Food Enzymes

Food Enzymes
Structure and Mechanism

Dominic W.S. Wong

CHAPMAN & HALL

I ⓣP® **International Thomson Publishing**
Thomson Science

New York • Albany • Bonn • Boston • Cincinnati • Detroit • London • Madrid • Melbourne
Mexico City • Pacific Grove • Paris • San Francisco • Singapore • Tokyo • Toronto • Washington

Art direction: Andrea Meyer, emDASH inc.
Cover design: Saeed Sayrafiezadeh, emDASH inc.

Printed in the United States of America

For more information contact:

Chapman & Hall
115 Fifth Avenue
New York, NY 10003

Chapman & Hall
2-6 Boundary Row
London SE1 8HN
England

Thomas Nelson Australia
102 Dodds Street
South Melbourne, 3205
Victoria, Australia

Chapman & Hall GmbH
Postfach 100 263
D-69442 Weinheim
Germany

International Thomson Editores
Campos Eliseos 385, Piso 7
Col. Polanco
11560 Mexico D.F.
Mexico

International Thomson Publishing - Japan
Hirakawacho-cho Kyowa Building, 3F
1-2-1 Hirakawacho-cho
Chiyoda-ku, 102 Tokyo
Japan

International Thomson Publishing Asia
221 Henderson Road #05-10
Henderson Building
Singapore 0315

ISBN 978-1-4419-4722-2

2 3 4 5 6 7 8 9 XXX 01 00 99 98 97

Library of Congress Cataloging-in-Publication Data

Wong, Dominic W. S.
        Food enzymes: structure and mechanism / Dominic W. S. Wong.
                p.   cm.
        Includes bibliographical references and index.

        Enzymes. 2. Food—Composition. I. Title.
        QP601.W734   1995   94-42926
        664`.024—dc20   CIP

Visit Chapman & Hall on the Internet http://www.chaphall.com/chaphall.html

To order this or any other Chapman & Hall book, please contact **International Thomson Publishing, 7625 Empire Drive, Florence, KY 41042.** Phone (606) 525-6600 or 1-800-842-3636. Fax: (606) 525-7778. E-mail: order@chaphall.com.

For a complete listing of Chapman & Hall titles, send your request to **Chapman & Hall, Dept. BC, 115 Fifth Avenue, New York, NY 10003.**

To My Parents

Mrs. Yuet-Mei Yue & Mr. Po-Luk Wong

Contents

About the Author

Dominic W. S. Wong is a research scientist at the Western Regional Research Center, United States Department of Agriculture. He has been a research chemist with the Beet Sugar Development Foundation, and spent two years doing research at Cornell University. Dr. Wong is the author of the book *Mechanism and Theory in Food Chemistry*. He received his undergraduate training at the University of San Francisco, his M.S. from Iowa State University, and his Ph.D. in Agricultural Chemistry from the University of California, Davis, in 1984. He is a member of the American Chemical Society, Institute of Food Technologists, American Society for Biochemistry and Molecular Biology, and the International Society for Plant Molecular Biology.

About the Contributor

John R. Whitaker is a professor emeritus in the Department of Food Science and Technology, University of California, Davis, following 37 years as professor and biochemist in the same department. His research and teaching have focused on enzymes, enzyme inhibitors, and other proteins in plants and animals and their role in the quality of foods. Dr. Whitaker is the author of *Principles of Enzymology for the Food Sciences* and *Paper Chromatography and Electrophoresis I. Electrophoresis in Stabilizing Media*, the editor or co-editor of 12 books, and the author or co-author of more than 200 primary publications. He received the B.A. in chemistry from Berea College (1951) and the Ph.D. degree in Agricultural Biochemistry from Ohio State University (1954). He is a member of the American Chemical Society, Institute of Food Technologists, American Society for Biochemistry and Molecular Biology and the American Society of Plant Physiologists.

Preface

Enzyme technology is an integral part of food processing in the production of foods and in the improvement of food quality. Better control of enzyme reactions in foods, the utilization of enzymes, and the development of new enzyme uses are dependent on our understanding of the fundamentals of enzyme structure and mechanism. This volume is the result of an attempt to put together current information on the structures and mechanisms for a selected list of important food enzymes. It is the intent of the author to describe the dynamic aspects of enzyme structures and their relationship to the chemistry of catalysis. This book is not meant to be a general review on enzymes. Rather it reflects the author's preference primarily based on some of the following criteria. (1) The enzyme plays an important role in food systems and/or processing. (2) The three-dimensional structure of the enzyme is known. (3) The reaction mechanism of the enzyme has been investigated in sufficient detail. (4) The chemistry of catalysis can be described at the molecular level.

There is a personal satisfaction in the completion of this volume for two reasons. In writing this book, the author had an opportunity to obtain a glimpse of the many remarkable developments in the chemistry of enzymology. Most importantly, it was both exciting and inspiring to witness the tremendous efforts and dedication of thousands of scientists in unraveling the mysteries of nature's wonderful molecules at work.

DOMINIC W. S. WONG

Acknowledgements

I am deeply indebted to Professor John R. Whitaker, who contributed to writing an excellent chapter on polyphenol oxidase and to the development of this project since its inception. Special thanks are due to Professor Donald R. Babin for his critical review and invaluable suggestions, which helped to clarify many points in the text. I am very grateful to Professor Robert E. Feeney, who first introduced me to the world of protein chemistry, and to Professor Carl A. Batt, who showed me the route to gene cloning and modification. Both had an indirect, but no less important, influence in the development of this book. My appreciation goes to the authors whose publications are referenced in this book. Special thanks are due to Dr. Pedro M. Alzari (Institut Pasteur), Dr. Christina Divne (Uppsala University), Professor Francis Jurnak (University of California, Riverside), and Dr. H. J. Hecht (Gesellschaft für Biotechnologische Forschung mbH) for their generosity in providing the original figures of their enzyme work.

I would like to thank the following publishers for the permissions to use their copyrighted materials: Academic Press, Inc., Academic Press, Ltd. (London), American Association for the Advancement of Science, American Chemical Society, American Oil Chemists' Society, American Society for Biochemistry and Molecular Biology, The Biochemical Society and Portland Press (London), Elsevier Trends Journals (Cambridge), Elsevier Science Publisher B.V. (Amsterdam), Federation of European Biochemical Societies, Garland

Publishing, Inc., The Japanese Biochemical Society, J. B. Lippincott Co., Society for General Microbiology (Redding), Macmillan Magazine Ltd. (London), The National Academy of Sciences, National Research Council of Canada, Royal Society of Chemistry (Cambridge), University of Geneva, VCH Verlagsgesellschaft mbH, Wiley-Liss, Inc. (John Wiley & Sons, Inc.)

Chapter 1

Food Enzymes and Future Development

Enzymes have been utilized for food processing since ancient times. The use of calf rennet in cheese making has been in practice long before the development of enzymology as a science. Fermentation in wine making, likewise, is an age-old practice that utilizes enzymes occurring naturally in raw materials. Proteolytic enzymes in the form of malt extract, koji, and papaya extract have been used for centuries.

The field of enzymology has expanded rapidly, particularly in the last two decades, because of intensive efforts carried out by numerous laboratories to elucidate the structures and mechanisms of a large number of enzymes. Tailoring enzymes for specific functions becomes a reality with the advent of protein engineering techniques. This is a golden era for agricultural/food sciences to capture the opportunity of transforming these recent developments into practical applications. One can envision a new phase of enzyme technology that may well revolutionize the ways foods are prepared and processed.

ENZYMES IN FOOD PROCESSING

There is not a single food system that does not involve enzyme reactions. In many processes, a cascade of complex enzyme-mediated rections is in operation. However, there are relatively few enzymes utilized in food processing. The following is a list of important industrial food enzymes and their applications. The majority of them are hydrolases and oxidoreductases.

(1) α-Amylase
   (a) Conversion of starch to dextrins (liquefaction) in the production of corn syrup.
   (b) Supplement to flour types low in α-amylase to ensure a continuous supply of fermentable sugar for yeast growth and gas production in dough making.
   (c) Solubilization of adjuncts (nonmalt carbohydrate materials from barley and other cereal grains) used in brewing.

(2) Glucoamylase
   (a) Conversion of dextrins to glucose (saccharification) in the production of corn syrup.
   (b) Conversion of residual dextrins to fermentable sugar in brewing for the production of "light" beer.

(3) β-Amylase
   Production of high-maltose syrup.

(4) Xylose (glucose) isomerase
   Isomerization of glucose to fructose in the production of high-fructose corn syrup.

(5) β-Glucanase
   Breakdown of β-glucans in malt and other raw materials to aid filtration of wort after mashing in brewing.

(6) Lipase
   (a) Enhancing flavor development and shortening the time for cheese ripening.
   (b) Production of specialty fats with improved qualities.
   (c) Production of enzyme-modified cheese/butter from cheese curd or butterfat.

(7) Papain
   (a) Used as meat tenderizer.
   (b) Used in brewing to prevent chill-haze formation by digesting the proteins that can otherwise react with tannic substances to form insoluble colloid particles.

(8) Chymosin
   Curding of milk by specific proteolytic action on caseins in cheese making.

(9) Microbial proteases
   Processing of raw plant and animal proteins. Production of fish meals, meat extracts, texturized proteins, and meat extenders.

(10) Pectinase
   Treatment of fruit pulp to facilitate juice extraction and for clarification and filtration of juice.

(11) Lactase
   (a) Additive for dairy products for individuals lacking lactase.
   (b) Breakdown of lactose in whey products for manufacturing polylactide.

(12) Acetolactate decarboxylase
   Reduce maturation time in wine making by converting acetolactate to acetoin.

In the absence of the enzyme, acetolactate is oxidized to diacetyl that requires a secondary fermentation to reduce it to acetoin.

(13) Lysozyme

Antimicrobial preservative.

(14) Glucose oxidase

(a) Conversion of glucose to gluconic acid to prevent Maillard reaction in egg products caused by high heating used in dehydration.

(b) Potential use to remove O_2 in food packaging for protection against oxidative deterioration.

(15) Cellulases

Conversion of cellulose wastes to fermentable feedstock for ethanol or single-cell protein production.

DEVELOPMENT OF RECOMBINANT ENZYMES

Much of the recent effort in improving enzyme use in food processing has been focused on the production of enzymes in quantities at economically affordable cost. Recombinant chymosin is a well-known example. The eukaryotic gene of chymosin is cloned into and expressed by microorganisms. Hence the enzyme can be produced by fermentation. The recombinant enzyme is commercially obtained from *Escherichia coli*, *Kluyveromyces lactis*, and *Aspergillus awamori* (Dornenburg and Lang-Hinrichs 1994). Another recombinant enzyme, acetolactate decarboxylase, produced by *Bacillus subtilus*, is now commercially available for use in reducing maturation time in brewing (Alder-Nissen 1987; Dornenburg and Lang-Hinrichs 1994). The enzyme catalyzes the direct transformation of α-acetolactate to acetoin without the formation of diacetyl intermediate. In yeast fermentation, the conversion of diacetyl to acetoin requires a secondary fermentation that can last several weeks. These recombinant products allow the replacement of their natural counterparts without alteration in their structures or functions.

ENGINEERING ENZYME PROPERTIES AND FUNCTIONS

Many ongoing investigations are directed to modify individual enzymes for specific functional properties. Enzymes used in food processing can be tailored to increase the efficiency of the process, and ultimately lower the cost of operations. Two key enzymes involved in corn syrup production can be improved, for example, to benefit the industry and the consumers. For complete liquefaction, starch granules need to be gelatinized by heating to about 105°C for α-amylase to act. All α-amylases used today require Ca^{++} ion for heat stability. In the following step, glucoamylase is used to further break down dextrins to glucose. Because glucoamylase has an acidic pH optimum and relatively low heat stability, the syrup from the liqufaction step must be cooled to about 60°C, and the pH adjusted to ~4.5 before saccharification. It is therefore highly desirable to have a glucoamylase that is heat-stable at the temperature range used in liquefaction and a higher pH optimum, so that both steps can be carried out in one process (Spradin 1989). Furthermore, glucoamylase catalyzes the hydrolysis of α-1,4 about 30–50 times faster than the branching (α-1,6) linkage. The yield of conversion in saccharifation is ~96%. It is desirable to add a debranching enzyme, or alternatively, glucoamylase may be modified to enhance its action on α-1,6 bonds.

For the production of high-fructose corn syrup, the glucose syrup feedstock needs to have the Ca^{++} ions removed. For glucose isomerase, Mg^{++} is added and a pH > 7.0 is required for optimum activity. Reaction in a high-glucose medium at an alkaline pH causes glycosylation and eventually inactivation of the enzyme (Pickersgill and Goodenough 1991). Intensive effort is now focused on shifting the pH optimum of the enzyme to neutral or slightly acidic, and changing the metal preference of the enzyme. In addition, the catalytic efficiency of glucose isomerase in the conversion of glucose to fructose is lower than that in the transformation of xylose to xylulose. The former reaction utilized in food processing is not the natural physiological function in microorganisms. There is great interest in modifying the enzyme to enhance its substrate specificity for glucose.

Lipases are used in the food oil industry for transesterification of inexpensive oils (e.g. palm oil) to produce substitutes for cocoa butter, which is a major ingredient in chocolate and confectionery manufacturing (Zaks et al. 1988). The characteristic melting property (a low and sharp melting point at 30–40°C) of cocoa butter is derived from the positions and compositions of the fatty acids in its triglyceride structure (primarily 65% SUS, with mostly 16:0, 18:0, and 18:1). Palm oil with most of its 1,3 position occupied by palmitic acid can be transesterified with stearic acid to produce a cocoa butter-like replacement, using a 1,3-specific lipase. Both the stability as well as the catalytic efficiency of lipases under commercial oil processing conditions need to be improved.

Lipase is also utilized in the production of enzyme-modified cheese/butterfat (Dziezak 1986). Direct development of selected cheese flavor is achieved by lipase-catalyzed hydrolysis of cheese curd or butterfat to generate free fatty acids, mostly short chains. Generation of specific flavor is dependent upon the specificity and property of the lipase employed. There has been continuous effort to screen for enzymes suitable for use in flavor development. Proteases can be utilized in this regard, but a major drawback is the requirement of precise control of hydrolysis in preventing the formation of bitter peptides. The addition of lipases and proteases to accelerate the cheese-ripening process is of economic significance. Again, the delicate control of enzyme action in the development of a proper flavor profile for the product remains a challenge to food scientists.

All these cases point to the need and potential in the utilization of enzymes in food processing. In nearly all the industrial enzymes currently used there are properties awaiting improvement. Traditionally, the primary focus is on the screening of microorganisms for enzymes with desired or useful properties. Many investigations today turn to the use of protein engineering for targeted properties. The work in this area is facilitated by the remarkable advancement in recent years in understanding enzyme structures and fine details of their mechanisms. Tailoring enzymes for industrial use is gaining increasing momentum, although it is still in its infancy. This topic will be treated thoroughly in the following chapter.

FUTURE DEVELOPMENT OF ENZYME TECHNOLOGY

Three research areas of emerging importance are capturing much excitement and imagination. The rapid progress in these researches will prove to be important. Undoubtedly, utilization of the new resources generated from these investigations will have a profound effect on the future of the food industry.

In Vivo Modification of Food Quality

Initial interests in the application of biotechnology in agriculture have generally been concentrated in controlling pests and diseases, and in developing pesticide/herbicide tolerance in crops. Increasing efforts are now directed to the improvement of food quality. A well-known case is the production of a genetically engineered tomato with the polygalacturonase gene silenced by antisense RNA (Lelen 1992). This process represents an example of in vivo manipulation of an endogenous enzyme to effect alteration in the quality of a food. The success of this type of research and development substantiates the potential use of modification of enzymatic pathways in plants for:

(1)   Improving functional properties of proteins and other food components more suitable for food formulation and processing.

(2)   Tailoring the chemical composition and physical properties of food components for value-added products.

(3)   Removing or reducing toxicants, or undesirable compounds in food crops to improve their nutrition value.

Many projects are currently in progress in the private sector, academia, and government research centers. For example, a potato with a high starch content and less moisture would absorb less oil on frying (Erickson 1992). The canola plant can be modified to produce oils of specifications, for example, those that are suitable for margarine. The modified oil thus eliminates hydrogenation and transesterification, and hence the problem of cis-trans isomerization of fatty acids in margarine production (Skillicorn 1994). In vivo modification of food quality traits in food crops is an area that will definitely change the food market.

Synthesis of Catalytic Antibodies

In the mid-1980s, two research groups (Peter G. Schultz and coworkers at the University of California, Berkeley, and Richard A. Lerner and coworkers at the Research Institute of Scripps Clinic) succeeded in generating antibodies with specific catalytic activities. The basic idea was to design transition state analogs to serve as antigens to raise antibodies that act like enzymes in binding to the reactants in their transition state (Fig. 1.1). In reactions catalyzed by proteolytic enzymes, the transition state is a tetrahedral intermediate. Lerner's group constructed a phosphonate ester transition state analog (substituting phosphorus for the carbon) to produce antibodies that accelerate ester hydrolysis by a factor of 1,000-fold (Tramontano et al. 1986). Schultz's group took a slightly different approach in that a reactant that would yield a transition state resembling *p*-nitrophenyl phosphoryl choline was synthesized. In this case, the reactant is *p*-nitrophenyloxycarbonylcholine, and antibody that specifically binds to phosphenyl choline hydrolyzes the reactant (Pollack et al. 1986). Since then, antibodies have been produced that catalyze many different types of reactions, for example: amide hydrolysis (Janda et al. 1988; Gallacher et al. 1992), ester hydrolysis (Janda et al. 1989; Guo et al. 1994), β-elimination (Shokat et al. 1989), lactonization (Napper et al. 1987; Benkovic et al. 1988), Claisen rearrangement (Jackson et al. 1988; Haynes et al. 1994), cationic cyclization (Janda et al. 1993), Diels-Alder reaction (Hilvert et al. 1989), peptide cleavage (Iverson and Lerner 1989), and transesterification (Wirsching et al. 1991).

Crystallographic study of a catalytic antibody that catalyzes the conversion of chorismic acid to prehenate, and its structural comparison with the natural enzyme, chorismate mutase, suggest an overall similarity between the mecha-

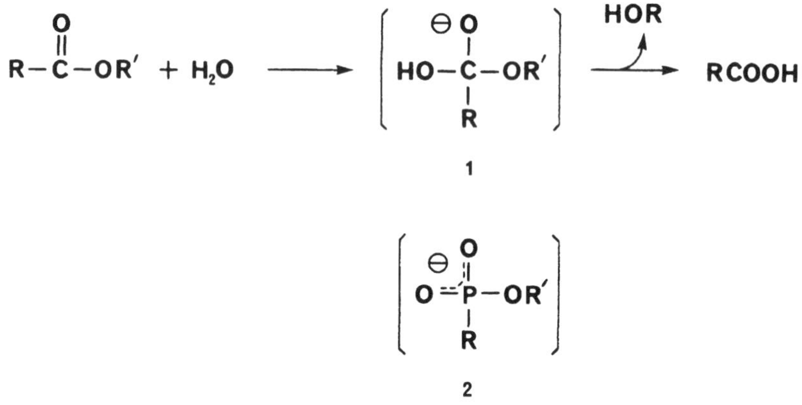

Figure 1.1.   Hydrolysis of ester proceeds via a tetrahedral transition state whose structure is comparable to that of a phosphonate ester.

nisms employed by the two proteins (Haynes et al. 1994). The three-dimensional structure of a hydrolytic antibody reveals features in the active site very similar to those of a serine protease (Zhou et al. 1994). The antibody active site contains a Ser-His diad (reminiscent of a catalytic triad in serine proteases), a Lys residue for oxyanion stabilization, and a hydrophobic pocket for substrate binding. Antibody-catalyzed ester hydrolysis shows a bell-shaped rate-pH profile suggesting two ionization groups with pKs of 8.9 and 10.1. Catalysis by the hydrolytic antibody involves a covalent acyl-antibody intermediate (Guo et al. 1994).

Along a similar line of strategy, antibodies can be chemically or genetically modified to acquire catalytic activity. One general scheme for creating semisynthetic antibodies is presented in Fig. 1.2A (Pollack et al. 1988; Pollack and Schultz 1989). An affinity labelling reagent is first synthesized that consists of three components: (1) a cleavable linkage group (e.g., disulfide) between (2) an antigen (to a specific antibody) that serves to direct the reagent to the binding site of the antibody, and (3) an affinity labelling group (e.g., aldehyde for alkalating Lys) that reacts with an amino acid at the binding site of the antibody. After the labelling group is introduced to the binding site, the disulfide can then be exchanged with a reactive thiol. The free thiol can further be derivatized with imidazole, cofactor, or metal complex, yielding antibodies possessing an array of biochemical functions.

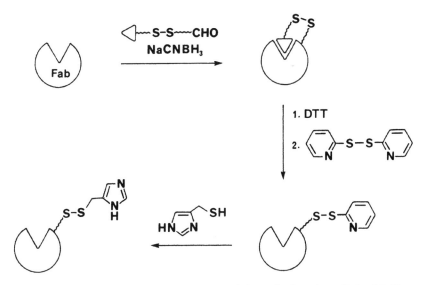

Figure 1.2A. Synthesis of semisynthetic catalytic antibody using affinity labelling method. (From Pollack and Schultz 1989. *J. Am. Chem. Soc. 111*, 1930, with permission. Copyright 1989 American Chemical Society.)

Antibodies can also be modified by altering the genetic code using site-directed mutagenesis, a technique with widespread use in protein engineering. This powerful technique has been extended to incorporating unnatural amino acids at specific sites in proteins utilizing acylated amber suppressor tRNA. The suppressor tRNA is recognized by the amber nonsense codon which acts as a blank to accept the desired unnatural amino acids (Fig. 1.2B) (Ellman et al. 1991; Anthony-Cahill et al. 1993). The methodology has been applied to engineering proteins with novel structures and functions. Most recently, a ribonuclease A is

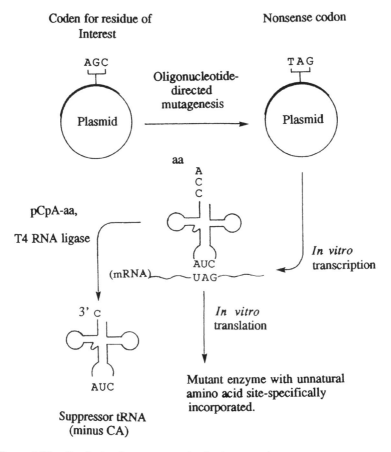

Figure 1.2B. Synthesis of mutant proteins by incorporating unnatural amino acids. (From Anthony-Cahill et al. 1989. *TIBS 14,* 401, with permission. Copyright 1989 Elsevier Trends Journals, Cambridge.)

synthesized by ligating peptide fragments containing catalytic 4-fluorohistidines (Jackson et al. 1994).

Catalytic antibodies generated by combinatorial libraries. Another breakthrough in antibody technology relates to the development of a bacterial-generated combinatorial library that allows for the synthetic construction of antibodies (Huse et al. 1989; McCafferty et al. 1990; Winter and Milstein 1991). One scheme involves the polymerase chain reaction (PCR) synthesis of randomized DNA segments of the hypervariable regions of heavy and light chains of Fab fragments. The PCR library of synthetic genes of heavy and light chains is inserted, separated into bacteriophage, and recombined to generate combinatorial constructs that can then be transformed into *Escherichia coli.* This and other similar methods allow the replacement of the hybridoma technology and immunological process for screening and selection of antibodies for catalytic activity. Large numbers of monoclonal Fab fragments against a transition state analog can be generated and selected in a relatively short time. The method has been successfully used to select for Fab fragments containing a free Cys that catalyzes the hydrolysis of thioester via a covalent acyl-antibody reminiscent of the mechanism in proteases (Janda et al. 1994). For a general discussion on the theory of combinatorial chemistry and its applications, see Clackson and Wells (1994), and Janda (1994).

Anti-idiotypic catalytic antibodies. In a recent report, the use of "molecular mimicry" in the construction of an anti-idiotypic antibody carrying acetylcholinesterase activity is described (Fiboulet et al. 1994). The technique is based on the binding interactions between antibody and antigen. In an idiotypic network, the antibody directed against an external antigen is an idiotope. Antibodies then raised against the idiotope are anti-idiotopes which carry the conformational image of the external antigen. The molecular structure of idiotope-anti-idiotope complex has been described for two different antigens: lysozyme (Bentley et al. 1990), and lipopolysaccharide A (Evans et al. 1994). The ability to raise anti-idiotope that mimics the antigen in binding interactions, has found widespread interest in its applications to diagnostic and therapeutic areas (Taub and Greene 1992). The most successful use has been in the identification of receptors for hormones, neurotransmitters and growth factors. The ligand to the receptor is the antigen, and the anti-idiotope mimics the binding site of the receptor (Fig. 1.3A).

The anti-idiotopic approach may open new routes in designing catalytic antibodies. In the scheme described by Fiboulet (1994), antibodies are raised against acetylcholinesterase followed by screening of antibodies that are structurally complementary to and thus recognized by the enzyme (Fig. 1.3B). The selected antibodies are used as immunogen to raise anti-idiotypes which will contain the image of the enzymatic active (binding) site. Selection of catalytic antibodies is achieved by screening the anti-idiotopes for enzyme activity.

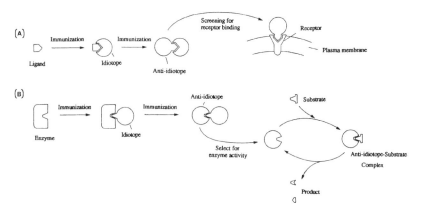

Figure 1.3. Generation of anti-idiotytic antibodies for (A) receptor binding (B) enzyme mimics (Fiboulet et al 1994.).

The ability to design a protein with specific catalytic functions has a profound implication on enzyme technology. Antibodies provide an enormous repertoire for creating novel catalytic functions that are not found in known enzymes. Catalytic antibodies could provide new ways to modify structures and functions of food proteins and other food components. The development of novel reactions may lead to conversion of agricultural raw materials to specialty food or nonfood products. Perhaps new reactions will be derived for effective removal/reduction of toxicants/antinutrients in foods, and for detoxification of processing wastes. Because catalytic antibodies have a combination of binding specificity and catalytic function, a potential application would be for targeted destruction of food pathogens.

Artificial Enzymes

The use of catalytic antibodies represents a novel approach in converting a nonenzyme protein to a semisynthetic enzyme with specific catalytic activity. Another active area of research focuses on the synthesis of miniature enzymes, without polypeptide backbones, that mimic the active site of an existing enzyme in catalysis. Enzyme models designed using synthetic polymers, crown ethers, cyclodextrins, or metal-steroid templates (Dugas 1989), as a matrix usually contain added reactive groups oriented in a geometry similar to that of the enzyme active site. Cyclodextrins have emerged as important enzyme models to use because of their ability for hydrophobic binding in the central cavity. A number of these miniature molecules have been constructed to mimic processes of hydrolytic enzymes.

An artifical enzyme with chymotrypsin-like activity (hydrolysis of esters) has been constructed by designing a catalytic group (an imidazolyl benzoic acid) to the secondary side (C2 or C3) of β-cyclodextrin (D'Souza et al. 1985), or by introducing an imidazole group at C2 (Rao et al. 1990). The former model hydrolyzes *m-t*-butylphenyl ester at twice the rate of chymotrypsin-catalyzed hydrolysis of *p*-nitrophenyl acetate at their respective pH optima of pH 10.7 and 8.0. The artificial enzyme shows increase in activity with increasing temperature to 80°C, and pH stability above pH 12 (D'Souza et al. 1987). Hydrolysis may proceed via a specific base catalysis assisted by imidazole nitrogen (Fig. 1.4A).

A synthetic transaminase designed by linking pyridoxamine to β-cyclodextrin via a thioether linkage has been shown to catalyze a transamination reaction (α-keto acid ⇌ amino acid) with stereochemical selectivity in substrates (Fig. 1.4B) (Breslow and Czarnik 1983). A later model has the pyridoxamine double linked by an additional thioester. These compounds show selectivity for the para or meta butyl substitution of the phenylpyruvic acid substrate (Breslow et al. 1990). Enzyme models of ribonuclease (Breslow et al. 1978) and carbonic anyhdrase (Tabushi and Kuroda 1984) have also been constructed by the same general approach. A water-soluble cyclodextrin sandwiched with porphyrin catalyzes epoxidation of hydrophobic alkenes (Kuroda et al. 1990). Cyclodextrin dimers linked by a metal-binding group have been constructed that mimic metalloenzymes (Fig. 1.4C) (Breslow and Zhang 1992). Ester substrates with hydrophobic substituents at both ends can then bind to the dimer, with the carbonyl group positioned close to the metal catalytic group. Hydrolysis of the ester involves the attack of a metal hydroxide species as observed in typical metalloenzymes.

Artificial enzymes certainly offer an attractive alternative for use in food processing. The highly complex structural features necessary for the functioning of a natural enzyme is reduced to a few functional groups attached onto a host molecule. Such molecular assemblies are less prone to the effects of temperature, pH, and ionic strength, parameters that are critical to conformational stability and biological functions of an enzyme. Artificial enzymes, devoid of amino acid backbones, are therefore particularly suitable for the harsh conditions required for their industrial applications.

The use of synthetic peptides to mimic the active site of an enzyme has also been the focus of many investigations. With the rapid accumulation of knowledge in protein structures and folding mechanisms, a number of successful attempts have been reported on de novo design of model proteins (DeGrado et al. 1989; O'Neil et al. 1990; Grove et al. 1993; Robertson et al. 1994). However, construction of synthetic peptides that can duplicate the catalytic activity, substrate specificity, and stability of an enzyme, presents additional challenges. Studies and reports today in this area can only be considered preliminary (Sasaki and Kaiser 1989; Hahn et al. 1990; Corey et al. 1994). There is definitely a need for more research in this area.

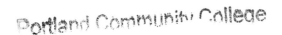

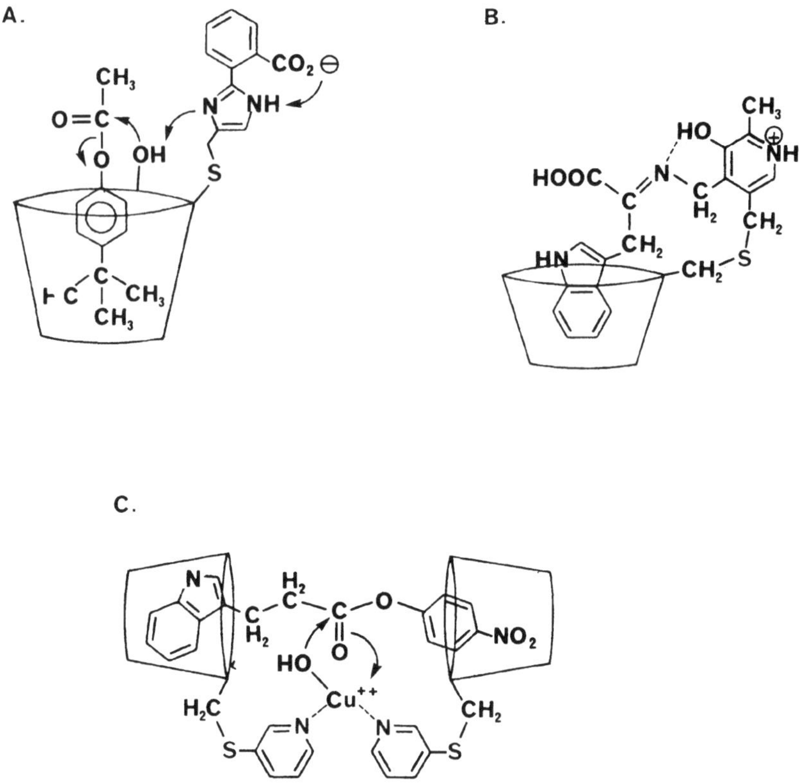

Figure 1.4. (A) Hydrolysis of *m-t*-butylphenyl ester by artificial enzyme with substituted imidazolyl-benzoic acid at secondary side of β-cyclodextrin. (From D'Souza and Bender 1987 *Acc. Chem. Res. 20,* 150, with permission. Copyright 1987 American Chemical Society.) (B) Conversion of indolepyruvic acid to tryptophan by artificial (β-cyclodextrin-pyridoxamine) transaminase. (From Breslow et al. 1980. *J. Am. Chem. Soc. 102,* 422, with permission. Copyright 1980 American Chemical Society.) (C) Cyclodextrin dimer linked by a metal-binding group catalyzes hydrolysis of esters with hydrophobic ends. (From Breslow and Zhang. 1992. *J. Am. Chem. Soc. 114,* 5883, with permission. Copyright 1992 American Chemical Society.)

NEEDS FOR ENZYME RESEARCH

All current developments amplify the importance of the basic understanding of structures and mechanisms of enzymes. The ability to manipulate enzymatic pathways, to construct artificial enzymes, to produce catalytic antibodies, depends on how well one knows the specific enzyme targeted for manipulation. The last decade has witnessed a marked progress in our knowledge of how enzymes work. The structures of a number of important food enzymes have been revealed by crystallographic studies, including α-amylase, β-amylase, glucoamylase, lipoxygenase, glucose oxidase, cellobiohydrolase, endoglucanase, xylose isomerase, horseradish peroxidase, and others. For example, it is revealing to learn that lipase contains a catalytic triad similar to that of serine proteases, and that interfacial activation is related to the movement of a lid located at the active site entrance. Digestion of cellulose is now known to require the cooperative action of a cellulose-binding domain and a catalytic domain in the enzyme, much like raw-starch digesting glucoamylases. The active site of soybean β-amylase is demonstrated to consist of Glu, and not Cys, as the catalytic residue. In lipoxygenase, the iron atom ligands to His residues. Xylose isomerase-catalyzed isomerization is mediated by two metal ions.

In spite of considerable progresses, many questions regarding enzyme mechanisms need to be addressed. For example, for amylolytic and cellulolytic enzymes, both an oxocarbonium intermediate mechanism and nucleophilic displacement have been proposed as possible models. Current evidences seem to indicate that xylose isomerase catalyzes isomerization via a hydride shift mechanism, but the base catalysts involved in the ring-opening step as well as the proton transfer have not yet been identified. Lipoxygenase shows a radical-mediated mechanism, but the activation of the native Fe(II)-enzyme to Fe(III)-enzyme has not been adequately explained. Both hydride transfer and one-electron transfer have been suggested as possible models for glucose oxidase-catalyzed reactions. Details on the reaction mechanisms of pectic enzymes are not yet available. The mechanism of synergism in cellulolytic enzyme systems is an area in need of further investigation.

The field of enzymology is expanding with new information on enzyme structures and mechanisms accumulating at a rapid pace. It is the synergistic activity of both basic and applied technology that will eventually benefit the general public. Fundamental knowledge on enzyme structures and mechanisms is important for effective utilization and improvement of existing enzymes and the development of new enzymes. A thorough understanding of enzyme reactions is critical to future success in the utilization of agricultural resources, and in the production of better foods.

REFERENCES

ALDER-NISSEN, J. 1987. Newer uses of microbial enzymes in food processing. *TIBTECH 5*, 170–174.

BENKOVIC, S. J.; NAPPER, A. D.; and LERNER, R. A. 1988. Catalysis of a stereospecific bimolecular amide synthesis by an antibody. *Proc. Natl. Acad. Sci. USA 85*, 5355–5358.

BENTLEY, G. A.; BOULOT, G.; RIOTTOT, M. M.; and POLJAK, R. J. 1990. Three-dimensional structure of an idiotope-anti-idiotope complex. *Nature 348*, 254–257.

BRESLOW, R.; CANARY, J. W.; VARNEY, M.; WADDELL, S. T.; and YANG, D. 1990. Artificial transaminases linking pyridoxamine to binding cavities: Controlling the geometry. *J. Am. Chem. Soc. 112*, 5212–5219.

BRESLOW, R., and CZARNIK, A. W. 1983. Transminations by pyridoxamine selectively attached at C-3 in β-cyclodextrin. *J. Am. Chem. Soc. 105*, 1390–1391.

BRESLOW, R.; DOHERTY, J. B.; GUILLOT, G.; and LIPSEY, C. 1978. β-Cyclodextrinyl-bisimidazole, a model for ribonucleases. *J. Am. Chem. Soc. 100*, 3229.

BRESLOW, R.; HAMMOND, M.; and LAUER, M. 1980. Selective transamination and optical induction by a β-cyclodextrin-pyridoxamine artificial enzyme. *J. Am. Chem. Soc. 102*, 421–422.

BRESLOW, R., and ZHANG, B. 1992. Very fast ester hydrolysis by a cyclodextrin dimer with a catalytic linking group. *J. Am. Chem. Soc. 114*, 5882–5883.

CLACKSON, T., and WELLS, J. A. 1994. In vitro selection from protein and peptide libraries. *TIBTECH 12*, 173–184.

COREY, M. J.; HALLUKOVA, E.; PUGH, K.; and STEWART, J. M. 1994. Studies on chymotrypsin-like catalysis by synthetic peptides. *Appl. Biochem. Biophys. 47*, 199–212.

DEGRADO, W. F.; WASSERMAN, Z. R.; and LEAR, J. D. 1989. Protein design, a minimalist approach. *Science 243*, 622–628.

DORNENBURG, H., and LANG-HINRICHS, C. 1994. Genetic engineering in food biotechnology. *Chem. & Industry 13*, 506–510.

D'SOUZA, V. T., and BENDER, M. L. 1987. Miniature organic models of enzymes. *Acc. Chem. Res. 20*, 146–152.

D'SOUZA, V. T.; HANABUSA, K.; O'LEARY, T.; GADWOOD, R. C.; and BENDER, M. L. 1985. Synthesis and evaluation of a miniature organic model of chymotrypsin. *Biochem. Biophys. Res. Comm. 129*, 727–732.

D'SOUZA, V. T.; LU, X. L.; GINGER, R. D.; and BENDER, M. L. 1987. Thermal and pH stability of "β-benzyme." *Proc. Natl. Acad. Sci. USA. 84*, 673–674.

Dziezak, J. D. 1986. Enzyme modification of dairy products. *Food Technol. 40*, 114–120.

DUGAS, H. 1989. *Bioorganic Chemistry. A Chemical Approach to Enzyme Action.* 2nd ed., Springer-Verlag, New York.

ELLMAN, J.; MENDEL, D.; ANTHONY-CAHILL, S.; NOREN, C. J.; and SCHULTZ, P. G. 1991. Biosynthetic method for introducing unnatural amino acids site-specifically into proteins. *Methods Enzymol. 202*, 301–337.

ERICKSON, D. 1992. Hot potato. *Scientific American 267(3)*, 160–161.

EVANS, S. V.; ROSE, D. R.; TO, R.; YOUNG, N. M.; and BUNDLE, D. R. 1994. Exploring the mimicry of polysaccharide antigens by anti-idiotypic antibodies. *J. Mol. Biol. 241*, 691–705.

FRIBOULET, A.; IZADYAR, L.; AVALLE, B.; ROSETO, A.; and THOMAS, D. 1994. Abzyme generation using an anti-idiotypic antibody as the internal image of an enzyme active site. *Appl. Biochem. Biophys. 47*, 229–239.

GALLACHER, G.; SEARCEY, M.; JACKSON, C. S.; and BROCKLEHURST, K. 1992. Polyclonal antibody-catalyzed amide hydrolysis. *Biochem. J. 284*, 675–780.

GROVE, A.; MUTTER, M.; RIVIER, J. E.; and MONTAL, M. 1993. Template-assembled synthetic proteins designed to adopt a globular, four-helix bundles conformation from ionic channels in lipid bilayer. *J. Am. Chem. Soc. 115*, 5915–5924.

GUO, J.; HUANG, W.; and SCANLAN, T. S. 1994. Kinetic and mechanistic characterization of an efficient hydrolytic activity: Evidence for the formation of an acyl intermediate. *J. Am. Chem. Soc. 116*, 6062–6069.

HAHN, K. W.; KLIS, W. A.; and STEWART, J. M. 1990. Design and synthesis of a peptide having chymotrypsin-like esterase activity. *Science 248*, 1544–1547.

HAYNES, M. R.; STURA, E. A.; HILVERT, D.; and WILSON, I. A. 1994. Routes to catalysis: Structure of a catalytic antibody and comparison with its natural counterpart. *Science 263*, 646–652.

HILVERT, D.; HILL, K. W.; NARED, K. D.; and AUDITOR, M.-T. M. 1989. Antibody catalysis of a Diels-Alder reaction. *J. Am. Chem. Soc. 111*, 9261–9262.

IVERSON, B. L., and LERNER, R. A. 1989. Sequence-specific peptide cleavage catalyzed by an antibody. *Science 243*, 1184–1188.

JACKSON, D. Y.; JACOBS, J. W.; SUGASAWARA, R.; REICH, S. H.; BARLETT, P. A.; and SCHULTZ, P. G. 1988. An antibody-catalyzed Claisen rearrangement. *J. Am. Chem. Soc. 110*, 4841–4842.

JANDA, K. D. 1994. Tagged versus untagged libraries: Methods for the generation and screening of combinatorial chemical libraries. *Proc. Natl. Acad. Sci. USA 91*, 10779–10785.

JANDA, K. D.; BENKOVIC, S. J.; and LERNER, R. A. 1989. Catalytic antibodies with lipase activity and R or S substrate selectivity. *Science 244*, 437–440.

JANDA, K. D.; LO, C.-H. L.; LI, T.; FARBAS, C. F. III; VIRSCHING, P.; and LERNER, R. A. 1994. Direct selection for a catalytic mechanism from combinatorial antibody libraries. *Proc. Natl. Acad. Sci. USA 91*, 2532–2536.

JANDA, K. D.; SCHLOEDER, D.; BENKOVIC, J.; and LERNER, R. A. 1988. Induction of an antibody that catalyzes the hydrolysis of an amide bond. *Science 241*, 1188–1191.

JANDA, K. D.; SHEVLIN, C. G.; and LERNER, R. A. 1993. Antibody catalysis of a disfavored chemical transformation. *Science 259*, 490–493.

KURODA, Y.; HIROSHIGE, T.; and OGOSHI, H. 1990. Epoxidation reaction catalyzed by cyclodextrin sandwiched porphyrin in aqueous buffer solution. *J. Chem. Soc. Chem. Comm. 1990*, 1594–1595.

LELEN, K. 1992. Ag-biotechnology companies move forward on heels of the FDA statement on biofoods. *Genetic Engineering News 12(11)*, 21–22.

MCCAFFERTY, J.; GRIFFITHS, A. D.; WINTER, G.; and CHISWELL, D. J. 1990. Phage antibodies: Filamentous phage displaying antibody variable domains. *Nature 348*, 552–554.

NAPPER, A. D.; BENKOVIC, S. J.; TRAMONTANO, A.; and LERNER, R. A. 1987. A stereospecific cyclization catalyzed by an antibody. *Science 237*, 1041–1043.

O'NEIL, K. T.; HOESS, R. H.; and DEGRADO, W. F. 1990. Design of DNA-binding peptides based on the leucine zipper motif. *Science 249*, 774–778.

PICKERSGILL, R. W., and GOODENOUGH, P. W. 1991. Enzyme Engineering. *Trends in Food Sci. & Technol. 9*, 122–126.

POLLACK, S. J.; JACOBS, J. W.; and SCHULTZ, P. G. 1986. Selective chemical analysis by an antibody. *Science 234*, 1570–1573.

POLLACK, S. J.; NAKAYAMA, G. R.; and SCHULTZ, P. G. 1988. Introduction of nucleophiles and spectroscopic probes into antibody combining sites. *Science 242*, 1038–1040.

POLLACK, S. J., and SCHULTZ, P. G. 1989. A semisynthetic catalytic antibody. *J. Am. Chem. Soc. 111*, 1929–1931.

RAO, K. R.; SRINIVASAN, T. N.; BHANUMATHI, N.; and SATTUR, P. B. 1990. Artificial enzymes: Synthesis of imidazole substituted at C(2) of β-cyclodextrin as an efficient enzyme model of chymotrypsin. *J. Chem. Soc. Chem. Comm. 1990*, 10–11.

ROBERTSON, D. E.; FARID, R. S.; MOSER, C. C.; URBAUER, J. L.; MULHOLLAND, S. E.; PIDIKITI, R.; LEAR, J. D.; WAND, A. J.; DEGRADO, W. F.; and DUTTON, P. L. 1994. Design and synthesis of multi-haem proteins. *Nature 368*, 425–432.

SASAKI, T., and KAISER, E. T. 1989. Helichrome: Synthesis and enzymatic activity of a designed hemeproteins. *J. Am. Chem. Soc. 111*, 380–381.

SHOKAT, K. M.; LEUMANN, C. J.; SUGASAWARA, R.; and SCHULTZ, P. G. 1989. A new strategy for the generation of catalytic antibodies. *Nature 338*, 269–271.

SKILLICORN, A. 1994. Oilseeds get a genetic makeover. *Food Processing 55(2)*, 48–52.

SPRADLIN, J. E. 1989. Tailoring enzyme systems for food processing. In: *Biocatalysis in Agricultural Biotechnology*, J. R. Whitaker, and P. E. Sonnet, eds., American Chemical Society Sym. Ser. 389, Washington, DC.

TABUSHI, I., and KURODA, Y. 1984. Bis(histamino)cyclodextrin-Zu-imidazole complex as an artificial carbonic anhydrase. *J. Am. Chem. Soc. 106*, 4580–4584.

TAUB, R., and GREENE, M. I. 1992. Functional validation of ligand mimicry by anti-receptor antibodies: Structural and therapeutic implications. *Biochemistry 31*, 7432–7435.

TRAMONTANO, A.; JANKA, K. D.; and LERNER, R. A. 1986. Catalytic antibodies. *Science 234*, 1566–1570.

WINTER, G., and MILSTEIN, C. 1991. Man-made antibodies. *Nature, 349*, 293–299.

WIRSCHING, P.; ASHLEY, J. A.; BENKOVIC, S. J.; JANDA, K. D.; and LERNER, R. A. 1991. An unexpected efficient catalytic antibody operating by ping-pong and induced fit mechanisms. *Science 252*, 680–685.

ZAKS, A.; EMPIE, M.; and GROSS, A. 1988. Potentially commercial enzymatic processes for the fine and specialty chemical industries. *TIBTECH 6*, 272–275.

ZHOU, G. W.; GUO, J.; HUANG, W.; FLETTERICK, R. J.; and SCANLAN, T. S. 1994. Crystal structure of a catalytic antibody with serine protease active site. *Science 265*, 1059–1064.

Chapter 2

Tailoring Enzyme Structures and Functions

The advantages of utilizing enzymes in producing food ingredients and in improving their functional properties have been recognized for many years. However, the application of enzymes in the food industry in general lags behind existing technology, in spite of the high expectations resulting from recent advances in the science of enzymology. Only a few enzymes are currently used in food processing; among these are glucose isomerase, amylases, chymosin, and papain. Catalase, pectolytic enzymes, and lactase are also used to a lesser extent. One of the major reasons for the slow growth in the use of enzymes in food processing is the cost. Enzymes are highly efficient and catalyze specific reactions often with high yield and minimum side effects, but they also need to function under rather stringent conditions, requiring a set range of pH, temperature, ionic strength, and in many cases, the addition of cofactors and coenzymes. Recovery and regeneration of enzymes in processing further add to the overall cost. Unlike developing pharmaceutical products, a slight addition to the manufacturing cost will have a significant impact on the market sales of a food product.

METHODS FOR MODIFICATION OF ENZYMES

Enzymes used in the food industry are traditionally obtained from microorganisms that have been screened and selected for improved properties. However, enzymes can also be tailored by in vitro methods—chemically, enzymatically, and genetically, for improvements in the following aspects: (1) increasing stability to pH and heat, (2) increasing resistance to solvents, chemicals, and inhibitors, (3) changing pH optimum, (4) changing substrate specificity and binding properties, (5) altering preference for metals/cofactors, and (6) modifying enzyme catalysis for new reactions.

Chemical modification has been the most widely used approach in the study of physical, chemical, and biological properties and in the elucidation of structure-function relationships in proteins. Much of our current understanding of enzyme actions comes from chemical studies. A number of monographs and texts are available on this subject (Kaiser et al. 1985; Feeney et al. 1985; Feeney 1987; Feeney and Whitaker 1982, 1986). On the other hand, the industrial aspects of chemical modification of enzymes have not been adequately treated, other than perhaps the topic on enzyme immobilization. The utilization of chemically modifed enzymes can be realized only if the following factors are considered: (1) The modification procedure should be reasonably specific to the targeted amino acid group(s) with minimum side reactions. (2) The specific modification should be achieved without employing harsh conditions that are deteriorative to the enzyme. (3) The modified enzyme should be separated and recovered from the reaction medium in a relatively simple and inexpensive procedure. (4) The reagents used should not impose toxicological problems, in particular, if the modified enzyme is to be used for food processing. (5) The process for large-scale use of the modified enzyme should be economically feasible.

The use of enzyme reactions to tailor the structure and function of an enzyme can be specific and may markedly affect the properties of the modified protein. Most research is directed to in vivo enzymatic modification of proteins in biological pathways. For food uses, the applications of enzymes to modify proteins that are of potential applications for nutritional and functional improvements have been thoroughly reviewed (Whitaker and Puigserver 1982; Feeney and Whitaker 1985; Matheis and Whitaker 1987). However, these studies are primarily focused on proteins that are not enzymes. Examples of the modification of enzymes by enzyme reactions are almost nonexistent. A number of factors work against the use of enzymes in modifying an enzyme. Unlike chemical reagents, which are small molecules, an enzyme, because of its structural size, does not have free access to a majority of residues in the native substrate enzyme. The potential sites accessible for modification are likely to be limited to the surface of the modified enzyme, unless the enzyme to be modifed is denatured. In modifying an enzyme, specificity may be at times a limiting factor rather than an advantage. Substrate binding involves a network of interactions between the active site residues of the modifying enzyme and the substrate molecule. Stereochemistry often plays a critical part in catalysis. A potential surface site therefore needs to satisfy a myriad of requirements before it can be acted upon by the modifying enzyme. An exception can be found in a recent study where phospholipase A_2 was modified by tissue transglutaminase. The modified enzyme containing intramolecular ε-(γ-glutamyl)lysine crosslinks involving the surface Gln4 showed a 3-fold increase in activity (Cordella-Miele et al. 1990). Transglutaminase-catalyzed incorporation of polyamine also caused activation in phospholipase A_2 (Cordella-Miele et al. 1993). In addition, enzymatic cleavage of carbohydrate chains is sometimes used for studying the functional role of glycosylation in enzymes. Proteolysis is occasionally performed as an activation step following gene expression in cloning.

Recent advances in recombinant DNA offer a new promise for tailoring enzymes for particular applications and their production in quantities for industrial uses. The procedure is the most specific, and least structurally destructive, compared with the other two described methods. By manipulating the genetic code of an enzyme, it is now possible to target a specific amino acid residue at any location for modification, and to examine the fine structure of interactions at a single locale in the molecule without the inference of a global effect. Although site-directed mutagenesis is fast becoming a routine procedure in many investigations, prediction of the effects of a particular mutation remains largely empirical. With few exceptions, mutation analyses are often done in hindsight, providing a rationale for the observed changes in mutation. There is no general model that allows a quantitative prediction regarding effective amino acid substitution. For site-directed mutagenesis, the sequence, the three-dimensional structure, and the reaction mechanism of the enzyme must be known. Subtilisin, lysozyme, and trypsin are among several model enzymes extensively studied in this respect.

Modification of one or more amino acids invariably causes perturbance of the various interacting forces in an enzyme molecule. The structural stability and catalytic function of an enzyme rely on a complex network of interactions of its amino acid residues and their interactions with the solvent. Quite often, a slight shift in the delicate balance of these forces may cause changes that are quite difficult to control. The more critical forces to be considered are: electrostatic interaction, hydrogen bonding, and hydrophobic effect. Covalent bonding, such as disulfide bridges, has also become an increasingly common target for protein engineering.

ELECTROSTATIC INTERACTIONS

Electrostatic interactions in proteins can be classified according to their functional effects into two major groups: long range or global, and short range or local (usually <5 angstroms (A) apart). Most charged residues are distributed on the surface of a protein, and less than 5% are buried. Buried charged residues are often catalytically or functionally important. Many enzyme reactions proceed via charge transition states. For example, electrostatic interaction contributes a stabilization energy of ~4 kcal/mol to the oxyanion intermediate in serine proteases (Graf et al. 1988; Warshel et al. 1989).

Replacement of charged residues involved in substrate binding may alter the stabilization energy of enzyme-substrate binding and subsequently cause a shift in the activity towards specific substrates. A mutation of Asp189->Ser located at the substrate binding pocket of trypsin shifts its substrate specificity towards hydrophobic tetrapeptide substrates (Graf et al. 1988). Increases in k_{cat}/K_m towards charged peptide substrates can be produced by complementary mutation of Glu156 and/or Gly166 in the substrate binding site of subtilisin. Based on the effect of alteration of charges in a binding site on substrate preference, the free energies, $\Delta\log(k_{cat}/K_m)$, for enzyme-substrate electrostatic interaction with charged residues at positions 156 and 166 have been determined to be -1.8 ± 0.5 and -2.3 ± 0.6 kcal/mol, respectively (Wells et al. 1987).

Modification of charged groups provides a useful means of changing the pH dependence of enzyme catalysis. The electrostatic effect of surface charges on the ionization constant of a catalytic active group can be analyzed by determining the pH dependence of k_{cat}/K_m for the wild type and the mutant. Substitution of Asp99 in the external loop of subtilisin with Ser lowers the pK_a of the active site His64 by 0.4 unit at low ionic strengths of 0.005–0.01 M. A similar mutation of Glu156->Ser also exhibits similar changes in pK_a, a result of destabilization of the protonated form of His64. The shift in pK_a corresponds to an effective dielectric constant of 40–50 between Asp99 (or Glu156) and His64 (Russell et al. 1987). Increasing ionic strength decreases the ΔpK_a because of the shielding effect of the counterions on charge interactions. The effect of destabilizing His64 by mutation of Asp99->Ser results in a decrease in k_{cat} and K_m for the hydrolysis of succinyl-Ala-Ala-Pro-Phe-p-nitroanilide (Russell and Fersht 1987).

Structural stability can be improved by interactions involving an α helix dipole. Synthetic polypeptides have been shown to increase stability by introducing charges at the end of the helix. The energy of interaction between the protonated His18 at the C–terminal end of an α helix and the peptide dipole contributed by the helix in the enzyme barnase (the small ribonuclease from *Bacillus amyloliquefaciens*) amounts to 1.4–2.1 kcal/mol (Sali et al. 1988). Engineering T4 lysozyme Ser38->Asp and Asn144->Asp at the N-terminal end of the helices B and J increases the enzyme stability with $\Delta\Delta G = 1.6$ kcal/mol at pH 3–5 relative to the native enzyme (Nicholson et al. 1988).

HYDROGEN BONDING

Hydrogen bonding contributes significantly in determining the specificity of enzyme catalysis. The energetic contribution of hydrogen bonds in the formation of the enzyme-substrate complex involves the competition of H_2O molecules for the hydrogen bond donors and acceptors.

$$E\text{-}AH\cdots OH_2 + S\text{-}B\cdots OH_2 \rightleftharpoons [E\text{-}A\cdots B\text{-}S] + H_2O\cdots HOH$$

The overall energy change comes from the interchange of hydrogen bonds in the above balance equation, and depends on the enthalpic difference in the formation of each bond, and predominantly the entropic change associated with the release of hydrogen-bonded H_2O (because H_2O hydrogen bonded to enzyme or substrate has lower entropy than bulk H_2O). The energetics describing these complex interactions in solution can be approximated by performing a hydrogen bond inventory (Fersht 1987). The net contribution of hydrogen bonds of different types can be quantitated by a systematic removal of amino acid side chains involved in hydrogen bonding interactions in substrate binding through site-specific mutation, and measurement of the affinities of the mutants.

Using tyrosyl-tRNA synthetase as a model enzyme, the overall energetics based on k_{cat}/K_m of deletion of a hydrogen bond between uncharged donor/acceptor has been estimated to be 0.5–1.5 kcal/mol. Removal of a hydrogen bond with a charged donor or acceptor reduces the binding affinity by 4 kcal/mol (Fersht et al. 1985). Replacement of Ser130 in the sulfate-binding protein with Gly and Ala causes a decrease of 1.6 and 2.7 kcal/mol in binding energy, corresponding to a 15- and 100-fold decrease in activity, respectively. Replacement of Ser130 with Cys, however, decreases the sulfate binding activity 3,200-fold, a binding energy loss of 4.8 kcal/mol, which can be attributed to the steric effect of the mutation (He and Quiocho 1991). The binding energy can also be established by modifying the ligand rather than the protein. Using matched pairs of either phosphonate or phosphonamidate inhibitors of thermolysin, a binding energy of 4.0 kcal/mol can be attributed to an amide hydrogen bond (Bartlett and Marlowe 1987).

The conformational stability of globular proteins is often attributed to hydrophobic effect as the major factor. However, in recent years, the importance of hydrogen bonding in protein structure has been recognized. Use of urea in thermal unfolding studies on the conformational stability of 12 ribonuclease T1 mutants suggests that the average contribution of a hydrogen bond to the conformation of the enzyme is 1.3 kcal/mol, which is comparable to that of the hydrophobic effect (Shirley et al. 1992). Substitution of Thr157 with Ile in T4 lysozyme disrupts the hydrogen bond with the buried main chain amide of Asp159. The δ-methyl group of Ile destabilizes the enzyme by ~1.3 kcal/mol while the Thr γ-methyl group contributes ~0.66 kcal/mol to the stability of the wild type enzyme (Alber et al. 1987). Other mutations that maintain certain kinds of hydrogen bond networks, but with different geometric arrangements, show lesser destabilization effects. The hydrogen bond in the folded state contributes to the dominant effect in all 13 amino acid substitutions.

In subtilisin, a distal hydrogen bonding of Thr220 has been found to be functionally important in stabilizing the oxyanion transition state. Site-directed substitution in removing the γ-hydroxyl and γ-methyl groups results in decrease of the transition energy, $\Delta\Delta G^{\neq} = \sim 2.0$ and 0.5 kcal/mol, respectively (Braxton and Wells 1991).

HYDROPHOBIC EFFECTS

Hydrophobic effect is often considered a major stabilizing factor in protein folding, although other forces also contribute to the net stability of a protein. For example, the stabilization of T4 lysozyme mutants has been shown to be directly related to the hydrophobicity of the substituted amino acids in replacing Ile3 (Matsumura et al. 1988). Even under optimum conditions, the net stability of a protein is marginal, with ΔG_N^0 typically -5 to -10 kcal/mol. In a mutation, one must also consider factors such as conformational changes, entropic loss, steric constraint, and bond distortion that are energetically unfavorable.

Hydrophobic effect has been traditionally estimated from model studies on the free energy ΔG_{tr} for transferring amino acid side chains from water to organic solvents (such as ethanol, dioxane, hexane, octanol, etc), or from macroscopic surface tension calculations (Sharp et al. 1991). The assumption that the microenvironment of the interior core of a protein can be adequately described by a scale using hydrocarbon or alcohol as the apolar phase is subject to debate (Eriksson et al. 1992; Sandberg and Terwilliger 1988, 1991). Replacement of Ile with Val/Ala in the hydrophobic core of the enzyme barnase gives a stabilization energy $\Delta(\Delta G)$ contributed by hydrophobic effect significantly higher than ΔG_{tr} in model studies (Kellis et al. 1989). It has been suggested that the hydrophobicity scales derived from H_2O/organic solvent models underestimate the magnitude of the effect without accounting for the dependence of volume entropy contribution (Sharp et al. 1991). The different values obtained by site-directed mutagenesis and solvent transfer model studies may be explained by the structural effect of substitutions imposed on the core-packing arrangement of the protein (Eriksson et al. 1992; Karpusas et al. 1989). These factors may include loss of van der Waals interactions, but most importantly, the addition of conformational entropy. The change in energy associated with substitution in mutants therefore contains an additional term that is proportional to the change in the cavity volume occupied by the substitution of the amino acid side chains. The corrected ΔG_{tr} values thus derived agree with the experimental $\Delta(\Delta G)$ values measured for 72 aliphatic side chain mutants from T4 lysoyme, *Staphylococcal* nuclease, barnase, and bacteriophage f1 gene 5 protein. The introduction of one methylene group contributes to a gain of ~ 1.3 kcal/mol in stability of globular proteins. The hydrophobic surface free energy amounts to ~ 50–60 cal/mol$\cdot A^2$ to protein stability.

DISULFIDE BONDS

A disulfide bond contributes to the stability of globular proteins by decreasing the conformational chain entropy in the unfolded state via restricting the conformational degree of freedom. Subtilisin mutants with disulfide bridges engineered into positions 24-87 or 22-87 show no detectable changes in the specific activities, autolysis stability, or the pH optima of the enzyme reaction. However, compared with the Cys22-Cys87 mutant, the Cys24-Cys87 disulfide is more resistant to reduction and exhibits a greater stability towards autolysis and aggregation. Apparently Cys22-Cys87 mutation creates destabilizing interactions in that the native Thr22-Ser87 intramolecular hydrogen bond is disrupted (Wells and Powers 1986). Crystallographic studies further reveal that the two disulfides assume atypical dihedral angles with the Cys22-Cys87 showing a higher dihedral energy (4.8 kcal/mol) than Cys24-Cys87 (2.5 kcal/mol) (Katz and Kossiakoff 1986). The disruptive effect of introduced disulfide bond on existing stabilizing interactions is underscored by another study of five mutants with disulfide bridges constructed at various locations of subtilisin; all these mutants are less stable than the native enzyme (Michinson and Wells 1989). In a similar work of engineering a disulfide bond (Cys22-Cys87) into subtilisin, Pantoliano et al. (1987) reported an enhanced stability of the mutant enzyme, as indicated by a ΔT_m of 3.1°C relative to the native enzyme.

Construction of disulfide bridges in T4 lysozyme takes a different approach. Unlike subtilisin, which does not contain free Cys residues, T4 lysozyme contains two free Cys residues at positions 54 and 97. Therefore substititution of an amino acid side chain proximal to one of the native Cys residues, followed by mild oxidation, can generate a new disulfide bond. Thus, a mutation of Ile3->Cys creates a disulfide between the new Cys3 and Cys97. The engineered disulfide does not affect the enzyme activity, but does enhance the stability against irreversible inactivation (Perry and Wetzel 1984) as well as reversible thermal unfolding (Wetzel et al. 1988). The stabilization effect contributed by each disulfide bond is additive. Mutants containing three constructed disulfide bonds at positions 3–97, 9–164, and 21–142 exhibit a ΔT_m of 23.4°C higher than the wild-type T4 lysozyme, approximating the sum of the ΔT_m values of the single disulfide mutants (Matsumura et al. 1989B). However, mutants with Cys90-Cys122 and Cys127-Cys154 show a decrease of stability due to the strain energy introduced by the new bonds (Matsumura et al. 1987A). All these results underline the importance in considering the balance between the favorable and unfavorable factors involved in engineering disulfide bonds. Stabilization conferred by the construction of new bonds can be offset by (1) the disruption of existing stabilizing interactions in the native structure, and (2) the steric constraint produced by the new bond formation.

An example of a novel use of engineering cysteine residues into an enzyme is given by the alteration of the active site of subtilisin to form a thiolsubtilisin. The modification of the subtilisin active site Ser221 to a Cys residue has been a focus of continuous interest (Neet et al. 1968; Philipp and Bender 1983). Thiolsubtilisin exhibits a large reduction in activity towards normal amide and especially ester substrates mainly because of a decreased rate in the acylation step (Neet and Koshland 1966). Interestingly enough, it has been demonstrated recently that thiolsubtilisin exhibits a shift in the preference for ammonolysis over hydrolysis of the acyl-enzyme intermediate (Abrahmsen et al. 1991; Nakatsuka et al. 1987). This unique property has been utilized in the synthesis of peptides in which the enzyme is acylated by a peptide segment of activated ester to form an acyl-enzyme intermediate that is then deacylated by the nucleophilic attack of the amino group of another peptide (Nakatsuka et al. 1987; Wu and Hilvert 1989).

The structural integrity of an engineered active site Cys into a serine protease has been examined in detail in thioltrypsin. Mutation of Ser195 → Cys in trypsin causes the k_{cat} to be lowered by a factor of 6.4 × 10⁵ without affecting the K_m (Higaki et al. 1989). The three-dimensional structure of the trypsin mutant reveals subtle conformational changes in the active site (McGrath et al. 1989). A rotation of 30° about the Cys195 C_α-C_β dihedral angle results in the sulfur atom pointing away from the essential His57, and intruding into the oxyanion hole. The Cys195 side chain thus imposes a steric interference in the formation of oxyanion intermediate during catalysis. These conclusions reaffirm the suggestion first noted by Koshland that the lack of reactivity of thiolsubtilisin must be ascribed to a change of geometry at the active site to accommodate the radius and bond angle of the sulfur atom (Neet and Koshland 1966).

ENGINEERING SUBSTRATE SPECIFICITY

Enzyme specificity for substrates is influenced in large part by residues located in the binding site, although distant residues may sometimes be involved to a certain extent. In fact, a change of substrate binding properties can be achieved by modifying the residues directly involved in substrate binding. Subtilisin from *Bacillus licheniformis* exhibits substantially higher catalytic efficiency than that from *Bacillus amyloliquefaciens* towards various substrates, because of two substitutions at positions 156 and 21. Mutation of Glu156->Ser and Tyr217->Leu in the *B. amyloliquefaciens* enzyme modifies the P_1 subsite preference towards negatively charged substrates, and results in an increased k_{cat}. In addition, mutation of Gly169->Ala located outside the direct binding site lowers the K_m for the substrates tested (Wells et al. 1987). The kinetic effects of this triple mutation are additive.

This design strategy for substrate specificity by the exchange of amino acid substitution based on a stable form has been applied to modify hen egg white lysozyme (Kumagai et al. 1992). Human and turkey lysozymes show significantly higher bacteriolytic activity than HEWL, which differs in three critical residues in the substrate-binding region. A triple mutation of these three residues (Trp62->Tyr, Asn37->Gly, and Asp101->Gly) in HEWL increases the lytic activity 3-fold without loss in hydrolytic activity.

Replacement of key multiple residues at the mobile loop and at the helix that folds around the active site in L-lactate dehydrogenase leads to an increase in k_{cat} and a broadened specificity for a number of α-keto acids including those with bulky side chains (Wilks et al. 1990). A shift in substrate specificity from pyruvate to oxaloacetate results from the Gln102->Arg mutant (Wilks et al. 1988), with the k_{cat}/K_m for oxaloacetate improved by three orders of magnitude and that for pyruvate decreased by four orders of magnitude.

Structural flexibility of the active site of α-lytic protease can be changed by shortening some residue side chains protruding into the binding pocket. The Met192->Ala mutation increases the range of substrates that are catalytically productive (Bone et al. 1989).

OTHER APPROACHES OF MODIFICATION

An enzyme can be engineered with Cys residues to act as a control element for activity. By introducing two Cys residues, each located on the opposite side of the active site cleft, the T4 lysozyme mutant becomes inactivated by oxidation, but its activity can be regenerated by a reducing reagent (Matsumura and Matthews 1989). A mutant of *Staphylococcal* nuclease with a Cys introduced into its hydrophobic binding pocket can be inactivated by mercuric or cupric ions, and reactivated by a chelating agent (Corey and Schultz 1989).

Another approach involves construction of hybrid proteins. Grafting a calcium-binding segment from the enzyme thermitase into one of the two sites of autoproteolysis in subtilisin confers a 10-fold stability to irreversible inactivation in the presence of 10 mM $CaCl_2$ (Braxton and Wells 1992). A mutant of *Bacillus subtilis* neutral protease containing a thermolysin calcium-binding segment shows a 2-fold increase in stability in 0.1 M $CaCl_2$ over that of the native enzyme (Toma et al. 1991).

Enzymes that contain Cys, Met, and Trp residues are subject to oxidation and subsequent inactivation. Subtilisin is among a group of enzymes that contain a sensitive Met222. Replacement of Met222 with Ala or Ser stabilizes the enzyme against oxidative inactivation, although the catalytic efficiency is lowered 5-fold compared with the wild type (Estell et al. 1985).

EXAMPLES OF MODIFICATION OF FOOD ENZYMES

The following discussion gives a number of specific examples of altering the stability and function of some important food enzymes.

Lipolytic Enzymes

Lipase is among an important group of enzymes under intensive investigation in recent years as catalysts in organic synthesis. Lipase has been shown to catalyze a wide variety of organic reactions of synthetic value in apolar media. Instead of engineering the enzyme directly, the reaction medium is altered to effect a particular change in properties. The chemistry and theory of nonaqueous enzymology has been covered thoroughly in many recent reviews (Klibanov 1989; Morihara 1987; Schneider 1991; Wong et al. 1991) and will not be elaborated upon in this discussion.

The lowering of water activity in the reaction system changes drastically the properties of the enzyme in the following aspects. (1) The reaction equilibrium is shifted to synthesis over hydrolytic action. This reversal in enzyme action has been well documented not only for lipase (Zaks and Klibanov 1986), but also for a number of other enzymes. These include chymotrypsin (Zaks and Klibanov 1986), trypsin (Sakurai et al. 1988), subtilisin (Zhong et al. 1991; Riva et al. 1988), thermolysin (Kitaguchi and Klibanov 1989), polyphenol oxidase (Kazandjian and Klibanov 1985), glucoamylase (Laroute and Willemot 1989), papain (Stehle et al. 1990), and chymosin (Abdel Malak 1992). (2) The lipase-catalyzed ester synthesis or ester exchange is highy stereospecific, producing optically active products (Cambau and Klibanov 1984; Kirchner et al. 1985). (3) The enzyme can catalyze new reactions not possible in H_2O. Lipase has been applied for synthesis of peptides in organic solvents using carboxylic esters and aliphatic amines (Margolin and Klibanov 1987). (4) The enzyme shows an increase of thermostability and storage stability. Lipase is catalytically active in organic solvents at temperatures as high as 100°C.

A number of enzymes modified chemically with polyethylene glycol (PEG) attached to lysyl residues via triazine groups become soluble and reactive in organic solvents (Inada et al. 1986). PEG-lipase catalyzes ester synthesis and transesterification, using a wide range of substrates (Takahashi et al. 1988). Because the enzyme is soluble in pure solvent, reaction kinetics of these reactions can be studied. In lipase-catalyzed ester synthesis, V_{max} increases with increasing carbon chain length of the fatty acid or the alcohol, with little change in K_m. PEG-lipase forms a very stable colloidal complex with magnetite (Fe_3O_4) which can be dispersed in both organic and aqueous solutions. This resulting magnetic lipase attains a maximum activity of 1770 μmol/min/mg protein in ester synthesis in benzene, and can be completely recovered after use by magnetic separation without loss of activity (Inada et al. 1988; Takahashi et al. 1987). In a similar modification of the lysyl residues in the *Candida cylindracea* lipase with decanol using a heterobifunctional photogenerated reagent, the enzyme retains 86% of the activity after heating at 50°C for 15 min, four times that of the native enzyme (Kawase et al. 1990).

An enzyme can also be engineered for increased stability in nonaqueous systems. A multiple mutant of subtilisin has been found to be 50 times more stable than the wild type in anhydrous dimethylformamide (Wong et al. 1990, 1991). Each of the six substitutions: Asn218->Ser, Gly169->Ala, Met50->Phe, Tyr217->Lys, Gln206->Cys, and Asn76->Asp, has a small and localized effect but an additive effect on the stability.

Phospholipase A_2 is an enzyme of both pharmaceutical and industrial importance. Most mutation studies are designed to probe the structure-function relationship of the enzyme. Results from these studies often provide insight into the effects of mutation on the structural and catalytic properties. Mutations of Asn89->Asp and Glu92->Lys which are close to the N-terminal end of helix 5 increase the stability by 1.0 kcal/mol, which comes close to the predicted -0.76 kcal/mol based on calculation of the interactions between the subtituents and the helix dipole (Pickersgill et al. 1991).

Phospholipase A_2 of porcine pancreas consists of an interface recognition (binding) site around the entrance of an active site cleft that is composed of the N-terminal helix and loop 62–72. The snake venom enzyme, lacking several residues in the loop, shows a higher activity and affinity for phospholipid micelles. Deletion of residues 62–66 in the porcine phospholipase A_2 and mutation of Asp56->Ser, Ser60->Gly, and Asn67->Tyr in the same loop, results in a snake venom enzyme-like surface loop. The mutant shows a 16-fold increase in activity on micellar substrates (Kuipers et al. 1989; Thunnissen et al. 1990). The mutant apparently has all the residues involved in the recognition site aligned in a smooth plane that facilitates improved interactions of the enzyme with aggregated substrates. In a similar work on the bovine phospholipase A_2, replacing the Lys56 involved in interface binding with Met results in a 3–4-fold increase in k_{cat} for phosphatidylcholine micelle substrate. Replacement of Lys56 reduces the repulsive electrostatic interaction with the charged choline moiety of the substrate, and causes conformational changes in the surface loop for accommodation of the substrate (Noel et al. 1991).

Xylose Isomerase

Xylose isomerase is used in the commercial process for the enzymatic conversion of glucose to fructose in the production of high-fructose corn syrup. This process is carried out in a continuous bioreactor at a temperature range of 60–65°C. Thermal inactivation becomes a limiting factor in the process operation. There is tremendous interest to develop mutants with the following properties: (1) increased thermostability, (2) lower pH optimum, (3) altered metal cation preference, and (4) shift in substrate specificity from xylose to glucose.

Replacement of a number of Lys residues in the *Actinoplanes missouriensis* enzyme enhances heat stability (Mrabet et al. 1992; Quax et al. 1991). A Lys253->Arg mutation at the interdimer interface of the enzyme molecule increases the half-life 6-fold in the presence of 2.5 mM glucose (60°C) and 30% in the absence of sugar (5 mM Mg^{++}, 84°C). A triple mutation (Lys309, 319, 323 -> Arg) at the protein surface also shows a stabilizing effect in the presence of glucose but to a lesser extent. The stabilizing effect of Lys->Arg mutant in the presence of glucose may be partially due to elimination of the Maillard type glycosylation of the lysyl residues. Most importantly, the increase in thermostability is the result of direct effects on electrostatic interactions. In the single mutant (Lys253->Arg) additional new intra- and intersubunit interactions are created through the guanidium group. Furthermore, a water-bridged hydrogen bond in the wild type is replaced by a direct bonding with the Arg. In the triple mutant, two of the mutations (Lys309->Arg and Lys319->Arg) located in helix α8 at the surface gain new or replacement bonding. Another triple mutation centered at helix 2 (Glu70->Ser, Ala73->Ser, and Glu74->Thr) also shows improvement in stability (Quax et al. 1991).

The *Actinoplanes missouriensis* enzyme has maximum stability between pH 5.5–6.5. The pH dependence profile shows a bell-shaped curve with less denaturation in this pH range. A single mutation of Lys253->Arg shifts the pH optimum for thermostability half a unit higher (Quax et al. 1991). In immobilized form, the mutant has a lifetime twice that of the native enzyme at 60°C.

Site-directed mutation of the conserved Glu186 located near the active site changes the specificity of the metal sites and pH activity profile of the *Actinoplanes* enzyme (Jenkins et al. 1992; van Tilbeurgh et al. 1992). The wild type enzyme is most active with Mg^{++} at pH 7.5, while the mutant Glu186->Gln exhibits the highest k_{cat} with Mn^{++} as the activation cation. The K_m values at pH 7.5 are also lowered 4-fold with either Mg^{++} or Mn^{++}. Furthermore, the mutant has the optimum pH for activity shifted to pH 6.25. The change in the metal preference can be explained by a conformational change involving Asp255 which, in the Mg^{++} activated mutant, forms hydrogen bonding with the mutated residue Gln186, weakening the metal binding at site 2. In the native enzyme, as well as the Mn^{++}-activated mutant, the Asp255 carboxylate side chain constitutes one of the ligands of metal site 2.

Studies on all the xylose isomerases investigated thus far confirm that the enzyme has a lower K_m and higher k_{cat} for xylose than for glucose, although the enzme also isomerizes glucose, and in some cases ribose. Modification of the native enzyme to achieve a preference for glucose would be of immense benefit to the corn sweetener industry. Replacing key amino acids that are in close contact with the glucose substrate with residues containing smaller side chains causes a kinetic change in substrate specificity (Meng et al. 1991). Such mutations are based on the assumption that the extra methyl carbon in glucose may impose a steric crowding effect on the binding site. Mutation of Trp139 in the *Clostridium thermosulfurogenes* enzyme, in particular, causes a shift of catalytic efficiency in favor of glucose over xylose. Most mutants show higher k_{cat}/K_m toward glucose than the wild type. The double mutants (Trp139->Phe and Val186->Ser; Trp139->Phe and Val186->Thr) show 2- and 5-fold increases. However, although the wild type shows a 13-fold difference in k_{cat}/K_m in flavor of the xylose substrate, all the mutants tested exhibit at best, only a 3-fold higher catalytic efficiency for glucose than xylose (as in the double mutant Trp139->Phe and Val186->Ser). The changes in K_m in these mutations may be related to the increase in volume of the substrate binding pocket, to better accommodate the glucose molecule. Finally, the discussion on xylose isomerase would not be complete without mentioning the application of the immobilized enzyme in manufacturing high-fructose corn syrup. Characteristics and properties of various commerical immobilized xylose isomerases are presented in Table 13.1 (Chapter 13). The subject has been reviewed thoroughly by Jensen and Rugh (1987).

Other Food Enzymes

The active site of amylases consists of a number of subsites, each of which interacts with a monomeric unit of the substrate polymer. Mutation of the conserved Lys210 to Arg with a longer positively charged side chain enhances the binding affinities of both subsites 7 and 8 in *Saccharomycopsis fibuligera* α-amylase. However, replacement with a shorter neutral side chain such as Asn causes a decrease in both (Matsui et al. 1992A, B). The change in affinities of the subsites results in a shift of bond cleavage pattern of the substrate maltotetraose. A lowering of hydrolytic activity toward both amylose and short chain substrates has been observed for the mutants. Crosslinking the subsite Lys with *o*-phthalaldehyde results in two modified species that show significant increase in activity for maltotriose but a k_{cat} decrease for soluble starch (Kobayashi et al. 1992). Mutation of Trp84 located at subsite 3 to Leu in the *Saccharomycopsis* α-amylase gives an increased transglycosylation in comparison with the native enzyme (Matsui et al. 1991).

Chymosin is a member of the aspartic proteinase family that cleaves specifically the Phe105-Met106 of κ-casein. Modification of the amino acids involved in the hydrogen bond network of the active site aspartates results in an increase of the pH optimum. Kinetic analyses indicate that mutants Asp300->Ala, Thr219->Ala, and Gly247->Asp have pH optima of 4.4, 4.2, and 4.0 compared with 3.7 for chymosin B (Mantafounis and Pitts 1990). Similarly, substitution of Lys221->Leu causes a marked shift of pH optimum to 3.5, with a k_{cat}/K_m three times higher in the hydrolysis of denatured hemoglobin (Suzuki et al. 1989). Interestingly, chemical modification of lysyl residues in chymosin results in an increased milk clotting and proteolytic activity (Smith et al. 1991). The hydrophobic residues located in the flap region that forms the S_1 subsite can be substituted to effect a marked change of specificity for peptide substrates. For example, the mutation Tyr77->Phe causes a decrease in k_{cat} but no change in K_m on a synthetic hexapeptide LeuSerPhe(NO$_2$)NleAlaLeuOMe. However, a marked increase in K_m, but no change in k_{cat}, has been observed for an octapeptide LysProIleGluPhePhe(NO$_2$)ArgLeuOH. Other mutations involving Val113 and Phe114 in the same region also affect both the catalytic and substrate binding. Mutant Val113->Ser shows a 2-fold increase in k_{cat} without change in K_m.

REFERENCES

ABRAHMSEN, L.; TOM, J.; BURNIER, J.; BUTCHER, K. A.; KOSSIAKOFF, A.; and WELLS, J. A. 1991. Engineering subtilisin and its subtrates for efficient ligation of peptide bonds in aqueous solution. *Biochemistry 30*, 4151–4159.

ALBER, T.; DAO-PIN, S.; WILSON, K.; WOZNIAK, J. A.; COOK, S. P.; and MATTHEWS, B. W. 1987. Contributions of hydrogen bonds of Thr157 to the thermodynamic stability of phage T4 lysozyme. *Nature 330*, 41–46.

ALDEL MALAK, C. A. 1992. Calf chymosin as a catalyst of peptide synthesis. *Biochem. J. 288*, 941–943.

BARTLETT, P. A., and MARLOWE, C. K. 1987. Evaluation of intrinsic binding energy from a hydrogen bonding group in an enzyme inhibitor. *Science 235*, 569–571.

BONE, R.; SILEN, J. L.; and AGARD, D. A. 1989. Structural plasticity broadens the specificity of an engineered protease. *Nature 339*, 191–195.

BRAXTON, S., and WELLS, J. A. 1991. The importance of a distal hydrogen bonding group in stabilizing the transition state in subtilisin BPN'. *J. Biol. Chem. 266*, 11797–11800.

———. 1992. Incorporation of a stabilizing Ca^{2+}-binding loop into subtilisin BPN'. *Biochemistry 31*, 7796–7801.

CAMBOU, B., and KLIBANOV, A. M. 1984. Preparative production of optically active esters and alcohols using esterase-catalyzed stereospecific transesterification in organic media. *J. Am. Chem. Soc. 106*, 2687–2692.

CORDELLA-MIELE, E.; MIELE, L.; and MUKHERJEE, A. B. 1990. A novel transglutaminase-mediated post-translational modification of phospholipase A$_2$ dramatically increases its catalytic activity. *J. Biol. Chem. 265*, 17180–17188.

CORDELLA-MIELE, E.; MIELE, L., BENINATI, S.; and MUKHERJEE, A. B. 1993. Transglutaminase-catalyzed incorporation of polyamines into phospholipase A$_2$. *J. Biochem. 113*, 164–173.

COREY, D. R., and SCHULTZ, P. G. 1989. Introduction of a metal-dependent regulatory switch into an enzyme. *J. Biol. Chem. 264*, 3666–3669.

ERIKSSON, A. E.; BAASE, W. A.; ZHANG, X.-J.; HEINZ, D. W.; BLABER, M.; BALDWIN, E. P.; and MATTHEWS, B. W. 1992. Response of a protein structure to cavity-creating mutations and its relation to the hydrophobic effect. *Science 255*, 178–183.

ESTELL, D. A.; GRAYCAR, T. P.; and WELLS, J. A. 1985. Engineering an enzyme by site-directed mutagenesis to be resistant to chemical oxidation. *J. Biol. Chem. 260*, 6518–6521.

FEENEY, R. E. 1987. Chemical modification of proteins: Comments and perspectives. *Int. J. Peptide Protein Res. 29*, 145–161.

FEENEY, R. E., and WHITAKER, J. R. 1982. Modification of Proteins: Food, Nutritional, and Pharmaceutical Aspects. *Adv. Chem. Ser. 198,* American Chemical Society, Washington, D.C.

———. 1985. Chemical and enzymatic modification of plant proteins. In: Seed Storage Proteins, A. M. Altschul and H. L. Wilcke, eds., *New Protein Foods,* vol. 5, Academic Press, New York, pp. 181–219.

———. 1986. *Protein Tailoring for Food and Medical Uses.* Marcel Dekker, New York.

FEENEY, R. E.; WHITAKER, J. R.; WONG, D. W. S.; OSUGA, D. T.; and GERSHWIN, M. E. 1985. Chemical reactions of proteins. In: *Chemical Changes in Food During Processing,* T. Richardson, and J. W. Finley, eds., AVI Publ. Co., Westport, CT, pp. 255–287.

FERSHT, A. R. 1987. The hydrogen bond in molecular recognition. *TIBS 12,* 301–304.

FERSHT, A. R.; SHI, J.-P.; KNILL-JONES, J.; LOWE, D. M.; WILKINSON, A. J.; BLOW, D. M.; BRICK, P.; CARTER, P.; WAYE, M. M. Y.; and WINTER, G. 1985. Hydrogen bonding and biological specificity analyzed by protein engineering. *Nature 314,* 235–238.

GRAF, L.; JANCSO, A.; SZILAGYI, L.; HEGYI, G.; PINTER, K.; NARAY-SZABO, G.; HEPP, J.; MEDZIHRADSZKY, K.; and RUTTER, W. 1988. Electrostatic complementarity within the substrate-binding pocket of trypsin. *Proc. Natl. Acad. Sci. 85,* 4961–4965.

HE, J. J., and QUIOCHO, F. A. 1991. A nonconservative serine to cysteine mutation in the sulfate-binding protein, a transport receptor. *Science 251,* 1479–1481.

HIGAKI, J. N.; EVNIN, L. B.; and CRAIK, C. S. 1989. Introduction of a cysteine protease active site into trypsin. *Biochemistry 28,* 9256–9263.

INADA, Y.; TAKAHASHI, K.; YOSHIMOTO, T.; AJIMA, A.; MATSUSHIMA, A.; and SAITO, Y. 1986. Application of polyethylene glycol-modified enzymes in biotechnological processes: Organic solvent-soluble enzymes. *TIBTECH 4,* 190–194.

INADA, Y.; TAKAHSAHI, K.; YOSHIMOTO, T.; KODERA, Y.; MATSUSHIMA, A.; and SAITO, Y. 1988. Application of PEG-enzyme and magnetite-PEG-enzyme conjugates for biotechnological process. *TIBTECH 6,* 131–134.

NKINS, J.; JANIN, J.; REY, F.; CHIADMI, M.; VAN TILBEURGH, H.; LASTERS, I.; DE MAEYER, M.; VAN BELLE, D.; WODAK, S. J.; LAUWEREYS, M.; STANSSENS, P.; MRABET, N. T.; SNAUWAERT, J.; MATTHYSSENS, G.; and LAMBEIR, A.-M. 1992. Protein engineering of xylose (glucose) isomerase from *Actinoplanes missouriensis.* 1. Crystallography and site-directed mutagenesis of metal binding site. *Biochemistry 31,* 5449–5458.

JENSEN, V. J., and RUGH, S. 1987. Industrial-scale production and application of immobilized glucose isomerase. *Methods in Enzymology 136,* 356–370.

KAISER, E. T.; LAWRENCE, D. S.; and ROKITA, S. E. 1985. The chemical modification of enzymatic specificity. *Ann. Rev. Biochem. 54,* 565–595.

KARPUSAS, M.; BAASE, W. A.; MATSUMURA, M.; and MATTHEWS, B. W. 1989. Hydrophobic packing in T4 lysozyme probed by cavity-filling mutants. *Proc. Natl. Acad. Sci. USA 86,* 8237–8241.

KATZ, B. A., and KOSSIAKOFF, A. 1986. The crystallographically determined structures of atypical strained disulfides engineered into subtilisin. *J. Biol. Chem. 261*, 15480–15485.

KAWASE, M.; SONOMOTO, K.; and TANAKA, A. 1990. Improvement of heat stability of yeast lipase by chemical modification with a heterobifunctional photogenerated reagent. *J. Ferment. Bioengineer. 70*, 155–157.

KAZANDJIAN, R. Z., and KLIBANOV, A. M. 1985. Regioselective oxidation of phenols catalyzed by polyphenol oxidase in chloroform. *J. Am. Chem. Soc. 107*, 5448–5450.

KELLIS, J. T. JR.; NYBERG, K.; and FERSHT, A. R. 1989. Energetics of complementary side-chain packing in a protein hydrophobic core. *Biochemistry 28*, 4914–4922.

KIRCHNER, G.; SCOLLAR, M. P.; and KLIBANOV, A. M. 1985. Resolution of racemic mixtures via lipase catalysis in organic solvents. *J. Am. Chem. Soc. 107*, 7072–7076.

KITAGUCHI, H., and KLIBANOV, A. M. 1989. Enzymatic peptide synthesis via segment condensation in the presence of water mimics. *J. Am. Chem. Soc. 111*, 9272–9273.

IBANOV, A. M. 1989. Enzymatic catalysis in anhydrous organic solvents. *TIBS 14*, 141–144.

KOBAYASHI, M.; MIURA, M.; and ICHISHIMA, E. 1992. Modification of subsite Lys residue induced a large increase in maltosidase activity of Taka-amylase A. *Biochem. Biophys. Res. Comm. 183*, 321–326.

KUIPERS, O. P.; THUNNISSEN, M. M. G. M.; DE GEUS, P.; DIJKSTRA, B. W.; DRENTH, J.; VERHEIJ, H. M.; and DE HAAS, G. H. 1989. Enhanced activity and altered specificity of phospholipase A₂ by deletion of a surface loop. *Science 244*, 82–85.

KUMAGAI, I.; SUNADA, F.; TAKEDA, S.; and MIURA, K.-I. 1992. Redesign of the substrate-binding site of hen egg white lysozyme based on the molecular evolution of C-type lysozyme. *J. Biol. Chem. 267*, 4608–4612.

LAROUTE, V., and WILLEMOT, R.-M. 1989. Glucose condensation by glucoamylase in organic solvents. *Biotechnology Letters 11*, 249–254.

MANTAFOUNIS, D., and PITTS, J. 1990. Protein engineering of chymosin; modification of the optimum pH of enzyme catalysis. *Protein Engineering 3*, 605–609.

MARGOLIN, A. L., and KLIBANOV, A. M. 1987. Peptide synthesis catalyzed by lipases in anhydrous organic solvents. *J. Am. Chem. Soc. 109*, 3802–3804.

MATHEIS, G., and WHITAKER, J. R. 1987. A review: Enzymatic cross-linking of proteins applicable to foods. *J. Food Biochem. 11*, 309–329.

MATSUI, I.; ISHIKAWA, K.; MIYAIRI, S.; FUKUI, S.; and HONDA, K. 1991. An increase in the transglycosylation activity of *Saccharomycopsis* α-amylase altered by site-directed mutagenesis. *Biochim. Biophys. Acta 1077*, 416–419.

————. 1992A. A mutant α-amylase with enhanced activity specific for short substrates. *FEBS Lett. 310*, 216–218.

————. 1992B. Alteration of bond-cleavage pattern in the hydrolysis catalyzed by Saccharomycopsis α-amylase altered by site-directed mutagenesis. *Biochemistry 31*, 5232–5236.

MATSUMURA, M.; BECKTEL, W. J.; LEVITT, M.; and MATTHEWS, B. W. 1989A. Stabilization of phage T4 lysozyme by engineered disulfide bonds. *Proc. Natl. Acad. Sci. USA 86*, 6562–6566.

MATSUMURA, M.; BECKTEL, W. J.; and MATTHEWS, B. W. 1988. Hydrophobic stabilization in T4 lysozyme determined directly by multiple substitutions of Ile3. *Nature 334*, 406–410.

MATSUMURA, M., and MATTHEWS, B. W. 1989. Control of enzyme activity by an engineered disulfide bond. *Science 243*, 792–794.

MATSUMURA, M.; SIGNOR, G.; and MATTHEWS, B. W. 1989B. Substantial increase of protein stability by multiple disulfide bonds. *Nature 342*, 291–293.

MCGRATH, M. E.; WILKE, M. E.; HIGAKI, J. N.; CRAIK, C. S.; and FLETTERICK, R. J. 1989. Crystal structures of two engineered thiol trypsins. *Biochemistry 28*, 9264–9270.

MENG, M.; LEE, C.; BAGDASARIAN, M.; and ZEIKUS, J. G. 1991. Switching substrate preference of thermophilic xylose isomerase from D-xylose to D-glucose by redesigning the substrate binding pocket. *Proc. Natl. Acad. Sci. USA 88*, 4015–4019.

MITCHINSON, C., and WELLS, J. A. 1989. Protein engineering of disulfide bonds in subtilisin BPN'. *Biochemistry 28*, 4807–4815.

MORIHARA, K. 1987. Using proteases in peptide synthesis. *TIBTECH 5*, 164–170.

MRABET, N. T.; VAN DEN BROECK, A.; VAN DEN BRANDE, I.; STANSSENS, P.; LAROCHE, Y.; LAMBEIR, A.-M.; MATTHIJSSENS, G.; JENKINS, J.; CHIADMI, M.; VAN TILBEURGH, H.; REY, F.; JANIN, J.; QUAX, W. J.; LASTERS, I.; DE MAEYER, M.; and WODAK, S. J. 1992. Arginine residues as stabilizing elements in proteins. *Biochemistry 31*, 2239–2253.

NAKATSUKA, T.; SASAKI, T.; and KAISER, E. T. 1987. Peptide segment coupling catalyzed by the semisynthetic enzyme thiolsubtilisin. *J. Am. Chem. Soc. 109*, 3808–3810.

NEET, K. E., and KOSHLAND, D. E., JR. 1966. The conversion of serine at the active site of subtilisin to cysteine: A "chemical mutation." *Proc. Natl. Acad. Sci. USA 56*, 1606–1611.

NEET, K. E.; NANCI, A.; and KOSHLAND, D. E., JR. 1968. Properties of thiol-subtilisin. *J. Biol. Chem. 243*, 6392–6401.

NICHOLSON, H.; BECKTEL, W. J.; and MATTHEWS, B. W. 1988. Enhanced protein thermostability from designed mutations that interact with α-helix dipoles. *Nature 336*, 651–656.

NOEL, J. P.; BINGMAN, C. A.; DENG, T.; DUPUREUR, C. M.; HAMILTON, K. J.; JIANG, R.-T.; KWAK, J.-G.; SEKHARUDU, C.; SUNDARALINGAM, M.; and TSAI, M.-D. 1991. Phospholipase A$_2$ engineering. X-ray structural and functional evidence for the interaction of lysine-56 with substrates. *Biochemistry 30*, 11801–11811.

PACE, C. N. 1992. Contribution of the hydrophobic effect to globular protein stability. *J. Mol. Biol. 226*, 29–35.

PANTOLIANO, M. W.; LADNER, R. C.; BRYAN, P. N.; ROLLENCE, M. L.; WOOD, J. F.; and POULOS, T. L. 1987. Protein engineering of subtilisin BPN': Enhanced stabilization through the introduction of two cysteines to form a disulfide bond. *Biochemistry 26*, 2077–2082.

PERRY, L. J., and WETZEL, R. 1984. Disulfide bond engineered into T4 lysozyme: Stabilization of the protein toward thermal inactivation. *Science 226*, 555–557.

PHILIPP, M., and BENDER, M. L. 1983. Kinetics of subtilisin and thiolsubtilisin. *Mol. Cell. Biochem. 51*, 5–32.

PICKERSGILL, R. W.; SUMMER, I. G.; COLLINS, M. E.; WARWICKER, J.; PERRY, B.; BHAT, K. M.; and GOODENOUGH, P. W. 1991. Modification of the stability of phospholipase A$_2$ by charge engineering. *FEBS Lett. 281*, 219–222.

QUAX, W. J.; MRABET, N. T.; LUITEN, R. G. M.; SCHUNRHUIZEN, P. W.; STANSSENS, P.; and LASTERS, I. 1991. Enhancing the thermostability of glucose isomerase by protein engineering. *Bio/Technology 9*, 738–742.

RIVA, S.; CHOPINEAU, J.; KIEBOOM, A. P. G.; and KLIBANOV, A. M. 1988. Protease-catalyzed regioselective esterification of sugars and related compounds in anhydrous dimethyl formamide. *J. Am. Chem. Soc. 110*, 584–589.

RUSSELL, A. J., and FERSHT, A. R. 1987. Rational modification of enzyme catalysis by engineering surface charge. *Nature 328*, 496–500.

RUSSELL, A. J.; THOMAS, P. G.; and FERSHT, A. R. 1987. Electrostatic effects on modification of charged groups in the active site cleft of subtilisin by protein engineering. *J. Mol. Biol. 193*, 803–813.

SAKURAI, T.; MARGOLIN, A. L.; RUSSELL, A. J.; and KLIBANOV, A. M. 1988. Control of enzyme enantioselectivity by the reaction medium. *J. Am. Chem. Soc. 110*, 7236–7237.

SALI, D.; BYCROFT, M.; and FERSHT, A. R. 1988. Stabilization of protein structure by interaction of α-helix dipole with a charged side chain. *Nature 335*, 740–743.

SANDBERG, W. S., and TERWILLIGER, T. C. 1988. Influence of interior packing and hydrophobicity on the stability of a protein. *Science 245*, 54–57.

———. 1991. Energetics of repacking a protein interior. *Proc. Natl. Acad. Sci. USA 88*, 1706–1710.

SCHNEIDER, L. V. 1991. A three-dimensional solubility parameter approach to nonaqueous enzymology. *Biotechnol. Bioengineer. 37*, 627–638.

SHARP, K. A.; NICHOLLS, A.; FRIEDMAN, R.; and HONIG, B. 1991. Extracting hydrophobic free energies from experimental data: Relationship to protein folding and theoretical models. *Biochemistry 30*, 9686–9697.

SHIRLEY, B. A.; STANSSENS, P.; HAHN, U.; and PACE, C. N. 1992. Construction of hydrogen bonding to the conformational stability of ribonuclease T1. *Biochemistry 31*, 725–732.

SMITH, J. L.; BILLINGS, G. E.; and YADA, R. Y. 1991. Chemical modification of amino groups in *Mucor miehei* aspartyl proteinase, porcine pepsin, and chymosin. I. Structure and function. *Agric. Biol. Chem. 55*, 2009–2016.

STEHLE, P.; BAHSITTA, H.-P.; MONTER, B.; and FURST, P. 1990. Papain-catalyzed synthesis of dipeptides: A novel approach using free amino acids as nucleophiles. *Enzyme Microb. Technol. 12*, 56–60.

SUZUKI, J.; SASAKI, K.; SASAO, Y.; HAMU, A.; KAWASAKI, H.; NISHIYAMA, M.; HORINOUCHI, S.; and BEPPU, T. 1989. Alteration of catalytic properties of chymosin by site-directed mutagenesis. *Protein Engineering 2*, 563–569.

TAKAHASHI, K.; SAITO, Y.; and INADA, Y. 1988. Lipase made active in hydrophobic media. *JAOCS 65*, 911–916.

TAKAHASHI, K.; TAMAURA, Y.; KODERA, Y.; MIHAMA, T.; SAITO, Y.; and INADA, Y. 1987. Magnetic lipase active in organic solvents. *Biochem. Biophys. Res. Comm. 142*, 291–296.

THUNNISSEN, M. M. G. M.; KALK, K. H.; DRENTH, J.; and DIJKSTRA, B. W. 1990. Structure of an engineered porcine phospholipase A_2 with enhanced activity at 2.1 A resolution. Comparison with the wild-type porcine and *Crotalus atrox* phospholipase A_2. *J. Mol. Biol. 216*, 425–439.

TOMA, S.; CAMPAGNOLI, S.; MARGARIT, I.; GIANNA, R.; GRANDI, G.; BOLOGNESI, M.; DE FILIPPIS, V.; and FONTANA, A. 1991. Grafting of a calcium-binding loop of thermolysin to *Bacillus subtilis* neutral protease. *Biochemistry 30*, 97–106.

VAN TILBEURGH, H.; JENKINS, J.; CHIADMI, M.; JANIN, J.; WODAK, S. J.; MRABET, N. T.; and LAMBEIR, A.-M. 1992. Protein engineering of xylose (glucose) isomerase from *Actinoplanes missouriensis*. 3. Changing metal specificity and the pH profile by site-directed mutagenesis. *Biochemistry 31*, 5467–5471.

WARSHEL, A.; NARAY-SZABO, G.; SUSSMAN, F.; and HWANG, J.-K. 1989. How do serine proteases really work? *Biochemistry 28*, 3629–3637.

WELLS, J. A.; CUNNINGHAM, B. C.; GRAYCAR, T. P.; and ESTELL, D. A. 1987. Recruitment of substrate-specificity properties from one enzyme into a related one by protein engineering. *Proc. Natl. Acad. Sci. USA 84*, 5167–5171.

WELLS, J. A., and POWERS, D. B. 1986. In vivo formation and stability of engineered disulfide bonds in subtilisin. *J. Biol. Chem. 261*, 6564–6570.

WETZEL, R.; PERRY, L. J.; BAASE, W. A.; and BECKTEL, W. J. 1988. Disulfide bonds and thermal stability in T4 lysozyme. *Proc. Natl. Acad. Sci. USA 85*, 401–405.

WHITAKER, J. R., and PUIGSERVER, A. J. 1982. Fundamentals and applications of enzymatic modifications of proteins: An overview. In: Modification of Proteins: Food, Nutritional, and Pharmaceutical Aspects. *Adv. Chem. Ser. 198*, American Chemical Society, Washington, D.C.

WILKS, H. M.; HALSALL, D. J.; ATKINSON, T.; CHIA, W. N.; CLARKE, A. R.; and HOLBROOK, J. J. 1990. Designs for a broad substrate specificity keto acid dehydrogenase. *Biochemistry 29*, 8587–8591.

WILKS, H. N.; HART, K. W.; FEENEY, R.; DUNN, C. R.; MUIRHEAD, H.; CHIA, W. N.; BARSTOW, D. A.; ATKINSON, T.; CLARKE, A. R.; and HOLBROOK, J. J. 1988. A specific, highly active malate dehydrogenase by redesign of a lactate dehydrogenase framework. *Science 242*, 1541–1544.

WONG, C.-H.; CHEN, S.-T.; HENNEN, W. J.; BIBBS, J. A.; WANG, Y.-F.; LIU, J. L.-C.; PANTOLIANO, M. W.; WHITLOW, M.; and BRYAN, P. N. 1990. Enzymes in organic synthesis: Use of subtilisin and highly stable mutant derived from multiple site-specific mutations. *J. Am. Chem. Soc. 112*, 945–953.

WONG, C.-H.; SHEN, G.-J.; PEDERSON, R. L.; WANG, Y.-F.; and HENNEN, W. J. 1991. Enzymatic catalysis in organic synthesis. *Meth. Enzymol. 202*, 591–620.

WU, Z.-P., and HILVERT, D. 1989. Conversion of a protease into an acyl transferase: selenolsubtilisin. *J. Am. Chem. Soc. 111*, 4513–4514.

ZAKS, A., and KLIBANOV, A. M. 1986. Substrate specificity of enzymes in organic solvents vs. water is reversed. *J. Am. Chem. Soc. 108*, 2767–2768.

ZHONG, Z.; LIU, J. L.-C.; DINTERMAN, L. M.; FINKELMAN, M. A. J.; MUELLER, W. T.; ROLLENCE, M. L.; WHITLOW, M.; and WONG, C.-H. 1991. Engineering subtilisin for reaction in dimethylformamide. *J. Am. Chem. Soc. 113*, 683–684.

Chapter 3

Amylolytic Enzymes

Amylolytic enzymes are a group of starch-degrading enzymes that include the industrial important amylases, and a number of enzymes with potential applications, such as pullulanase, α-glucosidase, and cyclodextrin glycosyltransferase. Amylases have found major applications in the starch sweetener industry. α-Amylase is used in the liquefaction step producing soluble dextrins, while glucoamylase further hydrolyzes the dextrins to glucose in the saccharification step. β-Amylase is used in the production of high-maltose syrups. These enzymes also play an important role in the brewing industry, in distilleries and in the baking process, as described in Chapter 1.

GENERAL CHARACTERISTICS OF AMYLASES

Amylases are glycosidases that catalyze the hydrolysis of α-D-1,4-glucosidic linkages of starch and related oligo- and polysaccharides by the transfer of a glucosyl residue (donor) to H_2O (acceptor).

α-Amylase

α-Amylase (EC 3.2.1.1, 1,4-α-D-glucan glucanohydrolase) is an endoglucosidase that cleaves the α-1,4-glucosidic bond of the substrate at internal positions to yield dextrins and oligosaccharides with the C1-OH in the α configuration. Although the endo-enzyme implies random cleavage, numerous experiments have suggested that the enzyme action follows a definite pattern depending on the source of the enzyme. (Refer to the section later in this chapter, Action Patterns.) An α-1,4 linkage neighboring an α-1,6 branching point in the substrate is resistant to attack by the enzyme (Takeda and Hizukuri 1981). The enzyme has a MW range of 50 kD, and requires Ca^{++} for stability and activity. The pH optimum varies depending on the enzyme source (6.0–7.0 for mammalian, 4.8–5.8 for *Aspergillus oryzae*, 5.85–6.0 for *Bacillus subtilis*, 5.5–7.0 for *Bacillus licheniformis*). Temperature optimum for activity varies—70–72°C for the α-amylase produced by *Bacillus subtilis*, and 90°C for the enzyme from *Bacillus licheniformis* (Anon 1984B). The applications of α-amylase in starch processing are at temperatures ≥ 105°C. The loss of enzyme activity in these operations, which is due to irreversible thermal inactivation (Brosnan et al. 1992), has prompted intensive use of genetic engineering to develop novel, thermostable enzymes. Commercially used α-amylase in particular for starch liquefaction is most often obtained from *Bacillus licheniformis* (Dawson and Allen 1984).

β-Amylase

β-Amylase (EC 3.2.1.2, 1,4-α-D-glucan maltohydrolase) occurs commonly in seeds of higher plants and in sweet potatoes. It is an exo-enzyme that successively cleaves maltosyl units from the nonreducing end of the polymer substrate to yield maltose with the C1-OH in the β-configuration (Thoma et al. 1971). Unlike α-amylase, which bypasses the α-1,6 branch point, β-amylase stops the action at such locations. The enzyme has a molecular weight of ∼50 kD with the exception of sweet potato β-amylase, which is a tetramer of 215 kD. The tetrameric structure is not required for catalytic function of the enzyme; an active monomer of sweet potato β-amylase has been isolated (Ann et al. 1990). The enzyme does not require Ca^{++} for activity. The pH optimum for activity ranges from 5.0 for wheat, malt, and sweet potato to 6.0 for soybean and pea. Plant β-amylases often consists of several isoforms. Soybean β-amylase exists predominantly as isozymes 2 and 4 with pI = 5.25 and 5.5, respectively (Ji et al. 1990). The maize enzyme comprises two isozymes with pI of 4.25 and 4.40 (Doyen and Lauriere 1992). In alfalfa seeds, five isozymes have been identified (pI = 5.05, 4.97, 4.85, 4.82, and 4.77) (Kohno et al. 1990). Barley and wheat consist of two isozymes, free and bound forms. The bound form is converted to the free form during germination (Lauriere et al. 1992). β-Amylase has also been found in several species of *Bacillus*, including *Bacillus polymyxa*, *Bacillus cereus*, and *Bacillus megaterium*, and in *Clostridium thermosulfurogenes*. These enzymes have MW of 30 to 160 kD; the highest value may represent association of subunits as in the case of the sweet potato enzyme.

Glucoamylase

Glucoamylase (EC 3.2.1.3, 1,4-α-D-glucan glucohydrolase), (also occasionally called γ-amylase or amyloglucosidase in some literature), is an exoglucosidase that catalyzes the hydrolysis of both α-D-1,4- and α-D-1,6-glucosidic linkages at the branch point, although hydrolysis of the latter occurs at a slower rate. The enzyme removes glucose units successively from the nonreducing end of the substrate. The rate of hydrolysis increases with the chain length of the substrate. The end product is exclusively glucose in the β configuration. Glucoamylase is a microbial enzyme found only in molds and yeasts. The enzyme is commercially produced from *Aspergillus niger*. The enzyme has a pH optimum of 4.0–4.4, with stability over a pH range of 3.5–5.5. The enzyme is stable at a temperature range of 40–65°C with optimum range 58–65°C in the hydrolysis of starch at pH 4.2 (Anon 1984A). Glucoamylase is a glycoprotein containing 5–20% carbohydrate, mainly mannose, glucose, galactose, and glucosamine. The molecular weight varies depending on the source of the enzyme, and many exist in multiple forms.

STRUCTURE OF α-AMYLASE

Amino Acid Sequences

Complete amino acid sequences of α-amylases have been established by sequencing the protein from *Aspergillus oryzae* (Toda 1982) and porcine pancreas (Kluh 1986; Pasero et al. 1986). Amino acid sequences are also deduced from the gene sequences of the following α-amylases: *Bacillus subtilis* (Yang et al. 1983; Yamazaki et al. 1983), *Bacillus licheniformis* (Yuuki et al. 1985), *Bacillus amyloliquefaciens* (Takkinen et al. 1983), *Bacillus stearothermophilus* (Ihara et al. 1985, Gray et al. 1986), *Bacillus polymyxa* (Uozumi et al. 1989), *Aspergillus oryzae* (Wirsel et al. 1989), *Saccharomycopsis fibuligera* (Itoh et al. 1987B), *Streptomyces hygroscopicus* (Hoshiko et al. 1987), *Micrococcus* sp. 207 (Kimura and Horikoshi 1990), barley (Rogers 1985B; Rogers and Williams 1983), wheat (Baulcombe et al. 1987), mouse pancreas (Gumucio et al. 1985), and human salivary and pancreas (Emi et al. 1988; Gumucio et al. 1988).

Three well-conserved regions are evident in comparing the amino acid sequences of α-amylases from plants (barley), mammals (porcine pancreas, mouse salivary) and bacteria (*Bacillus amyloliquefaciens* and *Bacillus subtilis*) (Rogers 1985A). Comparison of amino acid sequences of α-amylases from *Aspergillus oryzae*, porcine pancreatic, *Bacillus subtilis*, *Bacillus stearothermophilus*, *Bacillus amyloliquefaciens*, and *Bacillus licheniformis* reveals three short homologous sequences (Ihara et al. 1985). A comparison of the amino acid sequences of 11 α-amylases (*Bacillus stearothermophilus*, *Bacillus amyloliquefaciens*, *Bacillus subtilis*, *Aspergillus oryzae*, barley, pancreas (human, mouse, rat, hog), saliva (mouse, human) suggest an additional fourth highly homologous region (Nakajimi et al. 1986). The four regions of homologous sequences start at amino acid residues 117, 202, 230, and 292 in the α-amylase from *Aspergillus oryzae* (Fig. 3.1). Sequences in between the regions show very few discernable similar-

```
PO  αAMY                                               QYAPQTQ
HM  αAMY                                     MGKNGSLCCFSCWAQYSSNTQ
BA  αAMY             MIQKRKRTVSFRLVLMCTLLFVSLPITKTSAVNGTLMQYF
BL  αAMY             MKQQKRLYARLLTLFALIFLLPHSAAAAANLNGTLMQYF
AO  αAMY                          ATPADWRSQSIYFLLTDRFARTDGSTTA


PO  αAMY    8     SGRTDIVHLFEWRWVDIALECERYLGPKGFGGVQVSPPNE
HM  αAMY   11     QGRTSIVHLFEWRWVDIALECERYLAPKGFGGVQVSPPNE
BA  αAMY   10     EWYTPNDGQHWKRLQNDAEHLSDIGITAVWIPPAYKGLSQ
BL  αAMY   12     EWYMPNDGQHWKRLQNDSAYLAEHGITAVWIPPAYKGTSQ
AO  αAMY   29     TCNTADQKYCGGTWQGIIDKLDYIQGMGFTAIWITPVTAQ


PO  αAMY   48     NVVVTNPSRPWWERYQPVSYKLCTRSGNENEFRDMVTRCN
HM  αAMY   51     HVAIHNPFRPWWERYQPVSYKLCTRSGNEDEFRNMVTRCN
BA  αAMY   50     SDNGYGPYDLYDLGEFQQKGTVRTKYGTKSELQDAIGSLH
BL  αAMY   52     ADVGYGAYDLYDLGEFHQKGTVRTKYGTKGELQSAIKSLH
AO  αAMY   69     LPQTTAYGDAYHGYWQQDIYSLNENYGTADDLKALSSALH


PO  αAMY   88     NVGVRIYVDAVINHMCGSGAAAGTGTTC
HM  αAMY   91     NVGVRITV------MCGNAVSAGTSSTC
BA  αAMY   90     SRNVQVYG-V-L--KAGADATEDVTAVEVNPANRNQETSE
BL  αAMY   92     SRDINVYG-V----KGGADATEDVTAVEVDPADRNRVISG
AO  αAMY  109     ERGMYLMV-V-A--MGYDGAGSSVDYSV


PO  αAMY  116                              GSYCNPGNREFPAVPYSAWD
HM  αAMY  119                              CSYFNPGSRDFPAVPYSGWD
BA  αAMY  130     EYQIKAWTDFRFPGRGNTYSDFKWHWYHFDGADWDESRKI
BL  αAMY  132     EHRIKAWTHFHFPGRGSTYSDFKWHWYHFDGTDWDESRKL
AO  αAMY  137                                           FKPFSSQD


PO  αAMY  136     FNDGKCKTASGGIESYNDPYQVRDCQLVGLLDLALEKDYV
HM  αAMY  139     FNDGKCKTGSGDIENYNDATQVRDCRLSGLLDLALGKDYV
BA  αAMY  170     SRIFKFRGEGKAWDWEVSSENGNYDYLMYADVDYDHPDVV
BL  αAMY  172     NRIYKFQG  KAWDWEVSNENGNYDYLMYADIDYDHPDVA
AO  αAMY  145     YFHPFCFIQNYEDQTQVEDCWLGDNTVSLPDLDTTKDVVK
```

Figure 3.1. Comparison of amino acid sequences of α-amylases: PO (porcine pancreatic isozyme I, Kluh 1981, Pasero et al. 1986), HM (human salivary, 512 amino acids deduced from cDNA, 500 amino acids for the mature protein, 12 amino acids for the signal peptide, Nishida et al. 1986), BA (*Bacillus amyloliquefaciens*, 514 amino acids deduced from cDNA, mature protein with 483 amino acids, 31 amino acids for the signal peptide, Takkinen et al. 1983), BL (*Bacillus licheniformis*, 512 amino acids deduced from cDNA, mature protein with 483 amino acids, 29 amino acids for the signal peptide, Yuuki et al. 1985), AO (*Aspergillus oryzae*, Toda et al. 1982, Wirsel et al. 1989). Catalytic residues for the porcine pancreatic and the *Aspergillus oryzae* enzymes are indicated by boldface. The N-terminal amino acids are double underlined.

```
PO αAMY  176   RSMIADYLNKLIDIGVAGFRLDASKHMWPGDIKAVLDKLH
HM αAMY  179   RSKIAEYMNHLIDIGVA---------MWPCDIKAILDKLH
BA αAMY  210   AETKKWGIWYANELSLD---I--A--IKFS
BL αAMY  210   AEIKRWGTWYANELQLD------V--IKFS
AO αAMY  185   NEWYDWVGSLVSNYSID-L-I-TV--VQ


PO αAMY  216   NLNTNW    FPAGSRPFIFQEVIDLGGEAIKSGEYFSNG
HM αAMY  219   NLNSNW    FPECSKPFIYQ----LGGEPIKSSDYFGHG
BA αAMY  240   FLRDWVQAVRQATGKEMFTVA-YWQNNAGKLENYLNKTSF
BL αAMY  240   FLRDWVNHVREKTGKEMFTVA-YWQNDLGALENYLNKTNF
AO αAMY  213     KDFWPGYNKAAGVYCIGE-L-GDPAYTCPYQNVMDG


PO αAMY  252   RVTEFKYGAKLGTVVRKWSGEKMSYLKNWGEGWGFMPSDR
HM αAMY  255   RVTEFKYGAKLGTVIRKWTGEKMSYLKNWEEGWGFMPSDR
BA αAMY  280   NQSVFDVPLHFNLQAASSQGGGYDMRRLLDGTVVSRHPEK
BL αAMY  280   NHSVFDVPLHYQFHAASTQGGGYDMRKLLNSTVVSKHPLK
AO αAMY  249   VLNYPIYYPLLNAFKSTSGSMDDLYNMINTVKSDCPDSTL


PO αAMY  292   ALVFVDNHDNQRGHGAGG SSILTFWDAYRKLVAVGFMLA
HM αAMY  295   ALV--D---NRRGHGAGGGATTLTFWDARLYNHAVGFMLA
BA αAMY  320   AVT--E---TQPGQSLESTVQTWFKPLAYAFILTRESGYP
BL αAMY  320   AVT--D---TQPGQSLESTVQTWFKPLAYAFILTRESGYP
AO αAMY  289   LGT--E--DNPRFASYTNDIALAKNVAAFIILNDGIPIIY


PO αAMY  331   HPYGFTRVMSSYRWARNFVNGEDVNDWIGPPNNNGVI
HM αAMY  335   ILMDFTRVMSSYRWPRYFENCNDVNDWVGPPNDNGVT
BA αAMY  360   QVFYGDMYGTKGTSPKEIPSLKDNIEPILKARKEYAYGPQ
BL αAMY  360   QVFYGDMYGTKGDSQREIPALKHKIEPILKARKQYAYGAQ
AO αAMY  329   AGQEQHYAGGNDPANREATWLSGYPTDSELY KLIASA


PO αAMY  368   KEVTINADTTCGNDWVCEHRWREIRNMVWFRNVVDGEPFA
HM αAMY  372   KEVTIHPDTTCGNDWVCEHRWRQIRNMVNFRNVVDGQPFA
BA αAMY  400   HDYIDHPDVIGWTREGDSSAAKSGLAALITDGPGGSKRMY
BL αAMY  400   HDYFDHHDIVGWTREGDSSVANSGLAALITDGPGGAKRMY
AO αAMY  366   NAIRNYAISKDTGFVTYKNWPIYKDDTTIAMRKGTDGSQI


PO αAMY  408   NWWDNGSNQVAFGRGNRGFIVFNNDDWQLSSTLQTGLPAG
HM αAMY  412   NWWDNVSIQVAFGRGNRGFIVFNNGDWTFSLTLQTGLPAG
BA αAMY  440   AGLKNAGETWYDITGNRSDTVKIGSDGWGEFHVNDGSVSI
BL αAMY  440   VGRQNAGETWHDITGNRSEPVVINSEGWGEFHVNGGSVSI
AO αAMY  406   VTILSNKGASGDSYTLSLSGAGYTAGQQLTEVIGCTTVTV
```

Figure 3.1. (Continued).

```
PO αAMY   448   TYCDVISGDKVGNSCTGIKVYVSSDGKAQFSISNSAEDPF
HM αAMY   452   TYCDVISCDKINGHCTAIKIYVSDDGKAHFSISNSAIDPF
BA αAMY   480   YVQK    483
BL αAMY   480   YVQR    483
AO αAMY   446   GSDGNVPVPMAGGLPRVLYPTEKLAGSKICSSS    478


PO αAMY   488   IAIHAESKL   496
HM αAMY   492   IAIRAESKL   500
BA αAMY
BL αAMY
AO αAMY
```

Figure 3.1.   (Continued).

ities. In the homologous regions, amino acids analogous to His122, Asp206, His210, Glu230, His296, and Asp297 in the *Aspergillus oryzae* α-amylase are conserved in all other amylases.

Three-Diminensional Structures

Porcine pancreatic α-amylase. The tertiary structure of porcine pancreatic α-amylase is known (Fig. 3.2) (Payan et al. 1980; Buisson et al. 1987; Qian et al. 1993). The enzyme is composed of 496 amino acids with domains A (1–99 and 170–404), B (101–168), and C (405–496). Domain A has a structural arrangement of $(\alpha/\beta)_8$ barrel. Domain B is inserted in domain A between β strand 3 and α helix 3, forming an extended loop that contains two antiparallel β sheets (4 + 3 strands) plus a single 2-stranded parellel β sheet. This extended structure is tied to domain A via a disulfide bridge (Cys70-Cys115). Other disulfides are: 28–86, 141–160, 378–384, and 450–462 (Pasero et al. 1986). Domain C is found in the C terminus, which folds into two 4-stranded and one 2-stranded antiparallel β sheets.

Aspergillus oryzae α-amylase. The *Aspergillus oryzae* enzyme consists of 478 amino acid residues folded into two domains (Matsuura et al. 1979, 1980, 1984). Domain A, which contains the N-terminal 380 amino acid residues, consists of a $(\beta/\alpha)_8$ barrel with eight β strands alternating with eight helices joined by loops. The loop linking the β strand 3 and helix 3 contains an extended chain of three β strands. Domain B is an 8-stranded antiparallel β sheet linked to domain A via a single polypeptide chain. Four disulfide bonds are located at residues 30–38, 150–164, 240–283, and 439–474. The enzyme is N-glycosylated at Asn197.

Barley α-amylase. The barley isozyme 2 shows the same basic features in the overall structure (Kadziola et al. 1994). Domain A (1–88, 153–350) forms a $(\beta/\alpha)_8$ barrel. The loop protruding between β strand 3 and α helix 3 of domain

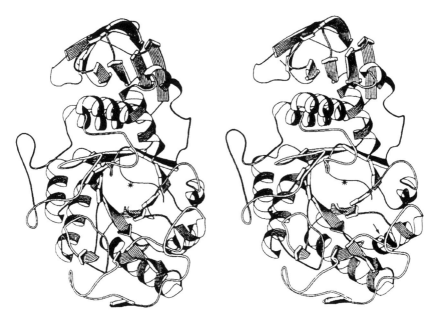

Figure 3.2. A stereo view of the pancreatic a-amylase structure. The chloride ion is shown as an asterisk near the axis of the β-barrel. The arrow points to the calcium ion binding site. (From Qian et al. 1993. *J. Mol. Biol. 231*, 791, with permission. Copyright 1993 Academic Press, Ltd.)

A constitutes domain B (89–152), which assumes an irregular fold stabilized by Ca⁺⁺ ions. The C-terminal domain C (351–403) forms a 5-stranded antiparallel β sheet.

Active site. The active site of the three α-amylases described is located in a cleft at the C-terminal side of the β strands of the $(\alpha/\beta)_8$ barrel of domain A. All the amino acid residues involved in the active site are found on the loops or β strands close to the C-terminal ends adjacent to the loops (MacGregor 1988). The location of active amino acids in these specific positions allows flexibility and variation among the α-amylases from different sources without changing the basic $(\beta/\alpha)_8$ structure. The two catalytic amino acid residues in the *Aspergillus oryzae* α-amylase, Glu230 and Asp297, are located ~3 A from the glucosidic oxygen of the substrate unit between subsites 4 and 5. Structural analysis of pancreatic α-amylase complexed with acarbose (a transition state analog inhibitor) reveals that the catalytic residues are Glu233 and Asp300 (corresponding to Glu230 and Asp297 in the *Aspergillus oryzae* enzyme) (Qian et al. 1994).

Substrate binding site. A substrate binding model for the *Aspergillus ory-*

zae α-amylase involving the seven subsites of the active site has been proposed (Matsuura et al. 1984). Bond cleavage occurs between subsites 4 and 5. The amino acid residues His122, His296, and Asp297 at subsite 4 interact with the hydroxyl groups at C6, C2, and C3 of the same glucose unit in the substrate. Among the other amino acid residues involved in the subsite binding, Lys209, Tyr155, Val231, His210, and Asp206 are conserved in both the *Aspergillus oryzae* and porcine pancreatic α-amylases. In the porcine enzyme, the region around subsite 3 (bond cleavage between subsite 3 and 4 in the pancreatic enzyme) is surrouded by the side chains of Glu233, Asp300, and Asp197. Both Glu233-$O_{\varepsilon1}$ and Asp300-$O_{\delta1}$ hydrogen bonded with the glycosidic NH of the inhibitor acarbose (corresponding to the glycosidic oxygen in substrate).

The role of a histidine in reactions catalyzed by the α-amylases has received considerable attention (Hoschke et al. 1980A; Kita et al. 1982; Dua and Kochhar 1985). The His residue is involved both in substrate and calcium binding, and these interactions maintain proper orientation of the substrate in the active site (Vihinen et al. 1990; Buisson 1987). The His201 located in subsite 4 in porcine and human pancreatic α-amylases interacts with the substrate residue at the subsite, and may induce a localized alteration of conformation that may assist in optimum catalysis. Because His201 and the two ligands for Cl^- (Arg195, Asp197) are located on the same peptide loop, the activation effect of Cl^- may also be related to this induce-fit process (Ishikawa et al. 1993). Studies on the human pancreatic α-amylase by site-directed mutagenesis confirm the multifunctional role of His201 in calcium binding, substrate binding, control of pH optimum, and activation by chloride ion (Ishikawa et al. 1992). The substrate units are bound to the subsites by hydrogen bonding, and also by dipole interaction between the amylose chain and the parallel β barrel.

Chemical modification studies indicate that Tyr and Trp residues are also involved in substrate binding (Hoschke et al. 1980B; Kochhar and Dua 1985A, B). In the pancreatic enzyme, the aromatic ring of this residue stacks onto the glucose ring in subsite 3, causing a conformational strain in the sugar substrate (Qian et al. 1994).

Site-directed mutagenesis of a thermostable α-amylase from *Bacillus stearothermophilus* reveals a possible function for the highly conserved Arg232 (Arg204 in the *Aspergillus oryzae* α-amylase). The Arg233 residue interacts with another conserved amino acid, Tyr63 (Tyr82 in the *Aspergillus oryzae* α-amylase), which lies in the active site region. The ionic interaction tends to hold the Tyr in place, so that the active center is accessible to substrate binding (Vihinen et al. 1990).

A second substrate binding site has been identified on the surface of porcine pancreatic α-amylase (Loyter and Schramm 1966; Payan et al. 1980), ~20 A from the active site, near the amino end of α helix 8 (Buisson et al. 1987). This binding site may play a role in anchoring the long chain starch substrate at a distal glucose unit to facilitate a specific action pattern.

Calcium binding site.   The calcium binding sites in both the porcine pancreatic enzyme and the *A. oryzae* enzyme are superimposible. Four amino acid residues in the porcine pancreatic α-amylase, Asn100, His201, Arg158, and Asp167 (Asn121, His210, Glu162, and Asp175 in the *Aspergillus oryzae* α-amylase) function to coordinate the calcium ion (Buisson et al. 1987; Boel et al. 1990; Qian et al. 1993). The first two amino acids belong to β strand 3 and 4 of domain A, and the latter two are from the β sheets in domain B. Asp167 acts as a bidentate ligand. Three H_2O molecules also are involved. The Ca^{++} binding to eight ligands assumes a coordinate geometry of a distorted pentagonal bipyramid. The function of the metal ion is to stabilize and maintain the active cleft in the correct conformation for enzyme activity by positioning the Asp197 (in β strand 4) in proximity to Asp300 (in β strand 7).

In the *Aspergillus oryzae* α-amylase, the corresponding amino acids Asn121, Glu162, and Asp175 are located in the extended loop, and the His210 in β strand 4. All are close to the Ca^{++} ion, and are likely involved in binding and stabilizing the structure (MacGregor 1988; Vihinen et al. 1990). Barley α-enzyme also contains an analogous Ca^{++} binding site with a distorted pentagonal bipyramid structure coordinated with seven ligands (Kadziola et al. 1994). The importance of His210 in Ca^{++} binding and substrate binding is also supported by mutation studies of the *Bacillus stearothermophilus* enzyme. A His210->Asp mutation reduced both the specific activity and thermal stability of the enzyme (Vihinen et al. 1990).

In addition to the high-affinity binding site, a second Ca^{++} binding site has been located at the bottom of the substrate binding site in the *Aspergillus niger* enzyme (Boel et al. 1990). The Ca^{++} ion has a distorted octahedral coordination with Glu230 and Asp206 and three water molecules, suggesting that this Ca^{++} binding site may possibly be involved in the catalytic step. In the barley enzyme, two additional Ca^{++} sites are located. All three Ca^{++} ions in the enzyme are bound to domain B, and probably function in stabilizing the structure fold of this domain.

Chloride binding site.   The porcine pancreatic enzyme has been known to be activated by chloride ion. The binding site for Cl^- is located close to the active site near the carboxyl end of the β barrel of domain A (Buisson et al. 1987). The Cl^- is coordinated by five charged ligands—Arg337 (bidentate), Arg195 (bidentate), Asn298, and H_2O molecule (Qian et al. 1993). Binding of Cl^- to mammalian α-amylases causes a shift in optimum pH to a neutral pH in addition to an increase in activity.

STRUCTURE OF β-AMYLASE

Amino Acid Sequences

Amino acid sequences are known for β-amylases from soybean (Mikami et al. 1989), sweet potato (Toda et al. 1993), barley (Kreis et al. 1987, Yoshigi et al. 1994), rye (Rorat et al. 1991), *Arabidopsis thaliana* (Monroe et al. 1991), *Bacillus polymyxa* (Kawazu et al. 1987; Uozumi et al. 1989), and *Clostridium thermosulfurogenes* (Kitamoto et al. 1988). Sweet potato β-amylase is homotetrameric, whereas most plant and bacterial enzymes are monomeric enzymes. The sweet potato enzyme (subunit) and the seed enzymes of soybean, barley, and rye share ~50–60% similarities in amino acid sequences (Fig. 3.3). The subunit of sweet potato mature enzyme is composed of 498 amino acids with a calculated MW = 55,880, compared to 495 for soybean, and 535 for barley. The β-amylases of *Bacillus polymyxa* and *Clostridium thermosulfurogenes* show 60% and 30% sequence similarity with the soybean enzyme, respectively. The gene from *Bacillus polymyxa* has been cloned into *Bacillus subtilis* and the nucleotide sequence coding the enzyme (Rhodes et al. 1987; Kawazu et al. 1987) translates into a MW of ~160 kD, which is proteolytically cleaved into enzymatically active fragments (42–70 kD).

```
SWP βAMY         MAP IPGVMP IGNYVSLYVMLP LGVVNADNVFPDKEKVEDE
SOY βAMY          MATSDSNMLL---PV------GV-NVD-V-EDPDGLKEQ
BAR βAMY           MEVNVKG---QV------PA-SVN-R-EKGDELRAQ

SWP βAMY    40   LKQVKAGGCDGVMVDVWWGIIEAKGPKQYDWSAYRELFQL
SOY βAMY    39   -LQLRAA-V----------II-L----Q---R--RS--Q-
BAR βAMY    37   -RKLVEA-V----------LV-G----A---S--KQ--E-

SWP βAMY    80   VKKCGLKIQAIMSFHQCGGNVADAVFIPIPQWILQIGDKN
SOY βAMY    79   -QEC--TL------------G-I-N------VLDI-ESN
BAR βAMY    77   -QKA--KL------------G-A-N------VRDV-TRD

SWP βAMY   120   PDIFYTNRAGNRNQEYLSLGVDNQRLFQGRTALEMYRDFM
SOY βAMY   119   H-----NRS-T--K---TV----EPI-H--T-IEI-S-Y-
BAR βAMY   117   P-----DGH-T--I---TL----QPL-H--S-VQM-A-Y-
```

Figure 3.3. Comparison of amino acid sequences of β-amylases: SWP (sweet potato subunit, 499 amino acids deduced from cDNA, mature protein with 498 amino acids, Yoshida and Nakamura 1991), SOY (soybean, Mikami et al. 1988), BAR (barley, 535 amino acids deduced from cDNA, mature protein with 535 amino acids, Kreis et al. 1987). Catalytic residue of the soybean enzyme is indicated with boldface. The N-terminal amino acids for the mature proteins are double underlined.

```
SWP βAMY 160   ESFRDNMADFLKAGDIVDIEVGCGAAGELRYPSYPETQGW
SOY βAMY 159   K---E--S---ES-L-I-----L-P--EL------QSQ--
BAR βAMY 157   T---E--K---DA-V-V-----L-P--EM------QSH--

SWP βAMY 200   VFPGIGEFQCYDKYMVADWKEAVKQAGNADWEMPGKGAGT
SOY βAMY 199   E--R----Q-----LK--F-A-VARA-HPE--L-DD AGK
BAR βAMY 197   S--G----I-----LQ--F-A-AAAV-HPE--F-ND VGQ

SWP βAMY 240   YNDTPDKTEFFRPNGTLQDGYGQVFLTWYSNKLIIHGDQV
SOY βAMY 238   ---V-ES-G--KS---YVTEK-KF--T----K-LN---QI
BAR βAMY 236   ---T-ER-Q--RD---YLSEK-RF--A----N-IK---RI

SWP βAMY 280   LEEANKVFVGLRVNIAAKVSGIHWWYNHVSHAAELTAGFY
SOY βAMY 278   -D----A-L-CK-KL-I-V------KVEN--------YY
BAR βAMY 276   -D----V-L-YK-QL-I-I------KVPS--------YY

SWP βAMY 320   NVAGRDGYRPIARMLARHHATLNFTCLEMRDSEQPAEAKS
SOY βAMY 318   -LND-----P-----S--H-IL----L----S--PSD-K-
BAR βAMY 316   -LHD-----T-----K--R-SI----A----L--SSQ-M-

SWP βAMY 360   APQELVQQVLSSGWKEYIDVAGENALPRYDATAYNQMLLK
SOY βAMY 358   G-Q--------G--R-DIR--G--------A----QILLN
BAR βAMY 356   A-E--------A--R-GLN--C--------P----TILRN

SWP βAMY 400   LRPNGVNLNGPPKLKMSGLTYLRLSDDLLQTDNFELFKKF
SOY βAMY 398   AK-Q-V-NN---KLSMF-V------DD-LQKS-FNI--K-
BAR βAMY 396   AR-H-I-QS---EHKLF-F------NQ-VEGQ-YVN--T-

SWP βAMY 440   VKKMHADLDPSPNAISP  AVLERSNSAITID E LMEAT
SOY βAMY 438   -LK---DQDYCANPQKYNH-ITPLKPSAPK-PIEV-LE-T
BAR βAMY 436   -DR---NLPRDPYVD P M-PLPRSGPEIS-E MI-QA-Q

SWP βAMY 476    KGSRPFPWYDVTDMPVDGSNPFD   498
SOY βAMY 478   -PTL---WLPE--MK-D-   495
BAR βAMY 473   P-LQ ---FQEH--LP- -PTGGMGGQAEGPTCGMGGQVK

SWP βAMY
SOY βAMY
BAR βAMY 511   GPTGGMGGQAEDPTSGIGGELPATM   535
```

Figure 3.3. (Continued).

Three-Dimensional Structures

The crystal structure of the soybean β-amylase (isozyme 2) has a dimension of 50 × 55 × 60 A, composed of a large domain and a small domain, separated by a cleft 20 A long and 15 A deep (Fig. 3.4) (Mikami et al. 1989, 1992, 1993). The large domain consists of a $(\beta/\alpha)_8$ structure with the parallel β strands wound into a barrel, each β strand alternating with an α helix. The interior wall of the β barrel is lined with residues with charged and polar side chains. There are 15 loops interconnecting these strands and helices. A small lobe is formed by three long loops (L3, L4, and L5) extending from β3, β4, and β5 from the small domain. A long tail loop (L8, ~50 residues) following the α8 helix, at the C-terminal end, covers the entrance of the barrel.

Active site.   The active site is situated in a cleft at the interface between the two domains. The catalytic Glu186, identified by affinity labeling (Nitta et al. 1989) is located at the base of the active site and is included in the highly conserved regions among both plant and bacterial β-amylases. The Cys95, which is the reactive cysteine identified in various chemical modification studies, is located in loop L3, close to the active site subsite 1 where the binding of the nonreducing end of a substrate occurs. Besides loop L3, where subsite 1 is located, interactions also involve loops L5 and L6. The active site can accommodate two maltose molecules, corresponding to a maltotetreose. Binding of substrate may cause a possible movement of some segments in loop L3.

STRUCTURE OF GLUCOAMYLASE

Amino Acid Sequences

The amino acid sequences of three *Aspergillus* glucoamylases are presented in Fig. 3.5. Amino acid sequences are also deduced from the cDNA gene of the following: *Aspergillus niger* (Svensson et al. 1983), *Aspergillus oryzae* (Hata et al. 1991), *Aspergillus awamori* (Nunberg et al. 1984), *Rhizopus oryzae* (Ashikari et al. 1989), *Saccharomycopsis fibuligera* (Itoh et al. 1987A), *Saccharomyces diastaticus* (Yamashita 1985; Lambrechts et al. 1991), *Saccharomyces cerevisiae* (Yamashita et al. 1987). Five segments of homologous sequences in these enzymes are shown in Fig. 3.6.

Multiple Forms

Fungal glucoamylases generally exist in multiple forms. For example, *Aspergillus niger* glucoamylase exists in two forms, GAI and GAII. The GAI form (616 amino acids, MW 71 kD) consists of a catalytic site region (1–440), a Thr- and Ser-rich linker region (441–512) heavily glcosylated, and a C-terminal region (513–616). The latter is missing in the GAII form (MW 61 kD), and is referred

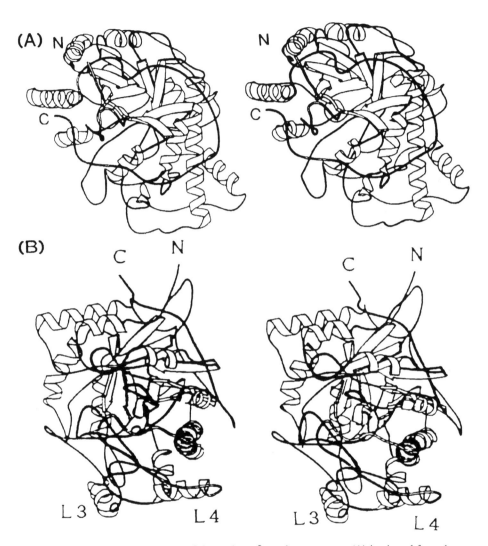

Figure 3.4. A stereo view of the soybean β-amylase structure. (A) is viewed from the N-terminal side of the (β/α)$_8$ barrel and (B) from the small domain. (From Mikami et al. 1992. *J. Biochem. 112*, 543, with permission. Copyright 1992 The Japanese Biochemical Society.)

```
AN GLA        M SFRSLL ALS GLVCTGLANVISKRATLDSWLSNEATV
AA GLA        M --R-L- --S -LVCTGLAN-ISKRAT--S---N--TV
AO GLA        MV--S-C-R--AL-SSVLAVQP-LRQATG--T---T--NF

AN GLA    14  ARTAILNNIGADGAWVSGADSGIVVASPSTDNPDYFYTWT
AA GLA    14  ART----------AWVS--DS-I-V----TDN--------
AO GLA    41  SRQ----------QSAQ--SP-V-I----KSD--------

AN GLA    54  RDSGLVLKTLVDLFRNGDTSLLSTIENYISAQAIVQGISN
AA GLA    54  ------L---------N--TS--ST--NY--A--IV-----
AO GLA    81  ------M---------G--AD--PI--EF--S--RI-----

AN GLA    94  PSGDLSSGAGLGEPKFNVDETAYTGSWGRPQRDGPALRAT
AA GLA    94  ---D----AGL----------Y--S--------------
AO GLA   121  ---A----G L----------F--A--------------

AN GLA   134  AMIGFGQWLLDNGYTSTATDIVWPLVRNDLSYVAQYWNQT
AA GLA   134  ---G--Q--LD-GY--T---I---L------------N-T
AO GLA   160  ---S--E--VE-SH--I---L---V------------S-S

AN GLA   174  GYDLWEEVNGSSFFTIAVQHRALVEGSAFATAVGSSCSWC
AA GLA   174  -Y---E--N-S--F-I--Q--------A--TA-----SW-
AO GLA   200  -F---E--Q-T--Y-V--S--------S--KT-----PY-

AN GLA   214  DSQAPEILCYLQSFWTGSFILANFDSSRSGKDANTLLGSI
AA GLA   214  -----EIL----------F-L---DSS-----A--L----
AO GLA   240  -----QVR----------Y-Q---GGG-----I--V----

AN GLA   254  HTFDPEAACDDSTFQPCSPRALANHKEVVDSFRSIYTLNN
AA GLA   254  -----E-A---S------PR------E-V-------TL-D
AO GLA   280  -----Q-T---A------AR------V-T-------AI-S

AN GLA   294  GLSDSEAVAVGRYPEDTYYNGNPWFLCTLAAAEQLYDALY
AA GLA   294  -LSDSE---------T---------C-------------
AO GLA   320  -RAENQ---------S---------T-------------

AN GLA   334  QWDKQGSLEVTDVSLDFFKALYSDAATGTYSSSSSTYSSI
AA GLA   334  ----Q---EV-----D-------D------S--SST-SS-
AO GLA   360  ----I---AI-----P-------S------A--TTV-KD-
```

Figure 3.5. Comparison of amino acid sequences of glucoamylases: AN (*Aspergillus niger*, 640 amino acids deduced from cDNA, 616 amino acids for the mature protein, signal peptide with 18 amino acids , 6 amino acids for the propeptide, Boel et al. 1984), AA (*Aspergillus awamori*, 637 amino acids deduced from cDNA, 616 amino acids for the mature protein, 21 amino acids signal peptide, Nunberg et al. 1984), AO (*Aspergillus oryzae*, 612 amino acids deduced from cDNA, Hata et al. 1991). Catalytic residues are indicated by boldface. The N-terminal amino acids of the proteins are double underlined.

```
AN GLA   374   VDAVKTFADGFVSIVETHAASNGSMSEQYDKSDGEQLSAR
AA GLA   374   -D---TF---F-S--E-H---N---SE--D-S--E-L---
AO GLA   400   -S---AY---Y-Q--Q-Y---T---AE--T-T--S-T---

AN GLA   414   DLTWSYAALLTANNRRNSVVPASWGETSASSVPGTCAATS
AA GLA   414   ----------------S----S----S-S-V-GT-AA--
SO GLA   440   ----------------A----P----A-T-I-SA-ST--

AN GLA   454   AIGTYSSVTVTSWPSIVATGGTTTTATPTGSGSVTSTSKT
AA GLA   454   -I------TV----S-VATGGTTTTATPTGSGSVTSTSKT
AO GLA   480   -S------VI----T-SGYPGAPDSP

AN GLA   494   TATASKTSTSTSSTSCTTPTAVAVTFDLTATTTYGENIYL
AA GLA   494   TATASKTSTSTSSTSCTT--A-A--FDLT---T---D-YL
AO GLA   506             CQV--T-S--FAVK---V---S-KI

AN GLA   534   VGSISQLGDWETSDGIALSADKYTSSDPLWYVTVTLPAGE
AA GLA   534   --------D-ET-DGI--S-DK--SSD---YV-VT----E
AO GLA   531   --------S-NP-SAT--N-DS--TDN---TG-IN----Q

AN GLA   574   SFEYKFIRIESDDSVEWESDPNREYTVPQACGTSTATVTD
AA GLA   574   --------IESDDS-E-------E----QA--TST-TVT-
AO GLA   571   -------- VQNGA-T-------K----ST--VKS-VQS-

AN GLA   614   TWR    616
AA GLA   614   T--    616
AO GLA   610   V--    612
```

Figure 3.5. (Continued).

to as the granular starch binding domain. The GAII form is synthesized by post-translational proteolytic cleavage (Svensson et al. 1983, 1986) or by differential splicing of the mRNA (Boel et al. 1984) from the C-terminal region. The glycosyl chains in the linker region contain predominantly mannose (mono-, di-, and tri-units) linked to Ser and Thr (Gunnarsson et al. 1984).

The *Aspergillus awamori* glucoamylase GAI (MW 90 kD) contains a segment of 45 amino acid residues that is missing from the GAII form. This linker segment is characterized by a high content of hydroxy amino acids (Thr$_8$, Ser$_4$, Asp$_5$, Glu$_3$, Gly$_4$, Ala$_4$, Val$_3$, His$_2$), and extensively glycosylated, in particular with mannose residues (Hayashida et al. 1982).

Three forms of glucoamylase are produced from *Rhizopus oryzae*—GAI (74 kD), GAII (58.6 kD), and GAIII (61.4 kD). The latter two forms are the

```
Aspergillus niger GLA          35    GIVVASPSTDNPDYFYTWTRDSGL
Aspergillus awamori GLA        35    -I-V----TDN-----T-----G-
Aspergillus oryzae GLA         62    -V-I----KSD-----T-----G-
Rhizopus oryzae GLA           172    -FIA--L-TAG---Y-A----AA-
Saccharomyces diastaticus GLA 380    -V-I----QTH-----Q-I---A-
Saccharomyces cerevisiae GLA  112    -V-I----QTH-----Q-I---A-
Saccharomycopsis fibuligera GLA 50   -T-I----TSN---Y-Q-----AI

Aspergillus niger GLA         104    LGEPKFNVDETAYTGSWGRPQRDGPALR
Aspergillus awamori GLA       104    ---------E--Y--SW----R------
Aspergillus oryzae GLA        130    ---------E--F--AW----R------
Rhizopus oryzae GLA           238    -------P-GSGY--AW----N----E-
Saccharomyces diastaticus GLA 450    --D--W---N--F-EPW----N------
Saccharomyces cerevisiae GLA  182    --D--W---N--F-EDW----N------
Saccharomycopsis fibuligera GLA 123  -------T-GS-Y--AW----N------

Aspergillus niger GLA         162    DLSYVAQYWNQTGYDLWEEVNGSSFFTIAVQ
Aspergillus awamori GLA       162    --S--AQY-NQT-Y---E----SS---IA--
Aspergillus oryzae GLA        188    --S--AQY-SQS-F---E--Q-TS-Y-VA-S
Rhizopus oryzae GLA           297    --D--VNV-SNGCF---E----VH-Y-LM-M
Saccharomyces diastaticus GLA 515    --RFIIDH-NSS-F---E----MH---LL--
Saccharomyces cerevisiae GLA  247    --RFIIDH-NSS-F---E----MH---LL--
Saccharomycopsis fibuligera GLA 193  --E--IGY-DST-F---E-NQ-RH---SL--

Aspergillus niger GLA         300    AVAVGRYPEDTYYNGN       PWFLC
Aspergillus awamori GLA       300    AV-V------T-YNGN       ----C
Aspergillus oryzae GLA        326    AV-V------S-YNGN       ----T
Rhizopus oryzae GLA           438    GNSI------T-NGNGNSQGNS---A
Saccharomyces diastaticus GLA 664    GI-L------V-DGYGVGE----V-A
Saccharomyces cerevisiae GLA  396    GI-L------V-DGYGFGE----V-A
Saccharomycopsis fibuligera GLA 340  GA-I------V-NGDGSSE------A

Aspergillus niger GLA         396    GSMSEQYDKSDGEQLSARDLTWSYAALLTA
Aspergillus awamori GLA       396    --MSE-YDKSD-EQLS---L---Y------
Aspergillus oryzae GLA        222    --MAE-YTKTD-SQTS--D----Y------
Rhizopus oryzae GLA           540    --LAEEFDRTT-LSTG--D----H-S-I--
Saccharomyces diastaticus GLA 771    ATW E-TGN
Saccharomyces cerevisiae GLA  503    ATW E-TGN
Saccharomycopsis fibuligers GLA 452  --LNE-LNRYT-YSTG-YS----SG---E-
```

Figure 3.6. Conserved regions in glucoamylases: *Aspergillus niger*, *Aspergillus awamori*, and *Aspergillus oryzae* GLA (described in Fig. 3.5 legend), *Rhizopus oryzae* (604 amino acids deduced from cDNA, 579 amino acids for the mature protein, 25 amino acids for the signal peptide, Ashikari et al. 1986), *Saccharomyces diastaticus* (778 amino acids deduced from cDNA, Yamashita et al. 1985), *Saccharomyces cerevisiae* (510 amino acids deduced from cDNA, Yamashita et al. 1987), *Saccharomycopsis fibuligera* (591 amino acids deduced from cDNA, 564 residues for the mature protein, 27 amino acids for the signal peptide, Itoh et al. 1987A).

products from proteolytic cleavage of GAI with the loss of its N-terminal segment (Takahashi et al. 1985).

Three-Dimensional Structures

The *Aspergillus awamori* enzyme consists of a catalytic domain (1–400), a glycosylated linker (~72 residues), and a starch-binding domain (513–616). The linker segment assumes a semirigid extended structure that wraps around the middle section of the α/α barrel. O-Glycosylation is essential for maximum stability for the two domains against thermal denaturation (Evans et al. 1990; Williamson et al. 1992A, B; Neustroev et al. 1993). In addition to the extensive O-glycosylation, there are two glycosyl chains N-linked to Asn171 and Asn395 that are hydrogen bonded to the surface of the enzyme molecule through their terminal sugars (Aleshin et al. 1992, 1994). The catalytic active fragment (minus the C-terminal binding domain) of glucoamylase from *Aspergillus awamori* var. X100 shows a crystal structure of an α/α barrel (in contrast to the α/β barrel in the other two amylases), with an inner core of six helices connected antiparallelly to six peripheral α helices by loops (Fig. 3.7A, B) (Aleshin et al. 1992, 1994). Wrapped around the barrel is the glycosylated linker segment, which is confirmed to be an extended conformation. The structure consists of 3 disulfides: 210-213, 262-270, and 222-449, the last acting as a linkage of the glycosylated segment to the main body.

Active site.   The active site, located at the interior cavity of the inner core, has a longitudinal shape with 15-A diameter at the open end, and a length of 10 A. The active site is lined with a number of polar side chains. Comparison of amino acid sequences indicates that four regions of homology exist among a number of glucoamylases (Tanaka et al. 1986; Itoh et al. 1987A; Sierks et al. 1989), and are conformationally similar. These conserved regions in the primary structures contain carboxyl groups and Trp, which seem to be involved in catalysis as determined by chemical modification studies (Inokuchi et al. 1982A, B; Ohnishi et al. 1990).

Raw starch binding site.   The GAI forms of the fungal glucoamylases described above exhibit a unique property in that they digest not only soluble starch, but also raw starch granules. Glucoamylases from *Aspergills phoenicis*, *Aspergillus cinnamomeus*, *Mucor rouxianus*, *Penicillium oxalicum*, and *Chalara paradoxa* are also able to digest starch granules. All forms (GAI, GAII, etc.) are equally active towards soluble starch and other suitable substrates; thus, it is believed that the segment cleaved from the GAI forms must contain the binding site for raw starch. The binding domain is located in the C-terminal end, except for the *Rhizopus* enzyme in which the N-terminus is the binding region. Sequence comparison reveals similar conserved segments of starch granule binding domain in other amylases as well. These include α-amylase from *Streptomyces limosus*,

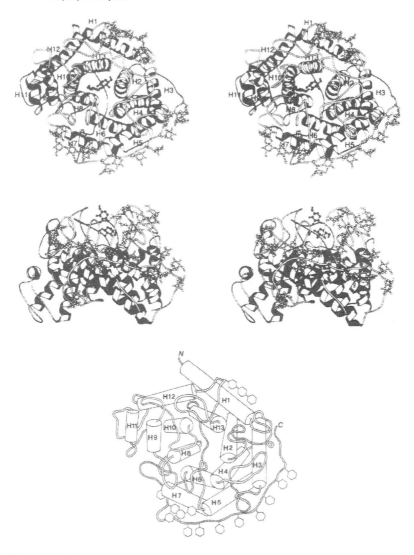

Figure 3.7. (A) A stereo view of glucamylase from *Aspergillus awamori* var. X100 as viewed (top) into the packing void of the α/α barrel and (bottom) perpendicular to the barrel. Atoms of the glycosyl chains are represented by open circles. Atoms of the bound 1-deoxynojirimycin are represented by filled circles. (From Aleshin et al. 1994. *J. Mol. Biol. 238*, 579, by courtesy of Academic Press.) (B) Schematic model of the glucoamylase. (From Aleshin et al. 1992. *J. Biol. Chem. 267*, 19294, with permission. Copyright 1992 American Society for Biochemistry and Molecular Biology.)

β-amylase from *Clostridium thermosulfurogenes*, maltogenic α-amylase from *Bacillus stearothermophilus*, and maltotetraose-forming amylase from *Pseudomonus stutzeri* (Svensson et al. 1989). In addition, porcine pancreatic α-amylase (Hayashida et al. 1991), *Lactobacillus amylovorus* α-amylase (Imam et al. 1991), and cereal (barley) α-amylases (Weselake and Hill 1983) contain binding sites for starch granules.

The C-terminal segment (residues 471–616) of the *Aspergillus niger* enzyme (GAI) has been produced by proteolytic hydrolysis and purified to homogeneity (Belshaw and Williamson 1990). The peptide binds to granular corn starch at 1.08 nmol/mg starch. Using malto-oligosaccharides of chain length 2–11, a binding stoichiometry of 1 mole substrate/mole protein was obtained (Belshaw and Williamson 1993). The affinity increases with increasing polymerization up to nine glucose units, a minimum chain length that can mimic the single-helix amlyase V form. The C-terminal binding domain also binds to linear α-1,6-linked glucose oligosaccharide apparently at a low-affinity site. The starch binding domain of *Aspergillus niger* GA has a high percentage of residues forming strand structure (Williamson et al. 1992A, B). The interaction at the interface consists primarily of hydrophobic forces, mostly involving the conserved Trp590 and Trp615 (Svensson et al. 1989). In a similar study, intact enzyme of *Aspergillus saitoi* has a binding constant of $1.6 \times 10^5 \ M^{-1}$ at pH 3.0 and 4°C, while the truncated enzyme shows a $K_b = 3.2 \times 10^3 \ M^{-1}$, practically little binding of raw starch (Takahashi et al. 1990). In a study on the binding mode of *Aspergillus awamori* enzyme with cyclodextrins, Trp562 is implicated in contributing to the formation of inclusion complexes (Goto et al. 1994).

It has been postulated that the adsorption of affinity site dissociates the water molecules clustered on the surface of the starch granule through the hydroxyl groups in the carbohydrate moiety via hydrogen bonding. This disruptive effect leads to alteration of the local structure, and penetration of water molecules into the micelles, so that the enzyme can then act on the localized soluble starch (Hayashida et al. 1982, 1989).

The degradation of a starch granule is dependent on several factors, such as amylose/amylopectin ratio, % crystallinity, accessible surface, particle size, and particular type of starch granule structure. α-Amylase has been known to digest starch granules from potato, corn, amylo-maize, waxy-maize, and wheat differently, as observed by electron microscopy (Gallant et al. 1972). Rice and tapioca starch are among the most susceptible to α-amylase digestion in a comparison of 11 starch types (Cone et al. 1990). The rate of hydrolysis may also be related to the crystal type of the starch granule with the A-type crystalline amylose being hydrolyzed more rapidly than the B-type (Williamson et al. 1992B), probably due to a greater surface/volume ratio found in the A-type, although the B-type molecules are more loosely bound.

NUCLEOPHILIC DISPLACEMENT OR OXOCARBONIUM ION?

α-Amylases hydrolyze the α-D-1,4 bond with retention of configuration, and for actions of β-amylases and glucoamylases, inversion of configuration occurs. Two types of mechanisms have been postulated to account for the difference in configuration of the products produced by these enzymes.

Nucleophilic Displacement Mechanism

Koshland (1959) first proposed nucleophilic displacement mechanisms for amylolytic enzymes. For α-amylase, which retains the configuration of the substrate, a double displacement reaction is envisioned (Fig. 3.8). A concerted protonation of the glucosidic oxygen by an acidic group of a general acid and the nucleophilic attack of a carboxylate anion at C1 cleaves the glycosidic bond, resulting in formation of a covalent glucosyl-enzyme (a carboxyl-acetal ester) intermediate. A second displacement by a general base catalysis hydrolyzes the ester intermediate to yield a product with an α-configuration (Koshland 1959; Matsuura et al. 1984; Robyt 1984).

For β-Amylase, which catalyzes hydrolysis with inversion of configuration, the reaction proceeds via a single displacement (Fig. 3.9A) (Koshland 1959; Hanozet et al. 1981). A general acid catalysis weakens the maltosyl C-O bond, and the attack of a water molecule on the β side of the substrate is assisted by a general base. The same type of mechanism applies to glucoamylase-catalyzed reactions. Inversion of configuration by β-amylase-catalyzed hydrolysis can also be explained by a mechanism analogous to the double displacement of α-amylase. In this case, the water molecule is directed to the carbon atom of the carboxyl group that makes the covalent acetal-ester bond to yield the β product (Fig. 3.9B) (Lai et al. 1974; Robyt 1984).

Oxocarbonium-mediated Mechanism

In this mechanism, protonation of the glucosidic oxygen by general acid catalysis, together with an enzyme-induced distortion of the substrate to a half-chair conformation, produces an enzyme-stabilized, glucosyl oxocarbonium ion

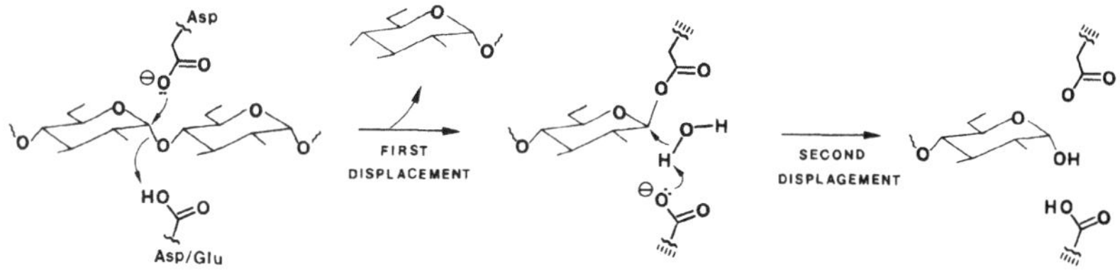

Figure 3.8.   Double displacement Mechanism for a-amylases. (Koshland 1959).

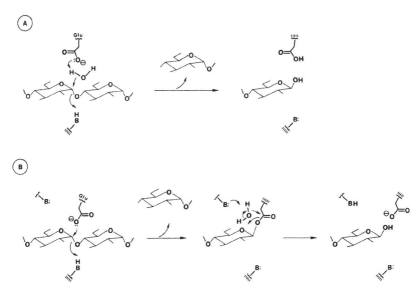

Figure 3.9. Reaction mechanism for β-amylases. (A) Single displacment (Koshland 1959), (B) Double displacement (From Robyt 1984. In: *Starch Chemistry and Technology*, p. 106, with permission. Copyright 1984 Academic Press, Inc.).

intermediate. The half-chair conformation relieves the steric strain at C1, and enhances the accessibility of the anomeric carbon to a front-side attack by H_2O as promoted by α-amylase. A back-side attack directed by β-amylase results in an inversion of the configuration (Fig. 3.10) (Thoma 1968; Wakim et al. 1969). Both α- and β-amylases are assumed to utilize a common mechanism, and the configuration of the product is governed by the direction of attack by the water molecule.

Glycosyl oxocarbonium ions have a short, but significant life time in aqueous solution with $k_{HOH} \sim 10^{-12}$ s (Amyes and Jencks 1989). Its stability is governed by the degree of nucleophilicity, the electrostatic interactions, and the positioning of catalytic groups in the active site. A complete formation of an oxocarbonium ion intermediate with an S_N1 reaction mechanism is preferred if bond breaking precedes the attack of H_2O facilitated by general base catalysis as described above. On the other hand, a stable oxocarbonium intermediate is unlikely to exist, if nucleophilic attack occurs prior to the complete formation of the cation. In this case, the reaction is "enforced" to undergo a concerted displacement S_N2 mechanism, possibly through an oxocarbonium ion-like transition state.

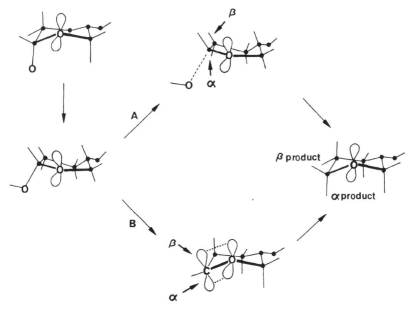

Figure 3.10. A schematic representation of possible conformations and intermediates along the reaction pathway for amylase-catalyzed hydrolysis of glucopyranoside. (From Thoma 1968. *J. Theoret. Biol. 19*, 306, with permission. Copyright 1968 Academic Press, Ltd.)

α-Amylase

The rate-pH profile for α-amylases show a bell-shaped curve. All α-amylases with structures elucidated are shown to contain Glu/Asp as catalytic residues. In the *Aspergillus oryzae* enzyme, the highly conserved Glu230 and Asp297 have been suggested to be the general acid and base, respectively (Matsuura et al. 1984). A refined structure of the porcine enzyme identifies Glu233 and Asp300 in proximity to one another at the active site (Qian et al. 1993). Homologous residues (Glu230, Asp297) are also found in the barley enzyme (Kadziola et al. 1994).

The formation of a covalent glycosyl-enzyme intermediate has been detected in porcine pancreatic α-amylase at cryogenic states using [1-^{13}C]maltotetraose as substrate (Tao et al. 1989). However, in another study of the crystal complex structure of pancreatic α-amylase and (transition state analog) inhibitor (acarbose), both the catalytic residues, Glu233 and Asp300, are positioned from the glycosidic linkage at a distance longer than that expected for a covalent bond (Qian et al. 1994). According to this study, formation of glycosyl-enzyme therefore is unlikely. Instead, hydrolysis occurs by general acid catalysis, as the two residues are suitably oriented for protonation of the glycosidic oxygen. The acid catalyst is likely to be Glu233, assuming a carbonium ion-mediated mechanism. If the Glu233 acts as an acid catalyst in protonating the glycosidic oxygen, it has to be protonated at the pH optimum of the enzyme, which is ~6–7. In the crystal structure of the native enzyme, Glu233-$O_{\varepsilon 1}$ is positioned close to the Cl$^-$ ion at a distance of ~4.8 A . The resulting electrostatic effect of the anion may shift the carboxyl group of Glu233 to a higher pK value (Qian et al. 1993, 1994).

β-Amylase

The rate-pH profile of a β-amylase-catalyzed reaction shows a bell-shaped curve, with pKs = 3.5 and 8.2 (Thoma and Koshland 1960B; Nitta et al. 1979). Affinity labelling has identified the carboxylate of Glu186 as the catalytic residue at the active site of soybean β-amylase (Nitta et al. 1989). Glu186 is located at the base of the cavity and is included in the highly conserved regions among both plant and bacterial β-amylases. Substitution of Glu186 in the soybean enzyme leads to complete loss of activity (Totsuka et al. 1994). The same result can be obtained by mutation of Asp101 to neutral and acidic residues, indicating that an Asp residue may also be essential to catalysis.

The conserved thiol groups in β-amylases have received considerable attention. Chemical modification of some reactive SH groups often results in a decrease of the enzyme activity and slight change in the Michaelis constant (Spradlin and Thoma 1970; Mikami et al. 1980). Site-directed mutagenesis of the three Cys in the *Bacillus polymyxa* enzyme demonstrates that the Cys residues are not directly involved in catalysis (Uozumi et al. 1991). The reactive Cys95 in the soybean enzyme is found in proximity to the binding site subsite 1, and it is believed that the attachment of alkylating agents used in the modification studies introduces steric stresses causing a slight structural distortion that prevents the substrate from productive binding (Spradlin and Thoma 1970; Mikami et al. 1980, 1992). Substitution of Cys95->Ser causes a considerable loss of thermal stability (Totsuka et al. 1994). Another cysteine in the soybean enzyme that is sensitive to chemical modification is identified to be Cys343. This Cys residue is located around subsite 3 and 4, near the entrance to the active site where the inhibitor α-cyclodextrin binds (Mikami et al. 1993, 1994). Alkylation of Cys343, a procedure commonly conducted in chemical modification, would therefore result in blocking the active site from substrate access, similar to the action of cyclodextrin.

Evidence for a single displacement for β-amylase comes from studies revealing that the enzyme hydrolyzes both α- and β-glycosyl fluorides to yield products of β configuration in each case (Hehre et al. 1979; Kitahata et al. 1981). The kinetics of the hydrolysis of β-maltosyl fluoride by the sweet potato β-amylase suggests the involvement of two molecules of substrates at the active site.

The fluoride is first displaced by the C4-OH of the second substrate molecule in a condensation forming a β-maltotetraosyl fluoride which then rapidly hydrolyzes to β-maltose (Fig. 3.11). Both processes involve a concerted acid-base catalysis. This finding is substantiated by labelling the sweet potato enzyme with 2,3-epoxy-propyl α-D-[U-^{13}C]glucopyranoside, identifying Glu187 specifically esterified in the reaction (Toda et al. 1993). In a later study, sweet potato and soybean β-amylases were found to convert maltal to the 1-deoxymaltose β anomer. The hydration process has been identified consistent with general acid catalysis, a carbonium ion-mediated mechanism (Fig. 3.12) (Kitahata et al. 1981; 1991). The reaction is conceived to proceed first by enzyme protonation of the double bond to yield a 2-deoxymaltosyl carbonium ion, and then followed by nucleophilic attack by H$_2$O. The hydration process in this case seems to be consistent with general acid catalysis in the oxocarbonium ion mechanism.

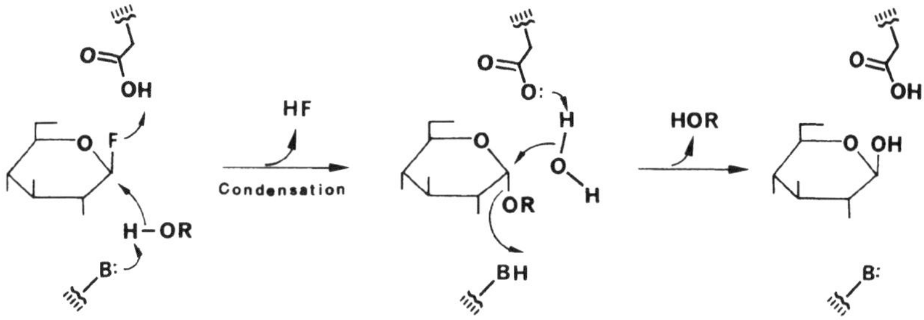

Figure 3.11.   Concerted displacement mechanism for the action of β-amylase on β-mal-tosyl fluoride. (From Hehre et al. 1979. *J. Biol. Chem. 254*, 5949, with permission. Copyright 1979 American Society for Biochemistry and Molecular Biology.)

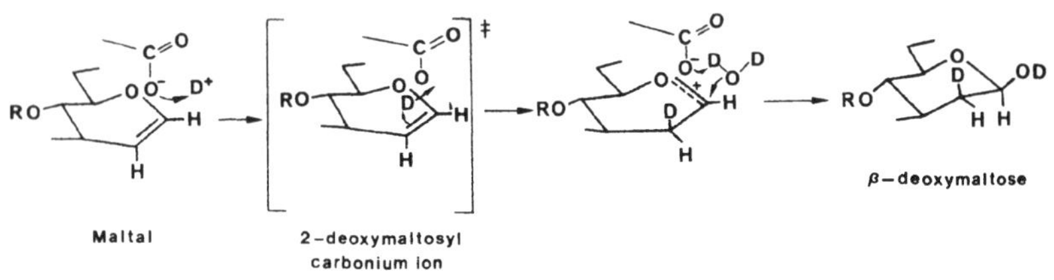

Figure 3.12.   Mechanism of maltal hydration catalyzed by β-amylase. (From Kitahata et al. 1991. *Biochemistry 30*, 6773, with permission. Copyright 1991 American Chemical Society.)

Glucoamylase

The pH-activity profile of glucoamylases indicates catalysis involving two essential carboxyl groups with pKs of 2.7 and 5.9. Inhibition and mutagenesis studies as well as crystallographic analysis reveal Glu179 and Asp176 as the general acid and base of pK = 5.9 and 2.7, respectively (Meagher and Reilly 1989; Svensson et al. 1990; Sierks et al. 1993; Aleshin et al. 1992, 1994). High-resolution structural studies of the *Aspergillus awamori* enzyme complexes with 1-deoxynojirimycin suggest that Glu179 and Glu400 as acid and base catalysts (Aleshin et al. 1994). Mutation of Glu179->Gln results in a 2,000-fold decrease in k_{cat} with maltoheptaose as substrate. The high pK of Glu179 is probably due to the electrostatic interaction with the charged Glu180 located in subsite 2. A mutation of Asp176->Asn leads to only a 12–20-fold decrease in k_{cat} but a 4-fold increase in k_m. It is possible that Asp176, which is located near subsite 1, functions to stabilize the transition, probably through interaction with Trp120 at subsite 4, although exact details are not clear. Crystallographic study shows that neither Glu179 nor Glu400 has accessibility to form a covalent bond with the C1 anomeric carbon in its complex with the inhibitor, 1-deoxynojirimycin. The former has its side chain blocked by the glycosidic linkage, and the latter is blocked by H$_2$O molecule (Aleshin et al. 1992, 1994). The Glu179, which is hydrogen bonded to Asp176 and Trp120, lies close to the aglycon, while the catalytic base Glu400 is located at the opposite side nearby the nucleophile (Harris et al. 1993). These results are taken as an argument in favor of a reaction pathway involving an oxocarbonium ion intermediate. Hydrolysis occurs by protonation of the aglycon by Glu179 followed by (or in concert with) nucleophilic attack by H$_2$O assisted by a general base (Glu400). α-Secondary tritium kinetics indicate that hydrolysis of α-glycosyl fluoride by glucoamylase involves a slow formation of carbonium ion character in the transition state (Matsui et al. 1989).

Based on kinetic studies of the *Rhizopus niveus* enzyme modified with *N*-bromosuccinimide, a Trp residue is found to be located at subsite 1 (Ohnishi et al. 1983; Ohnishi 1990). Another essential Trp120 at or near subsite 4 is found to function in stabilizing the enzyme-substrate complex at its transition state by inducing a conformational change at subsites 1 and 2 (Clarke and Svensson 1984; Sierks et al. 1989). Site-directed mutation of Trp120 of the *Aspergillus awamori* enzyme causes a ~10-fold increase in the inactivation energy compared to the wild type.

SUBSITE THEORY

Over the years, the subsite model has been applied to account for the action patterns of the amylases as well as other enzymes. The model assumes the active site of amylase consisting of a number of subsites, each of which interacts with a monomeric unit of the substrate. The measurable rate parameters, Michaelis constant (K_m), maximum velocity (V_{max}), and functions such as bond-cleavage frequency of a substrate, can be expressed in terms of affinity of each subsite.

Quantitation of the subsite model requires determination of (1) number of subsites, (2) location of the catalytic subsite, (3) subsite affinity (i.e. binding energy constant), and (4) rate of hydrolysis. There are basically two methods for subsite mapping.

The Kinetic Method

In a simple scheme (Hiromi 1970; Hiromi et al. 1973; Nitta et al. 1971; Hiromi et al. 1983), an enzyme (E) reacting with a substrate (S) to form a productive (p) or nonproductive (q) complex in the jth binding mode is represented by Eq. 1,

$$ES_q \overset{K_q}{\rightleftharpoons} E + S \overset{K_p}{\rightleftharpoons} ES_p \overset{k_2}{\longrightarrow} E + P \tag{1}$$

where the intrinsic rate of hydrolysis k_2 in every productive complex is assumed to be constant and independent of the degree of polymerization n of the substrate, and the binding mode. The binding constants K_p and K_q for a particular productive and nonproductive complex, respectively, are related to the Michaelis constant K_m as in Eq. 2.

$$1/K_m = \sum K_j = \sum K_p + \sum K_q \tag{2}$$

In addition, the maximum velocity (V_{max}) increases due to the increased probability of formation of productive binding between the substrate and the active site, as represented by Eq. 3, where E_0 = molar concentration of enzyme, k_{cat} = overall reaction rate constant, and k_p = breakdown rate constant of the productive complex.

$$v/E_0 = k_{cat} = \sum(k_p K_p) \Big/ \left(\sum K_p + \sum K_q \right) \tag{3}$$

The Michaelis constant K_m and the maximum rate V_{max} can be obtained by plotting $1/v$ versus $1/[S]$ according to $1/v = 1/V_{max} + K_m/(V_{max}[S])$, where the intercept is $1/V_{max}$ on the y axis as $1/[S]$ approaches zero, and $1/[S] = 1/k_m$ on the x axis.

By combining Eqs. 2 and 3,

$$k_{cat}/K_m = \sum(k_p K_p) \tag{4}$$

Assuming that the intrinsic rate constant for hydrolysis k_2 in the productive complexes is independent of the binding mode or chain length, Eq. 3 can be written as in Eq. 5.

$$k_{cat}/K_m = k_2 \sum K_p \tag{5}$$

$$\sum K_p = (k_{cat}/K_m)/k_2 \tag{6}$$

For substrates with chain length (n) equal to or greater than the number of subsites (m), the number of nonproductive complexes (ΣK_q) becomes negligible, and all the productive complexes could have all the subsites covered. In this case, all the K_p are identical, i.e., $\Sigma K_p = (n - m + 1) k_2$. Eqs. 2 and 3 become Eqs. 7 and 8.

$$1/K_m = \sum K_p = (n - m + 1)k_2 \tag{7}$$

$$k_{cat} = k_2 \tag{8}$$

The k_{cat} values therefore become constant when the substrate chain length $\geq$ the number of subsites. The number of subsites can then be estimated and k_{int} calculated by plotting $\log k_{cat}$ versus chain length of the substrate. The number of subsites is equal to n where k_{cat} becomes constant. From k_{cat} at constant value, k_2 can be estimated. Once k_2 is known, then ΣK_p can be calculated using Eq. 6.

The association constant K_j is related to subsite affinity as shown in the following way. The molecular binding affinity (B_j) is equal to the sum of the affinities of all covered individual subsites (A_i) and is directly related to the standard free energy change for the binding process (ΔG_j) and the free energy of mixing ($\Delta G_{mix} = 2.4$ kcal/mol, 25°C, aqueous solution).

$$B_j = \sum A_i = -\Delta G_j + \Delta G_{mix} = RT \ln K_j \tag{9}$$

$$K_j = \exp\left(\sum A_i/RT\right) \tag{10}$$

Substituting Eq. 10 in Eq. 5,

$$k_{cat}/K_m = k_2 \sum \exp\left(\sum A_i/RT\right) \tag{11}$$

For two substrates of n and $n - 1$ units, for which one additional subsite is covered by the substrate n,

$$\frac{(k_{cat}/K_m)_n}{(k_{cat}/K_m)_{n-1}} = \frac{\exp[(A_1 + A_2 + \cdots\cdots A_n)/RT]}{\exp[(A_1 + A_2 + \cdots\cdots A_{n-1})/RT]} \tag{12}$$

The Labelling Analysis

This method is based on the relationship between the cleavage mode as deduced from the end-labelling of the substrate, the association constant, and the covered subsite affinity (Allen and Thoma 1976A, B; MacGregor and MacGregor 1985). In hydrolysis of a polymeric substrate S of chain length n the rate of formation of product can be represented by Eq. 13, which has the form of a typical competitive inhibition.

$$\frac{d[P_m]}{dt} = \frac{[E][S]k_2/K_p}{1 + [S]/K_m} \tag{13}$$

If the substrate is labelled at the reducing end, so that P_m may be estimated, then the rate of product formation can be expressed from two adjacent productive modes, j and $j + 1$, of the same substrate. Then Eq. 13 can be rewritten into Eq. 14.

$$\frac{d[P_{j,m}]/dt}{d[P_{j+1,m}]/dt} = \frac{k_{2,j,m}\ k_{2,j+1,m}}{K_{j,m}\ K_{j+1,m}} \tag{14}$$

$$= \frac{\exp\left(\sum \Delta G_{j+1}\Big/ RT\right)}{\exp\left(\sum \Delta G_j\Big/ RT\right)}$$

Attempts have been made to refine the above methods and to minimize the discrepancy between the experimental results and theoretical predictions (Allen and Thoma 1976A, B; Matsuno et al. 1978; Suganuma et al. 1978; Kondo et al. 1980; Seigner et al. 1987). Two additional factors need to be considered for the refined equations. (1) The assumption that k_2 is constant may not be valid. Each occupied subsite lowers the activation energy and hence accelerates bond cleavage. This factor is referred to as the "acceleration factor,"

$$k_p = k_h \exp\left(\sum \Delta G_v\Big/ RT\right) \tag{15}$$

where k_h is the hypothetical rate constant as if the number of subsites occupied had no effect, and ΔG_v is the average value of each productive complex contributing to the acceleration of bond cleavage. (2) The methods assume only simple hydrolysis without considering more complicated mechanisms, involving multiple attack, transglycosylation, condensation, and shift binding, which are known to occur under certain conditions.

SUBSITES OF AMYLASES

Arrangement of subsites can be used to predict the predominant mode of productive binding and hence the action pattern on a given substrate. The most probable binding mode for a substrate should retain the highest molecular binding affinity. The *A. oryzae* α-amylase consists of seven subsites, with relatively high affinity at subsites 3 and 6, and negative affinities adjacent to the catalytic site (Table 3.1). The predominant mode of productive binding for maltopentaose is such that the substrate is hydrolyzed at G3-G4 (Fig. 3.13) (Nitta et al. 1971). The porcine pancreatic α-amylase contains five subsites with the catalytic attack between subsites 3 and 4. In this case, maltotriose is hydrolyzed predominantly at G1-G2 (from the reducing end), although two productive complexes can be formed (Fig. 3.13). A similar situation is observed with maltotetraose which is hydrolyzed at G2-G3 to yield mostly G2 units (Robyt and French 1970A). Barley α-amylases consist of 10 subsites, where hydrolytic cleavage of substrate occurs between subsites 6 and 7 (MacGregor and Morgan 1992).

Soybean β-amylase contains seven subsites with bond cleavage occurring between subsites 2 and 3 (Isoda and Nitta 1986; Kunikata et al. 1992). Studies of binding of cyclohexadextrin and maltose to the enzyme suggest that productive binding causes local conformational change at subsite 1 (Mikami et al. 1983; Nomura et al. 1987). Because β-amylase is exo-acting, the productive binding mode must have the nonreducing end of the substrate binding to the subsite 1 of the enzyme. Production of glucose is negligible because of the large affinity of subsite 1 compared to that of subsite 2. The large affinity of subsite 1 probably plays an important role for the exclusive maltose-forming action of this enzyme.

For glucoamylases from *Rhizopus niveus* and *Rhizopus delemar*, the active site consists of seven subsites; the affinity of subsite 1 is zero or negative, subsite 2 has the highest affinity, and the subsite affinity decreases with increasing distance from the scission bond (between subsite 1 and 2) toward the reducing end of the substrate (Ohnishi et al. 1983). Fluorescence stopped-flow kinetic studies suggest a two-step mechanism—a fast bimolecular association between the enzyme and substrate to form a loose complex followed by a unimolecular process to a tightly bound complex (E + S $\rightleftharpoons$ ES $\rightleftharpoons$ E*S $\rightarrow$ E + P) (Tanaka et al. 1983A, B). A maltose substrate binds the enzyme first with its nonreducing glucose residue associated with subsite 1 to form the ES complex. In the subsequent step, tight productive binding at subsite 2 occurs forming a tight productive complex with some change in the enzyme conformation in a formation of an induced-fit structure (Fig. 3.14) (Fagerstrom 1991; Olsen et al. 1992). This unimolecular step involves a Trp residue (Trp120) folding around the nonreducing end unit.

Table 3.1. Subsite Affinities of Amylases[a]

Enzymes[b]	No. of Subsites	Catalytic Sites	Subsite affinities (kcal/mol) 1	2	3	4	5	6	7	8	9	10
α-Amylases												
Bacillus amyloliquefaciens	10	6–7	−0.65	−1.90	0.25	−0.60	−1.30	2.00	−2.20	−0.90	−0.05	1.10
Porcine pancreatic	5	3–4	0.87	2.34	1.00	2.01	1.23					
Bacillus subtilis	8	6–7	1.1	2.4	0	0.6	2.4		−3.1	1.2		
Aspergillus oryzae	7	4–5	0.3	0.3	3.0	−1.7		3.2	1.3			
β-Amylases												
Soybean	7	2–3	7.09	−5.68	0.14	5.19	0.41	0.15	0.06			
Wheat bran	5	2–3	>2.7	< −1.4	−1.85	5.89	1.09					
Glucoamylases												
Aspergillus awamori	7	1–2	−0.62	5.11	1.51	0.43	0.38	0.24	0.07			
Rhizopus delemar	7	1–2	0	4.85	1.59	0.43	0.22	0.11	0.10			
Rhizopus niveus	7	1–2	−0.48	4.96	1.36	0.54	0.32	0.23	0.07			
Aspergillus niger I	7	1–2	−1.23	5.03	1.53	0.86	0.27	0.07	0.09			
II	7	1–2	−1.23	5.29	1.45	0.74	0.32	0.11	0.11			

[a]The subsites are numbered from the nonreducing end of the binding substrate.

[b]Source: α-Amylases: *Bacillus amyloliquefaciens* (Torgerson et al. 1979), Porcine pancreatic (Seigner et al. 1987), *Bacillus subtilis* (Iwasa et al. 1974), *Aspergillus oryzae* (Nitta et al. 1971); β-Amylases: Soybean (Kunikata et al. 1992), Wheat bran (Kato et al. 1974); Glucoamylases: *Aspergillus awamori* (Sierks et al. 1989), *Rhizopus delemar* (Hiromi et al. 1973), *Rhizopus niveus* (Tanaka et al. 1983B), *Aspergillus niger* (Meagher et al. 1989).

TAKA−AMYLASE

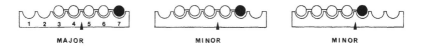

PANCREATIC α−AMYLASE

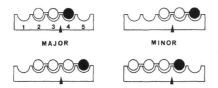

Figure 3.13. Hydrolysis of (A) maltopentaose by *Aspergillus oryzae* α-amylases (Nitta et al. 1971) and (B) maltotriose and maltotetraose by pancreatic α-amylase (Robyt and French 1970A).

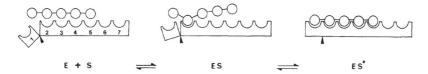

E + S ⇌ ES ⇌ ES*

Figure 3.14. Schematic representation of the proposed "induced-fit" model for binding of oligosaccharides to the subsites of glucoamylases. (From Fagerstrom 1991. *J. Gen. Microbiol. 137*, 1006, with permission. Copyright 1991 Society for General Microbiology.)

ACTION PATTERNS

Theoretically, three types of mechanisms are possible in describing the action of amylases.

(1) In a single-chain action pattern, the amylase, once a productive complex with a substrate molecule is formed, continuously acts on it to complete degradation.

(2) Another possible action is a multichain action pattern in which amylase, having cleaved a given substrate molcule once, dissociates and attacks another substrate molecule. This type of action leads to simultaneous shortening of all substrate molecules present.

(3) It is also conceivable that the enzyme, in a multiple attack pattern, cleaves a given substrate molecule more than once before it dissociates and forms a new productive complex with another substrate molecule. The direction of the multiple attack is from the reducing end toward the nonreducing end as was determined for

porcine pancreatic β-amylase. For exoenzymes, such as β-amylases and glucoamylases, cleavage occurs from the nonreducing end in a multiple attack pattern as expected (Suganuma et al. 1980).

Porcine pancreatic α-amylase at optimum pH 6.9 shows three times the degree of multiple attack that human salivary or *Aspergillus oryzae* amylases show at their optimum pH values (Robyt and French 1970B; Prodanov et al. 1984). At an alkaline pH of 10.5, the enzyme reaction proceeds through a multichain action. Glucoamylase from *Rhizopus niveus* has also been implicated in a multiple attack mechanism (Kondo et al. 1980). However, theoretical calculations suggest that only porcine pancreatic α-amylase exhibits a significant degree of multiple attack (Thoma 1976; Banks and Greenwood 1977; Mazur 1984). Using the malto-oligosaccharide mapping method, Pazur and Marchetti (1992) show that glucoamylase from *Aspergillus niger* and *Rhizopus niveus*, and β-amylase from sweet potato, hydrolyze malto-oligosaccharide predominantly by a multichain mechanism. In addition, bacterial α-amylases show little evidence of multichain action (Pazur and Marchetti 1992) or multiple attack (MacGregor and Morgan 1992).

The mechanism of multiple attack can be visualized as consisting of three routes (Fig. 3.15) (Robyt and French 1970B).

(1)   Complete dissociation and association. The substrate, after initial cleavage, dissociates completely and reassociates to a new position.

(2)   Flexible reattachment. The substrate, after initial cleavage, remains attached to the active site. A segment of the product substrate remains firmly attached, while another segment becomes detached and realigns to a new position. Such a mechanism requires a certain degree of flexibility either in the substrate or in the enzyme, or both.

(3)   Another possibility is for the product substrate to slide on the enzyme surface to a position of minimum potential energy where all of the subsites are completely filled.

If we consider that the molecular motion in each of the individual steps of dissociation, reassociation, and realignment in routes (1) and (2) is infinitesimally small, the product substrate can be visualized as "sliding" on the enzyme to a favorable alignment with minimal potential energy as in route (3).

The action of α-amylases on amylopectin differs in the hydrolysis of the α-D-1,4 linkages near the α-1,6 linkage. Both porcine pancreatic and malted-barley α-amylases produce 6^3-α-D-glucopyranosyl maltotetraose as the smallest limit dextrin (Hall and Manners 1978). With *Bacillus subtilis* α-amylase, the 6^2-α-maltosyl maltotriose is produced (Fig. 3.16) (French et al. 1972).

Dissociation plus Reassociation

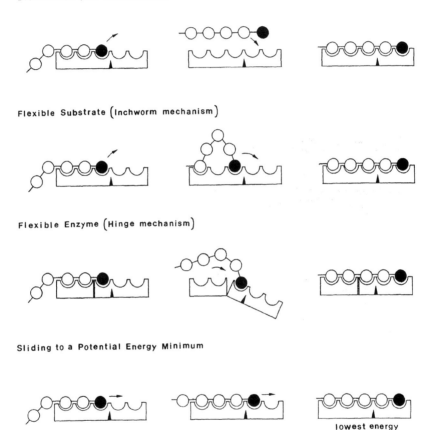

Flexible Substrate (Inchworm mechanism)

Flexible Enzyme (Hinge mechanism)

Sliding to a Potential Energy Minimum

lowest energy

Figure 3.15. Schematic representation of mechanisms for multiple attack. The site of cleavage is represented by arrow head. (From Robyt and French 1970B. *Arch. Biochem. Biophys. 138*, 668, with permission. Copyright 1970 Academic Press, Inc.)

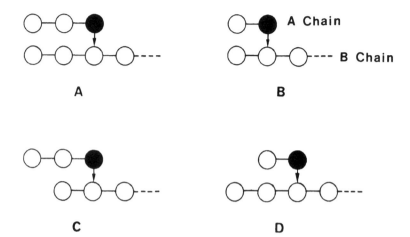

Figure 3.16. Possible structures of α-limit dextrins from the action of α-amylases. Product from (A) even A and B chain, (B) odd A and B chain, (C) even A and odd B chain, (D) odd A and even B chain (French et al. 1972; Hall and Manners 1978).

The β-amylases attack from the nonreducing end to release maltose units, until blocked by an α-D-1,6 linkage. For an amylopectin chain with an even number of glucose residues, a maltose unit is left next to the α-1,6 linkage. For chains having an odd number of glucose residues, the structure will contain one glycosyl residue next to the α-1,6 linkage. The same action patterns will be reflected on amylopectins with mixed A, B chains.

Multimolecular Reactions

α-Amylases do not exclusively catalyze simple unimolecular hydrolysis. Under certain conditions, the degradation of substrates proceeds via a multimolecular process, such as condensation, shifted binding, or transglycosylation (Fig. 3.17) (Robyt and French 1970A; Allen and Thoma 1978; Matsuno et al. 1978; Suganuma et al. 1978; Graber and Combes 1990; Matsui et al. 1991, 1994).

(1) Condensation: Two substrates polymerize to form a new glycosidic bond, followed by a rapid hydrolysis of the condensed product.

(2) Shifted binding: Inherent from the subsite model, two substrate molecules can bind to the same active site to yield a two-one complex. Depending on the binding affinities of the occupied subsites, the second substrate may bind in such a way that pushes the first bound substrate molecule into a new position promoting the hydrolysis at a different bond.

CONDENSATION

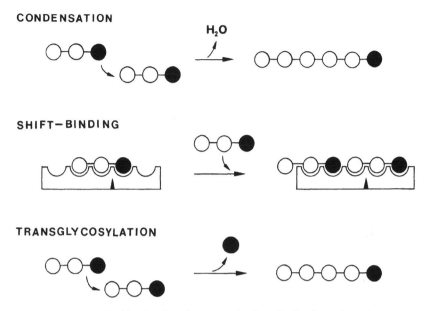

SHIFT—BINDING

TRANSGLYCOSYLATION

Figure 3.17. Possible bimolecular substrate mechanisms for the change in action pattern of porcine pancreatic a-amylase as the concentration of maltotriose increases. (From Robyt and French 1970A. *J. Biol. Chem. 245*, 3924, with permission. Copyright 1970 American Society for Biochemistry and Molecular Biology.)

(3) Transglycosylation: A glycosyl group of a substrate donor is transferred to another substrate acceptor, similar to hydrolysis, except that water is not the substrate acceptor.

These alternate mechanisms depend on substrate concentration and chain length. Substrates having a small chain length have an increasing chance of forming a nonproductive complex. At high substrate concentration, there is an increase in the probability of an interaction of a second substrate with the bound substrate molecule. Both factors contribute to an alternate pathway competitive to the hydrolytic process (Graber and Combes 1990).

Multimolecular mechanisms have been shown to be catalyzed by α-amylases, β-amylases, and glucoamylases. For porcine pancreatic α-amylase, condensation is the exclusive mechanism for the multimolecular degradation of high concentrations of maltotriose. However, the reaction for maltotetraose may involve both transglycosylation and condensation. With higher maltodextrins, unimolecular hydrolysis appears to be the dominant process. For the *Aspergillus oryzae* α-amylase, the major multimolecular mechanism involved is transglycosylation. Sweet potato β-amylase has been found to catalyze the synthesis of maltotetraose from β-maltose. Glucoamylase from *Rhizopus niveus* catalyzes the rapid synthesis of maltose from β-D-glucose (Hehre et al. 1969; Nikolov et al. 1989). Both are also found to catalyze transglycosylation with substrates having the correct anomeric form (Hehre et al. 1979; Kitahata et al. 1981).

REFERENCES

ALESHIN, A.; GOLUBEV, A.; FIRSOV, L. M.; and HONZATKO, R. B. 1992. Crystal structure of glucoamylase from *Aspergillus awamori* var. × 100 to 2.2-A resolution. *J. Biol. Chem. 267*, 19291–19298.

ALESHIN, A. E.; HOFFMAN, C.; FIRSOV, L. M.; and HONZATKO, R. B. 1994. Refined crystal structures of glucoamylase from *Aspergillus awamori* var. × 100. *J. Mol. Biol. 238*, 575–591.

ALLEN, J. D. 1980. Subsite mapping of enzymes: Application to polysaccharide depolymerases. *Methods in Enzymology 64*, 248–277.

ALLEN, J. D., and THOMA, J. A. 1976A. Subsite mapping of enzymes depolymerase computer modelling. *Biochem. J. 159*, 105–120.

———. 1976B. Subsite mapping of enzymes. Application of the depolymerase computer model to two α-amylases. *Biochem. J. 159*, 121–132.

———. 1978. Multimolecular substrate reactions catalyzed by carbohydrases. *Aspergillus oryzae* α-amylase degradation of maltooligosaccharides. *Biochemistry 17*, 2338–2344.

AMYES, T., and JENCKS, W. P. 1989. Lifetimes of oxocarbonium ions in aqueous solution from common ion inhibition of the solvolysis of α-azido ethers by added azide ion. *J. Am. Chem. Soc. 111*, 7888–7900.

ANN, Y.-G.; IIZUKA, M.; YAMAMOTO, T.; and MINAMIURA, N. 1990. Active monomer of sweet potato β-amylase: Stabilization and an improved preparation method using α-cyclodextrin. *J. Ferment. Bioengineer. 70*, 75–79.

ANON. 1984A. DIAZYME L200, fungal glucoamylase for starch hydrolysis. Miles Laboratories, Inc., Enzyme Products, Elkhart, IN.

ANON. 1984B. KINASE-HT, thermal stable bacterial alpha-amylase for the brewing industry. Miles Laboratories, Inc., Biotech Division, Elkhart, IN.

ASHIKARI, T.; KIUCHI-GOTO, N.; TANAKA, Y.; SHIBANO, Y.; AMACHI, T.; and YOSHIZUMI, H. 1989. High expression and efficient secretion of *Rhizopus oryzae* glucoamylase in the yeast Saccharomyces cerevisiae. *Appl. Microbiol. Biotechnol. 30*, 515–520.

ASHIKARI, T.; NAKAMURA, N.; TANAKA, Y.; KIUCHI, N.; SHIBANO, Y.; and TANAKA, T. 1986. *Rhizopus* raw-starch-degrading glucoamylase: Its cloning and expression in yeast. *Agric. Biol. Chem. 50*, 957–964.

BANKS, W., and GREENWOOD, C. T. 1977. Mathematical models for the action of alpha-amylase on amylose. *Carbohydr. Res. 57*, 301–315.

BAULCOMBE, D. C.; HUTTLY, A. K.; MARTIENSSEN, R. A.; BARKER, R. F.; and JARVIS,

M. C. 1987. A novel wheat α-amylase gene (α-Amy3). *Mol. Gen. Genet. 209*, 33–40.

BELSHAW, N. J., and WILLIAMSON, G. 1990. Production and purification of a granular-starch-binding domain of glucoamylase 1 from *Aspergillus niger. FEBS Lett. 269*, 350–353.

———. 1993. Specificity of the binding domain of glucoamylase 1. *Eur. J. Biochem. 211*, 717–724.

BOEL, E.; BRADY, L.; BRZOZOWSKI, A. M.; DEREWENDA, Z.; DODSON, G. G.; JENSEN, V. J.; PETERSEN, S. B.; SWIFT, H.; THIM, L.; and WOLDIKE, H. F. 1990. Calcium binding in α-amylases: An x-ray diffraction study at 2.1-A resolution of two enzymes from *Aspergillus. Biochemistry 29*, 6244–6249.

BOEL, E.; HJORT, I.; SVENSSON, B.; NORRIS, F.; NORRIS, K. E.; and FIIL, N. P. 1984. Glucoamylases G1 and G2 from *Aspergillus niger* are synthesized from two different but closely related mRNAs. *EMBO J. 3*, 1097–1102.

BROSNAN, M. P.; KELLY, C. T.; and FOGARTY, W. M. 1992. Investigation of the mechanisms of irreversible thermoinactivation of *Bacillus stearothermophilus* α-amylase. *Eur. J. Biochem. 203*, 225–231.

BUISSON, G.; DUEE, E.; HASER, R.; and PAYAN, F. 1987. Three dimensional structure of porcine pancreatic α-amylase at 2.9 A resolution. Role of calcium in structure and activity. *EMBO J. 6*, 3909–3916.

CLARKE, A. J., and SVENSSON, B. 1984. Identification of an essential tryptophanyl residue in the primary structure of glucoamylase G2 from *Aspergillus niger. Carlsberg Res. Comm. 49*, 559–566.

CONE, J. W., and WOLTERS, M. G. E. 1990. Some properties and degradability of isolated starch granules. *Starch/Starke 42*, 298–301.

DAWSON, H. G., and ALLEN, W. G. 1984. The use of enzymes in food technology. Miles Laboratories, Inc., Biotech Products Division, Elkhart, IN.

DOYEN, C., and LAURIERE, C. 1992. β-Amylases in germinating maize grains: Purification, partial characterization and antigen relationships. *Phytochemistry 31*, 3697–3702.

DUA, R. D., and KOCHHAR, S. 1985. Active site studies on *Bacillus amyloliquefaciens* α-amylase (I). *Mol. Cell. Biochem. 66*, 13–20.

EMI, M.; HORII, A.; TOMITA, N.; NISHIDE, T.; OGAWA, M.; MORI, T.; and MATSUBARA, K. 1988. Overlapping two genes in human DNA: a salivary amylase gene overlaps with a gamma-actin pseudogene that carries an integrated human endogenous retroviral DNA. *Gene 62*, 229–235.

EVANS, R.; FORD, C.; SIERKS, M.; NIKOLOV, Z.; and SVENSSON, B. 1990. Activity and thermal stability of genetically truncated forms of *Aspergillus glucoamylase. Gene 91*, 131–134.

FAGERSTROM, R. 1991. Subsite mapping of *Hormoconis resinae* glucoamylases and their inhibition by gluconolactone. *J. Gen. Microbiol. 137*, 1001–1008.

FOGARTY, W. M., and GRIFFIN, P. J. 1975. Purification and properties of β-amylase produced by *Bacillus polymyxa. J. Appl. Chem. Biotechnol. 25*, 229–238.

FRENCH, D.; SMITH, E. E.; and WHELAN, W. J. 1972. The structural analysis and enzymatic synthesis of a pentasaccharide alpha-limit dextrin formed from amylopectin by *Bacillus subtilis* alpha-amylase. *Carbohydr. Res. 22*, 123–134.

GALLANT, D.; MERCIER, C.; and GUILBOT, A. 1972. Electron microscopy of starch granules modified by bacterial α-amylase. *Cereal Chem. 49*, 354–365.

GOTO, M.; TANIGAWA, K.; KANLAYAKRIT, W.; and HAYASHIDA, S. 1994. The mechanism of binding of glucoamylase I from *Aspergillus awamori* var. *kawachi* to cyclodextrins and raw starch. *Biosci. Biotech. Biochem. 58*, 49–54.

GRABER, M., and COMBES, D. 1990. Action pattern of alpha-amylase from *Aspergillus oryzae* in concentrated media. *Biotechnol. Bioengineer. 36*, 12–18.

GRAY, G. L.; MAINZER, S. E.; REY, M. W.; LAMSA, M. H.; KINDLE, K. L.; CARMONA, C.; and REQUADT, C. 1986. Structural genes encoding the thermophilic α-amylases of *Bacillus stearothermophilus* and *Bacillus licheniformis*. *J. Bacteriol. 166*, 635–643.

GUMUCIO, D. L.; WIEBAUER, K.; CALDWELL, R. M.; SAMUELSON, C.; and MEISLER, M. H. 1988. Concerted evolution of human amylase genes. *Mol. Cell. Biol. 8*, 1197–1205.

GUMUCIO, D. L.; WIEBAUER, K.; DRANGINIS, A.; SAMUELSON, L. C.; TREISMAN, L. O.; CALDWELL, R. M.; ANTONUCCI, T. K.; and MEISLER, M. H. 1985. Evolution of the amylase multigene family. *J. Biol. Chem. 260*, 13483–13489.

GUNNARSSON, A.; SVENSSON, B.; NILSSON, B.; and SVENSSON, S. 1984. Structural studies on the O-glycosidically linked carbohydrate chains of glucoamylase G1 from *Aspergillus niger*. *Eur. J. Biochem. 145*, 463–467.

HALL, R. S., and MANNERS, D. J. 1978. The action of malted-barley alpha-amylase on amylopectin. *Carbohydr. Res. 66*, 295–297.

HANOZET. G.; PIRCHER, H.-P.; VANNI, P.; OESCH, B.; and SEMENZA, G. 1981. An example of enzyme hysteresis. *J. Biol. Chem. 256*, 3703–3711.

HARRIS, E. M. S.; ALESHIN, A. E.; FIRSOV, L. M.; and HONZATKO, R. B. 1993. Refined structure for the complex of 1-deoxynojirimycin with glucoamylase from *Aspergillus awamori* var. X100 to 2.4-A resolution. *Biochemistry 32*, 1618–1626.

HATA, Y.; KITAMOTO, K.; GOMI, K.; KUMAGAI, C.; TAMURA, G.; and HARA, S. 1991. The glucoamylase cDNA from *Aspergillus oryzae*: Its cloning, nucleotide sequence, and expression in *Saccharomyces cerevisae*. *Agric. Biol. Chem. 55*, 941–949.

HATA, Y.; TSUCHIYA, K.; KITAMOTO, K.; GOMI, K.; KUMAGAI, C.; TAMURA, G.; and HARA, S. 1991. Nucleotide sequence and expression of the glucoamylase-encoding gene (gla A) from *Aspergillus oryzae*. *Gene 108*, 145–150.

HAYASHIDA, S. 1975. Selective submerged productions of three types of glucoamylases by a black-koji mold. *Agric. Biol. Chem. 39*, 2093–2099.

HAYASHIDA, S.; KUNISAKI, S.; NAKAO, M.; and FLOR, P. Q. 1982. Evidence for raw starch-affinity site on *Aspergillus awamori* glucoamylase I. *Agric. Biol. Chem. 46*, 83–89.

HAYASHIDA, S.; NAKAHARA, K.; KURODA, K.; MIYATA, T.; and IWANAGA, S. 1989. Structure of the raw-starch-affinity site on the *Aspergillus awamori* var. *kawachi* glucoamylase I molecule. *Agric. Biol. Chem. 53*, 135–141.

HAYASHIDA, S.; TERAMOTO, Y.; and KIRA, I. 1991. Promotive and inhibitory effects of raw starch adsorbable fragments from pancreatic α-amylase on enzymatic digestions of raw starch. *Agric. Biol. Chem. 55*, 1–6.

HEHRE, E. J.; BREWER, C. F.; and GENGHOF, D. S. 1979. Scope and mechanism of carbohydrase action. Hydrolytic and nonhydrolytic actions of α-amylase on β-maltosyl fluoride. *J. Biol. Chem. 254*, 5942–5950.

HEHRE, E. J.; KITAHATA, S.; and BREWER, C. F. 1986. Catalytic flexibility of glycosylases. The hydration of maltal by β-amylase to form 2-deoxymaltose. *J. Biol. Chem. 261*, 2147–2153.

HEHRE, E. J.; OKADA, G.; and GENGHOF, D. S. 1969. Configurational specificity: Unappreciated key to understanding enzyme reversions and de Novo glycosidic bond synthesis. 1. Reversal of hydrolysis by α-, β- and glucoamylases with donors of correct anomeric form. *Arch. Biochem. Biophys. 135*, 75–89.

HIROMI, K. 1970. Interpretation of dependency of rare parameters on the degree of polymerization of substrate in enzyme-catalyzed reactions. Evaluation of subsite affinities of exo-enzyme. *Biochem. Biophys. Res. Comm. 40*, 1–6.

———. 1979. Amylase in food processing: the subsite theory and its application. *Proceedings of Fifth International Congress of Food Science and Technology*, ed. H. Chiba, Hodansha Ltd., and Elsevier Scientific Publ. Co., 1979.

HIROMI, K.; NITTA, Y.; NUMATA, C.; and ONO, S. 1973. Subsite affinities of glucoamylase: Examination of the validity of the subsite theory. *Biochim. Biophys. Acta 302*, 362–375.

HIROMI, K.; OHNISHI, M.; and TANAKA, A. 1983. Subsite structure and ligand binding mechanism of glucoamylase. *Mol. Cell. Biochem. 51*, 79–95.

HOSCHKE, A.; LASZLO, E.; and HOLLO, J. 1980A. A study of the role of histidine sidechains at the active centre of amylolytic enzymes. *Carbohydr. Res. 81*, 145–156.

———. 1980B. A study of the role of tyrosine groups at the active centre of amylolytic enzymes. *Carbohydr. Res. 81*, 157–166.

HOSHIKO, S.; MAKABE, O.; NOJIRI, C.; KATSUMATA, K.; SATOH, E.; and NAGAOKA, K. 1987. Molecular cloning and characterization of the *Streptomyces hygroscopicus* α-amylase gene. *J. Bacteriol. 169*, 1029–1036.

IHARA, H.; SASAKI, T.; TSUBOI, A.; YAMAGATA, H.; TSUKAGOSHI, N.; and UDAKA, S. 1985. Complete nucleotide sequence of a thermophilic α-amylase gene: Homology between prokaryotic and eukaryotic α-amylases at the active sites. *J. Biochem. 98*, 95–103.

IMAM, S. H.; BURGESS-CASSLER, A.; COTE, G. L.; GORDON, S. H.; and BAKER, F. L. 1991. A study of cornstarch granule digestion by an unusually high molecular weight α-amylase secreted by *Lactobacillus amylovorus*. *Current Microbiology 22*, 365–370.

INOKUCHI, N.; IWAMA, M.; TAKAHASHI, T.; and IRIE, M. 1982A. Modification of a glucoamylase from *Aspergillus saitoi* with 1-cyclohexyl-3-(2-morpholinyl-(4)-ethyl-carbodiimide. *J. Biochem. 91*, 125–133.

INOKUCHI, N.; TAKAHASHI, T.; YOSHIMOTO, A.; and IRIE, M. 1982B. N-Bromosuccinimide oxidation of a glucoamylase. *J. Biochem. 91*, 1661–1668.

ISHIKAWA, K.; MATSUI, I.; HONDA, K.; and NAKATANI, H. 1992. Multi-functional roles of a histidine residue in human pancreatic α-amylase. *Biochem. Biophys. Res. Comm. 183,* 286–291.

ISHIKAWA, K.; MATSUI, I.; KOBAYASHI, S.; NAKATANI, H.; and HONDA, K. 1993. Substrate recognition at the binding site in mammalian pancreatic α-amylases. *Biochemistry 32,* 6259–6265.

ISODA, Y., and NITTA, Y. 1986. Affinity labeling of soybean β-amylase with 2′,3′-epoxypropyl α-D-glucopyranoside. *J. Biochem. 99,* 1631–1637.

ITOH, T.; OHTSUKI, I.; YAMASHITA, I.; and FUKUI, S. 1987A. Nucleotide sequence of the glucoamylase gene GLU1 in the yeast *Saccharomycopsis fibuligera. J. Bacteriol. 169,* 4171–4176.

ITOH, T.; YAMASHITA, I.; and FUKUI, S. 1987B. Nucleotide sequence of the α-amylase gene (ALP1) in the yeast *Saccharomycopsis fibuligera.* FEBS Lett. 219, 339–342.

IWASA, S.; AOSHIMA, H.; HIROMI, K.; and HATANO, H. 1974. Subsite affinities of bacterial liquefying α-amylase evaluated from the rate parameters of linear substrates. *J. Biochem. 75,* 969–978.

JI, E.-S.; MIKAMI, B.; KIM, J.-P.; and MORITA, Y. 1990. Positions of substituted amino acids in soybean α-amylase isozymes. *Agric. Biol. Chem. 54,* 3065–3067.

KADZIOLA, A.; ABE, J.-I.; SVENSSON, B.; and HASER, R. 1994. Crystal and molecular structure of barley α-amylase. *J. Mol. Biol. 239,* 104–121.

KATO, M.; HIROMI, K.; and MORITA, Y. 1974. Purification and kinetic studies of wheat bran β-amylase. Evaluation of subsite affinities. *J. Biochem. 75,* 563–576.

KAWAZU, T.; NAKANISHI, Y.; UOZUMI, N.; SASAKI, T.; YAMAGATA, H.; TSUKAGOSHI, N.; and UDAKA, S. 1987. Cloning and nucleotide sequence of the gene coding for enzymatically active fragments of the *Bacillus polymyxa* β-amylase. *J. Bacteriol. 169,* 1564–1570.

KIMURA, T., and HORIKOSHI, K. 1990. The nucleotide sequence of an α-amylase gene from an alkalopsychrotrophic *Micrococcus* sp. *FEMS Microbiol. Lett. 71,* 35–42.

KITA, Y.; SAKAQUCHI, S.; NITTA, Y.; and WATANABE, T. 1982. Kinetic study on chemical modification of Taka-amylase A. II. Ethoxycarbonylation of histidine residues. *J. Biochem. 92,* 1499–1504.

KITAHATA, S.; BREWER, C. F.; GENGHOF, D. S.; SAWAI, T.; and HEHRE, E. J. 1981. Scope and mechanism of carbohydrase action. Stereocomplementary hydrolytic and glucosyl-transferring actions of glucoamylase and glucodextranase with α- and β-D-glucosyl fluoride. *J. Biol. Chem. 256,* 6017–6026.

KITAHATA, S.; CHIBA, S.; BREWER, C. F.; and HEHRE, E. J. i991. Mechanism of maltal hydration catalyzed by β-amylase: Role of protein structure in controlling the steric outcome of reactions catalyzed by α-glycosylase. *Biochemistry 30,* 6769–6775.

KITAMOTO, N.; YAMAGATA, H.; KATO, T.; TSUKAGOSHI, N.; and UDAKA, S. 1988. Cloning and sequencing of the gene encoding thermophilic β-amylase of *Clostridium thermosulfurogenes. J. Bacteriol. 170,* 5848–5854.

KLUH, I. 1981. Amino acid sequence of hog pancreatic α-amylase isoenzyme I. *FEBS Lett. 136*, 231–234.

KOCHHAR, S., and DUA, R. D. 1985A. Chemical modification of liquefying α-amylase: Role of tyrosine residues at its active center. *Arch. Biochem. Biophys. 240*, 757–767.

———. 1985B. An active center tryptophan residue in liquefying α-amylase from *Bacillus amyloliquefaciens. Biochim. Biophys. Res. Comm. 126*, 966–973.

KOHNO, A.; SHINKE, R.; and NANMORI, T. 1990. Features of the β-amylase isoform system in dry and germinating seeds of alfalfa (*Medicago sativa* L.). *Biochem. Biophys. Acta 1035*, 325–330.

KONDO, H.; NAKATANI, H.; MATSUNO, R.; and HIROMI, K. 1980. Product distribution in amylase-catalyzed hydrolysis of amylose. Comparison of experimental results with theoretical predictions. *J. Biochem. 87*, 1053–1070.

KOSHLAND, D. E., JR. 1959. Mechanisms of transfer enzymes. *The Enzymes 1*, 305–346.

KREIS, M.; WILLIAMSON, M.; BUXTON, B.; PYWELL, J.; HEIJGAARD, J.; and SVENDSEN, I. 1987. Primary structure and differential expression of β-amylase in normal and mutant barleys. *Eur. J. Biochem. 169*, 517–525.

KUNIKATA, T.; YAMANO, H.; NAGAMURA, T.; and NITTA, Y. 1992. Study on the interaction between soybean β-amylase and substrate by the stopped-flow method. *J. Biochem. 112*, 421–425.

LAI, H.-L.; BUTLER, L. G.; and AXELROD, B. 1974. Evidence for a covalent intermediate between α-glucosidase and glucose. *Biochem. Biophys. Res. Comm. 60*, 635–640.

LAMBRECHTS, M. G.; PRETORIUS, I. S.; SOLLITTI, P.; and MARMUR, J. 1991. Primary structure and regulation of a glucoamylase-encoding gene (STA 2) in *Saccharomyces diastaticus. Gene 100*, 95–103.

LAURIERE, C.; DOYEN, C.; THEVENOT, C.; and DAUSSANT, J. 1992. β-Amylases in cereals. A study of the maize β-amylase system. *Plant Physiol. 100*, 887–893.

LOYTER, A., and SCHRAMM, M. 1966. Multimolecular complexes of α-amylase with glycogen limit dextrin. *J. Biol. Chem. 241*, 2611–2617.

MACGREGOR, E. A. 1988. α-Amylase structure and activity. *J. Protein Chem. 7*, 399–415.

MACGREGOR, E. A., and MACGREGOR, A. W. 1985. A model for the action of cereal alpha-amylases on amylose. *Carbohydr. Res. 142*, 223–236.

MACGREGOR, A. W., and MORGAN, J. E. 1992. The action of germinated barley alpha-amylases on linear maltodextrins. *Carbohydr. Res. 227*, 301–313.

MATSUI, H.; BLANCHARD, J. S.; BREWER, C. F.; and HEHRE, E. J. 1989. α-Secondary tritium kinetic isotope effects for the hydrolysis of α-D-glucopyranosyl fluoride by exo-α-glucanases. *J. Biol. Chem. 264*, 8714–8716.

MATSUI, I.; ISHIKAWA, K.; MATSUI, E.; MIYAIRI, S.; FUKUI, S.; and HONDA, K. 1991. An increase in the transglycosylation activity of *Saccharomycopsis* α-amylase altered by site-directed mutagenesis. *Biochim. Biophys. Acta 1077*, 416–419.

MATSUI, I.; YONEDA, S.; ISHIKAWA, K.; MIYAIRI, S.; FUKUI, S.; UMEYAMA, H.; and HONDA, K. 1994. Roles of the aromatic residues conserved in the active center of *Saccharomycopsis* α-amylase for transglycosylation and hydrolysis activity. *Biochemistry 33*, 451–458.

MATSUNO, R.; SUGANUMA, T.; FUJIMORI, H.; NAKANISHI, K.; HIROMI, K.; and KAMIKUBO, T. 1978. Rate equation for amylase-catalyzed hydrolysis, transglycosylation and condensation of linear oligosaccharides and amylose. *J. Biochem. 83*, 385–394.

MATSUURA, Y.; KUSUNOKI, M.; HARADA, W.; and KAKUDO, M. 1984. Structure and possible catalytic residues of Taka-amylase A. *J. Biochem. 95*, 697–702.

MATSUURA, Y.; KUSUNOKI, M.; DATE, W.; HARADA, S.; BANDO, S.; TANAKA, N.; and KAKUDO, M. 1979. Low resolution crystal structures of Taka-amylase A and its complexes with inhibitors. *J. Biochem. 86*, 1773–1783.

MATSUURA, Y.; KUSUNOKI, M.; HARADA, W.; TANAKA, N.; IGA, Y.; YASUOKA, N.; TODA, H.; NARITA, K.; and KAKUDO, M. 1980. Molecular structure of Taka-amylase A. *J. Biochem. 87*, 1555–1558.

MAZUR, A. K. 1984. Mathematical models of depolymerization of amylose by α-amylases. *Biopolymers 23*, 1735–1756.

MEAGHER, M. M.; NIKOLOV, Z. L.; and REILLY, P. J. 1989. Subsite mapping of *Aspergillus niger* glucoamylases I and II with malto- and isomaltooligosaccharides. *Biotechnol. Bioengineer. 34*, 681–688.

MEAGHER, M. M., and REILLY, P. J. 1989. Kinetics of the hydrolysis of di- and trisaccharides with *Aspergillus niger* glucoamylases I and II. *Biotechnol. Bioengineer. 34*, 689–693.

MIKAMI, B.; AIBRA, S.; and MORITA, Y. 1980. Chemical modification of sulfhydryl groups in soybean β-amylase. *J. Biochem. 88*, 103–111.

MIKAMI, B.; HEBRE, E. J.; SATO, M.; KATSUBE, Y.; HIROSE, M.; MORITA, Y.; and SACCHETTINI, J. C. 1993. The 2.0-A resolution structure of soybean β-amylase complexed with α-cyclodextrin. *Biochemistry 32*, 6836–6845.

MIKAMI, B.; MORITA, Y.; and FUKAZAWA, C. 1988. Primary structure and function of β-amylase. *Seikagaku* (Japanese) *60*, 211–216.

MIKAMI, B.; NOMURA, K.; and MORITA, Y. 1983. Interaction of native and SH-modified β-amylase of soybean with cyclohexadextrin and maltose. *J. Biochem. 94*, 107–113.

————. 1994. Two sulfhydryl groups near the active site of soybean β-amylase. *Biosci. Biotech. Biochem. 58*, 126–132.

MIKAMI, B.; NOMURA, K.; MAJIMA, K.; and MORITA, Y. 1989. Structure of soybean β-amylase and the reactivity of its sulfhydryl groups. *Denpun Kagaku* (Japan) *36*, 67–72.

MIKAMI, B.; SATO, M.; SHIBATA, T.; HIROSE, M.; AIBARA, S.; KATSUBE, Y.; and MORITA, Y. 1992. Three-dimensional structure of soybean β-amylase determined at 3.0 A resolution: Preliminary chain tracing of the complex with α-cyclodextrin. *J. Biochem. 112*, 541–546.

MIKAMI, B.; SHIBATA, T.; HIROSE, M.; AIBARA, S.; SATO, M.; KATSUBE, Y.; and MO-

RITA, Y. 1991. X-ray crystal structure analysis of soybean β-amylase. *Denpun Kagaku* (Japan), *38*, 147–151.

MONROE, J. D.; SALMINEN, M. D.; and PREISS, J. 1991. Nucleotide sequence of a cDNA clone encoding a β-amylase from *Arabidopsis thaliana. Plant Physiol. 97*, 1599–1601.

NAKAJIMI, R.; IMANAKA, T.; and AIBA, S. 1986. Comparison of amino acid sequences of eleven different α-amylases. *Appl. Microbiol. Biotechnol. 23*, 355–360.

NEUSTROEV, K. N.; GOLUBER, A. M.; FIRSOV, L. M.; IBATULLIN, F. M.; PROTASEVICH, I. I.; and MAKAROV, A. A. 1993. Effect of modification of carbohydrate component on properties of glucoamylase. *FEBS Lett. 316*, 157–160.

NIKOLOV, Z. L.; MEAGHER, M. M.; and REILLY, P. J. 1989. Kinetics, equilibria, and modelling of the formation of oligosaccharides from D-glucose with *Aspergillus niger* glucoamylases I and II. *Biotechnol. Bioengin. 34*, 694–704.

NISHIDA, T.; EMI, M.; NAKAMURA, Y.; and MATSUBARA, K. 1986. Corrected sequences of cDNAs for human salivary and pancreatic α-amylases. *Gene 50*, 371–372.

NITTA, Y.; ISODA, Y.; TODA, H.; and SAKIYAMA, F. 1989. Identification of glutamic acid 186 affinity-labeled by 2,3-epoxypropyl α-D-glucopyranoside in soybean β-amylase. *J. Biochem. 105*, 573–576.

NITTA, Y.; KUNIKATA, T.; and WATANABE, T. 1979. Kinetic study of soybean β-amylase. The effect of pH. *J. Biochem. 85*, 41–45.

NITTA, Y.; MIZUSHIMA, M.; HIROMI, K.; and ONO, S. 1971. Influence of molecular structures of substrates and analogues on Taka-amylase A catalyzed hydrolyses. I. Effect of chain length of linear substrates. *J. Biochem. 69*, 567–576.

NOMURA, K.; MIKAMI, B.; NAGAO, Y.; and MORITA, Y. 1987. Effect of modification of sulfhydryl groups in soybean β-amylase on the interaction with substrate and inhibitions. *J. Biochem. 102*, 333–340.

NUNBERG, J. H.; MEADE, J. H.; COLE, G.; LAWYER, F. C.; MCCABE, P.; SCHWEICHART, V.; TAL, R.; WITTMAN, V. P.; FLATGAARD, J. E.; and INNIS, M. A. 1984. Molecular cloning and characterization of the glucoamylase gene of *Aspergillus awamori. Mol. Cell. Biol. 4*, 2306–2315.

OHNISHI, M. 1990. Subsite structure of *Rhizopus niveus* glucoamylase, estimated with the binding parameters for maltooligosaccharides. *Starch/Starke 42*, 311–313.

OHNISHI, M.; NAKAMURA, Y.; MURATA-NAKAI, M.; and HIROMI, K. 1990. A pH-induced change in state around active-site tryptophan residues of *Rhizopus niveus* glucoamylase, detected by stopped-flow studies of chemical modification with N-bromosuccinimide. *Carbohydr. Res. 197*, 237–244.

OHNISHI, M.; TANIGUCHI, M.; and HIROMI, K. 1983. Kinetic discrimination of tryptophan residues of glucoamylase from *Rhizopus niveus* by fast chemical modification with N-bromosuccinimide. *Biochim. Biophys. Acta 744*, 64–70.

OLSEN, K.; SVENSSON, B.; and CHRISTENSEN, U. 1992. Stopped-flow fluorescence and steady-state kinetic studies of ligand-binding reactions of glucoamylase from *Aspergillus niger. Eur. J. Biochem. 209*, 777–784.

PASERO, L.; MAZZEI-PIERRON, Y.; ABADIE, B.; CHICHEPORTICHE, Y.; and MARCHIS-MOUREN, G. 1986. Complete amino acid sequence and location of the five disulfide bridges in porcine pancreatic α-amylase. *Biochim. Biophys. Acta 869,* 147–157.

PAYAN, F.; HASER, R.; PIERROT, M.; FREY, M.; and ASTIER, J. P. 1980. The three-dimensional structure of α-amylase from porcine pancreas at 5 A resolution—The active-site location. *Acta Cryst. B36,* 416–421.

PAZUR, J. H., and MARCHETTI, N. T. 1992. Action patterns of amylolytic enzymes as determined by the [1-^{14}C]malto-oligosaccharide mapping method. *Carbohydr. Res. 227,* 215–225.

PAZUR, J. H.; KNULL, H. R.; and SIMPSON, D. L. 1970. Glycoenzymes: A note on the role for the carbohydrate moieties. *Biochem. Biophys. Res. Comm. 40,* 110–116.

PAZUR, J. H.; LIU, B.; PYKE, S.; and BAUMRUCKER, C. R. 1987. The distribution of carbohydrate side chains along the polypeptide chain of glucoamylase. *J. Protein Chem. 6,* 517–527.

PRODANOV, E.; SEIGNER, C.; and MARCHIS-MOUREN, G. 1984. Subsite profile of the active center of porcine pancreatic α-amylase. Kinetic studies using maltooligosaccharides as substrates. *Biochem. Biophys. Res. Comm. 122,* 75–81.

QIAN, M.; HASER, R.; and PAYAN, F. 1993. Structure and molecular model refinement of pig pancreatic α-amylase at 2.1A resolution. *J. Mol. Biol. 231,* 785–799.

QIAN, M.; HASER, R.; BUISSON, G.; DUEE, E.; and PAYAN, F. 1994. The active center of a mammalian α-amylase. Structure of the complex of a pancreatic α-amylase with a carbohydrate inhibitor refined to 2.2-A resolution. *Biochemistry 33,* 6284–6294.

RHODES, C.; STRASSER, J.; and FRIEDBERG, F. 1987. Sequence of an active fragment of *B. polymyxa* beta amylase. *Nucl. Acid Res. 15,* 3934.

ROBYT, J. F. 1984. Enzymes in the hydrolysis and synthesis of starch. In: *Starch: Chemistry and Technology,* R. L. Whistler, J. N. Bemiller, and E. F. Paschall, eds., Academic Press, New York.

ROBYT, J. F., and FRENCH, D. 1970A. The action pattern of porcine pancreatic α-amylase in relationship to the substrate binding site of the enzyme. *J. Biol. Chem. 245,* 3917–3927.

———. 1970B. Multiple attack and polarity of action of porcine pancreatic α-amylase. *Arch. Biochem. Biophys. 138,* 662–670.

ROGERS, J. C. 1985A. Conserved amino acid sequence domans in alpha-amylases from plants, mammals and bacteria. *Biochem. Biophys. Res. Comm. 128,* 470–476.

———. 1985B. Two barley α-amylase gene families are regulated differently in aleurone cells. *J. Biol. Chem. 260,* 3731–3738.

ROGERS, J. C., and WILLIAMS, C. 1983. Isolation and sequence analysis of a barley α-amylase cDNA clone. *J. Biol. Chem. 258,* 8169–8173.

RORAT, T.; SADOWSKI, J.; GRELLET, F.; DAUSSANT, J.; and DELSENY, M. 1991. Characterization of cDNA clones for rye endosperm β-amylase and analysis of β-amylase deficiency in rye mutant lines. *Theor. Appl. Genet. 83,* 257–263.

SEIGNER, C.; PRODANOV, E.; and MARCHIS-MOUREN, G. 1987. The determination of subsite binding energies of porcine pancreatic α-amylase by comparing hydrolytic activity towards substrates. *Biochim. Biophys. Acta 913,* 200–209.

SIERKS, M. R.; FORD, C.; REILLY, P. J.; and SVENSSON, B. 1989. Site-directed mutagenesis at the active site Trp 120 of *Aspergillus awamori* glucoamylase. *Protein Engineering 2*, 621–625.

————. 1993. Functional roles and subsite locations of Leu177, Trp178 and Asn182 of *Aspergillus awamori* glucoamylase determined by site-directed mutagenesis. *Protein Engineering 6*, 75–79.

SPRADLIN, J., and THOMA, J. A. 1970. β-Amylase thiol groups. *J. Biol. Chem. 245*, 117–127.

SUGANUMA, T.; MATSUNO, R.; OHNISHI, M.; and HIROMI, K. 1978. A study of the mechanism of action of Taka-amylase A on linear oligosaccharides by product analysis and computer simulation. *J. Biochem. 84*, 293–316.

SUGANUMA, T.; OHNISHI, M.; HIROMI, K.; and MORITA, Y. 1980. Evaluation of subsite affinities of soybean β-amylase by product analysis. *Agric. Biol. Chem. 44*, 1111–1117.

SVENSSON, B.; CLARKE, A. J.; SVENDSEN, IB.; and MOLLER, H. 1990. Identification of carboxylic acid residues in glucoamylase G2 from *Aspergillus niger* that participate in catalysis and substrate binding. *Eur. J. Biochem. 188*, 29–39.

SVENSSON, B.; JESPERSEN, H.; SIERKS, M. R.; and MACGREGOR, E. A. 1989. Sequence homology between putative raw-starch binding domains from different starch-degrading enzymes. *Biochem. J. 264*, 309–311.

SVENSSON B.; LARSEN, K.; and GUNNARSSON, A. 1986. Characterization of a glucoamylase G2 from *Aspergillus niger. J. Biochem. 154*, 497–502.

SVENSSON, B.; LARSEN, K.; SVENDSEN I.; and BOEL, E. 1983. The complete amino acid sequence of the glycoprotein, glucoamylase G1 from *Aspergillus niger. Carlsberg. Res. Comm. 48*, 529–544.

TAKAHASHI, T.; IDEGAMI, Y.; IRIE, M.; and NAKAO, E. 1990. Different behavior towards raw starch of two glucoamylases from *Aspergillus saitoi. Chem. Pharm. Bull. 38*, 2780–2783.

TAKAHASHI, T.; KATO, K.; IKEGAMI, Y.; and IRIE, M. 1985. Different behavior towards raw starch of three forms of glucoamylase from a *Rhizopus* sp. *J. Biochem. 98*, 663–671.

TAKAHASHI, T.; TSUCHIDA, Y.; and IRIE, M. 1982. Isolation of two inactive fragments of a Rhizopus sp. glucoamylase: Relationship among three forms of the enzyme and the isolated fragments. *J. Biochem. 92*, 1623–1633.

TAKASAKI, Y. 1989. Novel maltose-producing amylase from *Bacillus megaterium* G-2. *Agric. Biol. Chem. 53*, 341–347.

TAKEDA, Y., and HIZUKURI, S. 1981. Re-examination of the action of sweet-potato beta-amylase on phosphorylated (1–4)-α-D-glucan. *Carbohydr. Res. 89*, 174–178.

TAKEGAWA, K.; INAMI, M.; YAMAMOTO, K.; KUMAGAI, H.; TOCHIKURA, T.; MIKAMI, B.; and MORITA, Y. 1988. Elucidation of the role of sugar chains in glucoamylase using endo-β-N-acetylglucoaminidase from *Flavobacterium sp. Biochim. Biophys. Acta 955*, 187–193.

TAKKINEN, K.; PETTERSSON, R. F.; KALKKINEN, N.; PALVA, I.; SODERLUND, H.; and

KAARIAINEN, L. 1983. Amino acid sequence of α-amylase from *Bacillus amyloliquefaciens* deduced from the nucelotide sequence of the cloned gene. *J. Biol. Chem. 258*, 1007–1013.

TANAKA, A.; FUKUCHI, Y.; OHNISHI, M.; HIROMI, K.; AIBARA, S.; and MORITA, Y. 1983B. Fractionation of isozymes and determination of the subsite structure of glucoamylase from *Rhizopus niveus*. *Agric. Biol. Chem. 47*, 573–580.

TANAKA, A.; YAMASHITA, T.; OHNISHI, M.; and HIROMI, K. 1983A. Steady-state and transient kinetic studies on the binding of maltooligosaccharides to glucoamylase. *J. Biochem. 93*, 1037–1043.

TANAKA, Y.; ASHIKARI, T.; NAKAMURA, N.; KIUCHI, N.; SHIBANO, Y.; AMACHI, T.; and YOSHIZUMI, H. 1986. Comparison of amino acid sequences of three glucoamylases and their structure-function relationships. *Agric. Biol. Chem. 50*, 965–969.

TAO, B. Y.; REILLY, P. J.; and ROBYT, J. F. 1989. Detection of a covalent intermediate in the mechanism of action of porcine pancreatic α-amylase by using ^{13}C nuclear magnetic resonance. *Biochim. Biophys. Acta 995*, 214–220.

THOMA, J. A. 1968. A possible mechanism for amylase catalysis. *J. Theoret. Biol. 19*, 297–310.

———. 1976. Models for depolymerizing enzymes. Application to α-amylases. *Biopolymers 15*, 729–746.

THOMA, J. A., and KOSHLAND, D. E., JR. 1960A. Competitive inhibition by substrate during enzyme action. Evidence for the induce-fit theory. *J. Am. Chem. Soc. 82*, 3329–3333.

———. 1960B. Three amino acids at the active site of beta amylase. *J. Mol. Biol. 2*, 169–170.

THOMA, J. A.; SPRADLIN, J. E.; and DYGERT, S. 1971. Plant and animal amylases. *The Enzymes 5*, 115–189.

TODA, H.; KONDO, K.; and NARITA, K. 1982. The complete amino acid sequence of Taka-amylase A. *Proc. Japan Acad. 58*, Ser. B, 208–212.

TODA, H.; NITTA, Y.; ASANAMI, S.; KIM, J. P.; and SAKIYAMA, F. 1993. Sweet potato β-amylase. Primary structure and identification of the active-site glutamyl residue. *Eur. J. Biochem. 216*, 25–38.

TORGERSON, E. M.; BREWER, L. C.; and THOMA, J. A. 1979. Subsite mapping of enzymes. Use of subsite map to simulate complete time course of hydrolysis of a polymeric substrate. *Arch. Biochem. Biophys. 196*, 13–22.

TOTSUKA, A.; HONG, V. H.; KADOKAWA, H.; KIM, C.-S.; ITOH, Y.; and FUKAZAWA, C. 1994. Residues essential for catalytic activity of soybean β-amylase. *Eur. J. Biochem. 221*, 649–654.

UEDA, S. 1981. Fungal glucoamylases and raw starch digestion. *TIBS 6*, 89–90.

UOZUMI, N.; MATSUDA, T.; TSUKAGOSHI, N.; and UDAKA, S. 1991. Structural and functional roles of cysteine residues of *Bacillus polymyxa* β-amylase. *Biochemistry 30*, 4594–4599.

UOZUMI, N.; SAKURAI, K.; SASAKI, T.; TAKEKAWA, S.; YAMAGATA, H.; TSUKAGOSHI,

N.; and UDAKA, S. 1989. A single gene directs synthesis of a precursor protein with β- and α-amylase activities in *Bacillus polymyxa*. *J. Bacteriol. 171*, 375–382.

VIHINEN, M., and MANTSALA, P. 1990. Conserved residues of liquefying α-amylases are concentrated in the vicinity of active site. *Biochem. Biophys. Res. Comm. 166*, 61–65.

VIHINEN, M.; OLLIKKA, P.; NISKANEN, J.; MEYER, P.; SUOMINEN, I.; KARP, M.; HOLM, L.; KNOWLES, J.; and MANTSALA, P. 1990. Site-directed mutagenesis of a thermostable α-amylase from *Bacillus stearothermophilus*. Putative role of three conserved residues. *J. Biochem. 107*, 267–272.

WAKIM, J.; ROBINSON, M.; and THOMA, J. A. 1969. The active site of porcine-pancreatic alpha-amylase: Factors contributing to catalysis. *Carbohydr. Res. 10*, 487–503.

WESTLAKE, R. J., and HILL, R. D. 1983. Inhibition of alpha-amylase-catalyzed starch granule hydrolysis by cycloheptaamylase. *Cereal Chem. 60*, 98–101.

WILLIAMSON, G.; BELSHAW, N. J.; NOEL, T. R.; RING, S. G.; and WILLIAMSON, M. P. 1992A. O-Glycosylation and stability. Unfolding of glucoamylase induced by heat and guanidine hydrochloride. *Eur. J. Biochem. 207*, 661–670.

WILLIAMSON, G.; BELSHAW, N. J.; SELF, D. J.; NOEL, T. R.; RING, S. G.; CAIRNS, P.; MORRIS, V. J.; CLARK, S. A.; and PARKER, M. L. 1992B. Hydrolysis of A- and B-type crystalline polymorphs of starch by α-amylase, β-amylase and glucoamylase 1. *Carbohydr. Polymers 18*, 179–187.

WILLIAMSON, G.; BELSHAW, N. J.; and WILLIAMSON, M. P. 1992C. O-Glycosylation in *Aspergillus* glucoamylase. *Biochem. J. 282*, 423–428.

WIRSEL, S.; LACHMUND, A.; WILDHARDT, G.; and RUTTKNOWSKI, E. 1989. Three α-amylase genes of *Aspergillus oryzae* exhibit identical intron-exon organization. *Mol. Microbiol. 3*, 3–14.

YAMASHITA, I.; HATANO, T.; and FUKUI, S. 1984. Subunit structure of glucoamylase of *Saccharomyces diastaticus*. *Agric. Biol. Chem. 48*, 1611–1616.

YAMASHITA, I.; NAKAMURA, M.; and FUKUI, S. 1987. Gene fusion in a possible mechanism underlying the evolution of STA1. *J. Bacteriol. 169*, 2142–2149.

YAMASHITA, I.; SUZUKI, K.; and FUKUI, S. 1985. Nucleotide sequence of the extracellular glucoamylase gene STA1 in the yeast *Saccharomyces diastaticus*. *J. Bacteriol. 161*, 567–573.

YAMAZAKI, H.; OHMURA, K.; NAKAYAMA, A.; TAKEICHI, Y.; OTOZAI, K.; YAMASAKI, M.; TAMURA, G.; and YAMANE, K. 1983. α-Amylase genes (amyR2 and amyE +) from an α-amylase-hyperproducing *Bacillus subtilis* strain: Molecular cloning and nucleotide sequences. *J. Bacteriol. 156*, 327–337.

YANG, M.; GALIZZI, A.; and HENNER, D. 1983. Nucleotide sequence of the amylase gene from *Bacillus subtilis*. *Nucl. Acid Res. 11*, 237–249.

YASUDA, M.; KUWAE, M.; and MATSUSHITA, H. 1989. Purification and properties of two forms of glucoamylase from *Monascus* sp. No. 3403. *Agric. Biol. Chem. 53*, 247–249.

YOSHIDA, N.; HAYASHI, K.; and NAKAMURA, K. 1992. A nuclear gene encoding β-amylase of sweet potato. *Gene 120*, 255–259.

YOSHIDA, N., and NAKAMURA, K. 1991. Molecular cloning and expression in *Escherichia coli* of cDNA encoding the subunit of sweet potato β-amylase. *J. Biochem.* *110*, 196–201.

YOSHIGI, N.; OKADA, Y.; SAHARA, H.; and KOSHINO, S. 1994. PCR cloning and sequencing of the β-amylase cDNA from barley. *J. Biochem.* *115*, 47–51.

YUUKI, T.; NOMURA, T.; TEZUKA, H.; TSUBOI, A.; YAMAGATA, H.; TSUKAGOSHI, N.; and UDAKA, S. 1985. Complete nucleotide sequence of a gene coding for heat- and pH-stable α-amylase of *Bacillus licheniformis*: Comparison of the amino acid sequences of three bacterial liquefying α-amylases deduced from the DNA sequences. *J. Biochem.* *98*, 1147–1156.

Chapter 4

Cellulolytic Enzymes

Cellulolytic enzymes act synergistically to hydrolyze cellulose or its chemically modified polymers. Conversion of abundant cellulose waste materials, such as straw, husks, sawdust, paper, etc. to fuel-grade alcohol represents an important energy source. Many research efforts on these enzymes are devoted to the application of cellulolytic enzymes in making biomass conversion economically and technically feasible. The use of these enzymes, in conjunction with pectinases, is a potential alternative to chemical peeling in fruit and vegetable processing.

CLASSIFICATION

Cellulolytic enzymes are traditionally classified into three groups.

(1) 1,4-β-D-glucan cellobiohydrolase (EC 3.2.1.91)—Removal of cellobiose units from nonreducing ends of cellulose chains. Other names used include cellobiohydrolase, exoglucanase, exocellulase, cellobiosidase, and avicellase.

(2) 1,4-β-D-glucan 4-glucanohydrolase (EC 3.2.1.4)—Hydrolysis of internal 1,4-β-D-glycosidic linkages in cellulose, lichenin, and cereal β-D-glucans. Common names include endoglucanase, endo-1,4-β-glucanase, and CM-cellulase.

(3) β-D-glucoside glucohydrolase (EC 3.2.1.21)—Hydrolysis of cellobiose and removal of glucose from nonreducing ends of cello-oligosaccharides and glycosyl transfer to cellobiose. Commonly known as β-glucosidase and cellobiase.

It should be noted that β-glucosidase is not strictly a cellulase; the enzyme does not attack cellulose. It is part of the cellulolytic enzyme system because it removes end product inhibition caused by the accumulation of cellobiose formed by the combined action of the other two enzymes.

Comparison of amino acid sequences in the catalytic domains of cellulases by hydrophobic cluster analysis (based on the 2D structural elements that constitute the main 3D folding cores of globular proteins) (Henrissat et al. 1989; Beguin 1990; Gilkes et al. 1991) has led to a proposed classification of cellulases into nine families, A to I. The enzymes are grouped according to similarities of their catalytic domain alone rather than function. Consequently, exo- and endo-glucanases fall into the same families. Enzymes produced from the same organism, despite sharing the same type of cellulose-binding domain (CBD), appear in different families. Although hydrophobic-cluster analysis is effective in locating homologous sequences and domains that are difficult to assess by classical methods, classification of an enzyme according to structure rather than function requires careful consideration.

GENERAL CHARACTERISTICS

Most fungal enzymes are extracellularly secreted, while the enzymes synthesized by the bacterium *Clostridium thermocellum*, and possibly other bacteria, are organized into functional high molecular multiunit complexes called cellulosomes bound to the cell surface as postuberance-like structures (Lamed and Bayer 1988; Beguin et al. 1992).

It is well known that certain cellulolytic systems are able to hydrolyze crystalline cellulose completely. These are often known as "true" ("complete," "full-value") cellulase systems; they are found in a number of fungi, including *Trichoderma reesei, Trichoderma koningii, Trichoderma viride, Fusarium solani, Penicillium pinophilum*, and *Penicillium funiculosum*. These systems are distinguished in containing both exo- and endo-acting glucanases, in addition to β-glucosidase. Endoglucanase and cellobiohydrolase act in combination in the degradation of crystalline celluloses. In general, the enzymes acting alone have little capacity to hydrolyze crystalline cellulose, although they may effectively hydrolyze amorphous celluloses. Most bacteria, with the exception of *Clostridium thermocellum* and a few others, synthesize only endoglucanase and β-glucosidase (or cellobiose phosphorylase). Cellulolytic systems, which are inactive toward crystalline cellulose but hydrolyze only amorphous cellulose, are sometimes referred to as "low-value" cellulase systems.

STRUCTURES OF SUBSTRATES

Cellulases all cleave the same β-1,4-glycosidic bond, but there are variations in the structural details of cellulose substrates that inevitably affect enzyme activity.

Although all celluloses have the basic structure of poly-β(1–4)-D-glucans, there are at least four different crystalline forms, I, II, III, and IV (Blackwell 1982; Kamide et al. 1985). The native crystalline form, cellulose I, has the adjacent chains packed in parallel with two types of intramolecular hydrogen bonds, O_3-H—O_5' and O_2-H—O_6', and one type of intermolecular O_6-H—O_3 holding the chains along the *ac* plane. The three-dimensional structure assumes a series of hydrogen-bonded sheets of chains parallel to the *a* axis that are held by van der Waals forces. Cellulose I is converted irreversibly to cellulose II by mercerization (treatment with NaOH), a process used in modifying the properties of cotton fibers. Cellulose II has an antiparallel structure, with two types of intramolecular and three types of intermolecular hydrogen bonds. The high stability of this cellulose form is caused by an extra intermolecular bonding O_2-H—O_2' along the *b* axis. Therefore, unlike cellulose I, cellulose II has the chains linked along both the *a* and *b* axes. Celluloses III and IV can be generated by the treatment of cellulose I (or II) with liquid ammonia and hot glycerol, respectively. Both conversions are reversible.

Common crystalline celluloses used for enzyme studies include Avicel (a microcrystalline cellulose) and filter paper with relative crystallinity indexes (χ_c) of 47% and 45%, respectively, based on 100% for the highest crystalline cellulose from the algal *Valonia ventricosa*. The value of χ_c varies depending on the method of measurement. Native (untreated) cotton has $\chi_c = 73\%$ according to x-ray diffraction, but 65% by the density method. Using x-ray diffraction, Avicel and filter paper (Whatman) have $\chi_c = 69\%$ and 88%, respectively (Schultz et al. 1985). Substrates commonly used for measuring the degradation of amorphous cellulose such as H_3PO_4-swollen cellulose and other chemically modified celluloses are less orderly packed, with most of the intramolecular and intermolecular bonds disrupted. In addition to the degree of crystallinity, other physical and chemical features, such as specific surface area, degree of polymerization, and pore size, may all affect the efficiency of enzyme attack.

Because of the complexity involved in the use of these insoluble substrates, soluble substrates, such as carboxymethyl cellulose (CMC), *p*-nitrophenyl-oligosaccharides, cello-oligosaccharides (in the series from cellobiose to cellohexaose commonly used), and chromophoric glycosides are frequently used (Table 4.1). Although the use of these soluble substrates simplifies the interpretation in the characterization of enzyme actions, they are not natural substrates, and consequently, the effects of variables such as adsorption, synergism, and overlapping specificities cannot be adequately assessed. It is not unusual to find in the literature a wide range of substrates and conditions used by various research groups for studying the same enzyme. Consequently, meaningful interpretation and comparison of results become extremely difficult if not entirely impossible.

Table 4.1. Substrates of the Cellulolytic Enzymes and the Detection Methods Most Commonly Used

Product Measured	Soluble Substrates					
	CMC	HEC	β-Glucan	Cello-oligo-Saccharides	β-ME-umbeli-feryl-oligo-Saccharides	p-Nitro-phenyl-oligo-Saccharides
Formation of reducing sugars (biochemical assay)	+	+	+	+		
Liberation of chromophore (fluorescence, absorbance)	+				+	+
Oligosaccharides released (HPLC)	+	+		+	+	+
Decrease in degree of polymerization						
staining	+	+	+			
viscosity	+	+				

Product Measurement	Insoluble Substrate						
	Valonia Cellulose	Native Cotton	Bacterial micro-crystalline cellulose	Avicel	Filter paper	Amor-phous cellulose	Phosphoric Acid swollen cellulose
Formation of reducing sugars (biochemical assay)	+		+	+	+	+	+
Decrease in degree of polymerization							
turbidity		+	+		+		
staining				+			
Release of fibers							
light microscopy		+					
electron microscopy	+						

From Knowles et al. 1987. *TIBTECH* 5, 255, with permission. Copyright 1987 Elsevier Trends Journals (Cambridge).

CELLOBIOHYDROLASES

Trichoderma reesei

Two cellobiohydrolases (CbhI and CbhII) secreted by the fungus *Trichoderma reesei* have been purified from culture filtrates. CbhI comprises about 60% of the total extracellular proteins in culture filtrate. It has a MW of 65 kD and 10% carbohydrate content with a pI of 4.4. The enzyme attacks both amorphous and crystalline celluloses, but does not act on substituted celluloses such as CMC and hydroxyethylcellulose, nor on other soluble substrates, such as cellohexaose, *p*-nitrophenyl β-glucoside, and barley β-glucan (Nummi et al. 1983). CbhII, a minor enzyme, has a MW of 53 kD, 8% carbohydrate content, and a pI of 5.0. The enzyme, like CbhI, is not active toward CMC (Wood et al. 1989). It has also been reported that the enzyme hydrolyzes both insoluble and soluble substrates (Niku-Paavola et al. 1986), although the ability to produce reducing sugars from CMC is generally regarded as a property characteristic of endoglucanases. Some studies show that CbhI does not exhibit exclusively exo-actions (Vrsanska and Biely 1992). CbhI and CbhII from other fungal sources also show different substrate specificities.

The primary structures of *Trichoderma reesei* CbhI and CbhII are known. CbhI consists of 496 amino acids, with an N-terminal pyroglutamyl residue (Fagerstam et al. 1984). The primary structure agrees with the sequence deduced from the encoding gene with a nucleotide sequence corresponding to 513 amino acids, including a signal peptide of 17 residues (Shoemaker et al. 1983). Three Asn-X-Ser/Thr sites at positions 45, 270, and 384 are N-glycosylated with one to four hexose residues that are mostly mannose. Most of the carbohydrate is O-glycosylated to Ser and Thr, with structures $(Mann)_5(GlcNAc)_2$ and $(Mann)_9(GlcNAc)_2$, located in the Pro/Thr/Ser region near the C-terminus (Salovuori et al. 1987). The enzyme contains 12 disulfides—Cys4-Cys50, Cys19-Cys25, Cys61/67-Cys71/72, Cys138-Cys397, Cys172/176-Cys209/210, Cys230-Cys256, Cys238-Cys243, Cys261-Cys331, Cys469-Cys496, and Cys480-Cys486 (Bhikhabhai and Pettersson 1984). *Trichoderma reesei* CbhII contains 471 residues with a calculated MW of 49,712, as deduced from the nucleotide sequence. The CbhII protein, excluding the 24 amino acids in the signal peptide, consists of 447 amino acids. Comparison of the sequences of CbhI and CbhII reveals a common bifunctional structural organization that is representative of all the cellulases: a cellulose-binding domain and a core (consisting of the catalytic domain) interconnected by a Ser/Thr-rich linker (Fig. 4.1). A marked similarity exists between the C-terminal region 38 residues of CbhI and the N-terminus of CbhII, with about 70% of amino acid residues identical (Fig. 4.2) (Chen et al. 1987). These regions are cellulose-binding domains (CBDs). Adjacent to the CBD is the linker, a conserved region consisting of glycosylated Ser/Thr-rich sequences. The linker functions to join the CBD and the core segment of the enzyme where the

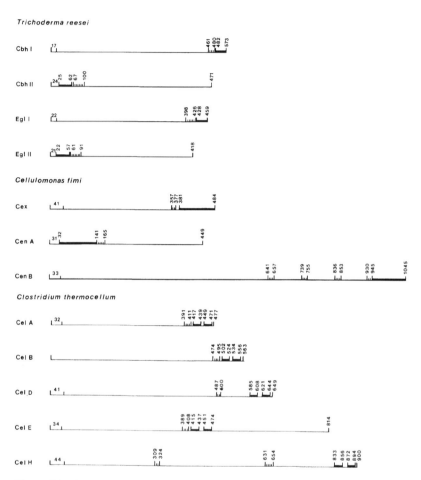

Figure 4.1. Schematic structures of different cellulolytic enzymes. Each enzyme consists of a catalytic domain, a linker region (shaded area), and a binding domain (solid bar). Sequences are numbered from Met of the leader peptides. Sources of information: *Trichoderma reesei*-CbhI (Shoemaker et al. 1983; Fagerstam et al. 1984), CbhII (Chen et al. 1987), EglI (Penttila et al. 1986), CenB (Meinke et al. 1991), Cex ()'Neill et al. 1986); *Clostridium thermocellum*-CelA (Beguin et al. 1985), CelB (Grepinet and Beguin 1986), CelD (Joliff et al. 1986), CelE (Hall et al. 1988), CelH (Yague et al. 1990). ALl amino acid sequences are deduced from cDNA sequencing.

```
Cbh I     482    YGQCGGIGYSGPTVCASGTTCQVLNPYYSQCL
Cbh II    31     W-----QNW--PTC-AS-ST-VYS-D--S--L
Egl I     428    W-----IGY--CKT-TS-TT-QYS-D--S--L
Egl II    26     W-----IGW--PTN-AP-SA-STL-P--A--I
```

Figure 4.2. Sequence alignment of conserved C-terminal domains of CbhI and EglI and the N-terminal domains of CbhII and EglII. Sequences from references listed for Fig. 4.1.

catalytic site is located. The core segment exhibits very little homology between the two enzymes.

Limited proteolysis of CbhI by papain cleaves the enzyme at the linker region into two segments—the CBD and the core. The core protein shows a decrease in activity and 50% loss of affinity on the crystalline cellulose, Avicel, but maintains full activity on amorphous cellulose (van Tilbeurgh et al. 1986; Johansson et al. 1989). For CbhII, loss of CBD results in the reduction of affinity and activity toward both types of celluloses (Tomme et al. 1988). The CbhI protein contains 21% helix, 42% β conformation, and 37% aperiodic (Bhikhabhai et al. 1985). Small-angle x-ray scattering studies show that molecules of both CbhI and CbhII have a tadpole-like structure, with an isotropic, ellipsoid head corresponding to the CBD, and a tail part coinciding with the core (Schmuck et al. 1986; Abuja et al. 1988A, B).

Cellulomonas fimi

For bacterial cellulase systems, the sequence of *cex* gene has been reported encoding an exoglucanase, Cex, in *Cellulomonas fimi* (O'Neill 1986). The gene encodes for 484 amino acids, including 41 residues in the leader peptide (Fig. 4.1). The enzyme consists of 443 amino acids and shows a bifunctional structure. The CBD (~100 amino acids) at the C-terminus and the core segment (~320 amino acids) at the N-terminal is joined by a linker (~20 amino acids) consisting of entirely Pro-Thr residues (Gilkes et al. 1988). Cex contains three disulfides; two in the catalytic domain (Cys208-Cys240 and Cys302-Cys308), and one in CBD (Cys382-Cys481). There is no disulfide bond between the two domains, nor are there free sulfhydryls. The *cex* gene has been cloned into *Escherichia coli*, and the nonglycosylated recombinant enzyme exhibits kinetic properties and pH and heat stabilities similar to the native enzyme, but is more susceptible to proteolysis with greatly reduced affinity for cellulose (Langsford et al. 1987). Although the enzyme is classified as an exo-β-1,4-glucanase, it shows significant activity toward soluble xylan in hydrolyzing p-nitrophenyl-β-D-xylobioside with a catalytic efficiency ~50 times that of p-nitrophenyl-β-D-cellobioside (K_m = 0.13 and 0.80 mM, k_{cat} = 1870 and 237 min^{-1}, V$_{max}$ = 30.4 and 3.8 μmol min^{-1}, k_{cat}/K_m = 14385 and 296 min^{-1} mM^{-1}, respectively) (Gilkes et al. 1991B).

Clostridium thermocellum

For the gram-positive, thermophilic, anaerobic *Clostridium thermocellum*, only one cellobiohydrolase has been isolated in a truncated form from the cellulosome of this bacterium, in spite of the fact that six endoglucanase genes have been isolated and sequenced (Morag et al. 1991). The active cellobiohydrolase fragment (MW = 68 kD) purified from a subunit (MW = 75 kD) has the highest activity on amorphous cellulose, followed by microcrystalline cellulose, and CMC. The subunit described may be identical to the S_s subunit purified earlier as a major protein from the bacterial cellulase system having a MW = 82 kD (Wu et al. 1988). The enzyme activity is unstable at temperatures above 50°C, but can be enhanced by the addition of calcium and thiol-reducing agents. It is interesting to note that the cellulase activity of the cellulosome of *Clostridium thermocellum* is also activated by the presence of calcium and dithiothreitol (Johnson et al. 1982; Johnson and Demain 1984).

ENDOGLUCANASES

Trichoderma reesei

A number of endoglucanases are found in *Trichoderma reesei* culture filtrates. The primary structures of EgII and EglII (formerly known as EglIII) are deduced from their respective nucleotide sequences (Saloheimo et al. 1988). EgII consists of 459 amino acid residues (including a leader sequence of 22 residues) with a calculated MW = 48,212, and like CbhI, the CBD is at the C-terminal region. The CBD of EgII is highly homologous (70%) with those of EglII, CbhI, and CbhII (Fig. 4.2). The core segment of EgII shows 40% similarity with that of CbhI, but not with EglII.

The amino acid sequence of the EglII protein, as deduced from the nucleotide sequence, consists of 418 amino acids (21 residues in the leader sequence) with MW = 42,200, and 15% carbohydrate content. The CBD of EglII is located at the N-terminus, similar to CbhII (Fig. 4.1). The linker region is rich in Ser/Thr and Pro residues. The function of the Pro residues in linkers probably serves to closely pack the heavily O-glycosylated chains. The core obtained after proteolytic removal of CBD remains fully active toward CMC, but shows decreased activity and adsorption with crystalline cellulose (Stahlberg et al. 1988).

EglII exhibits greater thermal stability than EgII, CbhI, and CbhII. The melting temperature T_m for EglII is 75°C, compared with T_m = 64°C for the other three enzymes (Baker et al. 1992). Differential scanning calorimetry also indicates the unfolding, which involves two transition stages.

Both EgII and EglII hydrolyze cello-oligosaccharides with the same efficiency, except that the latter does not hydrolyze cellobioside (Saloheimo et al.

1988). Apart from the CBD, EglII shows very little homology with the other three *T. reesei* enzymes.

Cellulomonas fimi

Two endoglucanases (CenA and CenB) have been isolated from the bacterium *Cellulomonas fimi*, and their primary structures deduced from the nucleotide sequences (Wong et al. 1986; Meinke et al. 1991). The nucleotide sequence encodes 449 amino acids (31 being in the leader peptide) with a calculated MW of 51,800. The CBD (~100 amino acids) at the N-terminus links with the C-terminal core segment (~300 amino acids) via a short Pro/Thr-rich linker (Gilkes et al. 1988). Two disulfide bonds are located in the CenA core—Cys248-Cys291, Cys390-Cys426—and one disulfide is identified in the CBD—Cys35-Cys134 (Gilkes et al. 1991B).

Similar to the Cex enzyme, the carbohydrate moieties of the CenA enzyme provide protection against proteolytic cleavage between the two functional domains (Langsford et al. 1987). The CBD of CenA shows 50% sequence homology with the C-terminal CBD of Cex. However, removal of up to 64 N-terminal amino acids does not affect its affinity to cellulose. Apparently, the remaining sequence in CBD is highly conserved in similar domains from other bacterial endoglucanases (Gilkes et al. 1989). The catalytic domain is tightly folded and resistant to proteolytic attack, whereas CBD shows a relatively loose conformation.

Similar to those described for CbhI and CbhII, the tertiary structure of CenA has a tadpole-like shape (Pilz et al. 1990) with an ellipsoid head (core protein containing catalytic domain) and an extended tail (linker plus CBD). A mutant of CenA obtained by in vitro deletion of the Pro/Thr-rich linker (23 amino acids) exhibits a tertiary structure with the relative orientation of the core and CBD altered. The conformational change causes a 50% decrease in enzyme activity, but the deletion does not affect adsorption with microcrystalline cellulose (Shen et al. 1991).

CenB consists of 1,045 amino acids encoded by the gene sequence (− 33 residues in the leader peptide) with a calculated MW of 105,905, making it the longest cellulase sequenced thus far (Meinke et al. 1991). It has an optimum pH of 8.5–9.0 for hydrolysis of CMC, in contrast to the optimal pH of 7.0–7.5 for CenA. The enzyme contains a C-terminal CBD (~100 amino acids) that is 50% identical to those of CenA and Cex and a core segment (~600 amino acids) that bears no similarity to CenA or Cex. The two domains are connected by three repeats of 98 amino acids, each with >60% homology (Fig. 4.1). Each repeat consists of a short Pro/Thr-rich region (~17 amino acids) analogous to the linkers of Cex and CenA. Between the third repeat and the CBD, there is a short sequence (13 amino acids) rich in Pro/Thr/Ser.

Clostridium thermocellum

Six endoglucanases from *Clostridium thermocellum* have had their genes (*celA, celB, celC, celD, celE, celH*) sequenced. CelA is a polypeptide of 477 amino acids corresponding to a MW of 52,503 (32 of the encoded amino acids are the signal sequence) (Beguin et al. 1985). CelB consists of 563 amino acid residues with a MW of 63,857. The N-terminal sequence does not resemble other signal peptides, suggesting that the protein is not secreted (Grepinet and Beguin 1986). CelC is a protein of 343 amino acids (14 amino acid signal peptide) with a calculated MW of 40,439 (Schwarz et al. 1988). The nucleotide sequence of CelD encodes 649 amino acids (41 amino acid signal sequence) with a calculated MW of 72,334. The purified native protein is found to have a MW of 65 kD (Joliff et al. 1986A, B). CelE contains 814 amino acids (34 in the signal peptide) and a calculated MW of 90,211 (Hall et al. 1988). The *CelH* gene encodes 900 amino acids (44 amino acids forming the signal peptide) with a calculated MW of 102,301. With the exception of CelC, all these endoglucanases contain a conserved region of 60–65 amino acids consisting of two repeats (Fig. 4.3). CelA, B, D, and H have their CBD at the C-terminus, while CelE has this region shifted more to the central position (Fig. 4.1). A short linker of Pro-Thr-Ser-rich sequence precedes the CBD in CelA, B, E, and H. CelH contains two Pro/Thr/Ser-rich regions; the sequence (306 amino acids) in between these two regions shows a strong similarity (30.2%) with the N-terminal region of CelE (where the core segment is located). CelC bears no similarity in sequence with the other four endoglucanases, although a short region (residues 76–88) contains the motif Glu-X_5-Asn-$X_{5/7}$-Asp found in the catalytic site of lysozyme (Schwarz et al. 1988). CelE also possesses xylanase activity although at a low rate of about 1/10 of that of the endoglucanase activity. Recombinant CelD produced by *Escherichia coli* has a pI of 5.4, and optimum pH of 6.0 for the hydrolysis of CMC. This enzyme

```
Cel A    417    DVNGDGNVNSTDLTMLKRYLLKS
Cel B    502    ---G--R---S-VAL----L-GL
Cel D    585    ---D--K------LTL----V-KA
Cel E    415    ---G--KI----CTM----I-RG
Cel H    833    -L-F-NA-----LLM----I-KS

Cel A    449    ---R--AI--S-MTI----LIKS
Cel B    534    ---VS-T-----LAIM---V-RS
Cel C    621    ---R--R---S-VTI-S--LIRV
Cel E    451    ---A-LKI----LVLM-K-L-RS
Cel H    872    -L-R-NK-D---LTI----L-KA
```

Figure 4.3. Sequence alignment of conserved terminal repeated segments of *Clostridium thermocellum* endoglucanases. Sequences from references listed for Fig. 4.1.

and CelA exhibit more random attack on CMC than CelB and CelC. The *Escherichia coli* expressed enzyme, however, has no activity toward Avicel, xylan, cellobiose, or *p*-nitrophenyl-β-D-glucoside (Joliff et al. 1986B).

β-GLUCOSIDASE

Trichoderma reesei

Multiple forms of β-glucosidase, both intracellular and extracellular, are reportedly produced by *Trichoderma reesei*, and have pI values ranging from acidic to alkaline (pH 4.4–8.4), and MW from 50 to 98 kD (Inglin et al. 1980; Enari et al. 1981; Umile and Kubicek 1986). Recent data suggest that the multiple β-glucosidases from *Trichoderma reesei* are the result of anomalous purification procedures and partial proteolysis of a single enzyme with alkaline pI (Hofer et al. 1989). The secreted and the cell wall-bound β-glucosidases are likely the same enzyme also (Jackson and Talburt 1988; Messner and Kubicek 1990). Less than 50% of the β-glucosidases is secreted into the culture medium; the rest remain tightly associated with a cell wall polysaccharide consisting of mannose, galactose, glucose, galacturonic and gluconic acid. This heteroglycan binds β-glucosidase in vitro, and the binding results in a 2-fold increase in the enzyme activity against *p*-nitrophenyl β-glycoside (Messner et al. 1990).

The enzyme is a polypeptide with a MW of 70 kD, and a pI of 8.4. It has an optimum activity at pH 4.5 and 70°C for the hydrolysis of cellobiose. The K_m for cellobiose and *p*-nitrophenyl β-glucoside is 0.5 mM and 0.3 mM, respectively (pH 4.5, 50°C). The enzyme also hydrolyzes cello-oligosaccharides (3–10 units) to glucose, but is inactive toward CMC or microcrystalline cellulose (Schmid and Wandrey 1987). Another study on apparently the same enzyme showed that the K_m for cellobiose is ~12 times that for *p*-nitrophenyl β-glucoside. Inhibition studies also reveal that the active site is comprised of at least three subsites with subsite 1 (numbering from the nonreducing end) having the highest binding energy (~ -7 kcal/mol). Binding energy for subsite 2 is slightly positive, and subsite 3 is -1.8 kcal/mol. The position of the catalytic cleavage is between subsites 1 and 2. The end product glucose binds to subsite 1 with $K_i = 700 \pm 60 \mu M$ (Chirico and Brown 1987).

The *Trichoderma reesei* β-glucosidase gene (*bglI*) encodes a protein of 713 amino acids with a calculated MW of 75,341 (Barnett et al. 1991). The primary structure contains four potential glycosylation sites with an Asn-X-Ser/Thr-X sequence at positions 208, 310, 417, and 566. The significance of glycosylation on the structure and function of the enzyme is not clear, because the purified protein has been reported to contain no carbohydrate chains, or very few (less than 1%).

β-Glucosidase represents about 1% of the total proteins secreted, which is insufficient for the *Trichoderma reesei* cellulase system to achieve practical saccharification of cellulose. The cellobiose, accumulated by the action of endoglucanases and cellobiohydrolases, exhibits end product inhibition. *Trichoderma reesei* mutants have been isolated with increased production (up to 3 times more) of β-glucosidase (Kawamori et al. 1986). Alternatively, complete conversion of cellulose to glucose can often be achieved by supplementing the cellulase system with β-glucosidase from a different source, for example, *Aspergillus niger*, which has a high production of the enzyme (about 10 times that of *Trichoderma reesei*) (Sternberg et al. 1977). The same is true for application in the removal of cellobiose in the *Clostridium thermocellum* cellulase complex (Ait et al. 1982; Kadam and Demain 1989; Lamed et al. 1991).

Clostridium thermocellum

Based on the characterization of two distinct genes *bgl*A and *bgl*B in *Clostridium thermocellum*, there is definitely more than one β-glucosidase produced by the bacterium (Schwarz et al. 1985; Grabnitz and Staudenbauer 1988). BglA, as deduced from the nucleotide sequence of the recombinant gene in *Escherichia coli*, consists of 448 amino acids and has a MW of 51,482. The enzyme has an optimum activity at pH 6.0–6.5 at 60°C. The enzyme hydrolyzes both cellobiose and *p*-nitrophenyl β-glucosidase, and also cello-oligosccharides, but not CMC. Competitive inhibition of β-glucosidase activity is observed by glucono-δ-lactone (K_i = 0.01 mM) and by high concentrations of glucose (K_i = 135 mM). Addition of thiol reagents (*p*-chloromercuricbenzoate and iodoacetamide) causes loss of activity. The primary structure of BglB, as encoded by the nucleotide sequence, consists of 754 amino acids with a calculated MW = 84,100 (Grabnitz et al. 1989). A recombinant enzyme expressed by *Escherichia coli* hydrolyzes cellobiose more efficiently than the long cello-oligosaccharides (3–5 units), but only at 43% of the rate of hydrolysis of aryl-β-glucosides (Kadam and Demain 1989). In this respect, the enzyme shows substrate specificity similar to that of BglA. Furthermore, neither enzyme hydrolyzes CMC. The N-termini of BglA and BglB contain no basic or hydrophobic regions characteristic of a leader peptide, suggesting that these enzymes are localized in the cytoplasmic region (Grabnitz et al. 1991).

The amino acid sequences of the *Trichoderma reesei* enzyme and the *Clostridium thermocellum* BglB consist of short segments homologous to the proposed active site sequence of β-glucosidase A3 of *Aspergillus wentii* (Bause and Legler 1980). Similar alignments can be identified in enzyme sequences of *Saccharomycopsis fibuligera* (Machida et al. 1988), *Candida pelliculosa* (Kohchi and Tohe 1985), *Ruminococcus albus* (Ohmiya et al. 1989), and *Kluyveromyces fragilis* (Raynal et al. 1987). The sequence around the catalytic Asp residue is highly conserved among all these homologous β-glucosidase sequences (Fig. 4.4). The Glu located nearby is also conserved. However, the deduced amino acid sequence

Trichoderma reesei Bgl 1	256	DELGFPGYVMT**D**WNAEHTTVESANSGLDMSMP
Clostridium thermocellum Bgl B	220	N-WMHD-F-VS**D**-G-VNDRVSGLDA---LE--
Aspergillus wentii Bgl A3		AZ---Z-F-MS**D**-A-HHAGVSG-LA--B-GSM
Saccharomycopsis fibuligera Bgl 1	284	E----Q-F-VS**D**-G-QLSGVYS-IS----S--
Saccharomycopsis fibuligera Bgl 2	288	E----Q-F-VS**D**-A-QMSGAYS-IS----S--
Candida pelliculosa	353	E----Q-F-MT**D**-G-LYSGIDA-NA----D--
Ruminococcus albus	685	KQW--D-FTMT**D**-W-NINDRGC-PDKNNFAAM

Figure 4.4. Sequence similarity of β-glucosidases with the active site sequence of the *Aspergillus wentii* enzyme A₃—*Trichoderma reesei* Bgl1 (Barnett et al. 1991), *Clostridium thermocellum* BglB (Grabnitz et al. 1989), *Aspergillus wentii* BglA3 (Bause and Legler 1980), *Saccharomycopsis fibuligera* Bgl1 and Bgl2 (Machida et al. 1988), *Candida pelliculosa* (Kohchi and Tohe 1985), *Ruminococcus albus* (Ohmiya et al. 1989).

of *Clostridium thermocellum* BglA shows no significant similarity to the BglB or fungal β-glucosidases noted above. Instead, strong similarities exist when it is compared to β-glucosidases of *Caldocellum saccharolyticum*, *Bacillus polymyxa*, and *Agrobacterium* (Grabnitz et al. 1991).

THREE-DIMENSIONAL STRUCTURE OF EXOGLUCANASES

Trichoderma reesei CbhI

The core protein of CbhI consists of a stack of two antiparallel β sheets enclosing a ~40-A-long tunnel (Fig. 4.5A) (Divne et al. 1994). The active site residues in the tunnel are contributed from several loops extending from the β sandwich. The cavity is lined with mostly hydrophobic side chains with small patches of polar residues. There are seven subsites with cleavage bond between subsites 2 and 3. Four Trp residues (376, 367, 38, and 40) interact by stacking with the pyranosyl rings of the glycosyl units at subsites 2, 4, 6, and 7, respectively. Located close to the glycosidic oxygen are three acidic residues. It has been suggested that Glu217 acts as the general acid catalyst in protonating the glycosidic oxygen, while Glu212 is the nucleophile attacking the anomeric carbon, based on a double displacement mechanism. CbhI catalyzes hydrolysis with retention in configuration. (Refer to the section later in this chapter, The Stereochemistry of Hydrolysis.)

Trichoderma reesei CbhII

The structure of the C-terminal domain (CBD) of *Trichoderma reesei* CbhII as determined by two-dimensional NMR, has a wedge-like structure (30 × 18 × 10 A) with one face mainly hydrophilic and the other hydrophobic (Kraulis et al. 1989). It is made up of a 3-stranded antiparallel β sheet, extensively hydrogen bonded. Two disulfides, Cys8-Cys25 and Cys19-Cys35, crosslink different parts

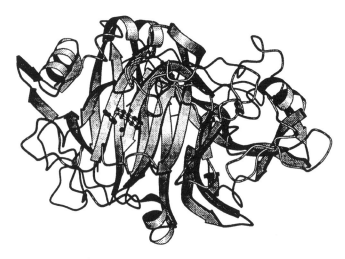

Figure 4.5A. A schematic MOLSCRIPT representation of the catalytic domain of cellobiohydrolase I from *Trichoderma reeser.* Arrows represent β strands, and α helices are
drawn as spirals. The active site is located at one end of the tunnel that spans the length
of a β sandwich motif. The ligand bound in the active site, *o*-iodo-benzyl-1-thio-β-D-
glucoside, is represented as a ball-and-stick model. (Reproduced from Divne C. (1994).
The three-dimensional structure of cellobiohydrolase I from *Trichoderma reesei,* Ph.D.
Thesis, Uppsala University, ReproHSC, Uppsala, Sweden. Courtesy of Dr. Christina
Divne.)

of the peptide chain. These four cysteine residues are conserved in the CBDs of
all four *Trichoderma reesei* enzymes.

The core protein of CbhII (residues 83–447) consists of a β barrel of seven
parallel strands with the first six strands connected by α helices; the connection
between the last two strands is irregular (Fig. 4.5B) (Rouvinen et al. 1990). Two
extensive loops (172–189, 394–429), each stabilized by a disulfide bridge
(Cys176-Cys235, Cys368-Cys415, respectively), form an enclosed tunnel about
20 A long (one-half the length of that in CbhII). The side chains that line the
tunnel are mainly hydrophilic, forming networks of hydrogen bonds and salt
bridges. The active site is located in the tunnel, with four subsites (numbered
from the nonreducing end of the cellulose chain). The OH groups of the sugar
substrate form an extensive hydrogen bond network with hydrophilic side chains
in the binding site. The indoles of three Trp residues (135, 169, and 367) in
subsites 1, 3, and 4 stack with the sugar rings of P_1, P_3, and P_4, respectively.
Cleavage of cellulose occurs between subsites 2 and 3, with inversion of configuration in the product, and involves Asp221 acting as a general-acid catalyst. The
pK of Asp221 is raised by the interaction with another acidic residue, Asp175,

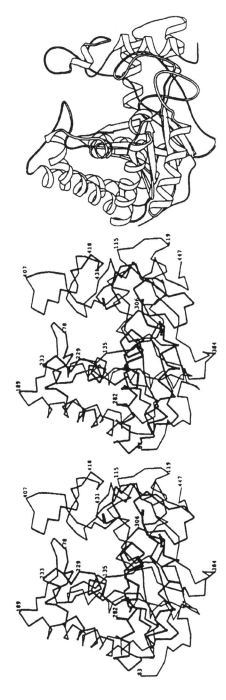

Figure 4.5B. The C_α skeleton and ribbon drawing of the CbhII core protein showing the active site tunnel and its position relative to the β barrel. (From Rouvinen et al. 1990. *Science 249*, 382, with permission. Copyright 1990 American Association for the Advancement of Science.)

which forms a hydrogen bond with Arg174. The general base that promotes nucleophilic attack of H_2O in the single displacement mechanism is likely to be Asp401 (Spezio et al. 1993), which forms salt bridges with Arg353 and Lys295.

Cellulomonas fimi Cex

The catalytic domain of the exo-enzyme from *cellulomonas fimi* consists of a central α/β barrel of eight parallel β strands connected by α helices (White et al. 1994). There are two disulfide bonds, Cys167-Cys199 joing β5 and β6, and Cys261-Cys267 connecting α7 to β8 (Fig. 4.5C). The active site is located at a crevice on the surface found at the C-terminal end of the barrel. The nucleophile, Glu233, is involved in hydrogen bonding with His205 and Asp235. The Glu-His-Asp network is believed to play a role in maintaining a deprotonated form of the Glu233 to facilitate attack at the anomeric center in a displacement reaction.

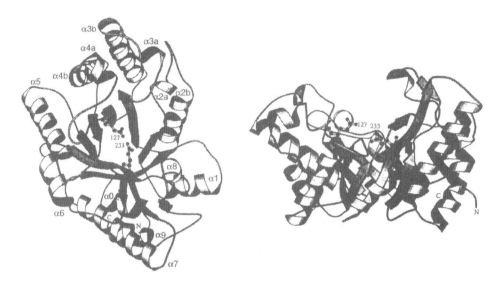

Figure 4.5C. Folding of the catalytic domain of cex from *Cellulomonas fimi*. (From White et al. 1994. Biochemistry *33*, 12549, with permission. Copyright 1994 American Chemical Society.)

THE THREE-DIMENSIONAL STRUCTURE OF ENDOGLUCANASES

Clostridium thermocellum CelD

The endoglucanase CelD is a globular molecule, slightly elongated with a dimension of $50 \times 50 \times 17$ A Fig. 4.6A (Juy et al. 1992). The N-terminal domain (residues 52–135) consists of seven antiparallel strands A–G, arranged into two β sheets that enclose a number of hydrophobic side chains. The topology of this domain resembles that of the immunologlobulin N-terminus, although the function of this domain is not known. The other domain (residues 136–574) contains 12 helices arranged in a "twisted" barrel shape with six of the helices oriented antiparallel to the other six helices. The opening end of the barrel contains extensive loops that form a long groove where the active site is located and substrate binding occurs. The action of CelD results in inversion of configuration in the product. It has been proposed that Glu555 acts as a general acid, and H_2O acts as nucleophile assisted by Asp201, assuming a double displacement mechanism.

Humicola insolens Endoglucanase V

The endo-enzyme consists of a 6-stranded β barrel of both anti-parallel and parallel arrangments (Fig. 4.6B) (Davies et al. 1993). The long groove on the surface of the molecule accommodates seven subsites with two catalytic Asp

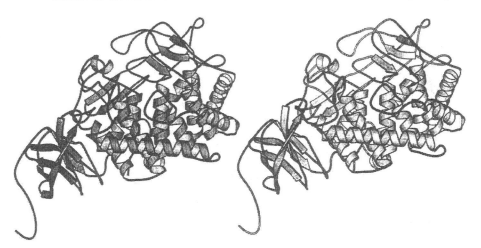

Figure 4-6A. Three-dimensional structure and topology of Cel D showing the α helices and β strands in the two domains. The catalytic site can be seen as an opening in the upper part of the ribbon diagram (From Juy et al. 1992. Nature *357*, 90, with permission. Copyright 1992 Macmillan Magazine Ltd.) Figure courtesy of Dr. Pedro M. Alzari, Institut Pasteur, Paris.)

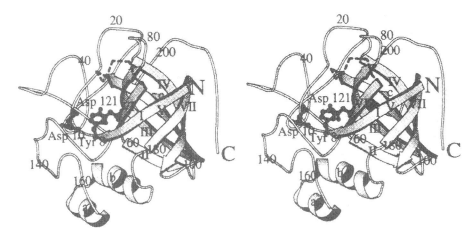

Figure 4.6B.   A stereo view showing the overall topology of the core protein of *Humicola insolens* endocellulase v. (From Davies et al. 1993. *Nature 365,* 362, with permission. Copyright 1993 *Macmillan Magazine* Ltd.)

residues located on either side. The active site topology bears a strong similarity to that of lysozyme, although the enzyme catalyzes reactions with inversion of configuration in the product.

Thermomonospora fusca E2

The thermophilic endoglucanase E2 from the *Actinomycete* consists of a catalytic domain that assumes an α/β barrel of eight strands connected by helices (Spezio et al. 1993). This structural arrangement is similar to that of CbhII. The β barrel contains extending loops forming a cleft that runs the entire length of the molecule. The tunnel can accommodate four subsites. Two Trp residues (41 and 162) are involved in stacking interaction with pyranosyl rings during substrate binding. Located close to the cleavage site is Asp117, which may act in protonating the glycosidc oxygen, while another acidic residue, Asp265 is the general base. The enzyme produces products with inversion in configuration. The pK of Asp117 is influenced by the interaction with Arg221 and Asp156.

All the exo- and endo-enzymes whose structures have been elucidated have the common feature of a large long groove or tunnel where the catalytic residues and the binding site are located, in spite of the fact that there is very little discernable relationship in the topological arrangements. In addition, compared with the exo-enzymes, the endoglucanases have a more open structure, and the active site is more accessible to substrates. This is the result of a deletion or movement of loops that enclose the active site in the tunnel of the exo-enzymes (Rouvinen et al. 1990; Spezio et al. 1993; Divne et al. 1994).

STEREOCHEMISTRY OF HYDROLYTIC ACTION

The stereochemical course of hydrolysis for a number of cellulolytic enzymes is known (Table 4.2). *Trichoderma reesei* CbhI and CbhII hydrolyze β-D-cellobiosyl fluoride with the release of β- and α-cellobiose, respectively (Knowles et al. 1988). CbhI catalyzes the hydrolysis of methyl-β-D-cellotetraoside yielding β-cellobiose (K_m = 4 μM and k_{cat} = 1.7 min^{-1}) while CbhII releases α-cellobiose (K_m = 7.2 μM, k_{cat} = 112 min^{-1}) (Claeyssens et al. 1990A). Similar stereochemistry is also obtained for the CbhI and CbhII purifed from *Talaromyces emersonii* (Brooks et al. 1992). The exoglucanase (Cex) and endoglucanases (CenA and CenB) from *Cellulomonas fimi* catalyze hydrolysis with retention and inversion of configuration, respectively (Withers et al. 1986; Meinke et al. 1991). The *Clostridium thermocellum* CelB, CelC, and CelH are configuration-retaining enzymes, whereas CelD produces inversion products (Gebler et al. 1992). The almond β-glucosidase produces hydrolyzed products with retention of the anomeric configuration (Withers et al. 1986).

There is increasing evidence that in reactions catalyzed by carbohydrases, including amylases, β-galactosidase, α- and β-glucosidases, and exo- and endoglucanases, the anomeric configuration of their reaction products is independent of the substrate configuration, but directed *de novo* by the inherent structural activity of the individual enzyme. The enzyme is not restricted to acting on a substrate with a glycosidic bond of appropriate α or β anomeric configuration. The β-glucosidase from sweet almond attacks D-glucal (which links the α or β anomeric configuration) to release deoxyglucose in the β form. This requires the enzyme to protonate D-glucal at C2 above the double bond, in contrast to the direction of protonation in the hydrolysis of β-D-glucosides. The α-glucosidase from *Candida tropicalis* protonates D-glucal from above the double bond and

Table 4.2. Stereochemical Courses of Hydrolysis for Cellulolytic Enzymes

Source	Enzyme	Configuration of product	
Clostridium	Cel B	Retention	Gebler et al. 1992
thermocellum	Cel C	Retention	Gebler et al. 1992
	Cel D	Inversion	Gebler et al. 1992
	Cel H	Retenton	Gebler et al. 1992
Cellulomonas	Cen A	Inversion	Withers et al. 1986
fimi	Cen B	Inversion	Meinke et al. 1991
	Cex	Retention	Withers et al. 1986
Trichoderma	Cbh I	Retention	Knowles et al. 1988
reesei	Cbh II	Inversion	Knowles et al. 1988
	Egl I	Retention	Claeyssens et al. 1990
	Egl II	Retention	Gebler et al. 1992
Almond	β-glucosidase	Retention	Withers et al. 1986

α-D-glucoside from below the glycosidic oxygen, to give the α-anomer of 2-deoxy-D-glucose (Hehre et al. 1977). Similar reactions occur using 2,6-anhydro-1-deoxy-D-gluco-hept-1-enitol (heptenitol) (Hehre et al. 1980). In a similar pattern, exo- and endoglucanases (from *Irpex lacteus* and *Aspergillus niger*, respectively) both protonate cellobial below the double bond, a direction opposite to that assumed for the protonation of the β-D-glycoside substrate or cellulose. In both enzymes, the products are the β-anomer of 2-deoxycellobiose (Kanda et al. 1986).

Trichoderma reesei CbhI catalyzes hydration of cellobial and lactal to yield β-anomer products, suggesting protonation of the double bond in a direction opposite to that for the hydrolysis of methyl β-cellotrioside although the product again maintains a β configuration (Claeyssens et al. 1990A). The different stereochemistry of protonation between D-glucal (and other similar substrates) and β-glucosides may also be explained by assuming that for the glucal reaction the carboxyl group donating the proton to the double bond is also the same group forming the 2-deoxy-β-glucose intermediate (Legler et al. 1979). The catalytic flexibility of glycosylases may suggest that the carboxylate/carboxylic acid groups can exchange in the role of a nucleophile or a proton donor.

REACTION MECHANISM

Two types of mechanisms have been proposed for enzyme hydrolysis of glycosides—via a stabilized oxocarbonium ion or a covalent glycosyl-enzyme intermediate. (See similar discussion in Chapter 3 on amylases.)

Oxocarbonium Ion Intermediate

In this mechanism, the glycosidic oxygen is protonated by an amino acid (carboxyl group) side chain (Fig. 4.7). This step of general acid catalysis gives a resonance-stabilized carbonium ion intermediate, assisted by steric strain. In lysozyme, an enzyme-induced ring distortion to the *N*-acetylglucosamine residue at site D assumes a twist-boat conformation that favors cleavage of the exocyclic C1–O bond and formation of a carbonium ion (Vernon 1967; Fife 1972; Yaguchi et al. 1983; Kanda et al. 1986). The resulting oxocarbonium ion intermediate is stabilized by interaction with a negatively charged (carboxylate) side chain. This is followed by a directed attack of H_2O at C1 to yield a product with specific anomeric configuration. An alternative pathway has also been proposed for the formation of an acyclic oxocarbonium ion intermediate via cleavage of the endocyclic C1–O5 bond (Post and Karplus 1986).

Displacement Mechanism

An alternative to the above described S_N1 mechanism is a displacement S_N2 process. A double displacement is envisioned for "retaining" enzymes. A con-

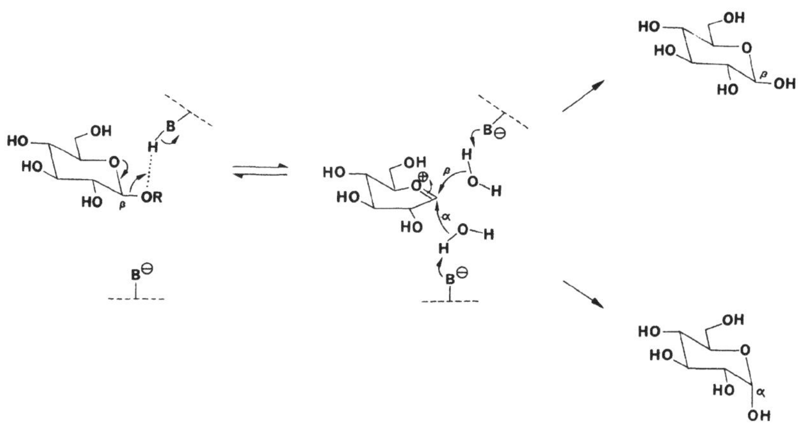

Figure 4.7.   Reaction mechanism of carbohydrases involving a stabilized oxocarbonium ion (Thoma 1968).

certed protonation of the glycosidic oxygen and a displacement of the O-acyl leaving group leads to the formation of a covalent glycosyl-enzyme intermediate (Koshland 1959). This is then followed by nucleophilic attack by a H_2O molecule assisted by a general-base catalyst. It is conceivable that the formation and breakdown of a glycosyl intermediate proceeds via an oxocarbonium-like transition state (Legler et al. 1980; Withers and Street 1988). A double displacement mechanism ensures a trans addition of the H–OH, resulting in a retention of the anomeric configuration of the substrate (Fig. 4.8A). For glycosidases that catalyze inversion of the anomeric configuration, the mechanism can be explained by a single displacement (Fig. 4.8B). With the O-aglycon linkage weakened by general acid catalysis, attack of a nucleophilic H_2O molecule assisted by a general base results in the inversion of configuration.

Cellulolytic enzymes are known to contain catalytic residues with carboxyl groups. For CbhI and EgII from *Trichoderma reesei*, Glu126 and Glu127 are plausible catalytic residues, respectively, as revealed by site-directed mutagenesis (Tomme and Claeyssens 1989). According to crystallographic studies of CbhI, however, Glu217 is the general acid and Glu212 is the nucleophile (Divine et al. 1994). For CbhII, Asp221 acts in protonating the glycosidic oxygen (Rouvinen et al. 1990). Affinity labelling of the EgIII enzyme and sequencing of the labeled peptide suggest Glu329 as an essential residue in the active site (Macarron et al. 1993). Catalytically important residues in the endoglucanase CelD from *Clostridium thermocellum* consist of Asp546 and Glu555 (Tomme et al. 1992). Studies on the crystal structure suggest Asp201 and Glu555 (Juy et al. 1992). Labelling the *Clostridium thermocellum* CelC enzyme with radioactive inactivator allows

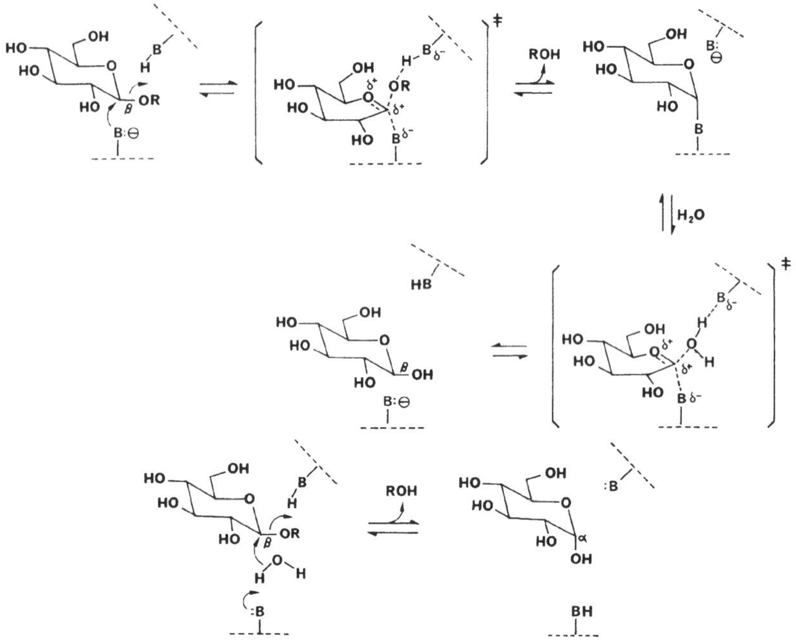

Figure 4.8. (A) A double displacement mechanism with the formation of a glycosyl enzyme intermediate via oxocarbonium-like transition states. (From Withers and Street 1988. *J. Am. Chem. Soc. 110*, 8552, with permission. Copyright 1988 American Chemical Society. (B) A single displacement mechanism (Koshland 1959).

the identification of Glu280 as the active site nucleophile (Wang et al. 1993). Site-directed mutagenesis studies of endoglucanases from *Bacillus polymyxa* and *Bacillus subtilis* also suggest an essential Glu residue in these enzymes (Glu194 and Glu169, respectively) (Baird et al. 1990). The active site nucleophile of the β-glucosidase of *Agrobacterium faecalis* has been identified to be Glu358 by directed mutagenesis (Trimbur et al. 1992).

The exoglucanase Cex from *Cellulomonas* catalyzes hydrolysis with retention of the anomeric configuration of the substrate, probably by the double displacement mechanism. Labeling the enzyme with 2',4'-dinitrophenyl 2-deoxy-2-fluoro-β-D-glucopyranoside allows the detection of accumulation of the covalent enzyme intermediate (2-deoxy-2-fluoro-α-D-glucopyranosyl enzyme) (Tull et al. 1991). In the glycosyl-enzyme structure, the Glu233 is esterified, demonstrating that Glu233 acts as the nucleophile in attacking the anomeric carbon of the substrate. In addition to Glu233 as the nucleophile, another catalytic residue functioning as an acid/base catalyst is required in a double displacement mechanism. This catalyst has been identified as Glu127 by kinetic studies of enzyme mutants with substitution at 127 (MacLeod et al. 1994; Tull and Withers 1994). Glu127 mutants show similar k_{cat} values for substrates with different leaving groups consistent with the fact that the general base catalyzed H_2O attack is the rate-limiting step. The k_{cat}/K_m values show significant reduction in substrate with poor leaving groups, suggesting that the protonation of the glycosidic oxygen involves Glu127.

In contrast to most studies revealing two Glu as catalytic residues, crystallographic studies identify Asp221 as the acid catalyst in CbhII (Kraulis et al. 1989). Endo-β-1,4-glucanase from *Schizophyllum commune* contains a short amino acid sequence (33–51) highly homologous with the active site sequence of lysozyme (Yaguchi et al. 1983). Kinetics and chemical modification studies suggest that Glu33 and Asp50 are the important residues analogous to those of lysozymes (Glu35 and Asp52 in HEWL) (Clarke and Yaguchi 1985; Clarke 1988).

A number of investigations demonstrate that β-glucosidase-catalyzed hydrolysis proceeds by a displacement reaction with retention of configuration. Measurement of the secondary α-deuterium kinetic isotope effect for β-glucosidase-catalyzed hydrolysis of phenyl-β-D-glucopyranoside gives $k_H/k_D = 1.010$, which is consistent with an S_N2 mechanism (Dahlquist et al. 1969). A rapid "burst" of *p*-nitrophenol from the hydrolysis of *p*-nitrophenyl-β-D-glucoside at subzero temperature experiments is also indicative of a glycosyl enzyme intermediate (Weber and Fink 1980). A glycosyl enzyme has been isolated from reactions of β-glucosidase A3 from *Aspergillus wentii* with *p*-nitrophenyl-β-2-deoxy-[^{3}H]-glycoside and the deoxyglycosyl residue identified binding to a carboxylate side chain (Roeser and Legler 1981). With 2-deoxy-2-fluoro-β-D-glucosyl fluoride as inhibitor, the existence of a covalent intermediate (2-deoxy-2-fluoro-α-D-glucopyranosyl-enzyme) involving the enzyme Glu358 residue with an α-anomeric linkage has been demonstrated by ^{19}F NMR (Withers and Street 1988; Withers et al. 1990).

MECHANISM OF ADSORPTION

The cellulolytic enzymes must diffuse to the surface of their insoluble substrates before catalytic action can take place. The adsorption process has been confirmed for *Trichoderma reesei* CbhI, CbhII, and EgII by electron microscopy (White and Brown 1981; Chanzy et al. 1983, 1984; Chanzy and Henrissat 1985) and for CenA from *Cellulomonas fimi* by scanning electron microscopy (Din et al. 1991).

The adsorption parameters of cellulase systems on cellulose may vary with the purity of the cellulose substrate, the extent of pretreatment, and the degree of crystallinity, and are also correlated to the changes in the substrate during hydrolysis, possible nonproductive binding of the enzyme, end product inhibition, and temperature and pH effects (Lee et al. 1983; Moloney and Coughlan 1983). Most studies involve action of a mixture of enzymes in culture filtrates, and therefore are difficult to interpret.

Intact enzymes of various endo- and exoglucanases have been shown to act by disruption or defibrillation of the crystalline structure of cellulose substrates (White and Brown 1981; White 1982; Chanzy and Henrissat 1985; Chanzy et al. 1983, 1984; Sprey and Bochem 1993). Studies with isolated fragments of CBD of CenA (from *Cellulomonas fimi*) suggest that, in addition to bringing the core segment of the enzyme close to sites accessible for hydrolysis, the CBD may also play a direct role in the disruption of the structural arrangement of the cellulose (Din et al. 1991). The isolated CBD binds to and penetrates cellulose fibers particularly at weak spots where discontinuities of microfibers or fibrils occur. Mutagenesis of Trp492 and Pro477 at the hydrophilic and hydrophobic surface, respectively, of CbhI of *Trichoderma reesei*, causes a decrease in activity and binding, indicating that both surfaces of the wedge-like shape CBD are required for the interaction with crystalline cellulose (Reinikainen et al. 1992). Although the disruptive action is revealed by scanning electron microscopy with the use of fluorescein-labelled CBD, the mechanism underlying this process and its implication in the rate of hydrolysis are not clear. Preincubation of Avicel with CBD of CbhI of *Trichoderma reesei* shows no increase in activity of either the core enzyme or the intact enzyme on the disrupted structure (Stahlberg et al. 1991). Both *Trichoderma reesei* CbhI and CbhII exhibit higher adsorption toward Avicel than amorphous cellulose, while the isolated core proteins show the same amount of adsorption, although specific activity toward amorphous is higher than that toward Avicel in all cases (Tomme et al. 1988). The core protein of *Trichoderma reesei* CbhII, obtained by papain digestion, also possesses the ability to adsorb on cellulose, and in addition, also shows a dispersion action on macrofibrils (Woodward et al. 1992).

The efficiency of adsorption can be described by the partition coefficient K_p of the enzyme between the cellulose substrate and the H_2O phase (amount of enzyme bound/concentration of free enzyme in solution). Endoglucanases in several microbial systems show that the rate of hydrolysis of crystalline cellulose is proportional to the endo-activity multiplied by K_p, while the hydrolysis of amorphous cellulose is determined only by the endo-activity (Klyosov et al. 1986). The effect of K_p on the hydrolysis of the crystalline structure may well be related to the action of CBD in disruption of the cellulose structure, rendering it more accessible to hydrolytic action. Furthermore, because the crystalline structure is composed of tightly packed chains, productive binding occurs only with certain accessible glycosidic bonds on the surface (Henrissat et al. 1988). Amorphous cellulose or soluble substrates apparently have all the surface readily accessible for enzyme attack, and hence the action of structural disruption by CBD becomes less important. However, the partition coefficient K_p corresponds to the initial slope of the Langmuir isotherm curve, and as such cannot be interpreted as an affinity constant of adsorption.

The adsorption behavior of cellulolytic enzymes generally obeys the Langmuir-type isotherm equation,

$$E_a = \frac{E_{max} K_{ad} E_f}{1 + K_{ad} E_f}$$

where E_a = amount of enzyme adsorbed (mg), E_f = free enzyme concentration (mg/ml), E_{max} = maximum amount of adsorbed enzyme (mg adsorbed enzyme/mg substrate), K_{ad} = adsorption equilibrium constant (ml/mg). Most studies obtain the K_{ad} and E_{max} values from the slope of a plot of E_f/E_a vs E_f according to the rearranged adsorption isotherm equation.

$$\frac{E_f}{E_a} = \frac{1}{K_{ad} E_{max}} + \frac{E_f}{E_{max}}$$

The initial rates of adsorption for six purified endoglucanases and two exoglucanases from *Trichoderma viride* indicate that the enzymes vary in their interactions with crystalline cellulose (Avicel) substrates: EndoIII > I > V > II > VI > IV and ExoIII > II (Beldman et al. 1987). The kinetic study also suggests that only a portion of the enzyme adsorbed on the substrate surface is hydrolytically active. The enzyme may be adsorbed on the substrate, but the catalytic domain is not positioned at the right site or orientation for action. In another study, the enzymes from *Trichoderma viride* can be divided into two groups: (EndoI, II and ExoII) and (EndoIII and IV) with the former having K_{ad} values 4–5 times that of the latter. The adsorption process is exothermic with negative ΔH values, and hence involves enthalpy-controlled reaction (Kim et al. 1994). However, the adsorption mode of the *Irpex lacteus* enzymes has been suggested to be endothermic (Hoshino et al. 1992).

The values of K_{ad} have been determined for the *Cellulomonas fimi* enzymes. With bacterial microcrytalline cellulose as substrate, the $K_{ad} = 0.40, 0.45$, and 0.33 for CenA, CBD of CenA (plus linker), and Cex, respectively (Gilkes et al. 1992). The results are consistent with the suggestion that the binding of the *Trichoderma reesei* enzyme is mediated mostly by the CBD and partly by the catalytic domain (Tomme et al. 1988). It has also been shown that core protein of *Trichoderma reesei* CbhI also exhibits very limited binding capacity. Total binding capacity equals 69, 9.3, and 53 mg/g Avicel for intact enzyme, core, and CBD, respectively. This corresponds to a maximal surface coverage of 19 m²/g Avicel for both intact enzyme and CBD, and only 2.5 m²/g Avicel for the core

(Stahlberg et al. 1991). Therefore, the binding process is controlled predominantly by the CBD.

The use of the Langmuir equation for the estimation of affinity of the enzymes is complicated by the fact that the adsorption is not always completely reversible in some cases. The binding interactions are heterogeneous in nature as shown by nonlinear (concave up) curves in Scatchard plots (E_a/E_f vs E_a) (Woodward et al. 1988; Gilkes et al. 1992; Reinikainen et al. 1992). Factors need to be considered include overlapping binding sites, differential affinity, and negative cooperativity. It is conceivable that the enzyme would bind predominantly to the high affinity sites in the initial phase, in particular when the enzyme is present in low concentration. The binding of lower affinity sites occur at a later stage or when the concentration of the enzyme is high.

In the adsorption/hydrolysis process, endoglucanases may, once adsorbed, continuously hydrolyzes the substrate before final desorption from the surface. Another possibility is that each adsorption is followed by hydrolysis, desorption, and readsorption. Some studies seem to support that endoglucanases act with multiple attack on a chain. The enzyme cleaves a given substrate many times before it dissociates (Klyosov 1990; Stahlberg et al. 1991).

MECHANISM OF SYNERGISM

Two types of synergism exist between enzyme components in cellulolytic systems—exo-endo and exo-exo. The synergistic action of the enzymes in the system determines to a large degree its ability to degrade crystalline celluloses.

Exo-Endo Synergism

The exo-endo synergism was demonstrated in the 1950s when the C_1-C_x activities were described (Reese et al. 1950). Since then, the synergistic action between endoglucanase and cellobiohydrolase has been well established for fungal cellulolytic systems (Table 4.3) (Coughlan et al. 1987). A mixture of *Trichoderma reesei* CbhI and EgII or EgIII produces synergistic actions on most crystalline celluloses (filter paper, Avicel, bacterial microcrystalline cellulose), but not on the highly crystalline *Valonia* cellulose. Mixtures of CbhII and EgII or EgIII also catalyze the synergistic degradation of the insoluble substrates, Avicel and bacterial MCC. Optimum ratios are 1:1 for CbhI: EgII or EgIII, and 19:1 for CbhII: EgII or EgIII (Henrissat et al. 1985). However, it has also been shown that the degree of synergism is not dependent on relative proportions but on the total concentration of the enzymes. With CbhI and EgIII at a ratio of 1:1, the synergistic hydrolysis of Avicel (10 mg/ml) reaches its maximum at an enzyme concentration of 40 mg/ml. Higher or lower concentrations of the enzymes, in spite of maintaining a 1:1 ratio, lead to a sharp decrease in hydrolysis (Woodward

Table 4.3. Synergistic Hydrolysis of Cellulose by Different Combinations of Endoglucanases and Cellobiohydrolases

Source of Exoglucanase	Source of Endoglucanase	Source of β-glucosidase	Amount of Reducing Sugar (mmol)	Sum by Individual Components	Ratio
T. emersonii			172		
P. funiculosum			823		
F. solani			869		
T. koningii			244		
	F. solani		615		
	T. koningii		1302		
	T. emersonii		145		
		F. solani	27		
P. funiculosum	T. emersonii	F. solani	2596	996	2.6
F. solani	F. solani	F. solani	5518	1511	3.6
F. solani	T. emersonii	F. solani	3084	1040	3.0
T. koningii	T. koningii	F. solani	4260	1547	2.7
T. koningii	T. emersonii	F. solani	1981	416	4.8
T. emersonii	F. solani	F. solani	1746	814	2.1
T. emersonii	T. koningii	F. solani	3157	1501	2.1
T. emersonii	T. emersonii	F. solani	896	344	2.6

From Coughlan et al. 1987. *Biochem. Soc. Trans. 15*, 263, with permisson. Copyright 1987 The Biochemical Society and Portland Press.

et al. 1988A, B). Apparently, optimum synergism occurs at nonsaturation concentrations of the combination of enzymes. This effect may be partially related to competitive adsorption between endoglucanase and cellobiohydrolase (Ryu et al. 1984). Adsorption reversibility and competition among *Trichoderma reesei* has been demonstrated by labelling the enzymes with ^{3}H and ^{14}C in sequential adsorption studies (Kyriacou et al. 1988). CbhI is adsorbed in preferential to the endoglucanases. Both CbhI and EglI cause displacement of EglII and EglIII from cellulose substrate.

Cross-synergistic actions between endoglucanase from one microbial source and cellobiohydrolase from another are often observed. Examples are the exo- and endo-acting enzymes from the fungi *Trichoderma koningii, Fusarium solani,* and *Penicillium funiculosum* (Table 4.3). However, the cellobiohydrolase from these fungi show little synergism with the endoglucanases of the fungi *Myrothecium verrucaria, Stachybotrys atra,* and *Memnoniella echinata,* nor with those of the bacteria *Ruminococcus albus, Ruminococcus flavefaciens,* and *Bacteroides succinogenes* (Wood et al. 1989). Purified *Penicillium pinophilum* CbhI can act synergistically with endoglucanases from *Trichoderma koningii* and *Fusarium solani.* CbhII shows little or no effect in this respect (Wood and McCrae 1986). *Talaromyces emersonii* enzymes are known to act synergistically (McHale and Coughlan 1980). The cellobiohydrolase combined with the endo-enzymes from *Fusarium solani or Trichoderma koningii* enhances the hydrolysis of Avicel to two times the additive action (non-synergistic sum) under the same reaction conditions. The endoglucanase combined with the exo-enzyme from *Penicillium funiculosum, Fusarium solani,* or *Trichoderma koningii* increases the hydrolytic action by a factor of 2.6, 30, and 4.8, respectively (Coughlan et al. 1987). *Trichoderma reesei* CbhI, when combined with either of the two endoglucanases from *Aspergillus niger,* results in no enhancement in hydrolysis (Lee et al. 1988). The two endoglucanases from *Aspergillus niger* acting alone show very little adsorption capacity or hydrolytic action on crystalline cellulose.

The bacterium *Clostridium thermocellum* cellulase complex requires calcium and reducing agent to effectively degrade crystalline cellulose as already described. It has been found that with the addition of *Trichoderma koningii* cellobiohydrolase, synergistic interaction occurs in the absence of calcium and DTT (Gow and Wood 1988).

It has been generalized from all these results that synergism occurs between fungal cellobiohydrolase and endoglucanase from another microbial source occurs only if the the endoglucanase used is part of a "true" cellulase system (Gow and Wood 1988); i.e., the endoglucanase must come from sources where the cellulase system contains both endoglucanase and cellobiohydrolase. The cellulase systems produced by *Myrothecium verrucaria, Stachybotrys atra,*

Memnoniella echinata, Ruminococcus albus, Ruminococcus flavefaciens, and *Bacteroides succinogenes* are not true cellulases.

Exo-Exo Synergism

Exo-exo synergism has been observed for *Trichoderma reesei* CbhI and CbhII (Fagerstam and Pettersson 1980; Henrissat et al. 1985). *Trichoderma reesei* CbhII alone does not hydrolyze crystalline cellulose to any significant extent, but incorporation of CbhI results in synergistic increase in activity against native cotton by a factor of 1.7 (Niku-Paavola et al. 1986). *Penicillium pinophilum* CbhI and CbhII in combination solubilize crystalline celluloses synergistically. A 1:1 ratio results in maximum synergism in the degradation of this substrate (Wood and McCrae 1986).

Cross-synergism is observed in combining *Penicillium pinophilum* CbhII with the cellobiohydrolase of *Fusarium solani* or *Trichoderma koningii*. However, the CbhI from *Penicillium pinophilum* does not exhibit this same effect. Synergistic action between two exo-enzymes in the degradation of crystalline celluloses is difficult to explain. It is speculated that the two cellobiohydrolases from *Penicillium pinophilum* catalyze hydrolytic actions stereospecifically in a complementary manner, although evidence for this hypothesis is lacking (Wood and McCrae 1986). To complicate the matter further, highly purified endoglucanases (I, II, III, IV, and V) isolated from *Penicillum pinophilum* do not act synergistically with either CbhI or CbhII alone. Any one of the five endoglucanases requires the addition of both CbhI and CbhII for synergism in degrading crystalline cellulose (Wood et al. 1989).

Klyosov (1990) proposes a hypothesis to explain the observations in both exo-endo and exo-exo and even endo-endo synergism that is based on the adsorption efficiency of the component enzymes. The efficiency of the degradation of crystalline cellulose is enhanced if the system contains enzymes with both high and low affinity to the substrate. Tightly adsorbed enzymes bind to cellulose and cause a disruption or defribrillation of the crystalline structure at the binding site, a concept also advanced by others (Stahlberg et al. 1991). The weakly adsorbed enzymes can now act effectively on the disrupted regions of the substrates. This is in line with the suggestion that a weakly adsorbed enzyme can only hydrolyze amorphous cellulose or soluble substrates. Inherent in this hypothesis, synergism occurs to a significant extent only when the disruption of the crystalline region is done by an endo-enzyme where the crystalline->amorphous change occurs randomly along the entire chain. Under this condition, the weakly adsorbed enzyme can now have access to a large number of newly created sites. This explanation is compatible with the suggestion by Gow and Wood (1988) as described above, because in general, "true" cellulase systems contain endoglucanases that are tightly adsorbed enzymes. Similarly, the existence of exo-exo synergism between two cellobiohydrolases can also be explained by the complementary action between tightly adsorbed and weakly-adsorbed enzymes.

REFERENCES

ABUJA, P. M.; PILZ, I.; CLAEYSSENS, M.; and TOMME, P. 1988A. Domain structure of cellobiohydrolase II as studied by small angle x-ray scattering close resemblance to cellobiohydrolase I. *Biochem. Biophys. Res. Comm. 156*, 180–185.

ABUJA, P. M.; SCHMUCK, M.; PILZ, I.; TOMME, P.; CLAEYSSENS, M.; and ESTERBAUER, H. 1988B. Structural and functional domains of cellobiohydrolase I from *Trichoderma reesei*. A small angle x-ray scattering study of the intact enzyme and its core. *Eur. Biophys. J. 15*, 339–342.

AIT, N.; CREUZET, N.; and CATTANEO, J. 1982. Properties of β-glucosidase purified from *Clostridum thermocellum. J. Gen. Microbiol. 128*, 569–577.

BAIRD, S. D.; HEFFORD, M. A.; JOHNSON, D. A.; SUNG, W. L.; YAGUCHI, M.; and SEILIGY, V. L. 1990. The Glu residue in the conserved Asn-Glu-Pro sequence of two highly divergent endo-β-1,4-glucanases is essential for enzymatic activity. *Biochem. Biophys. Res. Comm. 169*, 1035–1039.

BAKER, J. O.; TATSUMOTO, K.; GROHMANN, K.; WOODWARD, J.; WICHERT, J. M.; SHOE-MAKER, S. P.; and HIMMEL, M. E. 1992. Thermal denaturation of *Trichoderma reesei* cellulases studied by differential scanning calorimetry and tryptophan fluorescence. *Appl. Biochem. Biotechnol. 34/35*, 217–231.

BARNETT, C. C.; BERKA, R. M.; and FOWLER, T. 1991. Cloning and amplification of the gene encoding an extracellular β-glucosidase from *Trichoderma reesei*: Evidence for improved rates of saccharification of cellulosic substrates. *Bio/Technology 9*, 562–567.

BAUSE, E., and LEGLER, G. 1980. Isolation and structure of a tryptic glycopeptide from the active site of β-glucosidase A3 from *Aspergillus wentii. Biochim. Biophys. Acta 626*, 459–465.

BEGUIN, P. 1990. Molecular biology of cellulose degradation. *Ann. Rev. Microbiol. 44*, 219–248.

BEGUIN, P.; CORNET, P.; and AUBERT, J.-P. 1985. Sequence of a cellulase gene of the thermophilic bacterium *Clostridium thermocellum. J. Bacteriol. 162*, 102–105.

BEGUIN, P.; MILLET, J.; and AUBERT, J.-P. 1992. Cellulose degradation by *Clostridium thermocellum*: From manure to molecular biology. *FEMS Microbiol. Lett. 100*, 523–528.

BELDMAN, G.; VORAGEN, A. G. J.; ROMBOUTS, F. M.; SEARLE-VAN LEEUWEN, M. F.; and PILNIK, W. 1987. Adsorption and kinetic behavior of purified endoglucanases and exoglucanases from *Trichoderma viride. Biotechnol. Bioengineer. 30*, 251–257.

BHIKHABHAI, R.; JOHANSSON, G.; and PETTERSSON, G. 1985. Cellobiohydrolase from *Trichoderma reesei*. Internal homology and prediction of secondary structure. *Int. J. Peptide Protein Res. 25*, 368–374.

BHIKHABHAI, R., and PETTERSSON, G. 1984. The disulfide bridges in a cellobiohydrolase and an endoglucanase from *Trichoderma reesei. Biochem. J. 222,* 729–736.

BLACKWELL, J. 1982. The macromolecular organization of cellulose and chitin. In: *Cellulose and Other Natural Polymer Systems. Biogenesis, Structure, and Degradation.* R. Malcolm Brown, Jr., ed., Plenum Press, New York and London.

BROOKS, M. M.; TUOKY, M. G.; SAVAGE, A. V.; CLAEYSSENS, M.; and COUGHLAN, M. P. 1992. The stereochemical course of reactions catalyzed by the cellobiohydrolases produced by *Talaromyces emersonii. Biochem. J. 283,* 31–34.

CHANZY, H., and HENRISSAT, B. 1985. Unidirectional degradation of *Valonia* cellulose microcrystals subjected to cellulase action. *FEBS Lett. 184,* 285–288.

CHANZY, H.; HENRISSAT, B.; and Vuong, R.; and Schulein, M. 1983. The action of 1,4-β-D-glucan cellobiohydrolase on *Valonia* cellulose microcrystals. *FEBS Lett. 153,* 113–118.

CHANZY, H.; HENRISSAT, B.; and VUONG, R. 1984. Colloid gold labelling of 1,4-β-D-glucan cellobiohydrolase adsorbed on cellulose substrates. *FEBS Lett. 172,* 193–197.

CHEN, C. M.; GRITZALI, M.; and STAFFORD, D. W. 1987. Nucleotide sequence and deduced primary structure of cellobiohydrolase II from *Trichoderma reesei. Bio/Technology 5,* 274–278.

CHIRICO, W. J., and BROWN, R. D., JR. 1987. Purification and characterization of a β-glucosidase from *Trichoderma reesei. Eur. J. Biochem. 165,* 333–341.

CLAEYSSENS, M.; TOMME, P.; BREWER, C. F.; and HEHRE, E. J. 1990A. Stereochemical course of hydrolysis and hydration reactions catalyzed by cellobiohydrolases I and II from *Trichoderma reesei. FEBS Lett. 263,* 89–92.

CLAEYSSENS, M.; VAN TILBEURGH, H.; KAMERLING, J. P.; BERG, J.; VRSANSKA, M.; and BEILY, P. 1990B. Studies of the cellulolytic system of the filamentous fungus *Trichoderma reesei* QM9414. Substrate specificity and transfer activity of endoglucanase I. *Biochem. J. 270,* 251–256.

CLARKE, A. J. 1988. Active-site-directed inactivation of *Schizophyllum commune* cellulase by 4′,5′-epoxypentyl-4-D-(β-D-glucopyranosyl)-β-D-glucopyranoside. *Biochem. Cell Biol. 66,* 871–879.

CLARKE, A. J., and YAGUCHI, M. 1985. The role of carboxyl groups in the function of endo-β-1,4-glucanase from *Schizophyllum commune. Eur. J. Biochem. 149,* 233–238.

COUGHLAN, M. P.; MOLONEY, A. P.; McCRAE, S. I.; and WOOD, T. M. 1987. Cross-synergistic interactions between components of the cellulase systems of *Talaromyces emersonii, Fusarium solani, Penicillium funiculosum* and *Trichoderma koningii. Biochem. Soc. Trans. 15,* 263–264.

DAHLQUIST, F. W.; RAND-MEIR, T.; and RAFTERY, M. A. 1969. Application of secondary α-deuterium kinetic isotope effects to studies of enzyme catalysis. Glycoside hydrolysis by lysozyme and β-glucosidase. *Biochemistry 8,* 4214–4221.

DAVIES, G. J.; DODSON, G. G.; HUBBARD, R. E.; TOLLEY, S. P.; DAUTER, Z.; WILSON, K. S.; HJORT, C.; MIKKELSEN, J. M.; RASMUSSEN, G.; and SCHULEIN, M. 1993. Structure and function of endoglucanase V. *Nature 365,* 362–364.

DIN, N.; GILKES, N. R.; TEKANT, B.; MILLER, R. C., JR.; WARREN, R. A. J.; and

KILBURN, D. G. 1991. Non-hydrolytic disruption of cellulose fibres by the binding domain of a bacterial cellulase. *Bio/Technology 9*, 1096–1099.

DIVNE, C.; STAHLBERG, J.; REINIKAINEN, T.; RUOHONEN, L.; PETTERSSON, G.; KNOWLES, J. K. C.; TEERI, T. T.; and JONES, T. A. 1994. The three-dimensional crystal structure of the catalytic core of cellobiohydrolase I from *Trichodermi reesei*. *Science 265*, 524–528.

ENARI, T.-M.; NIKU-PAAVOLA, M.-L.; HARJU, L.; LAPPALAINEN, A.; and NUMMI, M. 1981. Purification of *Trichoderma reesei* and *Aspergillus niger* β-glucosidase. *J. Appl. Biochem. 3*, 157–163.

FAGERSTAM, L. G., and PETTERSSON, L. G. 1980. The 1,4-β-glucan cellobiohydrolases of *Trichoderma reesei* QM9414. *FEBS Lett. 119*, 97–100.

FAGERSTAM, L. G.; PATTERSSON, L. G.; and ENGSTROM, J. A. 1984. The primary structure of a 1,4-β-glucan cellobiohydrolase from the fungus *Trichoderma reesei* QM9414. *FEBS Lett. 167*, 309–315.

FIFE, T. H. 1972. General acid catalysis of acetal, ketal, and ortho ester hydrolysis. *Acc. Chem. Res. 5*, 264–272.

GEBLER, J.; GILKES, N. W.; CLAEYSSENS, M.; WILSON, D. B.; BEGUIN, P.; WAKARCHUK, W. S.; KILBURN, D. G.; MILLER, R. C.; WARREN, R. A. J. JR.; and WITHERS, S. G. 1992. Stereoselective hydrolysis catalyzed by related β-1,4-glucanases and β-1,4-xylanases. *J. Biol. Chem. 267*, 12559–12561.

GILKES, N. R.; CLAEYSSENS, M.; AEBERSOLD, R.; HENRISSAT, B.; MEINKE, A.; MORRISON, H. D.; KILBURN, D. G.; WARREN, A. J.; and MILLER, R. C., JR. 1991A. Structural and functional relationships in two families of β-1,4-glycanases. *Eur. J. Biochem. 202*, 367–377.

GILKES, N. R.; HENRISSAT, B.; KILBURN, D. G.; MILLER, R. C.; and WARREN, R. A. J. 1991B. Domains in microbial β-1,4-glycanases: Sequence conservation, function, and enzyme families. *Microbiol. Rev. 55*, 303–315.

GILKES, N. R.; JERVIS, E.; HENRISSAT, B.; TEKANT, B.; MILLER, R. C., JR.; WARREN, R. A. J.; and KILBURN, D. G. 1992. The adsorption of a bacterial cellulase and its two isolated domains to crystalline cellulose. *J. Biol. Chem. 267*, 6743–6749.

GILKES, N. R.; KILBURN, D. G.; MILLER, R. C. JR.; and WARREN, R. A. J. 1989. Structural and functional analysis of a bacterial cellulase by proteolysis. *J. Biol. Chem. 264*, 17802–17808.

GILKES, N. R.; WARREN, A. J.; MILLER, R. C.; and KILBURN, D. G. 1988. Precise excision of the cellulose binding domains from two *Cellulomonas fimi* cellulases by a homologous protease and the effect on catalysis. *J. Biol. Chem. 263*, 10401–10407.

GOW, L. A., and WOOD, T. M. 1988. Breakdown of crystalline cellulose by synergistic action between cellulase components from *Clostridium thermocellum* and *Trichoderma koningii*. *FEMS Microbiol. Lett. 50*, 247–252.

GRABNITZ, F.; RUCKNAGEL, K. P.; SEIB, M.; and STAUDENBAUER, W. L. 1989. Nucleotide sequence of the *Clostridium thermocellum bgl*B gene encoding thermostable β-glucosidase B: Homology to fungal β-glucosidases. *Mol. Gen. Genet. 217*, 70–76.

GRABNITZ, F.; SEISS, M.; RUCKNAGEL, K. P.; and STAUDENBAUER, W. L. 1991. Structure of the β-glucosidase gene bgl A of *Clostridium thermocellum*. *Eur. J. Biochem. 200*, 301–309.

GRABNITZ, F., and STAUDENBAUER, W. L. 1988. Characterization of two β-glucosidase genes from *Clostridium thermocellum*. *Biotechnol. Lett. 10*, 73–78.

GREPINET, O., and BEGUIN, P. 1986. Sequence of the cellulase gene of *Clostridium thermocellum* coding for endoglucanase B. *Nucl. Acids Res. 14*, 1791–1799.

HALL, J.; HAZLEWOOD, G. P.; BARKER, P. J.; and GILBERT, H. J. 1988. Conserved reiterated domains in *Clostridium thermocellum* endoglucanases are not essential for catalytic activity. *Gene 69*, 29–38.

HEHRE, E. J.; BREWER, C. F.; UCHIYAMA, T.; SCHLESSELMANN, P.; and LEHMANN, J. 1980. Scope and mechanism of carbohydrase action. Stereospecific hydration of 2,6-anhydro-1-deoxy-D-gluco-hept-1-enitol catalyzed by α- and β-glucosidases and an inverting exo-α-glucanase. *Biochemistry 19*, 2557–2564.

HEHRE, E. J.; GENGHOF, D. S.; STERNLICHT, H.; and BREWER, C. F. 1977. Scope and mechanism of carbohydrase action: Stereospecific hydration of D-glucal catalyzed by α- and β-glucosidase. *Biochemistry 16*, 1780–1787.

HENRISSAT, B.; CLAEYSSENS, M.; TOMME, P.; LEMESLE, L.; and MORNON, J.-P. 1989. Cellulase families revealed by hydrophobic cluster analysis. *Gene 81*, 83–95.

HENRISSAT, B.; DRIGUEZ, H.; VIET, C.; and SCHULEIN, M. 1985. Synergism of cellulases from *Trichoderma reesei* in the degradation of cellulose. *Bio/Technology 3*, 722–726.

HENRISSAT, B.; VIGNY, B.; BULEON, A.; and PEREZ, S. 1988. Possible adsorption sites of cellulases on crystalline cellulose. *FEBS Lett. 231*, 177–182.

HOFER, F.; WEISSINGER, E.; MISCHAK, H.; MESSNER, R.; MEIXNER-MONORI, B.; BLASS, D.; VISSER, J.; KUBICEK, C. P. 1989. A monoclonal antibody against the alkaline extracellular β-glucosidase from *Trichoderma reesei*: Reactivity with the *Trichoderma* β-glucosidases. *Biochim. Biophys. Acta 992*, 298–306.

HOSHINO, E.; KANDA, T.; SASAKI, Y.; and NISIZAWA, K. 1992. Adsorption mode of exo- and endo-cellulases from *Irpex lacteus* (*Polyporus tulipiferae*) on cellulose with different crystallinities. *J. Biochem. 111*, 600–605.

INGLIN, M.; FEINBERG, B. A.; and LOEWENBERG, J. R. 1980. Partial purification and characterization of a new intracellular β-glucosidase of *Trichoderma reesei*. *Biochem. J. 185*, 515–519.

JACKSON, M. A., and TALBURT, D. E. 1988. Purification and partial characterization of an extracellular β-glucosidase of *Trichoderma reesei* using cathodic run, polyacrylamide gel electrophoresis. *Biotechnol. Bioengineer. 32*, 903–909.

JOHANSSON, G.; STAHLBERG, J.; LINDEBERG, G.; ENGSTROM, A.; and PETTERSSON, G. 1989. Isolated fungal cellulase terminal domains and a synthetic minimum analogue bind to cellulose. *FEBS Lett. 243*, 389–393.

JOHNSON, E. A., and DEMAIN, A. L. 1984. Probable involvement of sulfhydryl groups and a metal as essential components of the cellulase of *Clostridium thermocellum*. *Arch. Microbiol. 137*, 135–138.

JOHNSON, E. A.; SAKAJOH, M.; HALLIWELL, G.; MADIA, A.; and DEMAIN, A. L. 1982. Saccharification of complex cellulosic substrates by the cellulase system from *Clostridium thermocellum*. *Appl. Environ. Microbiol. 43*, 1125–1131.

JOLIFF, G., BEGUIN, P., and AUBERT, J.-P. 1986A. Nucleotide sequence of the cellulase gene *cel*D encoding endoglucanase D of *Clostridium thermocellum*. *Nucl. Acids Res. 14*, 8605–8613.

JOLIFF, G.; BEGUIN, P.; JUY, M.; MILLET, J.; RYTER, A.; POLJAK, R.; and AUBERT, J.-P. 1986B. Isolation, crystallization and properties of a new cellulase of *Clostridium thermocellum* overproduced in *Escherichia coli*. *Bio/Technology 4*, 896–900.

JUY, M.; AMIT, A. G.; ALZARI, P. M.; POLJACK, R. J.; CLAEYSSENS, M.; BEGUIN, P.; and AUBERT, J.-P. 1992. Three-dimensional structure of a thermostable bacterial cellulase. *Nature 357*, 89–91.

KADAM, S. K., and DEMAIN, A. L. 1989. Addition of cloned β-glucosidase enhances the degradation of crystalline cellulase by the *Clostridium thermocellum* cellulase complex. *Biochim. Biophys. Res. Comm. 161*, 706–711.

KAMIDE, K.; OKAJIMA, K.; KOWSAKA, K.; and MATSUI, T. 1985. CP/MASS ^{13}C NMR spectra of cellulose solids: An explanation by the intramolecular hydrogen bond concept. *Polymer J. 17*, 701–706.

KANDA, T.; BREWER, C. F.; OKADA, G.; and HEHRE, E. J. 1986. Hydration of cellobial by exo- and endo-type cellulases: Evidence for catalytic flexibility of glycosylases. *Biochemistry 25*, 1159–1165.

KAWAMORI, M.; ADO, Y.; and TAKASAWA, S. 1986. Preparation and application of *Trichoderma reesei* mutants with enhanced β-glucosidase. *Agric. Biol. Chem. 50*, 2477–2482.

KIM, D. W.; JEONG, Y. K.; and LEE, J. K. 1994. Adsorption kinetics of exoglucanase in combination with endoglucanase from *Trichoderma viride* on microcrystalline cellulose and its influence on synergistic 152 degradation. *Enzyme Microb. Technol. 16*, 649–658.

KLYOSOV, A. A. 1990. Trends in biochemistry and enzymology of cellulose degradation. *Biochemistry 29*, 10577–10585.

KLYOSOV, A. A.; MITKEVICH, O. V.; and SINITSYN, A. P. 1986. Role of the activity and adsorption of cellulases in the efficiency of the enzymatic hydrolysis of amorphous and crystalline cellulose. *Biochemistry 25*, 540–542.

KNOWLES, J.; LEHTOVAARA, P.; and TEERI, T. 1987. Cellulase families and their genes. *TIBTECH 5*, 255–261.

KNOWLES, J. K. C.; and LENTOVAARA, P.; MURRAY, M.; and SINNOTT, M. L. 1988. Stereochemical course of the action of the cellobioside hydrolases I and II of *Trichoderma reesei*. *J. Chem. Soc. Chem. Commun. 1988*, 1401.

KOHCHI, C., and TOHE, A. 1985. Nucleotide sequence of *Candida pelliculosa* β-glucosidase gene. *Nucl. Acids Res. 13*, 6273–6282.

KOSHLAND, D. E. JR. 1959. Mechanisms of transfer enzymes. *The Enzymes 1*, 305–346.

KRAULIS, P. J.; CLORE, G. M.; NILGES, M.; JONES, T. A.; PETTERSSON, G.; KNOWLES, J.; and GRONENBORN, A. M. 1989. Determination of the three-dimensional solution structure of the C-terminal domain of cellobiohydrolase I from *Trichoderma reesei*. A study using nuclear magnetic resonance and hybrid distance geometry-dynamical simulated annealing. *Biochemistry 28*, 7241–7257.

KYRIACOU, A.; NEUFELD, R. J.; and MACKENZIE, C. R. 1989. Reversibility and competition in the adsorption of *Trichoderma reesi* cellulase components. *Biotechnol. Bioengineer. 33*, 631–637.

LAMED, R., and BAYER, E. A. 1988. The cellulosome of *Clostridium thermocellum*. *Adv. Appl. Microbiol. 33*, 1–46.

LAMED, R.; KENIG, R.; MORAG, E.; CALZADA, J. F.; DE MICHEO, F.; and BAYER, E. A. 1991. Efficient cellulose solubilization by a combined *cellulosome*-β-glucosidase system. *Appl. Biochem. Biotechnol. 27*, 173–183.

LANGSFORD, M. L.; GILKES, N. R.; SINGH, B.; MOSER, B.; MILLER, R. C., JR.; WARREN, R. A. J.; and KILBURN, D. G. 1987. Glycosylation of bacterial cellulases prevents proteolytic cleavage between functional domains. *FEBS Lett. 225*, 163–167.

LEE, N. E.; LIMA, M.; and WOODWARD, J. 1988. Hydrolysis of cellulose by a mixture of *Trichoderma reesei* cellobiohydrolase and *Aspergillus niger* endoglucanase. *Biochim. Biophys. Acta 967*, 437–440.

LEE, S. B.; SHIN, H. S.; RYU, D. D. Y.; and MANDELS, M. 1983. Adsorption of cellulase on cellulose: Effect of physiochemical properties of cellulose on adsorption and rate of hydrolysis. *Biotechnol. Bioengineer. 24*, 2137–2153.

LEGLER, G.; ROESER, K.-R.; and ILLIG, H.-K. 1979. Reaction of β-D-glucosidase A3 from *Aspergillus wentii* with D-glucal. *Eur. J. Biochem. 101*, 85–92.

LEGLER, G.; SINNOTT, M. L.; and WITHERS, S. G. 1980. Catalysis by β-glucosidase A3 of *Aspergillus wentii*. *J. Chem. Soc. Perkin II 9*, 1376–1383.

LEI, S.-P.; LIN, H.-C.; WANG, S.-S.; CALLAWAY, J.; and WILCOX, G. 1987. Characterization of the *Erwinia carotovora pel B* gene and its product pectate lyase. *J. Bacteriol. 169*, 4379–4383.

MACARRON, R.; VAN BEEUMEN, J.; HENRISSAT, B.; DE LA MATA, I.; and CLAEYSSENS, M. 1993. Identification of an essential glutamate residue in the active site of endoglucanase III from *Trichoderma reesei. FEBS Lett. 316*, 137–140.

MACHIDA, M.; OHTSUKI, I.; FUKUI, S.; and YAMASHITA, I. 1988. Nucleotide sequences of *Saccharomycopsis fibuligera* genes for extracellular β-glucosidases as expressed in *Saccharomyces cerevisiae. Appl. Environ. Microbiol. 54*, 3147–3155.

MACLEOD, A. M.; LINDHORST, T.; WITHERS, S. G.; and WARREN, R. A. J. 1994. The acid/base catalyst in the exoglucanase/xylanase from *Cellulomonas fimi* is glutamic acid 127: Evidence from detailed kinetic studies of mutants. *Biochemistry 33*, 6371–6376.

MCHALE, A., and COUGHLAN, M. P. 1980. Synergistic hydrolysis of cellulose by components of the extracellular cellulase system of *Talaromyces emersonii. FEBS Lett. 117*, 319–322.

MEINKE, A.; BRAUN, C.; GILKES, N. R.; KILBURN, D. G.; MILLER, R. C., JR.; and WARREN, R. A. J. 1991. Unusual sequence organization in CenB, an inverting endoglucanase from *Cellulomonas fimi. J. Bacteriol. 173*, 308–314.

MESSNER, R.; HAGSPIEL, K.; and KUBICEK, C. P. 1990. Isolation of a β-glucosidase binding and activating polysaccharide from cell walls of *Trichoderma reesei. Arch. Microbiol. 154*, 150–155.

MESSNER, R., and KUBICEK, C. P. 1990. Evidence for a single, specific β-glucosidase in cell walls from *Trichoderma reesei* QM9414. *Enzyme Microb. Technol. 12,* 685–690.

MOLONEY, A., and COUGHLAN, M. P. 1983. Sorption of *Talaromyces emersonii* cellulase on cellulosic substrates. *Biotechnol. Bioengineer. 25,* 271–280.

MORAG, E.; HALEVY, I.; BAYER, E. A.; and LAMED, R. 1991. Isolation and properties of a major cellobiohydrolase from the cellulosome of *Clostridium thermocellum. J. Bacteriol. 173,* 4155–4162.

NIKU-PAAVOLA, M.-L.; LAPPALAINEN, A.; ENARI, T.-M.; and NUMMI, M. 1986. *Trichoderma reesei* cellobiohydrolase II. Purification by immunoadsorption and hydrolytic properties. *Biotechnol. Appl. Biochem. 8,* 449–458.

NUMMI, M.; NIKU-PAAVOLA, M.-L.; LAPPALAINEN, A.; ENARI, T.-M.; and RAUNIO, V. 1983. Cellobiohydrolase from *Trichoderma reesei. Biochem. J. 215,* 677–683.

OHMIYA, K.; TAKANO, M.; and SHIMIZU, S. 1989. DNA sequence of a β-glucosidase from *Ruminococcus albus. Nucl. Acids Res. 18,* 671.

O'NEILL, G. P.; GOH, S. H.; WARREN, R. A. J.; KILBURN, D. G.; and MILLER, R. C., JR. 1986. Structure of the gene encoding the exoglucanase of *Cellulomonas fimi. Gene 44,* 325–330.

PENTTILA, M.; LEHTOVAARA, P.; NEVALAINEN, H.; BHIKHABHAI, R.; and KNOWLES, J. 1986. Homology between cellulase genes of *Trichoderma reesei*: Complete nucleotide sequence of the endoglucanase I gene. *Gene 45,* 253–263.

PILZ, I.; SCHWARZ, E.; KILBURN, D. G.; MILLER, R. C., JR.; WARREN, R. A. J.; and GILKES, N. R. 1990. The tertiary structure of a bacterial cellulase determined by small-angle x-ray-scattering analysis. *Biochem. J. 271,* 277–280.

POST, C. B., and KARPLUS, M. 1986. Does lysozyme follow the lysozyme pathway? An alternative based on dynamic, structural, and stereoelectronic considerations. *J. Am. Chem. Soc. 108,* 1317–1319.

RAYNAL, A.; GERBAUD, C.; FRANCINGUES, M. C.; and GUERINEAU, M. 1987. Sequence and transcription of the β-glucosidase gene of *Kluyveromyces fragilis* cloned in *Saccharomyces cerevisiae. Curr. Genet. 12,* 175–184.

REESE, E. T.; SIU, R. G. H.; and LEVINSON, H. S. 1950. The biological degradation of soluble cellulose derivatives and its relationship to the mechanism of cellulose hydrolysis. *J. Bacteriol. 59,* 485–497.

REINIKAINEN, T.; RUOHONEN, L.; NEVANEN, T.; LAAKSONEN, L.; KRAULIS, P.; JONES, T. A.; KNOWLES, J. K. C.; and TEERI, T. T. 1992. Investigation of the function of mutated cellulose-binding domains of *Trichoderma reesei* cellobiohydrolase I. *Proteins: Structure, Function, and Genetics 14,* 475–482.

ROESER, K.-R., and LEGLER, G. 1981. Role of sugar hydroxyl groups in glycoside hydrolysis. Cleavage mechanism of deoxyglucosides and related substrates by β-glucosidase A3 *Aspergillus wentii. Biochim. Biophys. Acta 657,* 321–333.

ROUVINEN, J.; BERGFORS, T.; TEERI, T.; KNOWLES, K. C.; and JONES, T. A. 1990. Three-dimensional structure of cellobiohydrolase II from *Trichoderma reesei. Science 249,* 380–386.

RYU, D. D. Y.; KIM, C.; and MANDELS, M. 1984. Competitive adsorption of cellulase components and its significance in a synergistic mechanism. *Biotechnol. Bioengineer. 26,* 488–496.

SALOHEIMO, M.; LEHTOVAARA, P.; PENTTILA, M.; TEERI, T. T.; STAHLBERG, J.; JO-HANSSON, G.; PETTERSSON, G.; CLAEYSSENS, M.; TOMME, P.; and KNOWLES, J. K. C. 1988. EGIII, a new endoglucanase from *Trichoderma reesei*: The characterization of both gene and enzyme. *Gene 63*, 11–21.

SALOVUORI, I.; MAKAROW, M.; RAUVALA, H.; KNOWLES, J.; and KAARIAINEN, L. 1987. Low molecular weight high-mannose type glycans is a secreted protein of the filamentous fungus *Trichoderma reesei*. *Bio/Technology 5*, 152–156.

SCHMID, G., and WANDREY, Ch. 1987. Purification and partial characterization of a cellodextrin glucohydrolase (β-glucosidase) from *Trichoderma reesei* strain QM9414. *Biotechnol. Bioengineer. 30*, 571–585.

SCHMUCK, M.; PILZ, I.; HAYN, M.; and ESTERBAUER, H. 1986. Investigation of cellobiohydrolase from *Trichoderma reesei* by small angle x-ray scattering. *Biotechnol. Lett. 8*, 397–402.

SCHULTZ, T. P.; McGINNIG, G. D.; and BERTRAN, M. S. 1985. Estimation of celulose crystallinity using Fourier transform-infrared spectroscopy and dynamic thermogravimetry. *J. Wood Chemistry and Technology 5*, 543–557.

SCHWARZ, W.; BRONNENMEIER, K.; and STAUDENBAUER, W. L. 1985. Molecular cloning of *Clostridium thermocellum* genes involved in β-glucan degradation in bacteriophage lambda. *Biotechnol. Lett. 7*, 859–864.

SCHWARZ, W. H.; SCHIMMING, S.; RUCKNAGEL, K. P.; BURGSCHWAIGER, S.; KREIL, G.; and STAUDENBAUER, W. L. 1988. Nucleotide sequence of the *celC* gene encoding endoglucanase C of *Clostridium thermocellum*. *Gene 63*, 23–30.

SHEN, H.; SCHMUCK, M.; PILZ, I.; GILKES, N. R.; KILBURN, D. G.; MILLER, R. C., JR.; and WARREN, R. A. J. 1991. Deletion of the linker connecting the catalytic and cellulose-binding domains of endoglucanase A (Cen A) of *Cellulomonas fimi* alters its conformation and catalytic activity. *J. Biol. Chem. 266*, 11335–11340.

SHOEMAKER, S.; SCHWEICKART, V.; LADNER, M.; GELFAND, D.; KWOK, S.; MYAMBO, K.; and INNIS, M. 1983. Molecular cloning of exo-cellobiohydrolase I derived from *Trichoderma reesei* strain L27. *Bio/Technology 1*, 691–696.

SPEZIO, M.; WILSON, D. B.; and KARPLUS, P. A. 1993. Crystal structure of the catalytic domain of a thermophilic endocellulase. *Biochemistry 32*, 9906–9916.

SPREY, B., and BOCHEM, H.-P. 1993. Formation of cross-fractures in cellulose microfibril structure by an endoglucanase-cellobiohydrolase complex from *Trichoderma reesei*. *FEMS Microbiol. Lett. 106*, 239–244.

STAHLBERG, J.; JOHANSSON, G.; and PETTERSSON, G. 1988. A binding-site-deficient catalytically active, core protein of endoglucanase III from the culture filtrate of *Trichoderma reesei*. *Eur. J. Biochem. 173*, 179–183.

———. 1991. A new model for enzymatic hydrolysis of cellulose based on the two-domain structure of cellobiohydrolase I. *Bio/Technology 9*, 286–290.

STERNBERG, D.; VIJAYAKUMAR, P.; and REESE, E. T. 1977. β-Glucosidase: Microbial production and effect on enzymatic hydrolysis of cellulose. *Can. J. Microbiol. 23*, 139–147.

THOMA, J. A. 1968. A possible mechanism for amylase catalysis. *J. Theoret. Biol. 19*, 297–310.

TOMME, P., and CLAEYSSENS, M. 1989. Identification of a functionally important carboxyl group in cellobiohydrolase I from *Trichoderma reesei. FEBS Lett. 243*, 239–243.

TOMME, P.; VAN BEEUMEN, J.; and CLAEYSSENS, M. 1992. Modification of catalytically important carboxy residues in endoglucanase D from *Clostridium thermocellum. Biochem. J. 285*, 319–324.

TOMME, P.; VAN TILBEURGH, H.; PETTERSSON, G.; VAN DAMME, J.; VANDEKERCKHOVE, J.; KNOWLES, J.; TEERI, T.; and CLAEYSSENS, M. 1988. Studies of the cellulolytic system of *Trichoderma reesei*: QM9414. Analysis of domain function in two cellobiohydrolases by limited proteolysis. *Eur. J. Biochem. 170*, 575–581.

TRIMBUR, D. E.; WARREN, R. A. J.; and WITHERS, S. G. 1992. Region-directed mutagenesis of residues surrounding the active site nucleophile in β-glucosidase from *Agrobacterium faecalis. J. Biol. Chem. 267*, 10248–10251.

TRIPP, V. W. 1971. In: *Cellulose and Cellulose Derivatives.* Part IV, N. M. Bikales, and L. Segal, eds., John Wiley & Sons, New York, Chapter XIII-G, p. 305.

TULL, D., and WITHERS, S. G. 1994. Mechanisms of cellulases and xylanases: A detailed kinetic study of the exo-β-1,4-glycanase from *Cellulomonas fimi. Biochemistry 33*, 6363–6370.

TULL, D.; WITHERS, S. G.; GILKES, N. R.; KILBURN, D. G.; WARREN, R. A. J.; and AEBERSOLD, R. 1991. Glutamic acid 274 is the nucleophile in the active site of a "retaining" exoglucanase from *Cellulomonas fimi. J. Biol. Chem. 266*, 15621.

UMILE, C., and KUBICEK, C. P. 1986. A constitutive plasma-membrane bound β-glucosidase in *Trichoderma reesei. FEMS Microbiol. Lett. 34*, 291–295.

VAN TILBEURGH, H.; TOMME, P.; CLAEYSSENS, M.; BHIKHABHAI, R.; and PETTERSSON, G. 1986. Limited proteolysis of the cellobiohydrolase I from *Trichoderma reesei. FEBS Lett. 204*, 223–226.

VERNON, C. A. 1967. The mechanisms of hydrolysis of glycosides and their relevance to enzyme-catalyzed reactions. *Royal Soc. Proc. Ser. B. 167*, 389–401.

VRSANSKA, M., and BIELY, P. 1992. The cellobiohydrolase I from *Trichoderma reesei* QM9414: Action on cello-oligosaccharides. *Carbohydr. Res. 227*, 19–27.

WANG, Q. P.; TULL, D.; MEINKE, A.; GILKES, N. R.; WARREN, R. A. J.; AEBERSOLD, R.; and WITHERS, S. G. 1993. Glu280 is the nucleophile in the active site of *Clostridium thermocellum* CelC, a family A endo-β-1,4-glucanase. *J. Biol. Chem. 268*, 14096–14102.

WEBER, J. P., and FINK, A. L. 1980. Temperature-dependent change in the rate-limiting step of β-glucosidase catalysis. *J. Biol. Chem. 255*, 9030–9032.

WHITE, A.; WITHERS, S. G.; GILKES, N. R.; and ROSE, D. R. 1994. Crystal structure of the catalytic domain of the β-1,4-glycanase Cex from *Cellulomonas fimi. Biochemistry 33*, 12546–12552.

WHITE, A. R. 1982. Visualization of cellulases and cellulose degradation. In: *Cellulose and Other Natural Polymer Systems: Biogenesis, Structure, and Degradation*, R. M. Brown, Jr. ed., Plenum Press, New York and London, chapter 23, pp. 489–509.

WHITE, A. R., and BROWN, R. M. 1981. Enzymatic hydrolysis of cellulose: Visual characterization of the process. *Proc. Natl. Acad. Sci. USA 78*, 1047–1051.

WITHERS, S. G.; DOMBROSKI, D.; BERVEN, L. A.; KILBURN, D. G.; MILLER, R. C. JR.; WARREN, R. A. J.; and GILKES, N. R. 1986. Direct ¹H N.M.R. determination of the stereochemical course of hydrolyses catalyzed by glucanase components of the cellulase complex. *Biochem. Biophys. Res. Comm. 139*, 487–494.

WITHERS, S. G., and STREET, I. P. 1988. Identification of a covalent α-D-glucopyranosyl enzyme intermediate formed on a β-glucosidase. *J. Am. Chem. Soc. 110*, 8551–8553.

WITHERS, S. G.; WARREN, R. A. J.; STREET, I. P.; RUPITZ, K.; KEMPTON, J. B.; and AEBERSOLD, R. 1990. Unequivocal demonstration of the involvement of a glutamate residue as a nucleophile in the mechanism of a "retaining" glycosidase. *J. Am. Chem. Soc. 112*, 5887–5889.

WONG, W. K. R.; GERHARD, B.; GUO, Z. M.; KILBURN, D. G.; WARREN, R. A. J.; and MILLER, R. C., JR. 1986. Characterization and structure of an endoglucanase gene *cen*A of *Cellulomonas fimi. Gene 44*, 315–324.

WOOD, T. M., and MCCRAE, S. I. 1986. The cellulase of *Penicillium pinophilum.* Synergism between enzyme components in solubilzing cellulose with special reference to the involvement of two immunologically distinct cellobiohydrolases. *Biochem. J. 234*, 93–99.

WOOD, T. M.; MCCRAE, S. I.; and BHAT, K. M. 1989. The mechanism of fungal cellulase action. Synergism between enzyme components of *Penicillium pinophilum* cellulase in solubilizing hydrogen bond-ordered cellulose. *Biochem. J. 260*, 37–43.

WOODWARD, J.; AFFHOLTER, K. A.; NOLES, K. K.; TROY, N. T.; and GASLIGHTWALA, S. F. 1992. Does cellobiohydrolase II core protein from *Trichoderma reesei* disperse cellulose macrofibrils? *Enzyme Microb. Technol. 14*, 625–630.

WOODWARD, J.; HAYES, M. K.; and LEE, N. E. 1988A. Hydrolysis of cellulose by saturating and non-saturating concentration of cellulase: Implications for synergism. *Bio/Technology 6*, 301–304.

WOODWARD, J.; LIMA, M.; and LEE, N. E. 1988B. The role of cellulase concentration in determining the degree of synergism in the hydrolysis of microcrystalline cellulose. *Biochem. J. 255*, 895–899.

WU, J. H. D.; ORME-JOHNSON, W. H.; and DEMAIN, A. L. 1988. Two components of an extracellular protein aggregate of *Clostridium thermocellum* together degrade crystalline cellulose. *Biochemistry 27*, 1703–1709.

YAGUCHI, M.; ROY, C.; ROLLIN, C. F.; PAICE, M. G.; and JURASEK, L. 1983. A fungal cellulase shows sequence homology with the active site of hen egg-white lysozyme. *Biochem. Biophys. Res. Comm. 116*, 408–411.

YAGUE, E.; BEGUIN, P.; and AUBERT, J.-P. 1990. Nucleotide sequence and deletion analysis of the cellulase-encoding gene *cel*H of *Clostridium thermocellum. Gene 89*, 61–67.

Chapter 5

Proteolytic Enzymes

The term protease refers to all enzymes that hydrolyze peptide bonds. Other names include peptidase and peptide hydrolase. This group of enzymes can be subdivided into exopeptidases and endopeptidases for exo-acting and endo-acting patterns. Endopeptidase is used synonymously with proteinase.

Proteinases are classified into four groups according to the catalytic residue involved in the nucleophilic attack at the carbonyl carbon of the scissile bond: serine (EC 3.4.21), cysteine (EC 3.4.22), aspartic (EC 3.4.23), and metalloproteinases (EC 3.4.24). Physiologically, proteases have a major function in cellular catabolism and protein turnover. These are the proteins used in digestion and in processes of the immune system. In food processing, the major uses of proteases occur in meat tenderizing, cheese manufacturing, beer chill-proofing, and textural modification.

SUBTILISIN

Serine proteinases are a class of proteolytic enzymes that have been studied in more detail than any other group of enzymes. Two families of serine proteinases receive particular attention: the trypsin family and the subtilisin family. Both groups consist of a catalytic triad of Ser-His-Asp, and an oxyanion hole in the active site, with the catalytic residues showing a high degree of similarity in their conformational arrangements. However, enzymes in these two families show little similarity in their amino acid sequence or their overall three-dimensional structure. These two families of enzymes are generally considered to have a convergent origin in that they probably evolved independently from unrelated ancestral proteins.

The scope of this discussion will be limited to subtilisin, which has gained considerable importance due to the recent advent in recombinant DNA technology (Wells and Estell 1988). Unlike trypsin and chymotrypsin, which are mammalian enzymes, subtilisin is a group of alkaline serine proteinases secreted by species of *Bacillus*. Over the years, various names have been designated for subtilisin enzymes. Subtilisin BPN' (also known as Nagarase and subtilopeptidase C) and subtilisin Novo (also referred to as subtilopeptidase B), originally obtained from different commercial sources, are now known to be identical enzymes produced by *Bacillus amyloliquefaciens* (Olaitan et al. 1968; Robertus et al. 1971; Drenth et al. 1972). Subtilisin Carlsberg is produced by *Bacillus licheniformis* (Keay and Moser 1969). Other less known subtilisin enzymes include subtilisin E and DY isolated from *Bacillus subtilis* strains, and subtilisin Amylosacchariticus from *Bacillus amylosacchariticus* (Ikemura et al. 1987).

Subtilisins (EC 3.4.21.14) have a MW range of 26 to 28 kD. The enzymes are in general more stable at acid pH than at alkaline pH where autolysis may occur. Subtilisins, similar to trypsin and chymotrypsin, are synthesized as preproenzymes with a prosequence between the signal peptide and the mature protein. Subtilisin Carlsberg contains a pro-peptide of 76 residues in addition to the 274 amino acid sequence (Jacobs et al. 1985). Likewise, subtilisin BPN' contains a pro-peptide sequence of 75 amino acids (Wells et al. 1983).

AMINO ACID SEQUENCES

Two subtilisins were sequenced in the late 1960s (Markland and Smith 1967; Smith et al. 1968). Subtilisin Carlsberg and BPN' consist of 274 and 275 residues, respectively. The two sequences are 70% identical, showing differences in 84 residues, with the Pro56 missing in Carlsberg as compared to BPN' (Fig. 5.1). The amino acid sequence deduced from the DNA sequence differs from the protein sequence at residues 102, 128, 157, 160, and 211, in the Carlsberg enzyme (Jacobs et al. 1985), which could reflect mutational variations existing in different strains. Sequencing of the cDNA clone of subtilisin BPN' also shows discrepancies with the protein sequence in six amino acid residues (Wells et al. 1983; Vasantha et al. 1984). Subtilisin Amylosacchariticus contains 275 residues. The enzyme differs from BPN' in 35 residues (Kurihara et al. 1972). The calculated MW for subtilisin Carlsberg, BPN', and Amylosacchariticus are 27,277, 27,537, and 27,671, respectively. Subtilisin DY produced by *Bacillus subtilis* strain DY consists of 274 residues with 82 substititions compared to the primary structure of BPN' (Nedkov et al. 1985). Subtilisin E from *Bacillus subtilis* strain I168

```
BPN'                  AQSVPYGVSQIKAPALHSQGYTGSNVKVAVIDSGIDSSHP
Carlsberg             --T----IPL---DKVQA--FK-AD-----LDT--QASHP
Amylosacchariticus    -------IS---------------------D--------
E                     -------IS---------------------D--------
DY                    --T----IPL---DKVQA---K-A----GI-DT--AA--T

BPN'             41   DLKVAGGASMVPSETPNFQDDNSHGTHVAGTVAALNNSIG
Carlsberg        41   --N-V----F-AG-A YNT-G-GH-----------D-TT-
Amylosacchariticus 41  --N-R----F-----PNY--GS-H-------I--------
E                41   --N-R----F-----NPY--GS-H-------I--------
DY               41   --K-V----F-SGES YNT-G-GH-----------D-TT-

BPN'             81   VLGVAPSSALYAVKVLGDAGSGQYSWIINGIEWAIANNMD
Carlsberg        80   -------VS-------NSS---SY-G-VS-----TT-G--
Amylosacchariticus 81  -------SA-------DST----------------S----
E                81   ----S--AS-------DST----------------S----
DY               80   ------NVS---I---NSS---TY-A-VS-----TQ-GL-

BPN'             121  VINMSLGGPSGSAALKAAVDKAVASGVVVVAAAGNEGSTG
Carlsberg        120  --------A---T-M-Q---N-Y-R-V--------S-NS-
Amylosacchariticus 121  ------------T---TV-----S--I--A--------S-
E                121  ---------T--T---TV-----S--I--A--------S-
DY               120  ------------T---Q-----Y---I--------S--S-

BPN'             161  SSSTVGYPGKYPSVIAVGAVDSSNQRASFSSVGPELDVMA
Carlsberg        160  -TN-I---A--D----------NSN-------A--E---
Amylosacchariticus 161  -S------A----T------N---------A-S------
E                161  -T------A----T------N---------A-S------
DY               160  -QN-I---A--D-V--------NKN-------AG-G---

BPN'             201  PGVSIQSTLPGNKYGAYNGTSMASPHVAGAAALILSKHPN
Carlsberg        200  --AGVY--Y-T-T-ATL---S------------------
Amylosacchariticus 201  -----------GT-------S--T---------------T
E                201  -----------GT-------S--T---------------T

BPN'             241  WTNTQVRSSLQNTTTKLGDSFYYGKGLINVQAAAQ    275
Carlsberg        240  LSAS---NR-SS-A-Y--S----------------    274
Amylosacchariticus 241  --NA---DR-ES-A-Y-------------------    275
E                241  --NA---DR-ES-A-Y--N----------------    275
DY               240  LSAS---NR-SS-A-N------------G---N      274
```

Figure 5.1. Comparison of amino acid sequences of subtilisins: BPN' (Markland and Smith 1967), Carlsberg (Smith et al. 1968), Amylosacchariticus (Kurihara et al. 1972), E (sequence deduced from cDNA, Stahl and Ferrari 1984), DY (Nedkov et al. 1985). Catalytic residues are indicated by boldface.

contains 275 residues with 85% similarity to the protein sequence of subtilisin BPN' (Stahl and Ferrari 1984).

Subtilisin shares little similarity with mammalian serine proteinases in their primary structures. In particular, the sequence around the reactive Ser in subtilisin (Thr-Ser-Met-Ala) is markedly different from that in trypsin or chymotrypsin (Markland and Smith 1967). The primary structures in subtilisin are noted for many di- and tri-repetitions of the same amino acids, and several repeated sequences in different segments of the molecule. Subtilisin contains no Cys but high contents of Gly and hydrophobic residues that are largely conserved ((Smith et al. 1968; Kurihara et al. 1972).

THREE-DIMENSIONAL STRUCTURES

The refined structures of subtilisin Carlsberg and BPN' are known (Fig. 5.2) (Wright et al. 1969; Kraut 1971; Wright 1972; Bott et al. 1988). In addition to this, information on the structural characteristics of subtilisins is available from analyses of complexes between subtilisin and various types of inhibitors (Hirono et al. 1984; Bode et al. 1986; McPhalen and James 1988; Heinz et al. 1991).

Subtilisin BPN' has an overall globular structure with 42-A diameter (Kraut 1971). The enzyme consists of a central 5-stranded parallel β sheet formed by five segments: Val28-Asp32, Ala45-Met50, Ala89-Lys94, Asp120-Met124, and Val148-Ala152. Running antiparallel to the sheet are eight helical segments (A to D), which make up 31% of the total residues. Helices C and F are buried, whereas the other helices are amphipathic and found on the surface of the enzyme molecule. The enzyme thus has an α/β type structure that is quite different from that of chymotrypsin, which contains two antiparallel β barrel domains.

Subtilisin Carlsberg has a very similar structure. The molecule is composed of a twisted β sheet of seven parallel β strands: Val26-Asp32, Asn43-Phe50, Ser89-Val95, Asp120-Met124, Val148-Ala153, Ile175-Val180, and Glu197-Gly202. Located on the surface are helices A (Tyr6-Ile11), B (Lys12-Gln19), C (Gly63-Ala74), D (Ser103-Asn117), E (Ser132-Gly146), F (Tyr220-His238), G (Ser242-Thr253), and H (Asn269-Ala274) (Bode et al. 1987; McPhalen and James 1988).

Active Site

The active site is located near the C-terminal edge of the β sheet connecting to two helices (C and F), in a shallow depression on the enzyme surface. The catalytic Ser221 is conformationally restricted by a Type 3.0_{10} reverse turn forming a hydrogen-bonded loop (Wright 1972). The essential His64 is also held in a rigid conformation by a type 3.0_{10} loop. The His imidazole is located close to the

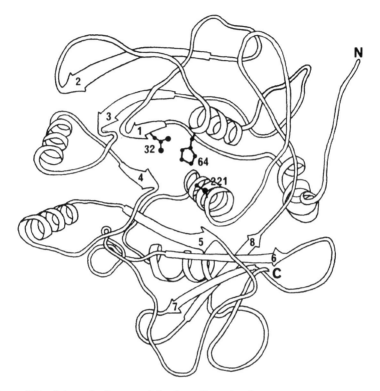

Figure 5.2. Schematic diagram of the three-dimensional structure of subtilisin viewed down the central parallel β sheet. (From Branden and Tooze 1991. *Introduction to Protein Structure*, p. 242, with permission. Copyright 1991 Garland Publishing, Inc.)

reactive Ser221-OH, but a hydrogen bond is not formed in the native enzyme. The Ser-O_γ is 3.7 A from His-$N_{\epsilon 2}$, and displaced 2.5 A from its ideal position for hydrogen bonding (Matthews et al. 1977). The $N_{\delta 1}$ atom, however, is in a suitable location to hydrogen bond with $O_{\delta 2}$ of the buried side chain of Asp32, with N. . .O distance of 2.4 A. A H_2O molecule is located near this Ser-His that can form hydrogen bonds with Ser221-OH, His64-$N_{\epsilon 2}$, and Asn155-NH_2

The Substrate Binding Site

The binding site is an extended cleft consisting of at least six subsites (S_4 to S_2') (Morihara et al. 1971; Robertus et al. 1972A), or even eight sub-fjsites (S_5 to S_3') (Gron et al. 1992). According to subsite/substrate nomenclature in the scheme used by Schechter and Berger (1967) for cysteine proteinases, the catalytic site is located between S_1 and S_1' (Fig. 5.3). The substrate

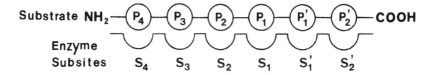

Figure 5.3.   Schematic representation of complex of enzyme-substrate binding at subsites (Schechter and Berger 1967).

$(NH_2-P_n--P_1-P_1'---P_n'-COOH)$ therefore has the N-terminal segment of the scissile bond as the acylating group (or specificity side), and the carboxyl segment as the leaving group. An extended backbone segment of Ser125-Gly127 on one side of the binding cleft assumes an antiparallel β sheet structure by forming an array of hydrogen bonds with the acylating side of the substrate polypeptide chain. The other wall is comprised of side chains and backbone of Ala152-Glu156 (Robertus et al. 1972A; Wells et al. 1987A). The P_1' and P_2' binding subsites are less defined, and the enzyme-substrate interactions are less extensive than on the acylating side (Robertus et al. 1972B; Morihara and Oka 1977). All the residues lining the substrate binding region are located in external loops connecting the β strands and helical segments (Bott et al. 1988).

The Cation Binding Site

Subtilisins are stabilized against autolysis by divalent cations such as calcium. Two calcium binding sites are apparent from the x-ray crystal structure of subtilisin Carlsberg-eglic C complex (Bode et al. 1986, 1987). The two sites are located with the loop segments Ala74-Leu82 and Pro168-Ile175, respectively. In the former site, the calcium is positioned to ligand octahedrally six oxygen atoms contributed by the carbonyl groups of Leu75, Thr79 and Val81, and the carboamide groups of Gln2 and Asn77. The second site which has a less well-defined geometry, is surrounded by carbonyl oxygens of Ala167, Tyr171, and Val174. Similar coordination for the first site is also identified in the crystal structure of the *Bacillus amyloliquefaciens* enzyme (Bott et al. 1988). The ligand distances for the second site, however, are not consistent with Ca^{++} as cation, but more typical of a K^+ ion (Bott et al. 1988; McPhalen and James 1988).

MECHANISM OF CATALYSIS

Subtilisin catalyzes the hydrolysis of esters and amides. However, the hydrolysis of esters is at a rate several orders of magnitude higher than that of the corresponding amides in oligopeptide substrates (Glazer 1967). The high reactivity of ester substrates is most useful for characterization of binding sites in subtilisin. Kinetic values of k_{cat} and K_m of a wide range of reactive esters and amides are available (Philipp and Bender 1983). The hydrolysis reaction catalyzed by serine proteases can be described by the following kinetic scheme:

$$E + S \underset{k_{-1}}{\overset{k_1}{\rightleftharpoons}} E{\cdot}S \xrightarrow{k_2} EP_2 \xrightarrow{k_3} E + P_2$$

$$\searrow P_1$$

where ES = Michaelis complex, and EP_2 = acyl-enzyme intermediate. The Michaelis-Menton parameters are obtained from the steady state reaction according to:

$$K_m = K_s k_3/(k_2 + k_3)$$

$$K_{cat} = k_2 k_3/(k_2 + k_3)$$

For reactive ester substrates, the rate of acylation is greater than deacylation ($k_2 \gg k_3$), $K_m \sim K_s k_3/k_2$, and the rate-limiting step is that controlled by k_3 (deacylation). For amide and peptide substrates, $k_2 \ll k_3$. Therefore, k_2 approximates k_{cat} and $K_m \sim K_s$.

Subtilisin has in common with other serine proteases the characteristic that all are inhibited by diisopropyl phosphorofluoridate. The k_{cat}-pH profile of serine proteinase-catalyzed reactions shows a bell-shaped curve reflecting a maximum activity close to pH 8. The acid and alkaline pH limbs correspond to two kinetically important ionizations with $\sim$pKs of 7 and 9, respectively (Bender et al. 1964). Both acylation and deacylation steps show a sigmoidal k_{cat}-pH profile, corresponding to a protoptropic group with $\sim$pK of 7. This group is identified as His that is involved in k_{cat} (k_3 for esters, k_2 for amide substrate). The ionization group with pK $\sim$9 affects the K_m—a plot of $1/K_m$ versus pH shows a sigmoid relationship with pK $\sim$9. The increase in K_m at high pH values is related to a conformational change. In chymotrypsin this is caused by the interconversions between the inactive and the active form of the enzyme, with the latter consisting of the formation of an Ile16-Asp194 salt bridge (Fersht and Requena 1971). However, in subtilisin-catalyzed reaction, K_m does not change with pH (Polgar and Bender 1967), suggesting no conformational conversion in the alkaline pH range.

The general catalytic mechanism of serine proteinases is represented by two half-reactions (Fig. 5.4).

(1) Acylation: Transfer of the acylating group (specificity side) of the substrate to the reactive Ser of the enzyme.

(2) Deacylation: The acyl group in the acyl-enzyme replaced by H_2O (or some other nucleophile) to become the leaving group.

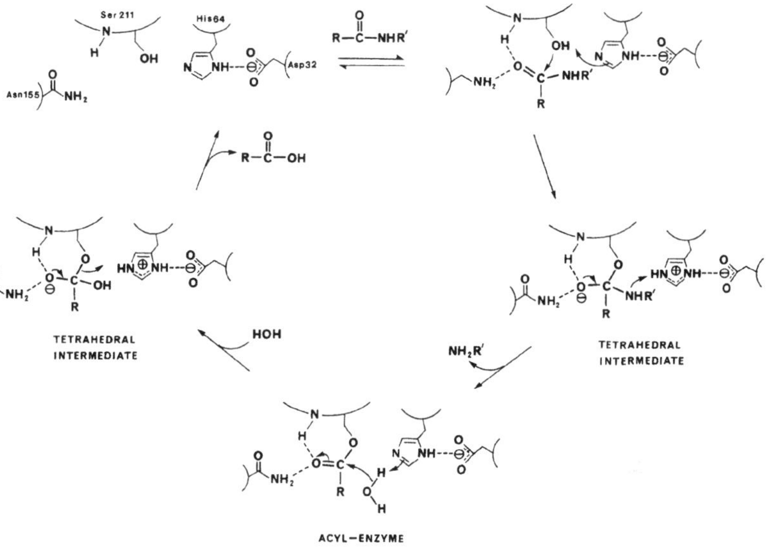

Figure 5.4. Reaction mechanism of subtilisin showing the catalytic pathways of acylation and deacylation via the formation of stable tetrahedral intermediates.

The two half-reactions consist of the following steps.

(a) Formation of Michaelis complex (ES) between the enzyme and substrate that is noncovalent and reversible.

(b) Formation of a covalent tetrahedral intermediate via nucleophilic attack by the reactive Ser221-O_γ on the carbonyl carbon of the scissile bond of the substrate. This step is facilitated by general base catalysis by His64.

(c) A collapse of the intermediate to an acyl-enzyme via the protonation of the leaving group (C-terminal segment) of the substrate by His64.

(d) Nucleophilic attack by a H_2O molecule, assisted by general-base catalysis involving His64. This results in the formation of another tetrahedral intermediate.

(e) Breakdown of the intermediate via His64-catalyzed protonation of the Ser220-O_γ, liberating the acylating side (N-terminal segment) of the substrate as an acid product.

CATALYTIC TRIAD

The catalytic triad in serine proteases synergistically functions to enhance the rate of hydrolysis by a factor of $\sim 2 \times 10^6$ as compared to non-enzymatic reaction as demonstrated by a site-mutagenesis study of subtilisin (Carter and Wells 1988).

Individual replacement of the catalytic Ser221, His64, and Asp32 results in significant decreases in k_{cat} by factors of 2×10^6, 2×10^6, and 3×10^4, respectively, whereas the K_m values show little change.

Protonation and Charge Distribution

The role of Ser221 as a nucleophile in the formation of a covalent bond between the hydroxyl oxygen and the carbonyl carbon of the substrate has been well established. Subtilisin has in common with other serine proteinases the characteristics that all are inhibited by diisopropylfluorophosphate. The charge distribution and protonation state of the essential His as well as that of Asp32 has been a subject of continuous debate since the pioneer work on the charge relay system of chymotrypsin (Matthews et al. 1967; Hunkapillar et al. 1973). The charge-relay mechanism involves a concerted double transfer of protons—from Ser221 to His64-$N_{\varepsilon 2}$ and from His64-$N_{\delta 1}$ to Asp32-$O_{\delta 2}$ with a schematic representation [-COOH ... N-Im-NH ... $^-$O-CH$^{2-1}$] (Fig. 5.5). Some molecular calculations also give support to this scheme of double proton transfer (Scheiner and Lipscomb 1976; Stamato et al. 1984). NMR and neutron diffraction studies establish that the latter transfer does not occur, and that the proton is at the imidazole and not at the aspartate (Jordan and Polgar 1981; Kossiakoff and Spencer 1981). Theoretical calculations of electrostatic potentials at the active site of chymotrypsin and subtilisin subtituting Asp with Asn also suggest that the Asp in the triad must remain ionized (Kollman and Hayes 1981). A transfer of proton from His to Asp would cause an unfavorable increase of ~10 kcal/mol in the free energy of the transition state (Warshel and Russel 1986).

The Role of Asp32

The carboxylate of Asp32 in the triad is believed to stabilize the positive charge developed in the essential His in the transition state. The energy difference in having an Asp-His ion pair as compared to a noncharged residue amounts to ~4 kcal/mol in agreement with the observed stabilization effect in site-directed mutation studies (Warshel et al. 1989). The charge interaction among residues in the triad [Asp$^-$His$^+$Ser$^-$] plays a significant role in lowering the activation energy of the reaction in the transition state (Kossiakoff and Spencer 1981). The estimated stabilization energy is 27 kcal/mol (Daggett et al. 1991).

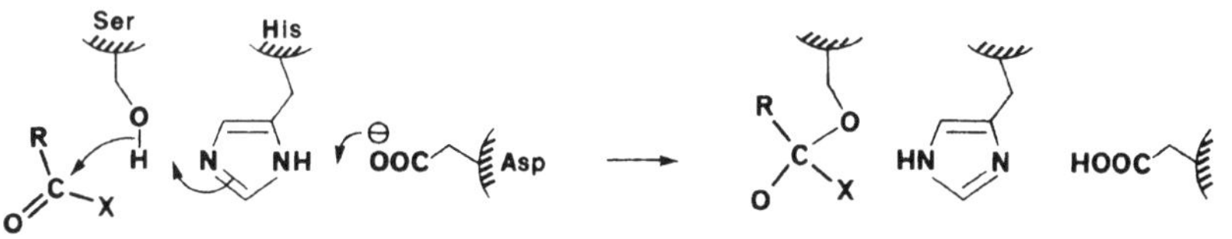

Figure 5.5.   The charge-relay system.

In addition to the above-described functions, Asp32 may assist in orienting the His64 imidazole in the correct stereochemistry for proton transfer to occur during catalysis. More importantly, the Asp carboxylate, in the formation of a charged interaction in the triad [Asp$^-$His$^+$Ser$^-$] together with the development of the negatively charged tetrahedral intermediate, changes the local electrostatic and dielectric environment of the essential His imidazole. The pK of His64 is raised from its normal value of 7.2 in the native enzyme (Jordon et al. 1985; Bycroft and Fersht 1988) to an effective pK of about 9.5 comparable to that of the leaving group nitrogen of the substrate in the tetrahedral intermediate (Robillard and Schulman 1974; Kossiakoff and Spencer 1981).

The pK of His64

The change in the pK and hence the proton affinity of the His64 in the transition to the formation of tetrahedral intermediate has implications in the mechanism of catalysis. If the pK of the His imidazole remains at 7.2 in the transition state, the Ser-O$_\gamma$ proton will be transferred to the leaving group nitrogen, which has a much higher pK of 8–11, before a tetrahedral intermediate can form (i.e., before the bond between the carbonyl carbon of the amide and the attacking Ser-O$_\gamma$ is completed) (Komiyama and Bender 1979).

His64 has a bifunctional role, acting as a catalytic base to abstract a proton from Ser221-OH, and in a following step, as an acid to protonate the leaving group of the substrate. The former is required for the covalent attack of Ser221 oxygen on the carbonyl carbon of the substrate. The protonation step is needed because unprotonated amine is a poor leaving group. For analogous reasons, the tetrahedral intermediate has the carbonyl oxygen in an anion form.

OXYANION HOLE

The functional role of oxyanion hole in the transition state stabilization is supported by numerous studies. In subtilisin, the substrate carbonyl oxygen anion developed in the tetrahedral intermediate is hydrogen-bonded to the Asn155-N$_{\delta 2}$ and the backbone NH of Ser221 (Matthews et al. 1975). Mutation of Asn155 to Ala results in a reduced k_{cat} by a factor of 200–300. The Asn155-oxyanion hydrogen bond contributes a 3.7 kcal/mol to transition stabilization (Bryan et al. 1986). Free energy calculations support this stabilization effect (Rao et al. 1987; Hwang and Warshel 1987). A distal hydrogen bonding is also found between the oxyanion and the γ-hydroxyl group of Thr220 (Braxton and Wells 1991), which provides 1.8–2.0 kcal/mol of stabilization in the oxyanion. The total contribution by hydrogen bonding in the oxyanion hole is estimated to be ~14 kcal/mol (Daggett et al. 1991). The overall rate acceleration of serine protease-catalyzed hydrolysis can be accounted for by the stabilization energy contributed by the oxyanion hole and by the Asp32 carboxylate forming ion pair interaction with His64 (Warshel et al. 1989; Daggett et al. 1991).

MICHAELIS COMPLEX

In the binding complex between subtilisin and substrate, there is no hydrogen bonding between Ser221 and the imidazole of His64 (Robertus et al. 1972B). Hydrogen bonding between the active site His and Asp does not exist in the native enzyme (Jordan and Polgar 1981). The H_2O molecule in the free enzyme blocking the oxyanion hole and locating close to $Ser221-O_\gamma$ and $His64-N_\gamma$ is displaced in the complex. However, hydrogen bonds do not exist between carbonyl oxygen of the substrate and the backbone NH_2 of Ser221 or $Asn155-N_{\delta2}$ that constitute part of the oxyanion hole. In the complex, there is no S_1-P_1 hydrogen bond interaction between the P_1 amido nitrogen and the carboxyl oxygen of Ser125, although the two groups are positioned to form such a bond. However, the carbonyl carbon of the scissile bond in the substrate is positioned for the nucleophilic attack by $Ser221-O_\gamma$ and nitrogen protonation by His64.

TETRAHEDRAL INTERMEDIATE

For the transition from Michaelis complex to tetrahedral intermediate, a number of key steps need to be fulfilled (Robertus et al. 1972A, 1972B). (1) Transfer of $Ser221-O_\gamma$ proton to $His64-N_{\epsilon2}$ increases the nucleophilicity of the hydroxyl oxygen for atttack at the carbonyl carbon of the scissile bond of the substrate. It is likely that the nucleophilic attack by $Ser-O_\gamma$ and protein transfer to $His-N_{\epsilon2}$ occur simultaneously (Kollman and Hayes 1981; Daggett et al. 1991). (2) The covalent bond formation between $Ser221-O_\gamma$ and the carbonyl carbon causes conformational change at the carbon center (from the sp^2 to sp^3 geometry). Consequently, the carbonyl oxygen is shifted into a position, allowing it to hydrogen bond with the backbone NH of Ser221 and $Asp155-NH_{\delta2}$. The formation of this oxyanion hole stabilizes the carbonyl oxyanion in the tetrahedral intermediate. (3) The substrate forms a system of hydrogen bonds of antiparallel β structure with an extended segment in the binding cleft of the enzyme.

It should be noted that direct detection of the *accumulation* of tetrahedral intermediate in the hydrolysis of amides by serine proteinases is not available. Low-temperature stopped-flow experiments as well as ^{13}C NMR fail to give conclusive evidence (Hunkapiller et al. 1979; Markley et al. 1981; O'Connell et al. 1993). Whereas the accumulation of tetrahedral intermediate has not been observed directly, numerous data implicating its occurrence are found in structure-reactivity studies in substrate catalysis, crystallographic studies, and spectroscopic analyses of the binding of a number of inhibitors, including boronic acid derivatives and chloromethyl ketone substrate analogs (Robertus et al. 1972A; Bender and Philipp 1973; Matthews et al. 1975; Poulos et al. 1976; Kossiakoff and Spencer 1981). Regardless of the existence of a stable tetrahedral adduct, the transition state must structurally resemble closely a tetrahedral intermediate, and it is stabilized by hydrogen bonding formed in the oxyanion hole as already discussed. Under the assumption that a stable tetrahedral is formed, there must be two transition states—TS1 and TS2 for bond forming and bond breaking, respectively (Fig. 5.6).

ACYL-ENZYME

Conversion of the tetrahedral intermediate to an acyl-enzyme requires the protonation of the amine nitrogen of the scissile peptide bond of the leaving group. This leads to cleavage of the peptide bond, and the carbonyl carbon resumes its planar conformation. Indoleacryloyl chymotrypsin, an analog of acyl-enzyme, has been investigated at high resolution. The indolyl portion binds deep in the hydrophobic pocket, and the carboxyl is covalently attached to the Ser-OH (Henderson 1970). A similar conclusion was obtained by structural investigation of carbamy-chymotrypsin (Robillard et al. 1972). The linkage in the acyl-enzyme shows absorption spectra and raman resonance frequencies corresponding to an aldehyde rather than an ester (Argade et al. 1984). This conformational distortion may increase the nucleophilic susceptibility of the acyl-enzyme carbonyl and facilitate the breakdown of the acyl-enzyme. In fact, acyl-enzymes are known to hydrolyze at a rate $\sim 10^9$ times faster than serine esters of similar structures (Stein et al. 1983).

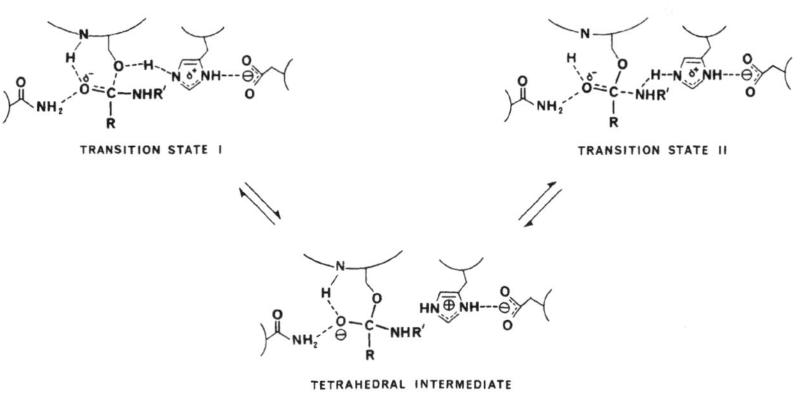

Figure 5.6. The transition states of tetrahedral intermediate showing the dynamics of bond forming and bond breaking in the triad and oxyanion hole in the acylation step.

Resonance raman spectroscopy has been used to study a series of acyl-subtilisins (arylacryloyl acyl groups) and subtilisin BPN′ with mutations at oxyanion hole residues (Tonge and Carey 1990). Analysis of the vibrational spectra of the acyl carbonyl groups in these acyl-enzymes demonstrates an increase in single bond character in the carbonyl bond. The increase in length in the more reactive acyl-enzymes studied reaches 7% of the expected change from C=O to C−O bond. Such distortion in the ground-state structure lowers the free energy for the transformation to the next transition state, and contributes to a decrease in activation energy for the deacylation step. Most importantly, all these changes that ultimately reflect in the observed rate increase can be attributed to the formation of hydrogen bonds of the oxyanion (Tonge and Carey 1992). Recent results in a combined approach of cryoenzymology and electron nuclear double resonance spectroscopy demonstrate that the formation of an acyl-enzyme of chymotrypsin in solution is accompanied by a significant torsional alteration of the substrate conformation during binding (Wells et al. 1994). Such development of an unstable conformation in the substrate in enzyme binding is consistent with the transition-state theory in that the enzyme active site is complementary not to the ground state structure of the substrate, but to an altered form approaching that of the transition state.

SUBSTRATE SPECIFICITY

The characteristics of enzyme subsites and their interactions with substrates are revealed by the investigations of enzyme complexing with various substrates and inhibitors—including chloromethyl ketone derivatives (Morihara et al. 1971; Robertus et al. 1972; Poulos et al. 1976), boronic acid derivatives (Lindquist and Terry 1974; Matthews et al. 1975), fluorogenic substrates (Morihara and Oka 1977; Gron et al. 1992), and proteinaceous inhibitors (Wright 1972; Bode et al. 1987; McPhalen and James 1988; Takeuchi et al. 1991).

The binding cleft of subtilisin enzymes consists of at least six subsites—S_4, S_3, S_2, S_1, $S_1′$, $S_2′$ (Morihara and Oka 1970; Morihara et al. 1971). Binding interactions involving eight subsites have also been described (Hirono et al. 1984; Gron et al. 1992). The acylating group residues have significantly more interactions with the subsites than residues of the leaving-group side (Morihara and Oka 1977). The primary binding subsite S_1 together with subsite S_4 account for about half of all the intermolecular contacts (McPhalen et al. 1985).

In all binding studies of both subtilisin and trypsin family enzymes, the S_1, S_2, and S_3 subsites complex with P_1, P_2, and P_3 of oligopeptide substrates in forming a system of hydrogen bonds of antiparellel β sheet structure. In subtilisin BPN′, three subsites involve the chain segment of residues Ser125-Leu126-Gly127 (Robertus et al. 1972A). In the interaction with protein inhibitors, a more elaborated system can be identified. An additional antiparellel β sheet is formed between P_4, P_5, and P_6 of *Streptomyces* subtilisin inhibitor and another enzyme chain segment of residues Gly102-Gln103-Tyr104 in subtilisin BPN' (Hirono et al. 1984). However, in the complex between subtilisin Carlsberg and eglin-C, and between subtilisin Novo and chymotrypsin inhibitor 2, the P_1-P_4 residues in these inhibitors form a 3-stranded antiparallel β sheet with two enzyme segments Ser125-Gly127 and Gly100-Gly102 as flanking strands (McPhalen et al. 1985; McPhalen and James 1988).

Subtilisin shows a preference for aromatic and apolar groups at P_1, a characteristic also exhibited in chymotrypsin (Morihara et al. 1970). Chymotrypsin prefers large hydrophobic residues, whereas trypsin cleaves peptides with Arg/Lys residue at P_1. The P_1 specificity in subtilisin is less restrictive compared with that of chymotrypsin. Subsite S_1 in subtilisin is a large open crevice lined with a high number of hydophobic side chains. In the case of the *Bacillus amyloliquefaciens* enzyme, these include Ala152-Glu156, Met124-Leu126, and Thr164-Tyr167 (Wells and Estell 1988). The side chain of aromatic P_1 substrates extends towards the α carbon of Gly166 at the bottom of the P_1 binding site. Site-directed mutation of Gly166 to 12 different nonionic residues reveals large changes in the specificity towards substrates of varying size and hydrophobicity (Estell et al. 1986). The catalytic efficiency k_{cat}/K_m increases with increasing hydrophobicity of the substitution, but is also adversely affected by steric repulsion created by the S_1-P_1 residues. Increasing side chain volume of residue 166 causes a decrease in k_{cat}/K_m, which is also proportional to the size of the P_1 substrate side chain. Amino acids with β-branched side chains are poor P_1 substrates for this reason (Morihara and Tsuzuki 1970). The optimum combined side chain volume for a productive S_1-P_1 binding is $\sim$160 A^3 (Estell et al. 1986).

In addition to aromatic or apolar residues, basic amino acid Lys at P_1 also interacts with S_1 in a productive binding mode (Morihara et al. 1974). Crystallographic analysis of subtilisin BPN' shows that the P_1 Lys peptide NH forms a hydrogen bond with Ser125 main chain oxygen (Robertus et al. 1972A) while the ϵ-amino nitrogen is twisted to a position near the enzyme surface to form an ion pair with Glu156-$O_{\epsilon2}$ at the entrance of the binding cleft (Poulos et al. 1976). In the Carlsberg enzyme, Glu156 is substituted by Ser. A Glu156 $\rightarrow$ Ser mutation in the *Bacillus amyloliquefaciens* enzyme causes an increase in the k_{cat}/K_m towards Glu P_1 substrate, and a decrease toward Lys substrate, due primarily to changes in K_m (Wells et al. 1987B). The average free energy for ion-pair interactions at Glu156 and Gly166 has been estimated by site mutation studies to be in the range of $\sim$2 kcal/mol (Wells et al. 1987A).

The P_2 binding site is situated in a hydrophobic pocket comprised of the imidazole ring of His64 and the side chain of Leu96 (McPhalen and James 1988). Subtilisin exhibits a P_2 preference for aliphatic amino acids, and Ala is the optimum substrate (Morihara et al. 1974). The P_2 Ala-C_β interacts with His64 and Leu96-$C_{\delta 1}$ (Robertus et al. 1972A). Residues with bulkier side chains may cause steric interference.

The S_3-P_3 interaction is unique in that the P_3 substrate side chain points towards the solvent (Robertus et al. 1972A), and shows no specific contact with the enzyme (McPhalen and James 1988; Gron et al. 1992)

The P_4 binding site exhibits a very strong preference for aromatic residues with an affinity even exceeding that of the S_1 subsite. The high affinity for aromatic residues at P_4 complements the rather broad specificity at P_1. In great contrast are trypsin and chymotrypsin where the interactions involving P_1 substrate contribute the dominant effect in substrate specificity. The side chain of P_4 fits into the S_4 hydrophobic pocket bound by side chains of Tyr104, Ile107, Leu126, and Leu135, and main chain atoms of Ser101, Gly102, Gly127, and Gly128 (McPhalen and James 1988; Takeuchi et al. 1991). The binding of P_4 substrate induces movement of Tyr104 resulting in an alignment with the aromatic ring of the substrate (Robertus et al. 1972A, 1972B). This also imparts flexibility in the size of the S_4 pocket, allowing it to accept P_4 substrates with bulky side chains (Hirono et al. 1984). However, subtilisin 309 which has a Val at position 104, also exhibits a strong preference for aromatic residues at P_4. Mutational replacements of Val104 to Ala, Arg, Asp, Phe, Ser, Trp, and Tyr show that all the mutants have the same order of preference for aromatic residues (Bech et al. 1992). The catalytic efficiency towards hydrophilic P_4 substrates is not affected by the replacement of Val104 with hydrophobic residues. Apparently the Val104 side chain is mobile so that it may direct towards the P_4 binding only when favorable interactions become possible.

Subsite $S_1{}'$ is a shallow pocket lined with side chains of Tyr217, His64, and Met222 in subtilisin BPN' (Robertus 1972B; Morihara and Oka 1977). The pocket accepts small amino acids quite well—Ala and Gly show high k_{cat}/K_m. Peptides with Asp at $P_1{}'$ are hydrolyzed at low rates. Amino acids with β-branched side chains cause steric hindrance. Both Val and Ile cause a decrease in k_{cat} (Bromme et al. 1986). Introduction of Pro at $P_1{}'$ results in inhibition of hydrolysis, due to the fact that Pro cannot be protonated in the acylation step (Morihara and Oka 1977; Bromme et al. 1986).

The $P_2{}'$ binding site is bound by extended chains of Asn218, Gly219, and Phe189 in subtilisin BPN' (Robertus et al. 1972B). Substrates with Leu at $P_2{}'$ exhibit a high k_{cat}/K_m due mainly to the high binding affinity (Morihara and Oka 1977). The $P_2{}'$ amido nitrogen and the carbonyl oxygen form hydrogen bonds with the backbone carbonyl oxygen and the amido group of Asn218, respectively (Robertus et al. 1972B). The enzyme $S_2{}'$ shows a preference for aromatic and alipathic amino acids. The positively charged Arg is hydrolyzed well, whereas Pro causes a significant decrease in k_{cat}/K_m (Gron et al. 1992).

PAPAIN

Papain (EC 3.4.22.2) is a proteolytic enzyme found in the latex of the unripe papaya fruit of the *Carica papaya* tree. It belongs to a family of cysteine (sulfhydryl) proteinases that also includes other plant enzymes—ficin (fig), actinidin (kiwifruit), and bromelain (pineapple), as well as mammalian enzymes—calpains (cytoplasmic Ca^{++} dependent) and cathepsins (lysosomal). The enzyme is exceptionally stable to high temperatures at neutral pH, but denatured at acidic pH < 4. The pH optimum of papain is 5.5–7.0 and its pI is 9.6 (Glazer and Smith 1971). It is also resistant to denaturation in concentrated (8 M) urea solution at neutral pH range. Papain retains its activity in various organic media. In fact, the enzyme is extensively investigated for organic synthesis of peptides and other chemical compounds. The high stability of papain toward various extreme conditions and its availability at affordable cost has made it widely used as a meat tenderizer and in chill haze prevention in the brewery industry. Physiologically, lysosomal cysteine proteinases play a role in intracellular protein degradation. Imbalance of these enzymes is believed to be a possible cause of diseases such as muscular dystrophy, osteoporosis, inflammatory heat diseases, and tumor invasion (Poole et al. 1978; Delaisse et al. 1984; Falanga and Gordon 1985). The development of therapeutic inhibitors to control proteolysis by these enzymes has drawn considerable attention.

AMINO ACID SEQUENCES

Papain is a single polypeptide that contains 212 amino acids with a calculated MW of 23,350 (Mitchel et al. 1970). The sequence consists of high Gly_{28}, Tyr_{19}, and Val_{18} (Fig. 5.7). There are 3 disulfide bonds: Cys22-Cys63, Cys56-Cys95, and Cys153-Cys200, but only one free thiol group, which is the catalytic Cys25. Actinidin from kiwi consists of 220 residues with a MW of 23,500 (Carne and Moore 1978) and shares 48% similarity with papain. The three-dimensional structures of the two enzymes are virtually identical (Baker 1980; Kamphuis et al. 1985).

THE THREE-DIMENSIONAL STRUCTURE

The enzyme molecule of papain is folded in an ellipsoid with a dimension of $50 \times 37 \times 37$ A (Fig. 5.8) (Drenth et al. 1971). A number of electrostatic interactions involving Arg and Asp/Glu are observed on the surface of the molecule: Arg8-Asp6, Arg41-Asn212, Arg96-Glu89, and Arg188-Glu183. The polypeptide chain is folded into two domains. The L domain consists of residues 10–100, forming three α helices, A (24–42), B (50–57), and C (67–78) (Kamphuis

```
Papain              IPEYVDWRQKGAVTPVKNQGSCGSCWAFSAVVTIEGIIKI
Actinidin           L-S-----SA---VDI-S-GE--GC-A-S-IA-V-GINK-
Bromelain           V-QSI---DY---TSV-N-NP--AC-G-A-IA-V-SVAS-

Papain         41   RTGNLNQYSEQELLDCDR  RSYGCDGGYPWSALQLVAQY
Actinidin      41   TS-S-ISL-E-ELI--GRTQNTR------ITDGF-FIIND
Bromelain      41   YK-I-QPL-Q-QVD--AK

Papain         79   GIHYRNTPYYEGVQRYCRSREKGPYAAKTDGVRQVQPYNQ
Actinidin      81   -GINTDEN-PYTA-DGDCDVALQDQKYV-IDTYDNV-Y-N
Bromelain      59                                     KARV-R-N

Papain        119   GALLYSIAN QPVSVVLQAAGKDFQLYRGGIFVGPCGNKV
Actinidin     121   EWALQTAVTY--VS-ALDA-GDAFKQ-AS-IFT-P-GTAV
Bromelain      66    ESSMYAVSK--IT-AVVS- ANFZL-KS-VGD-Y-KDKL

Papain        158   DHAVAAVGYNPGYILIKNSWGTG     WGENGYIRIKRGT
Actinidin     161   DH-IVIV--GTEGGVDYWIVKNSWDTT---E--M-IL-NV
Bromelain     104   NH-VTAI--NKAEFGDGSGKKAR     ---A--I-MA-DV

Papain        194   GNSYGVCGLYTSSFYPVKN    212
Actinidin     201    GGA-T--IATMPS--VKYNN    220
Bromelain     140    SSSS-I--IAIDPL--TEG    158
```

Figure 5.7. Amino acid sequences of papain (Mitchel et al. 1970), actinidin (Carne and Moore 1978), and stem bromelain (Goto et al. 1976, 1980).

et al. 1984). Helix A lies antiparallel to B, while C is perpendicular to B and at a 60° angle to A. The folding of the α helices forms a central core of hydrophobic residues. The L domain also contains a short C-terminal segment (208–212) that folds over from the R domain (Baker and Drenth 1987). The R domain (113–207) consists primarily of a twisted antiparallel β sheet forming a barrel with a large hydrophobic core. The opposite ends of the barrel are closed by short helices (117–127 and 138–143). In addition, the N-terminal end crosses over from the L domain. The N- and C-termini therefore are attached, respectively, to the R and L domain, and their cross folding assists in connecting the two domains together. Some of the Arg-carboxylate interactions and a number of hydrogen bonds are also involved in this function.

The active site is located in a groove between the two domains. At the bottom are the two catalytic residues, Cys25 and His159, contributed by the L and R domains, respectively. The Cys-S_γ lies adjacent to the $N_{\delta1}$ of the His imidazole. The other His nitrogen, $N_{\epsilon2}$, is hydrogen bonded to the side chain oxygen of Asn175, which lies above the His imidazole ring. During the covalent addition of acyl-enzyme, the imidazole ring rotates about 30°, moving the His159-$N_{\delta1}$ to

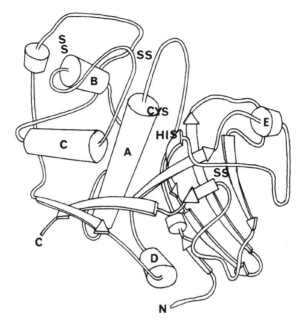

Figure 5.8. Schematic diagram showing the polypeptide chain folding in papain and actinidin. Domain L contains helices A, B, and C, whereas the other domain is based on a double β sheet structure with helices D and E at opposite ends. The catalytic Cys and His are shown. (From Baker and Drenth 1987. In: *Biological Macromolecules and Assemblies,* vol. 3, p. 324, with permission. Copyright 1987 Wiley-Liss, Inc., a subsidiary of John Wiley & Sons, Inc.)

a position coplanar with the leaving group nitrogen (Kamphuis et al. 1984). The His is thus ideally positioned to protonate the leaving group. For bromelain, the function of Asn175 in papain and in actinidin is replaced by Ser-O_γ, which can also orient the His imidazole by forming hydrogen bonds (Kamphius et al. 1985). The Asp158 residue in papain and actinidin found at a distance of 6.7 A from the His imidazole has been implicated in forming charge interactions as observed in the catalytic triad in serine proteinases. However, the Asp residue is not conserved in a number of cysteine proteinases, making it unlikely to have such a functional role (Kamphius et al. 1985).

IONIZATION STATE OF THE ACTIVE SITE

For papain-catalyzed reaction, the pH dependence of k_{cat}/K_m gives a bell-shaped curve with optimum activity at pH 6.0. The acid and alkaline limbs of the curve indicate the involvement of ionization groups with pKs of 4.3 and 8.5 (Glazer and Smith 1971). These two groups are assigned to Cys25 and His159, respectively. Deacylation of acyl-papain shows a sigmoidal pH rate profile indicating the dependence on a basic group with pK of 4.3. This ionization group is also attributed to His159. The pK values of Cys-SH and His-Im in proteins are normally in the range of ~8 and 6, respectively. The anomalous low pK of Cys25 and the shifting of the pK of His159 in papain are attributed to the formation of an ion pair between thiolate anion and imidazolium cation (Polgar 1973, 1974).

Coupled ionization in the active site of papain has been demonstrated by the solvent isotope effect (Creighton and Schamp 1980; Greighton et al. 1980; Frankfater and Kuppy 1981), proton titration (Sluyterman and Wijdenes 1976), potentiometric determination (Lewis et al. 1976; Migliorini and Creighton 1986; Roberts et al. 1986), NMR (Lewis et al. 1981), biomimetic model studies (Skorey and Brown 1985), and fluorescence spectroscopy (Johnson et al. 1981). All these investigations support the theory that the catalytic Cys and His residues exist predominantly as ion pairs at physiological pH range, and the unusual pK values for the Cys and His are caused by the effect of interactive ionization of these two groups.

The presence of the negative charge on the Cys25 thiolate anion causes an increase in the pK of His159 by 4.2 units. An analogous effect on the positive charge of His159 imidazolium results in a 4.2-unit decrease in the pK of Cys25 (Fig. 5.9) (Lewis et al. 1976, 1981). In the ion-pair form, the thiolate anion acts as an efficient nucleophile to attack the carbonyl carbon of the substrate, and the

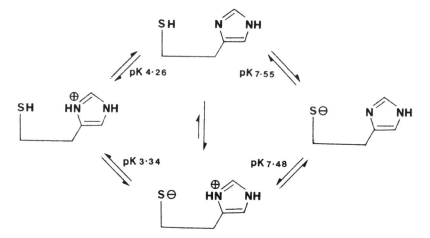

Figure 5.9. Interactive ionization between thiolate anion and imidazolium cation in the active site of papain.

His imidazolium can participate in protonating the leaving group. In the acylation step, the Cys25 thiolate anion is neutralized in the formation of a covalent bond with the substrate carbonyl carbon. The removal of the negative charge in Cys25 causes the pK of His159 to decrease from 8.5 to 4.3. The effect of this decrease in the pK of His159 facilitates proton transfer from the imidazole to the leaving group and subsequent bond breaking in deacylation. In this step, the His159 becomes an effective acid catalyst.

Some studies suggest that two acid ionization groups are involved. This additional group in the acid limb of the rate-pH profile is attributed to Asp158 (Schack and Kaarsholm 1984; Lewis et al. 1978). Replacement of Asp158 with Asn by site-directed mutagenesis shows no significant effect on k_{cat}/K_m or k_{cat}. The activity-pH profile, however, is shifted to a lower pH. The results of this study suggest that the negative charge of the Asp residue may have some effect on the pKs associated with ion-pair formation (Menard et al. 1990). In another investigation, removing the hydrogen bonding network involving the Asp158 by mutational substitution with Gly or Ala causes significant drop in k_{cat}/K_m values (Menard et al. 1991B). The hydrogen bonds involving the Asp158 side chain may function in the stabilization of the ion-pair formaion.

MECHANISM OF CATALYSIS

Cysteine proteinases share a number of similar features in reaction mechanism with serine proteinases. The hydrolysis reaction follows a similar kinetic scheme, in that the rate-limiting step is determined by k_2 and k_3, respectively, for amides and esters (Glazer and Smith 1971; Lowe and Yuthavong 1971). The rates for hydrolysis of esters and related amides, however, are comparable, in contrast to the large differences observed in subtilisin, trypsin, and chymotrypsin. The general mechanism of both proteinases involves acylation and deacylation with similar basic chemistry. Papain-catalyzed reaction consists of the following steps (Fig. 5.10) (Baker and Drenth 1987).

(1) The enzyme binds reversibly with the substrate to form a Michaelis complex.

(2) Nucleophilic attack by the Cys25 thiolate anion on the carbonyl carbon of the scissile bond of the substrate leads to formation of a tetrahedral intermediate.

(3) A collapse of the tetrahedral intermediate to an acyl-enzyme is facilitated by protonation of the leaving group (amide nitrogen in the case of a peptide substrate) by the His159 imidazolium.

(4) Nucleophilic attack by a H_2O molecule assisted by His159 acting as base catalyst results in deacylation of the acyl-enzyme (which is a thioester in this case).

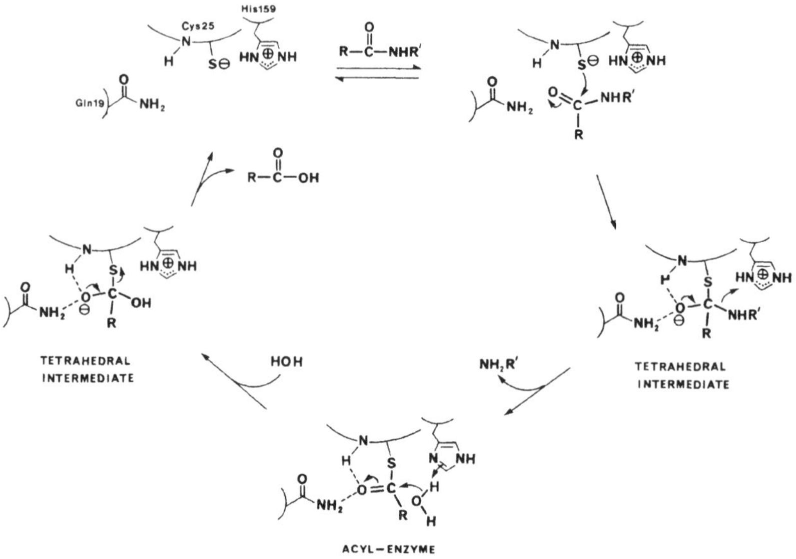

Figure 5.10. Reaction mechanism of papain showing acylation and deacylation via the formation of stable tetrahedral intermediates.

TETRAHEDRAL INTERMEDIATE

In the acylation step involving the conversion of the Michaelis complex to an acyl-enzyme as well as in the deacylation step leading to product formation, it is generally assumed that a covalent tetrahedral intermediate is formed, which resembles the transition state structure. Evidence for the presence of an intermediate with a tetrahedral configuration in papain-catalyzed hydrolysis reaction comes from inhibition studies on the hemithioacetal formation between the active site thiol and aldehyde analogs of peptide substrates (Westerik and Wolfenden 1972). Crystallographic analyses and NMR studies of similar inhibition complexes support tetrahedral structures (Bendall et al. 1977; Gamcsik et al. 1983). Investigation of the binding mode between chloromethyl ketone transition state analogs and papain also suggest tetrahedral configurations stabilized by hydrogen bonding (Drenth et al. 1976). Other indirect support is also provided by nitrogen isotope effects on hydrolysis of *N*-benzoyl-L-argininamide (O'Leary et al. 1974) and solvent deuterium isotope studies on binding *N*-acetyl-L-phenylalanylglycinal to papain (Frankfater and Kuppy 1981). However, direct detection for the *accumulation* of a tetrahedral intermediate in papain-catalyzed hydrolysis mechanism is not available (Mackenzie et al. 1985).

OXYANION HOLE

The negative charge developed at the carbonyl oxygen of the substrate P_1 residue in the tetrahedral intermediate is stabilized by hydrogen bonding with Gln19-NH_2 and the main chain NH of Cys25 (Drenth et al. 1976; Schroder et al. 1993). This oxyanion interaction resembles that suggested for serine proteinases. Replacement of Gln19 with Ala or Ser by site-directed mutagenesis causes a decrease in k_{cat}/K_m for the hydrolysis of peptide aldehyde analogs by a factor of 60 and 610, respectively (Menard et al. 1991A). These values correspond to a contribution of 2.4 and 3.8 kcal/mol for transition state stabilization by the Gln19 side chain in the oxyanion hole in papain.

A number of results tend to argue against the importance of an oxyanion in the reaction catalyzed by papain. Specific thiono esters (containing a sulfur atom in place of a carbonyl oxygen) that are not substrates for serine proteinases, are shown to be hydrolyzed by papain at rates comparable to those of the corresponding esters (Asboth et al. 1985). Furthermore, the thiohemiacetal OH group does not orient towards the oxyanion hole where hydrogen bond stabilization can occur (Mackenize et al. 1986). Solvent deuterium isotope studies and the proton inventory technique conclude that the hydrolyses of esters and thiono esters follow the same hydrolytic pathway (Storer and Carey 1985).

ACYL-ENZYME

Trans-cinnamoyl-papain, an acyl-enzyme analog, has been prepared and its properties characterized (Brubacher and Bender 1966). The formation of acyl-enzyme is also supported by kinetic studies on papain-catalyzed reaction of thiono esters (Lowe and Williams 1965). ^{13}C-NMR studies on the inhibition of papain by peptide nitriles indicate formation of a covalent thioimidate ester (Moon et al. 1986). Direct structural evidence for the acyl-enzyme formation is achieved by the application of ^{13}C-NMR in detecting a productive thioacyl intermediate prepared from papain and N-benzoylimidazole under cryoenzymological conditions (Malthouse et al. 1982).

The deacylation step is first order in acyl-enzyme with the rate dependent on a basic group of pK = 4.3, corresponding to His159. The His159 imidazole acts in a general base catalysis in the conversion of the acyl-enzyme to a tetrahedral intermediate with subsequent hydrolysis of the thioester bond. The solvent deuterium isotope effect causes a rate reduction of 2–3 fold, and a linear correlation of k_{cat} for ester hydrolysis to ^{2}H composition (Brubacher and Bender 1966; Szawelski and Wharton 1981).

SUBSTRATE SPECIFICITY

Papain has a fairly broad substrate specificity. The enzyme consists of seven subsites extending over 25 A in a groove at the interface of the L and R domains— $S_4S_3S_2S_1S_1'S_2'S_3'$—that interact with substrate amino acid residues— NH_2-$P_4P_3P_2P_1P_1'P_2'P_3'$-COOH (Schechter and Berger 1967). Substrate specificity of the enzyme is largely determined in the acylating side. The S_1 subsite is relatively nonselective. The binding pocket of S_1 is unable to accommodate P_1 substrates with β-branched side chains such as Val. The carbonyl group of P_1 lies close to the backbone chain of Gly23-Ser24, while the carbonyl oxygen is oriented for hydrogen bonding with Gln19-$N_{\epsilon 2}$ and backbone NH of Cys25 (Drenth et al. 1976). The S_2-P_2 interaction plays a dominant role in substrate binding. The S_2 subsite has a hydrophobic surface composed mainly of side chains of Tyr67, Pro68, Trp69, Phe207, Ala160, Val133, and Val157 (Lowe and Yuthavong 1971). This is consistent with the fact that the S_2 subsite has a strong preference for hydrophobic amino acids Phe, Tyr, Val, and Leu (Schechter and Berger 1968; Lynn 1983). Side chains of Phe or Tyr fit particularly well in the S_2 product. The P_2 peptide NH and carbonyl oxygen form hydrogen bonds with the backbone oxygen and NH of Gly66, respectively (Drenth et al. 1976). The S_2-P_2 hydrogen bonds may also be associated with proper alignment of the substrate. Replacement of the P_2 carbonyl oxygen by a sulfur atom lowers the reaction by 2–3 orders of magnitude compared to the corresponding oxo substrate (Asboth et al. 1988). Electrostatic interactions seem to play an insignificant part in S_2-P_2 binding (Williams et al. 1972). The S_2 subsite shows high stereospecificity for L-amino acid residues (Schechter and Berger 1968; Lowe and Yuthavong 1971). The ratio of k_{cat}/K_m values determined for the *N*-acetyl-L-phenylalanylglycine4-nitroanilide and its D-enantiomer is 330, indicating a high selectivity towards the L-Phe at P_2 (Kowlessur et al. 1990). The C_β of the L-amino acid is oriented near Ala160, whereas that of the D-enantiomer points more towards the solvent (Drenth et al. 1976; Kowlessur et al. 1990). The S_3-P_3 interaction involves hydrophobic contact between the P_3 side chain and Tyr61 ring and C_α of Gly65 (Drenth et al. 1976). The S_1' specificity, as demonstrated in serine proteinases, is expressed in the deacylation step. The S_1' subsite has a preference for hydrophobic residues, Leu and Trp, and the S_1'-P_1' interaction makes considerable contribution to binding and to the rate of hydrolysis (Alecio et al. 1974; Williams et al. 1972). Substrates having side chains longer than three methylene groups do not fit well in the S_1'subsite (Schuster et al. 1992). The hydrophobicity of the side chains of P_1' substrate shows no correlation with K_m or k_{cat}/K_m (Garcia-Echeverria and Rich 1992A). For the S_2' subsite, the hydrophobicity of P_2' versus k_{cat}/K_m and k_{cat} correlation follows a parabolic curve pattern, and the extent of binding interaction at the leaving group may affect product release in acylation (Garcia-Echeverria and Rich 1992B).

Crystallographic studies of papain complexed with E64 (1[*N*-[*N*-(L-3-trans-carboxyoxirane-2-carbonyl)-L-leucyl]-amino]-4-guanidino butane), a potent irreversible inhibition of cysteine proteases, yield further insight into the various interactions in the binding sites. The reactive Cy25-S$_\gamma$ is covalently linked to the C2 atom of the inhibitor. An extensive network of hydrogen bonding is involved between the inhibitor and the S subsites, including Gln19, Gly66, His159, Cys25, Asp158, Tyr61, and Tyr67 (Varughese et al. 1989; Yamamoto et al. 1990). Another group of well-studied cysteine proteinase inhibitors belongs to a superfamily of protein inhibitors that includes three families—cystatins, stefins, and kininogens (Muller-Esterl et al. 1985). Unlike serine proteinase protein inhibitors, these protein inhibitors form reversible complexes with simple bimolecular kinetics. The crystal structures of chicken egg white cystatin, and (recombinant) human stefin B are known (Bode et al. 1988; Stubbs 1990). Both molecules consist of a 5-stranded β sheet with a 5-turn α helix. The conserved residues form a wedge edge, consisting of the N-terminal segment, a primary hairpin loop that contains a QLVSG variation of the conserved consensus sequence QVVAG in all members of the superfamily, and a second hairpin loop with the highly conserved Trp104. The inhibitor binds to papain by having this wedge edge slot into the enzyme active site, with $K_i \sim 6$ pM (Machleidt et al. 1989). The interactions are predominantly hydrophobic involving major contacts with Gly23, Gln19, Trp177, and Ala136 around the reactive Cys25. The first hairpin loop (QLVSG segment) makes extensive surface contact with the S$_1$' subsite of the enzyme. The N-terminal Leu7-Leu8-Gly9-Ala10 segment adopts a 1–4 turn conformation so that the side chain of Leu8 interacts with Gly66 and with the S$_2$ subsite (Bode et al. 1988; Machleidt et al. 1989). The Gly9-Ala10 peptide lies in the vicinity of Cys25 but is not correctly positioned for cleavage. The binding mode of these protein inhibitors is in sharp contrast to the "standard mechanism" exhibited by protein inhibitors of serine proteinases (Laskowski and Kato 1980). The majority of the serine proteinase protein inhibitors consist of a surface binding loop that fits into the active-site cleft to form a typical enzyme-substrate stable complex. These inhibitors resemble true substrates. The peptide bond in the loop that orients close to the active site Ser195 is subject to potential albeit extemely slow hydrolysis. The inhibitor in this case becomes irreversibly bound. In most systems, an equilibrium exists between unreacted inhibitor and its modified form (with peptide bond hydrolyzed).

CHYMOSIN

Chymosin (EC 3.4.23.4), also commonly known as "rennin," is the major proteinase used in cheese making. The enzyme is found in the fourth stomach of a calf, although other animal rennets can be used as substitutes. The chymosin gene has been cloned and expressed in *Escherichia coli* (Nishimori et al. 1982A; Emtage et al. 1983), *Saccharomyces cerevisiae* (Goff et al. 1984), *Aspergillus nidulans*, and *Aspergillus awamori* (Hayenga et al. 1988). Recombinant chymosin produced in genetically modified *E. coli* (Flamm 1991), as well as enzyme preprations from (fungus) *Aspergillus awamori* and (yeast) *Kluyveromyces lactis* are commercially available. Proteinases from microbial sources are also used in replacing calf chymosin in cheese making. These proteinases, which are commercially available, are obtained from the fungus *Mucor pusillus* or *Mucor miehei* (also called *Rhizomucor miehei*). The *Mucor pusillus* enzyme has an optimal pH of 3.5, and like calf chymosin, possesses low proteolytic activity, but high milk-clotting activity (Iwasaki et al. 1967). It is more heat resistant with an optimum temperature of 56°C and more stable in pH range 3–8.

AMINO ACID SEQUENCES

The primary structure of calf chymosin is known from both direct amino acid sequence and cDNA nucleotide sequence. The enzyme consists of a single polypeptide of 323 amino acids with 3 disulfides, and MW of 35 kD (Fig. 5.11). Chymosin A and B differ by a single amino acid with Asp and Gly at position 244, respectively (Foltmann et al. 1977, 1979). The former has a pH optimum of 4.2 and the latter 3.7. The amino acid sequence deduced from the nucleotide sequence of cDNA shows slight discrepancies with Asn for Asp at positions 160 and 172 (Harris et al. 1982; Nishimori et al. 1982B).

The enzyme, like other acid proteinases (pepsin, cathepsin D, renin) is secreted in the form of zymogens. This prochymosin is activated at pH 4.7 by an autocatalytic process, resulting in removal of the N-terminal segment of 42 amino acid residues (Pedersen and Foltmann 1975; McCaman and Cummings 1986). At pH below 2.5, the prochymosin is cleaved at residues 27–28 to form pseudochymosin via an intermolecular reaction between two prochymosin molecules (Pedersen et al. 1979).

The *Mucor* proteinases are synthesized as preproenzymes. The *Mucor pusillus* enzyme contains 361 amino acids with an extra 66 amino acids at the N-terminal sequence (Fig. 5.11) (Tonouchi et al. 1986; Beppu 1988). The first 22 N-terminal amino acids of the preproenzyme function as a signal peptide for secretion (Yamashita et al. 1987; Hiramatsu et al. 1989). The proenzyme (inactive zymogen) is then converted to the mature enzyme under acidic pH probably by an autocatalytic process. The mature enzyme from *Mucor pusillus* consists of two glycosylation sites at Asn79 and 188, both located in the Asn-X-Ser sequence (Aikawa et al. 1990). The *Mucor miehei* enzyme contains 361 amino acid residues with a calculated MW of 38,701. The enzyme is synthesized in its precursor form containing an extra 69 amino acid residues in the N-terminal, the first 22 amino

```
BCH                     AEITRIPLYKGKSLRKAL
RMP     MLFSQITSAILLTAASLSLTTARPVSKQSESKD-L-ALP-
MPP     MLFSKISSAILLTAASFALTSRRPVSKQSDADD-L-ALP-
PPA            MKWLLLLSLVVLSECLVKVPLVRK-S-RQN-


BCH     KEHGLLEDFLQKQQYGISSKYSGF       GEVASVP
RMP     TSVSRKFSQTKFG-QQLAE-LA-LK PFSEAAADGSVDT-
MPP     TSVNRKYSQTKHG-QA AE-LG-IKAF  AEGDGSVDT-
PPA     IKDGKLKDFLKTHKHNPAS-YFPEAAAL   IGDE-


BCH   8  LTNYLD SQYFGKIYLGTPPQEFTVLFDTGSSDFWVPSIY
RMP  11  GYYDF-LEE-AIPVSI---G-D-LL--D-----T---HKG
MPP  11  GLYDF-LEE-AIPVSI---G-D-YL--D-----T---HKG
PPA   6  LENYL- TE-FGTIGI---A-D-TVI-D----NL---SVY


BCH  47  C KSNACKNHQRFDPRKSSTFQNLGKPLSIHYGTGS MQG
RMP  51  -TK-EG-VGSRF---SA--A-KATNYN-N-T----GANGL
MPP  51  -DN-EG-VGKRF---SS----KETDYN-N-T----GANGI
PPA  45  - S-LA-SDHNQ-N-DD----EATSQE-S-T----S MTG


BCH  85  ILGYDTVTV   SNIVDIQQTVGLSTQEPGDVFTYAEFD
RMP  91  YFEDSIAIGDITVTKQILAYVDNVRGPTAEQSPNADI-L-
MPP  91  YFRTSITVGGATVKQQTLAYVDNVSGPTAEQSPDSEL-L-
PPA  83  ILGYDTVQV   GGISDTNQIFGLSETEPGSFLYYAPF-


BCH 121  GILGMAYPSLASEYSIPV   FDNMMN   RHLVAQDLFS
RMP 131  -LF-A---DNTAMEAEYGSTYNTVHV-LYKQG-ISSPL--
MPP 131  --F-A---DNTAMEAEYGDTYNTVHV-LYKQG-ISSPV--
PPA 119  --L-L---SISASGATPV   FDNLWD   QG-VSQDL--


BCH 155  VYMDR DGQESMLTLGAIDPSYYTGSLHWVPV    TVQQ
RMP 171  ---NT NSGTGEVVF-GVNNTLLG-DIAYTD-MSRYGGYY
MPP 171  ---NT NDGGGQVVF-GVNNTLLG-DIQYTD-LKSRGGYF
PPA 153  --LSSNDDSGSVVLL-GIDSSYYT-SLNWVP-    SVEG


BCH 190  YWQFTVDSVTISGVVV ACEGGCQAILDTGTSKLVGPSSD
RMP 210  F-DAP-TGI-VD-SAAVRFSRPQAFTID---NFFIM---A
MPP 210  F-DAP-TGVKID-SDAVSFDGAQAFTID---NFFIA---F
PPA 189  Y-QITLDSI-MD-ETI ACSGGCQAIVD---SLLTG-T-A
```

Figure 5.11. Amino acid sequences of bovine chymosin (mature protein with 323 amino acids, prosequence with 42 amino acids, prochymosin with 365 amino acids, Foltmann et al. 1977, 1979), *Rhizomucor miehei* protease (430 amino acids deduced from cDNA, mature protein with 361 residues, signal peptide with 22 amino acids, presegment with 47 amino acids, Boel et al. 1986), *Mucor pusillus* protease (445 amino acids deduced from cDNA, mature protein with 361 residues, signal peptide = 18 amino acids, presegment = 66 amino acids, Tonouchi et al. 1986), porcine pepsin (385 amino acids deduced from cDNA, mature pepsin with 326 residues, signal peptide = 14 amino acids, prosequence = 44 amino acids, Lin et al. 1989).

```
BCH   229   IL NIQQAIGATQNA YGEFDIDCDNLSYMPTVVFEINGK
RMP   250   ASKIVKA--PDATETQQ- WVVPCASYQNSKSTISIVMQ-
MPP   250   AEKVVKA--PDATESQQ- YTVPCSKYQDSKTTFSLDLQ-
PPA   228   IA NIQSDIGASENS Y-EMVISCSSIDSLPDIVFTINGV


BCH   267   MYPLTPSAYTSQ            DQGFCTSGFQS     EN
RMP   289   SGSSSDTIEI-VPVSKMLLPVDQSNETCMFIILP     D
MPP   289   SGSSSDTIDV-VPISKMLLPVDKSGETCMFIVLP     D
PPA   266   QYPLSPSAYILQ           DDDSCTSGFEGMDVPTS


BCH   292   HSQKWILGDVFIREYYSVFDRANNLVGLAKAI 323
RMP   324   GGNQY-V-NL-L-FFVN-Y-FG--RI-F-PLASAYENE 361
MPP   324   GGNQF-V-NL-L-FFVN-Y-FGK-RI-F-PLASGYENN 361
PPA   295   SGELW-L-DV-I-QYYT-F-RA--KV-L-PVA 326
```

Figure 5.11.   (Continued).

acids composing the signal peptide, and the following 47 contributing to the pro-segment (Boel et al. 1986).

THREE DIMENSIONAL STRUCTURES

The three-dimensional structure has been characterized for recombinant bovine chymosin (Gilliland et al. 1988, 1990), chymosin mutant (Strop et al. 1990; Watson et al. 1988), and the wild type bovine chymosin (Newman et al. 1991).

Chymosin, like other aspartic proteinases, consists of two domains separated by a deep substrate binding cleft, with a pseudo 2-fold symmetry (Fig. 5.12A, B). The N-terminal (1–175) and C-terminal (176–323) domains are topographically equivalent. The enzyme molecule contains a repeating motif, each consisting of four antiparallel β strands (A, B, C, D) and a helix. Two motifs pair to form two interlocked four-stranded sheets that are packed orthogonally in each domain. The N- and C-terminal segments also contribute to a β sheet to each domain, giving rise to a total of six antiparallel β sheets (Newman et al. 1991). The sequence identity between the two domains is ~9%, but the sequence around the active site residues is highly (~80%) conserved (Newman et al. 1991).

Catalytic Site

Located at the base of the cleft separating the N- and C-terminal domains are the two essential Asp34 and Asp216 with their side chains oriented towards the cleft. The distance between the carboxyl oxygens of the two Asp is 3.1 A. The two Asp residues are located at the ends of the B_1 and B_3 loops formed by two homogenous segments, N-terminal 33–37, and C-terminal 215–219. The carboxyl groups of the Asp residues are connected via a complex network of hydro-

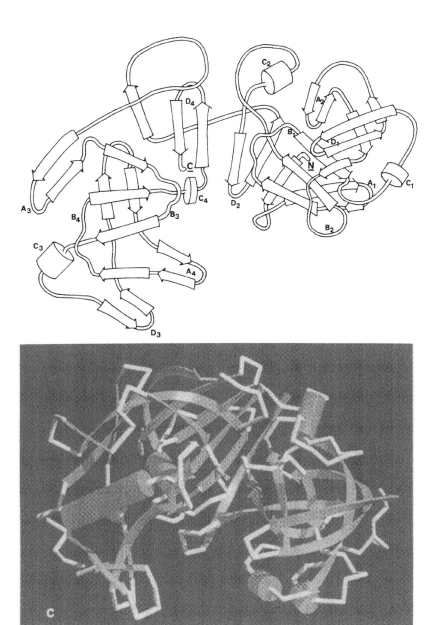

Figure 5.12. (A) The general structure of acid proteases showing the compositions of first β hairpin A, wide loop B, helix C, and second β hairpin D. (From Andreeva and Gustchina 1979. *Biochem. Biophys. Res. Comm. 87,* 36, with permission. Copyright 1979 Academic Press, Inc.) (B) Recombinant bovine chymosin (From Gilliland et al. 1990. *Proteins: Structure, Function, and Genetics 8,* 87, with permission. Copyright 1990 Wiley-Liss, Inc., a subsidiary of John Wiley & Sons, Inc.)

gen bonds and a number of water molecules in the immediate environment of the active site cleft (Mantafounis and Pitts 1990; Gilliland et al. 1990). The two adjacent Thr35 and 217 are hydrogen bonded to each other via the side chain oxygen and main chain nitrogen, while their side chains are directed in a hydrophobic pocket. This interaction, a common feature in aspartic proteinase, is known as the "fireman's grip" (the side chain oxygen of one Thr bonded to the main chain nitrogen of the corresponding Thr as shown in Fig. 5.13). This provides additional structural stability to the catalytic site.

A β loop (residues 71–82) forms a "flap" projecting over the cleft, such that the substrate is enclosed between the active site and the flap. It is evident that some movement and possibly conformational change of the flap region is required before the binding of substrate. Studies on the *Rhizopus chinensis* pepsin suggest that the flap movement probably plays a role in forming an induced fit of the hydrophobic cleft for a substrate chain (Bott et al. 1982). The formation of a flap closing over the substrate bound to the active site cleft also serves to ensure seclusion of solvent from the S_1-S_1' region (Blundell et al. 1990). The immediate environment of this flap region is more hydrophobic in chymosin than in other aspartic proteinases, due to the Tyr77 residue, forming part of the S_1-S_1' subsite and its interactions with Phe119 and Leu32.

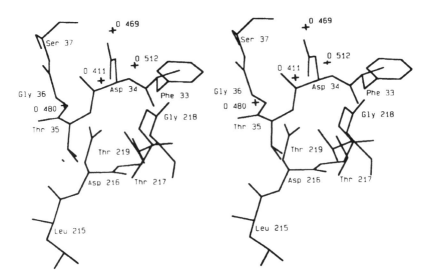

Figure 5.13.   The "fireman's grip" formed by hydrogen bonding of two Thr residues (one from each domain) with the main chain atoms associated with the Thr in the opposite domain. (From Gilliland 1990. *Proteins: Structure, Function, and Genetics 8,* 91, with permission. Copyright 1990 Wiley-Liss, Inc., a subsidiary of John Wiley & Sons, Inc.)

Substrate Binding Site

In addition to the active site, the rest of the cleft serves as an extended substrate binding site containing possibly seven subsites designated S_4, S_3, S_2, S_1, S_1', S_2', S_3', where the catalytic site is located between S_1 and S_1'. Hence the corresponding substrate units are NH_2-P_4, P_3, P_2, P_1, P_1', P_2', P_3'-COOH. The substrate binding residues in each individual subsite have not been precisely located. The Lys221 residue may be involved in electrostatic interactions with the P_2 side chain (Dunn et al. 1987), although site-directed mutagenesis studies cannot confirm this (Suzuki et al. 1989). The S_4 subsite is located close to the entrance of the active site cleft. For a synthetic substrate [LysProLeuGlu-PhePhe(NO_2)ArgLeu], the P_1(Phe) and P_3(Leu) side chains are in close contact with Tyr77 and Gln15, respectively (Strop et al. 1990). The Tyr77 in the flap and Val110 in the adjacent region participate in forming the S_1 subsite. The Tyr77 side chain interacts with a hydroxyl group of the cleavage bond in the substrate for stabilizing the transition state (Blundell et al. 1987; Suzuki et al. 1989). Alteration of Val113 and Phe114 by site-directed mutagenesis causes a marked change in k_{cat} and K_m, indicating the involvement of these hydrophobic residues in both the catalytic function and substrate binding.

SUBSTRATE SPECIFICITY

All known acid proteinases have similar substrate specificity, in that the enzyme acts preferentially at peptide bonds flanked by hydrophobic amino acid residues. In the clotting of milk, the Phe105-Met106 bond is preferentially cleaved by chymosin, although this cleavage is not related to the specificity of the enzyme on the Phe105-Met106 peptide per se. Based on the kinetic studies of hydrolysis of oligopeptides, the preferential cleavage of Phe105-Met106 is believed to be a consequence of secondary enzyme-substrate interactions (Voynick and Fruton 1971). A minimum chain length of five amino acids including -SerPheMetAla- is required for hydrolytic action to occur. Addition of Leu103 and Ile108 causes an increase in k_{cat} and decrease in K_m. Addition of Pro further on either side increases K_{cat}/K_m to 22 mM^{-1}s^{-1} (Visser et al. 1976). The Ser104 contributes strongly to the enzyme-substrate interaction, making the Phe-Met bond accessible to the enzyme action (Visser et al. 1977; Raap 1983). Lengthening the C-terminal end of the peptide increases the k_{cat}/K_m to 67 mM^{-1}s^{-1} (pH 4.7, 30°C).

Using the tryptic peptide from κ-casein, which contains residues 98–112 (-HisProHisProHisLeuSerPheMetAlaIleProProLysLys-) as substrate, the kinetic action of bovine chymosin has shown a catalytic efficiency k_{cat}/K_m of 4,715 mM^{-1}s^{-1} at pH 5.3–5.5, and 2,123 mM^{-1}s^{-1} at pH 6.6 (Visser et al. 1980). Kinetic studies on the enzyme action on bovine k-casein also yield similar results with k_{cat} = 68.5 s^{-1}, K_m = 0.048 mM, and k_{cat}/K_m = 1,413 mM^{-1}s^{-1} (pH 6.2,

30°C) (Carles and Martin 1985). These results substantiate the notion that the fragment containing 98–112 fulfills the requirement for maximal rate of cleavage of the Phe105-Met106 by chymosin.

A chemical modification study of the tryptic 98–112 fragment and synthetic peptides suggests that, in κ-casein, the segment containing residues 103–108 fits into the active site cleft of the enzyme with Leu103 and Ile108 directed towards the hydrophobic environment, and Ser104 forming hydrogen bonds. Electrostatic interactions probably occur between the His-Pro cluster sequence of the κ-casein and a negatively charged region on the active site cleft of the enzyme consisting of Glu244, Asp246, and Asp248. Proper positioning of the substrate is reinforced and stabilized by Pro99, 101, 109, and 110 (Visser 1987). These bindings provide a proper conformation of the substrate in the enzyme-substrate complex and exact positioning of the Phe-Met relative to the catalytic residue (Safro and Andreeva 1990). Interactions provided by the negatively charged region on the enzyme molecule (which is not found in pepsin and other aspartic proteinases) may account for the high milk-clotting acting/low proteolytic activity observed in chymosin. The lower k_{cat}/K_m at pH 6.6 compared to that at pH 5.5 may be related to the partial deprotonation of the His weakening the interaction.

MECHANISM OF CATALYSIS

All aspartic proteinases display a bell-shaped pH-activity profile, suggesting two catalytic Asp residues, with ~pKs of 2 and 5. For calf chymosin the two ionization groups have pKs 3.2 and 4.7 using LysProAlaGluPhePhe(NO$_2$)AlaLeu as substrate (Mantafounis and Pitts 1990). The pK values are 2.3 and 4.7 if acid-denatured hemoglobin serves as the substrate (Suzuki et al. 1989). The enzyme has a pH optimum of 3.7, where one of the Asp residues is deprotonated.

The abnormal pK values of the two Asp residues (normal pK = 3.9 for Asp) in aspartic proteinases are attributed to interactive ionization of the two carboxyl oxygens (Fig. 5.14). Deprotonation of the first Asp residue is relatively fast, and once the hydrogen bonding is formed, removing the second proton from a dianion becomes more difficult (Pearl and Blundell 1984). It has not been firmly established which of the Asp carboxyl groups has the lower pK. For *Rhizopus* pepsin and penicillopepsin, the proton is believed to reside on Asp220 and Asp215 (corresponding to 216 in chymosin), respectively.

The exact mechanism of aspartic proteinases is not available. It is generally agreed that nucleophilic catalysis with the formation of acyl-enzyme and amino-enzyme intermediates does not occur. Attempts to detect a covalently bonded enzyme-substrate intermediate have been unsuccessful (Dunn and Fink 1984; Hofmann et al. 1984; Hofmann and Hodges 1982). Current evidence seems to

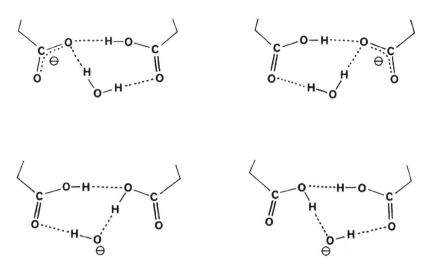

Figure 5.14. Possible hydrogen-bonded arrangements between the carboxylates of the two catalytic Asp residues in the diad of aspartic proteases. (From Pearl and Blundell 1984. *FEBS Lett. 174*, 98, with permission. Copyright 1984 Federation of European Biochemical Societies.)

suggest general acid-base catalysis with a noncovalent intermediate. A number of models have been proposed along this line, and all contain several common features (Fig. 5.15) (James and Sielecki 1985; Sunguna et al. 1987; Bott et al. 1982; Blundell et al. 1987).

(1) The peptide carbonyl oxygen of the substrate is protonated by the Asp diad proton. The effect is an increase in polarization of the carbonyl bond $[>C = O \cdots\cdots H^+]$, resulting in a more positive carbon center for nucleophilic attack.

(2) One of the H_2O molecules located in the immediate environment of the Asp residues acts as the nucleophile. Nucleophilic attack occurs at the peptide carbonyl carbon of the substrate by base catalysis as assisted by the Asp carboxylate anion. This results in the formation of a tetrahedral intermediate.

(3) The peptide nitrogen atom in the tetrahedral intermediate accepts a proton from bulk solvent or the Asp diad. The result of N-protonation facilitates cleavage of the C-N bond of the tetrahedral intermediate.

Peptide bonds are known to be planar and *trans* due to delocalization of the lone pair electron of the nitrogen to the carbonyl oxygen. The C-N bond acquires a double bond character. It has been suggested that binding of the sub-

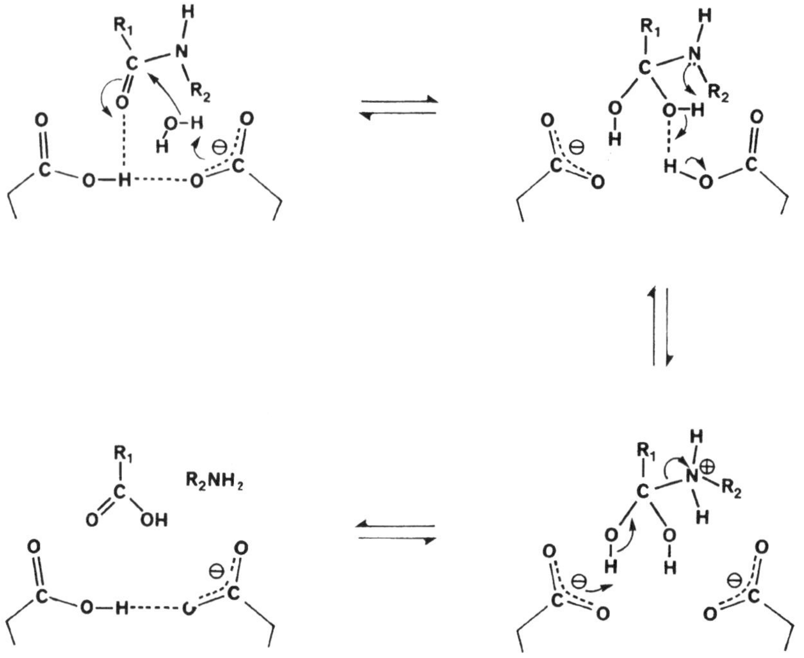

Figure 5.15. The proposed reaction mechanism of chymosin (Bott et al. 1982; Suguna et al. 1987).

strate with side chain and main chain interactions causes distortion of the scissile peptide bond from planarity. This rotation about the peptide bond further reduces the double bond character, rendering the nitrogen atom more acceptable to protonation (Suguna et al. 1987).

Protonation of the nitrogen atom can occur in concert with or subsequent to the protonation of the carbonyl oxygen. A protonated amine yields a good leaving group, allowing the breakdown of the C-N bond.

The location of the protons has not been firmly established. As mentioned previously, the proton of Asp215 in the penicillopepsin enzyme participates in hydrogen bonding with Asp32. In x-ray analysis of binding inhibition of the *Rhizopus chinensis* pepsin, the proton is not shared by the two Asp residues. Instead the carboxyl oxygens are hydrogen bonded to a centrally positioned H_2O molecule. The Asp35 acts as the electrophile in donating the proton to the peptide carbonyl oxygen. The central H_2O molecule provides the nucleophile, which attacks the carbonyl carbon.

REFERENCES

Subtilisn

ARGADE, P. V.; GERKE, G. K.; WEBER, J. P.; and PETICOLAS, W. L. 1984. Resonance raman carbonyl frequencies and ultraviolet absorption maxima as indicators of the active site environment in native and unfolded chromophoric acyl-α-chymotrypsin. *Biochemistry 23*, 299–304.

BACHOUCHIN, W. W., and ROBERTS, J. D. 1978. Nitrogen-15 nuclear magnetic resonance spectroscopy. The state of histidine in the catalytic triad of α-lytic protease. Implications for the charge-relay mechanism of peptide-bond cleavage by serine proteases. *J. Am. Chem. Soc. 100*, 8041–8047.

BECH, L. M.; SORENSEN, S. B.; and BREDDAM, K. 1992. Mutational replacements in subtilisin 309. Val104 has a modulating effect on the P_4 substrate preference. *Eur. J. Biochem. 209*, 869–874.

BENDER, M. L.; CLEMENT, G. E.; KEZDY, F. J.; and HECK, H. D. A. 1964. The correlation of the pH (pD) dependence and the stepwise mechanism of α-chymotrypsin-catalyzed reaction. *J. Am. Chem. Soc. 86*, 3680–3689.

BENDER, M. L., and PHILIPP, M. 1973. Subtilisin catalysis of nonspecific anilide hydrolyses. *J. Am. Chem. Soc. 95*, 1665–1666.

BODE, W.; PAPAMOKOS, E.; and MUSIL, D. 1987. The high-resolution x-ray crystal structure of the complex formed between subtilisin Carlsberg and eglin c, an elastic inhibitor from the leech *Hirudo medicinalis*. Structural analysis, subtilisin structure and interface geometry. *Eur. J. Biochem. 166*, 673–692.

BODE, W.; PAPAMOKOS, E.; MUSIL, D.; SEEMUELLER, U.; and FRITZ, H. 1986. Refined 1.2 Å crystal structure of the complex formed between subtilisin Carlsberg and the inhibitor eglin c. Molecular structure of eglin and its detailed interaction with subtilisin. *EMBO J. 5*, 813–818.

BOTT, R.; ULTSCH, M.; and KOSSIAKOFF, A. 1988. The three-dimensional structure of *Bacillus amyloliquefaciens* subtilisin at 1.8 Å and an analysis of the structural consequences of peroxide inactivation. *J. Biol. Chem. 263*, 7895–7906.

BRANDEN, C., and TOOZE, J. 1991. Introduction to Protein Structure. Garland Publishing, Inc., New York and London.

BRAXTON, S., and WELLS, J. A. 1991. The importance of a distal hydrogen bonding group in stabilizing the transition state in subtilisin BPN'. *J. Biol. Chem. 266*, 11797–11800.

BROMME, D.; PETERS, K.; FINK, S.; and FITTKAU, S. 1986. Enzyme-substrate interactions in the hydrolysis of peptide substrate by thermitase, subtilisin BPN', and proteinase K. *Arch. Biochem. Biophys. 244*, 439–446.

BRYAN, P.; PANTOLIANO, M. W.; QUILL, S. G.; HSIAO, H.-Y.; and POULOS, T. 1986. Site-directed mutagenesis and the role of the oxyanion hole in subtilisin. *Proc. Natl. Acad. Sci. USA 83*, 3743–3745.

BYCROFT, M., and FERSHT, A. R. 1988. Assignment of histidine resonances in the H NMR (500 MHz) spectrum of subtilisin BPN' using site-directed mutagenesis. *Biochemistry 27*, 7390–7394.

CARTER, P., and WELLS, J. A. 1988. Dissecting the catalytic triad of a serine protease. *Nature 332*, 564–568.

DAGGETT, V.; SCHRODER, S.; and KOLLMAN, P. 1991. Catalytic pathway of serine proteases: Classical and quantum mechanical calculations. *J. Am. Chem. Soc. 113*, 8926–8935.

DRENTH, J.; HOL, W. G. J.; JANSONIUS, J. N.; and KOEKOEK, R. 1972. Subtilisin Novo. The three-dimensional structure and its comparison with subtilisin BPN'. *Eur. J. Biochem. 26*, 177–181.

ESTELL, D. A.; GRAYCAR, T. P.; MILLER, J. V.; POWERS, D. B.; BURNIER, J. P.; NG, P. G.; and WELLS, J. A. 1986. Probing steric and hydrophobic effects on enzyme-substrate interactions by protein engineering. *Science 233*, 659–663.

FERSHT, A. R., and REQUENA, Y. 1971. Equilibrium and rate constants for the interconversion of two conformations of α-chymotrypsin. *J. Mol. Biol. 60*, 279–290.

GRON, H.; MELDAL, M.; and BREDDAM, K. 1992. Extensive comparison of the substrate preferences of two subtilisins as determined with peptide substrates which are based on the principle of intramolecular quenching. *Biochemistry 31*, 6011–6018.

HEINZ, D. W.; PRIESTLE, J. P.; RAHUEL, J.; WILSON, K. S.; and GRUTTER, M. G. 1991. Refined crystal structures of subtilisin Novo in complex with wild-type and two mutant eglins. Comparison with other serine proteinase inhibitor complexes. *J. Mol. Biol. 217*, 353–371.

HENDERSON, R. 1970. Structure of crystalline α-chymotrypsin IV. The structure of indoleacryloyl-α-chymotrypsin and its relevance to the hydrolytic mechanism of the enzyme. *J. Mol. Biol. 54*, 341–354.

HIRONO, S.; AKAGAWA, H.; MITSUI, Y.; and IITAKA, Y. 1984. Crystal structure at 2.6 A resolution of the complex of subtilisin BPN' with *Streptomyces* subtilisin inhibitor. *J. Mol. Biol. 178*, 389–413.

HUNKAPILLER, M. W.; FORGAC, M. D.; YU, E. H.; and RICHARDS, J. H. 1979. ¹³C NMR studies of the binding of soybean trypsin inhibitor to trypsin. *Biochem. Biophys. Res. Comm. 87*, 25–31.

HUNKAPILLAR, M. W.; SMALLCOMBE, S. H.; WHITAKER, D. R.; and RICHARD, J. H. 1973. Carbon nuclear magnetic resonance studies of the histidine residue in α-lytic protease. Implications for the catalytic mechanism of serine proteases. *Biochemistry 12*, 4732–4743.

HWANG, J.-K., and WARSHEL, A. 1987. Semiquantitative calculations of catalytic free energies in genetically modified enzymes. *Biochemistry 26*, 2669–2673.

IKEMURA, H.; TAKAGI, H.; and INOUYE, M. 1987. Requirement of pro-sequence for the production of active subtilisin E in *Escherichia coli. J. Biol. Chem. 262*, 7859–7864.

JACOBS, M.; ELIASSON, M.; UHLEN, M.; and FLOCK, J.-I. 1985. Cloning, sequencing and expression of subtilisin Carlsberg from *Bacillus licheniformis. Nucl. Acids. Res. 13*, 8913–8926.

JORDAN, F., and POLGAR, L. 1981. Proton nuclear magnetic resonance evidence for the absence of a stable hydrogen bond between the active site aspartate and histidine residues of native subtilisins and for its presence in thiolsubtilisins. *Biochemistry 20*, 6366–6370.

JORDAN, F.; POLGAR, L.; and TOUS, G. 1985. Proton magnetic resonance studies of the states of ionization of histidines in native and modified subtilisins. *Biochemistry 24*, 7711–7717.

KEAY, I., and MOSER, P. W. 1969. Differentiation of alkaline proteases from *Bacillus* species. *Biochem. Biophys. Res. Comm. 34*, 600–605.

KOLLMAN, P. A., and HAYES, D. M. 1981. Theoretical calculations on proton-transfer energetics: Studies of methanol, imidazole, formic acid, and methanethiol as models for the serine and cysteine proteases. *J. Am. Chem. Soc. 103*, 2955–2961.

KOMIYAMA, M., and BENDER, M. L. 1979. Do cleavages of amides by serine proteases occur through a stepwise pathway involving tetrahedral intermediates? *Proc. Natl. Acad. Sci. USA 76*, 557–560.

KOSSIAKOFF, A. A., and SPENCER, S. A. 1981. Direct determination of protonation states of aspartic acid-102 and histidine-57 in the tetrahedral intermediate of the serine proteases: Neutron structure of trypsin. *Biochemistry 20*, 6462–6473.

KRAUT, J. 1971. Subtilisin: X-Ray structure. *The Enzymes 3*, 547–560.

KURIHARA, M.; MARKLAND, F. S.; and SMITH, E. L. 1972. Subtilisin amylosacchariticus III. Isolation and sequence of the chymotryptic peptides and the complete amino acid sequence. *J. Biol. Chem. 247*, 5619–5631.

LINDQUIST, R. N., and TERRY, C. 1974. Inhibition of subtilisin by boronic acids, potential analogs of tetrahedral reaction intermediates. *Arch. Biochem. Biophys. 160*, 135–144.

MARKLAND, F. S., and SMITH, E. L. 1967. Subtilisin BPN'. VII. Isolation of cyanogen bromide peptides and the complete amino acid sequence. *J. Biol. Chem. 242*, 5198–5211.

MARKLEY, J. L.; TRAVERS, F.; and BALNY, C. 1981. Lack of evidence for a tetrahedral intermediate in the hydrolysis of nitroanilide substrate by serine proteinases. *Eur. J. Biochem. 120*, 477–485.

MATTHEWS, B. W.; SIGLER, P. B.; HENDERSON, R.; and BLOW, D. W. 1967. Three-dimensional structure of tosyl-α-chymotrypsin. *Nature 214*, 652–656.

MATTHEWS, D. A.; ALDEN, R. A.; BIRKTOFT, J. J.; FREER, S. T.; and KRAUT, J. 1975. X-Ray crystallographic study of boronic acid adducts with subtilisin BPN' (Novo). *J. Biol. Chem. 250*, 7120–7126.

———. 1977. Re-examination of the charge relay system in subtilisin and comparison with other serine proteases. *J. Biol. Chem. 252*, 8875–8883.

MCPHALEN, C. A., and JAMES, M. N. G. 1988. Structural comparison of two serine proteinase-protein inhibitor complexes: Eglin-C-subtilisin Carlsberg and CI-2-subtilisin Novo. *Biochemistry 27*, 6582–6598.

MCPHALEN, C. A.; SVENDSEN, I.; JONASSEN, I.; and JAMES, M. N. G. 1985. Crystal and molecular structure of chymotrypsin inhibitor 2 from barley seeds in complex with subtilisin Novo. *Proc. Natl. Acad. Sci. USA 82*, 7242–7246.

MORIHARA, K., and OKA, T. 1970. Subtilisin BPN': Inactivation by chloromethyl ketone derivatives of peptide substrates. *Arch. Biochem. Biophys. 138*, 526–531.

———. 1977. A kinetic investigation of subsites S_1' and S_2' in α-chymotrypsin and subtilisin BPN. *Arch. Biochem. Biophys. 178*, 188–194.

MORIHARA, K.; OKA, T.; and TSUZUKI, H. 1970. Subtilisin BPN': Kinetic study with oligopeptides. *Arch. Biochem. Biophys. 138*, 515–525.

MORIHARA, K.; OKA, T.; and TSUZUKI, H. 1974. Comparative study of various serine alkaline proteinases from microorganisms. *Arch. Biochem. Biophys. 165*, 72–79.

MORIHARA, K.; TSUZUKI, H.; and OKA, T. 1971. Comparison of various types of subtilisins in size and properties of the active site. *Biochem. Biophys. Res. Comm. 42*, 1000–1006.

NEDKOV, P.; OBERTHUR, W.; and BRAUNITZER, G. 1985. Determination of the complete amino acid sequence of subtilisin DY and its comparison with the primary structures of the subtilisin BPN', Carlsberg and amylosacchariticus. *Biol. Chem. Hoppe-Seyler 366*, 421–430.

O'CONNELL, T. P.; FINUCANE, M. D.; and MALTHOUSE, J. P. G. 1993. The use of ^{13}C n.m.r. and saturation transfer to detect tetrahedral intermediates in reactions catalyzed by chymotrypsin and also in an amide inhibitor complex. *Biochem. Soc. Trans. 22*, 30S.

OLAITAN, S. A.; DELANGE, R. J.; and SMITH, E. L. 1968. The structure of subtilisin Novo. *J. Biol. Chem. 243*, 5296–5301.

PHILIPP, M., and BENDER, M. L. 1983. Kinetics of subtilisin and thiolsubtilisin. *Mol. Cell. Biochem. 51*, 5–32.

POLGAR, L., and BENDER, M. L. 1967. The reactivity of thiol-subtilisin, an enzyme containing a synthetic functional group. *Biochemistry 6*, 610–620.

POULOS, T. L.; ALDEN, R. A.; FREER, S. T.; BIRKTOFT, J. J.; and KRAUT, J. 1976. Polypeptide halomethyl ketones bind to serine proteases as analogs of the tetrahedral intermediate. *J. Biol. Chem. 251*, 1097–1103.

RAO, S. N.; SINGH, U. C.; BASH, P. A.; and KOLLMAN, P. A. 1987. Free energy perturbation calculation on binding and catalysis after mutating Asn155 in subtilisin. *Nature 328*, 551–554.

ROBERTUS, J. D.; ALDEN, R. A.; and KRAUT, J. 1971. On the identity of subtilisins BPN' and Novo. *Biochem. Biophys. Res. Comm. 42*, 334–339.

ROBERTUS, J. D.; ALDEN, R. A.; BIRKTOFT, J. J.; KRAUT, J.; POWERS, J. C.; and WILCOX, P. E. 1972A. An x-ray crystallographic study of the binding of peptide chloromethyl ketone inhibitors to subtilisins BPN'. *Biochemistry 11*, 2439–2449.

ROBERTUS, J. D.; KRAUT, J.; ALDEN, R. A.; and BIRKTOFT, J. J. 1972B. Subtilisin; a stereochemical mechanism involving transition-state stabilization. *Biochemistry 11*, 4293–4303.

ROBILLARD, G. T.; POWERS, J. C.; and WILCOX, P. E. 1972. A chemical and crystallographic study of carbamyl-chymotrypsin A. *Biochemistry 11*, 1773.

ROBILLARD, G. T., and SHULMAN, R. G. 1974. High resolution nuclear magnetic resonance studies of the active site of chymotrypsin. II. Polarization of histidine 57 by substrate analogues and competitive inhibitors. *J. Mol. Biol. 86*, 541–558.

SCHECHTER, I., and BERGER, A. 1967. On the size of the active site in proteases. I. Papain. *Biochem. Biophys. Res. Comm. 27*, 157–162.

SCHEINER, S., and LIPSCOMB, W. N. 1976. Molecular orbital studies of enzyme activity: Catalytic mechanism of serine proteinases. *Proc. Natl. Acad. Sci. USA 73*, 432–436.

SMITH, E. L.; DELANGE, R. J.; EVANS, W. H.; LANDON, M.; and MARKLAND, F. 1968. Subtilisin Carlsberg. V. The complete sequence; comparison with subtilisin BPN'; evolutionary relationship. *J. Biol. Chem. 243*, 2184–2191.

STAHL, M. L., and FERRARI, E. 1984. Replacement of the *Bacillus subtilis* subtilisin structural gene with an in vitro-derived deletion mutation. *J. Bacteriol. 158*, 411–418.

STAMATO, F. M. L. G.; LONGO, E.; and YOSHIOKA, L. M. 1984. The catalytic mechanism of serine proteases: Single proton versus double proton transfer. *J. Theor. Biol. 107*, 329–338.

STEIN, R. L.; ELROD, J. P.; and SCHOWEN, R. L. 1983. Correlative variations in enzyme-derived and substrate-derived structures of catalytic transition states. Implications for the catalytic strategry of acyl-transfer enzymes. *J. Am. Chem. Soc. 105*, 2446–2452.

TAKEUCHI, Y.; SATOW, Y.; NAKAMURA, K. T.; and MITSUI, Y. 1991. Refined crystal structures of the complex of subtilisin BPN' and *Streptomyces* subtilisin inhibitor at 1.8 Å resolution. *J. Mol. Biol. 221*, 309–325.

TONGE, P. J., and CAREY, P. R. 1990. Length of the acyl carbonyl bond in acyl-serine proteases correlates with reactivity. *Biochemistry 29*, 10723–10727.

———. 1992. Forces, bond lengths, and reactivity: Fundamental insight into the mechanism of enzyme catalysis. *Biochemistry 31*, 9122–9125.

VASANTHA, N.; THOMPSON, L. D.; RHODES, C.; BANNER, C.; NAGLE, J.; and FILPULA, D. 1984. Genes for alkaline protease and neutral protease from *Bacillus amyloliquefaciens* contain a large open reading frame between the regions coding for signal sequence and mature protein. *J. Bacteriol. 159*, 811–819.

WARSHEL, A.; NARAY-SZABO, G.; SUSSMAN, F.; and HWANG, J.-K. 1989. How do serine proteases really work? *Biochemistry 28*, 3629–3637.

WARSHEL, A., and RUSSELL, S. 1986. Theoretical correlation of structure and energetics in the catalytic reaction of trypsin. *J. Am. Chem. Soc. 108*, 6569–6579.

WELLS, J. A.; CUNNINGHAM, B. C.; GRAYCAR, T. P.; and ESTELL, D. A. 1987B. Recruitment of substrate-specificity properties from one enzyme into a related one by protein engineering. *Proc. Natl. Acad. Sci. USA 84*, 5167–5171.

WELLS, J. A., and ESTELL, D. A. 1988. Subtilisin- an enzyme designed to be engineered. *TIBS 13*, 291–297.

WELLS, J.; FERRARI, E.; HENNER, D. J.; ESTELL, D. A.; and CHEN, E. Y. 1983. Cloning, sequencing, and secretion of *Bacillus amyloliquefaciens* subtilisin in *Bacillus subtilis*. *Nucl. Acids Res. 11*, 7911–7925.

WELLS, G. B.; MUSTAFI, D.; and MAKINEN, M. W. 1994. Structure at the active site of an acylenzyme of α-chymotrypsin and implications for the catalytic mechanism. *J. Biol. Chem. 269*, 4577–4586.

WELLS, J. A.; POWERS, D. B.; BOTT, R. R.; GRAYCAR, T. P.; and ESTELL, D. A. 1987A. Designing substrate specificity by protein engineering of electrostatic interactions. *Proc. Natl. Acad. Sci. USA 84*, 1219–1223.

WRIGHT, C. S. 1972. Comparison of the active site stereochemistry and substrate conformation in α-chymotrypsin and subtilisin BPN'. *J. Mol. Biol. 67*, 151–163.

WRIGHT, C. S.; ALDEN, R. A.; and KRAUT, J. 1969. Structure of subtilisin BPN' at 2.5 A resolution. *Nature 221*, 235–242.

Papain

ALECIO, M. R.; DANN, M. L.; and LOWE, G. 1974. The specificity of the S_1' subsite of papain. *Biochem. J. 141*, 495–501.

ASBOTH, B.; MAJER, Z.; and POLGAR, L. 1988. Cysteine proteases: The S_2P_2 hydrogen bond is more important for catalysis than is the analogous S_1P_1 bond. *FEBS Lett. 233*, 339–341.

ASBOTH, B.; STOKUM, E.; KHAN, I. U.; and POLGAR, L. 1985. Mechanism of action of cysteine proteinases: Oxyanion binding site is not essential in the hydrolysis of specific substrate. *Biochemistry 24*, 606–609.

BAKER, E. N. 1980. Structure of actinidin, after refinement at 1.7 A resolution. *J. Mol. Biol. 141*, 441–484.

BAKER, E. N., and DRENTH, J. 1987. The thiol proteases: structure and mechanism. In: *Biological Macromolecules and Assembles*. vol. 3. *Active Sites of Enzymes*. F. A. Jurnak and A. McPherson, eds., Wiley & Sons, New York.

BENDALL, M. R.; CARTWRIGHT, I. L.; CLARK, P. I.; LOWE, G.; and NURSE, D. 1977. Inhibition of papain by *N*-acyl-aminoacetaldehydes and *N*-acyl-aminopropanones. *Eur. J. Biochem. 79*, 201–209.

BODE, W.; ENGH, R.; MUSIL, D.; THIELE, U.; HUBER, R.; KARSHIKOV, A.; BRZIN, J.; KOS, J.; and TURK, V. 1988. The 2.0 A x-ray crystal structure of chicken egg white cystatin and its possible mode of interaction with cysteine proteinases. *EMBO J. 7*, 2593–3599 (1988).

BRUBACHER, L. J., and BENDER, M. L. 1966. The preparation and properties of trans-cinnamoyl-papain. *J. Am. Chem. Soc. 88*, 5871–5880.

CARNE, A., and MOORE, C. H. 1978. The amino acid sequence of the tryptic peptides from actinidin, a proteolytic enzyme from the fruit of *Actinidia chinensis*. *Biochem. J. 173*, 73–83.

CREIGHTON, D. J., and SCHAMP, D. J. 1980. Solvent isotope effects on tautomerization equilibria of papain and model thiolamines. *FEBS Lett. 110*, 313–318.

DELAISSE, J.-M.; EECKHOUT, Y.; and VAES, G. 1984. In vivo and in vitro evidence for the involvement of cysteine proteinases in bone resorption. *Biochem. Biophys. Res. Comm. 125*, 441–447.

DRENTH, J.; JANSONIUS, J. N.; KOEKOEK, R.; and WOLTHERS, B. G. 1971. Papain, x-ray structure. *The Enzymes 3*, 485–499.

DRENTH, J.; KALK, K. H.; and SWEN, H. M. 1976. Binding of chloromethyl ketone substrate analogues to crystalline papain. *Biochemistry 15*, 3731–3739.

FALANGA, A., and GORDON, S. G. 1985. Isolation and characterization of cancer procoagulant: A cysteine proteinase from malignant tissue. *Biochemistry 24*, 5558–5567.

FRANKFATER, A., and KUPPY, T. 1981. Mechanism of association of *N*-acetyl-L-phenylalanylglycinal to papain. *Biochemistry 20*, 5517–5524.

GAMCSIK, M. P.; MALTHOUSE, J. P. G.; PRIMROSE, W. U.; MACKENZIE, N. E.; BOYD, A. S. F.; RUSSELL, R. A.; and SCOTT, A. I. 1983. Structure and stereochemistry of tetrahedral inhibitor complexes of papain by direct NMR observation. *J. Am. Chem. Soc. 105*, 6324–6325.

GARCIA-ECHEVERRIA, C., and RICH, D. H. 1992A. New intramolecularly quenched fluorogenic peptide substrates for the study of the kinetic specificity of papain. *FEBS Lett. 297*, 100–102.

———. 1992B. Effect of P_2' substituents on kinetic constants for hydrolysis by cysteine proteinases. *Biochem. Biophys. Res. Comm. 187*, 615–619.

GLAZER, A. N., and SMITH, E. L. 1971. Papain and other plant sulfhydryl proteolytic enzymes. *The Enzymes 3*, 501–547.

GOTO, K.; MURACHI, T.; and TAKAHASHI, N. 1976. Structural studies on stem bromelain isolation, characterization and alignment of the cyanogen bromide fragments. *FEBS Lett. 62*, 93–95.

GOTO, K.; TAKAHASHI, N.; and MURACHI, T. 1980. Structural studies on stem bromelain. *Int. J. Peptide Protein Res. 15*, 335–341.

GREIGHTON, D. J.; GESSOUROUN, M. S.; and HEAPES, J. M. 1980. Is the thiolate-imidazolium ion pair the catalytically important form of papain? *FEBS Lett. 110*, 319–322.

HUBER, R., and BODE, W. 1978. Structural basis of the activation and action of trypsin. *Acc. Chem. Res. 11*, 114–122.

JOHNSON, F. A.; LEWIS, S. D.; and SHAFER, J. A. 1981. Perturbations in the free energy and enthalpy of ionization of histidine-159 at the active site of papain as determined by fluorescence spectroscopy. *Biochemistry 20*, 52–58.

KAMPHUIS, I. G.; DRENTH, J.; and BAKER, E. N. 1985. Thiol proteases. Comparative studies based on the high-resolution structures of papain and actinidin, and on amino acid sequence information for cathepsins B and H, and stem bromelain. *J. Mol. Biol. 182*, 317–329.

KAMPHUIS, I. G.; KALK, K. H.; SWARTE, M. B. A.; and DRENTH, J. 1984. Structure of papain refined at 1.65 A resolution. *J. Mol. Biol. 179*, 233–256.

KOWLESSUR, D.; THOMAS, E. W.; TOPHAM, C. M.; TEMPLETON, W.; and BROCKLE-HURST, K. 1990. Dependence of the P_2-S_2 stereochemical selectivity of papain on the nature of the catalytic-site chemistry. *Biochem. J. 266*, 653–660.

LASKOWSKI, M., and KATO, I. 1980. Protein inhibitors of proteinases. *Ann. Rev. Biochem. 49*, 593–626.

LEWIS, S. D.; JOHNSON, F. A.; OHNO, A. K.; and SHAFER, J. A. 1978. Dependence of the catalytic activity of papain on the ionization of two acidic groups. *J. Biol. Chem. 253*, 5080–5086.

LEWIS, S. D.; JOHNSON, F. A.; and SHAFER, J. A. 1976. Potentiometric determination of ionizations at the active site of papain. *Biochemistry 15*, 5009–5017.

———. 1981. Effect of cysteine-25 on the ionization of histidine-159 in papain as determined by proton nuclear magnetic resonance spectroscopy. Evidence for a His-159-Cys-25 ion pair and its possible role in catalysis. *Biochemistry 20*, 48–51.

LOWE, G., and WILLIAMS, A. 1965. Direct evidence for an acylated thiol as an intermediate in papain- and ficin-catalyzed hydrolyses. *Biochem. J. 96*, 189–193.

LOWE, G., and YUTHAVONG, Y. 1971. Kinetic specificity in papain-catalyzed hydrolyses. *Biochem. J. 124*, 107–115.

LYNN, K. G. 1983. Definition of the site of reactivity of the ancestral protease of the papain type. *Phytochemistry 22*, 2485–2487.

MACHLEIDT, W.; THIELE, U.; LABER, B.; ASSFALG-MACHEIDT, I.; ESTERL, A.; WIEGAND, G.; KOS, J.; TURK, V.; and BODE, W. 1989. Mechanism of inhibition of papain by chicken egg white cystatin. Inhibition constants of N-terminally truncated forms and cyanogen bromide fragments of the inhibitor. *FEBS Lett. 243*, 234–238.

MACKENZIE, N. E.; GRANT, S. K.; SCOTT, A. I.; and MALTHOUSE, P. G. 1986. ^{13}C NMR study of the stereospecificity of the thiohemiacetals formed on inhibition of papain by specific enantiomeric aldehydes. *Biochemistry 25*, 2293–2298.

MACKENZIE, N. E.; MALTHOUSE, P. G.; and SCOTT, A. I. 1985. Chemical synthesis and papain-catalyzed hydrolysis of N-α-benzyloxycarbonyl-L-lysine p-nitroanilide. *Biochem. J. 226*, 601–606.

MALTHOUSE, J. P. G.; GAMCSIK, M. P.; BOYD, A. S. F.; MACKENSIZE, N. E.; and SCOTT, A. I. 1982. Cryoenzymology of proteases: NMR detection of a productive thioacyl derivative of papain at subzero temperature. *J. Am. Chem. Soc. 104*, 6811–6813.

MENARD, R.; CARRIERE, J.; LAFLAMME, P.; PLOUFFE, C.; KHOURI, H. E.; VERNET, T.; TESSIER, D. C.; THOMAS, D. Y.; and STORER, A. C. 1991A. Contribution of the glutamine 19 side chain to transition-state stabilization in the oxyanion hole of papain. *Biochemistry 30*, 8924–8928.

MENARD, R.; KHOURI, H. E.; PLOUFFE, C.; DUPRAS, R.; RIPOLL, D.; VERNET, T.; TESSIER, D. C.; LALIBERTE, F.; THOMAS, D. Y.; and STORER, A. C. 1990. A protein engineering study of the role of aspartate 158 in the catalytic mechanism of papain. *Biochemistry 29*, 6706–6713.

MENARD, R.; KHOURI, H. E.; PLOUFFE, C.; LAFLAMME, P.; DUPRAS, R.; VERNET, T.; TESSIER, D. C.; THOMAS, D. Y.; and STORER, A. C. 1991B. Importance of hydrogen-bonding interactions involving the side chain of Asp158 in the catalytic mechanism of papain. *Biochemistry 30*, 5531–5538.

MIGLIORINI, M., and CREIGHTON, D. J. 1986. Active-site ionization of papain. An evaluation of the potentiometric difference titration method. *Eur. J. Biochem. 156*, 189–192.

MITCHEL, R. E. J.; CHAIKEN, I. M.; and SMITH, E. L. 1970. The complete amino acid sequence of papain. *J. Biol. Chem. 245*, 3485–3492.

MOON, J. B.; COLEMAN, R. S.; and HANZLIK, R. P. 1986. Reversible covalent inhibition of papain by a peptide nitrile. ^{13}C NMR evidence for a thioimidate ester adduct. *J. Am. Chem. Soc. 108*, 1350–1351.

MULLER-ESTERL, W.; FRITZ, H.; KELLERMANN, J.; LOTTSPEICH, F.; MACHLEIDT, W.; and TURK, V. 1985. Genealogy of mammalian cysteine proteinase inhibitors.

Common evolutionary origin of stefins, cystalins and kininogens. *FEBS Lett. 191*, 221–226.

O'LEARY, M. H.; URBERG, M.; and YOUNG, A. P. 1974. Nitrogen isotope effects on the papain-catalyzed hydrolysis of *N*-benzoyl-L-argininamide. *Biochemistry 13*, 2077–2081.

POLGAR, L. 1973. On the mode of activation of the catalytically essential sulfhydryl group of papain. *Eur. J. Biochem. 33*, 104–109.

———. 1974. Mercaptide-imidazolium ion-pair: The reactive nucleophile in papain catalysis. *FEBS Lett. 47*, 15–18.

POOLE, A. R.; TILTMAN, K. J.; RECKLIES, A. D.; and STOKER, T. A. M. 1978. Differences in secretion of the proteinase cathepsin B at the edges of human breast carcinomas and fibroadenomas. *Nature 273*, 545–547.

ROBERTS, D. D.; LEWIS, S. D.; BALLOU, D. P.; OLSON, S. T.; and SHAFER, J. A. 1986. Reactivity of small thiolate anions and cysteine-25 in papain toward methyl methanethiosulfonate. *Biochemistry 25*, 5595–5601.

SCHACK, P., and KAARSHOLM, N. C. 1984. Absence in papaya peptidase A catalyzed hydrolysis of a pK_a ~ 4 present in papain-catalyzed hydrolyses. *Biochemistry 23*, 631–635.

SCHECHTER, I., and BERGER, A. 1967. On the size of the active site in proteases. I. Papain. *Biochem. Biophys. Res. Comm. 27*, 157–162.

———. 1968. On the active site of proteases. III. Mapping the active site of papain. *Biochem. Biophys. Res. Comm. 32*, 898–902.

SCHRODER, E.; PHILLIPS, C.; GARMAN, E.; HARLOS, K.; and CRAWFORD, C. 1993. X-Ray crystallographic structure of a papain-leupeptin complex. *FEBS Lett. 315*, 38–42.

SCHUSTER, M.; KASCHE, V.; and JAKUBKE, H.-D. 1992. Contributions to the S'-subsite specificity of papain. *Biochim. Biophys. Acta 1121*, 207–212.

SLUYTERMAN, L. A., and WIJDENES, J. 1976. Proton eqilibria in the binding of Zn^{2+} and of methymercuric iodide to papain. *Eur. J. Biochem. 71*, 383–391.

SKOREY, K. I., and BROWN, R. S. 1985. Biomimetic models for cysteine proteases. 2. Nucleophilic thiolate-containing zwitterions produced from imidazole-thiol pairs. A model for the acylation step in papain-mediated hydrolysis. *J. Am. Chem. Soc. 107*, 4070–4072.

STORER, A. C., and CAREY, P. R. 1985. Comparison of the kinetics and mechanism of the papain-catalyzed hydrolysis of esters and thiono esters. *Biochemistry 24*, 6808–6818.

STUBBS, M. T.; LABER, B.; BODE, W.; HUBER, R.; JERALA, R.; LENARCIC, B.; and TURK, V. 1990. The refined 2.4 A x-ray crystal structure of recombinant human stefin B in complex with the cysteine proteainase papain: A novel type of proteinase inhibitor interaction. *EMBO J. 9*, 1939–1947.

SZAWELSKI, R. J., and WHARTON, C. W. 1981. Kinetic solvent isotope effects on the deacylation of specific acyl-papains. *Biochem. J. 199*, 681–692.

VARUGHESE, K. I.; AHMED, F. R.; CAREY, P. R.; HASNAIN, S.; HUBER, C. P.; and STORER, A. C. 1989. Crystal structure of a papain-E-64 complex. *Biochemistry 28*, 1330–1332.

WESTERIK, J. O., and WOLFENDEN, R. 1972. Aldehydes as inhibitors of papain. *J. Biol. Chem. 247*, 8195–8197.

WILLIAMS, A.; LUCAS, E.; RIMMER, A. R.; and HAWKINS, H. C. 1972. Proteolytic enzymes. Nature of binding forces between papain and its substrates and inhibitors. *J. Chem. Soc. Perkin Trans II*, 627–633.

YAMAMOTO, O.; OHISHI, H.; ISHIDA, T.; INOUE, M.; SUMIYA, S., and KITAMURA, K. 1990. Molecular dynamics simulation of papain-E-64 (*N*-[*N*-(L-3-*trans*-carboxyoxirane-2-carbonyl)-L-leucyl]agmatine) complex. *Chem. Pharm. Bull. 38*, 2339–2343.

Chymosin

AIKAWA, J.-(i).; YAMASHITA, T.; NISHIYAMA, M.; HORINOUCHI, S.; and BEPPU, T. 1990. Effects of glycosylation on the secretion and enzyme activity of *Mucor* rennin, an aspartic proteinase of *Mucor pusillus*, produced by recombinant yeast. *J. Biol. Chem. 265*, 13955–13959.

ANDREEVA, N. S., and GUSTCHINA, A. E. 1979. On the supersecondary structure of acid proteases. *Biochem. Biophys. Res. Comm. 87*, 32–42.

ANDREEVA, N.; DILL, J.; and GILLILAND, G. L. 1992. Can enzymes adopt a self-inhibited form? Results of x-ray crystallographic studies of chymosin. *Biochem. Biophys. Res. Comm. 184*, 1074–1081.

ANDREEVA, N. S.; ZDANOV, A. S.; GUSTCHINA, A. E.; and FEDOROV, A. A. 1984. Structure of ethanol-inhibited porcine pepsin at 2-Å resolution and binding of the methyl ester of phenylalanyl-diiodotyrosine to the enzyme. *J. Biol. Chem. 259*, 11353–11365.

BEPPU, T. 1988. Site-directed mutagenesis on chymosin and *Mucor pusillus* rennin. The 18th Linderstrom-Lang conference, Elsinore, Denmark, 4–8 July 1988.

BLUNDELL, T. L.; COOPER, J.; and FOUNDLING, S. I. 1987. On the rational design of renin inhibitors: X-ray studies of aspartic proteinases complexed with transition-state analogues. *Biochemistry 26*, 5585–5590.

BLUNDELL, T. L.; JENKINS, J. A.; SEWELL, B. T.; PEARL, L. H.; COOPER, J. B.; TICKLE, I. J.; VEERAPANDIAN, B.; and WOOD, S. P. 1990. X-ray analysis of aspartic proteinases. The three-dimensional structure at 2.1 Å resolution of endothiapepsin. *J. Mol. Biol. 211*, 919–941.

BOEL, E.; BECH, A.-M.; RANDRUP, K.; DRAEGER, B.; FIIL, N. P.; and FOLTMANN, B. 1986. Primary structure of a precursor to the aspartic proteinase from *Rhizomucor miehei* shows that the enzyme is synthesized as a zymogen. *Proteins: Structure, Function, and Genetics 1*, 363–369.

BOTT, R.; SUBRAMANIAN, E.; and DAVIES, D. R. 1982. Three-dimensional structure of the complex of the *Rhizopus chinensis* carboxyl proteinase and pepstatin at 2.5-Å resolution. *Biochemistry 21*, 6956–6962.

CARLES, C., and MARTIN, P. 1985. Kinetic study of the action of bovine chymosin and pepsin A on bovine κ-casein. *Arch. Biochem. Biophys. 242*, 411–416.

CREAMER, L. K.; RICHARDSON, T.; and PARRY, D. A. D. 1981. Secondary structure of bovine α_{s1}- and β-casein in solution. *Arch. Biochem. Biophys. 211*, 689–696.

DUNN, B. M., and FINK, A. L. 1984. Cryoenzymology of porcine pepsin. *Biochemistry 23*, 5241–5247.

DUNN, B. M.; VALLER, M. J.; ROLPH, C. E.; FOUNDLING, S. I.; JIMENEZ, M.; and KAY, J. 1987. The pH dependence of the hydrolysis of chromogenic substrates of the type, Lys-Pro-Xaa-Yaa-Phe-(NO$_2$)Phe-Arg-Leu, by selected aspartic proteinases: Evidence for specific interactions in subsites S$_3$ and S$_2$. *Biochim. Biophys. Acta 913*, 122–130.

EMTAGE, J. S.; ANGEL, S.; DOEL, M. T.; HARRIS, T. J. R.; JENKINS, B.; LILLEY, G.; and LOWE, P. A. 1983. Synthesis of calf prochymosin (prorennin) in *Escherichia coli. Proc. Natl. Acad. Sci. USA 80*, 3671–3675.

FLAMM, E. L. 1991. How FDA approved chymosin: A case history. *Bio/Technology 9*, 349–351.

FOLTMANN, B. 1988. Aspartic proteinases: Alignment of amino acid sequence. The 18th Linderstrom-Lang Conference, Elsinore, Denmark, 4–8 July, 1988.

FOLTMANN, B.; PEDERSEN, V. B.; JACOBSEN, H.; KAUFFMAN, D.; and WYBRANDT, G. 1977. The complete amino acid sequence of prochymosin. *Proc. Natl. Acad. Sci. USA 74*, 2321–2324.

FOLTMANN, B.; PEDERSEN, V. B.; KAUFFMAN, D.; and WYBRANDT, G. 1979. The primary structure of calf chymosin. *J. Biol. Chem. 254*, 8447–8456.

Gilliland, G. L.; WINBORNE, E. L.; NACHMAN, J.; and WLODAWER, A. 1988. The three-dimensional structure of recombinant bovine chymosin at 2.3 A resolution. The 18th Linderstrom-Lang Conference, Elsinore, Denmark, 4–8 July 1988.

————. 1990. The three-dimensional structure of recombinant bovine chymosin at 2.3 A resolution. *Proteins: Structure, Function, and Genetics 8*, 82–101.

GOFF, C. G.; MOIR, D. T.; KOHNO, T.; GRAVIUS, T. C.; SMITH, R. A.; YAMASAKI, E.; and TAUNTON-RIGBY, A. 1984. Expression of calf prochymosin in *Saccharomyces cerevisiae. Gene 27*, 35–46.

HARRIS, T. J. R.; LOWE, P. A.; LYONS, A.; THOMAS, P. G.; EATON, M. A. W.; MILLICAN, T. A.; PATEL, T. P.; BOSE, C. C.; CAREY, N. H.; and DOEL, M. T. 1982. Molecular cloning and nucleotide sequence of cDNA for calf preprochymosin. *Nucl. Acid Res. 10*, 2177–2187.

HAYENGA, K. J.; CRABB, D.; CARLOMAGNO, L.; ARNOLD, R.; HEINSOHN, H.; and LAWLIS, B. 1988. Protein chemistry and recovery of calf chymosin from *Aspergillus nidulans* and *Aspergillus awamori*. The 18th Linderstrom-Lang Conference, Elsinore, Denmark, 4–8 July 1988.

HIRAMATSU, R.; AIKAWA, J.-i., HORINOUCHI, S.; and BEPPU, T. 1989. Secretion by yeast of the zymogen form of *Mucor* rennin, an aspartic proteinase of *Mucor pusillus*, and its conversion to the mature form. *J. Biol. Chem. 264*, 16862–16866.

HOFMANN, T.; DUNN, B. M.; and FINK, A. L. 1984. Cryoenzymology of penicillopepsin. Appendix: Mechanism of action of aspartyl proteinases. *Biochemistry 23*, 5247–5256.

HOFMANN, T., and HODGES, R. S. 1982. A new chromophoric substrate for penicillopepsin and other fungal aspartic proteinases. *Biochem. J. 203*, 603–610.

IWASAKI, S.; TAMURA, G.; and ARIMA, K. 1967. Milk clotting enzyme from microorganisms. Part II. The enzyme production and the properties of crude enzyme. *Agric. Biol. Chem. 31*, 546–551.

JAMES, M. N. G., and SIELECKI, A. R. 1985. Stereochemical analysis of peptide bond hydrolysis catalyzed by the aspartic proteinase penicillopepsin. *Biochemistry 24*, 3701–3713.

LIN, X.-L.; WONG, R. N. S.; and TANG, J. 1989. Synthesis, purification, and active site mutagenesis of recombinant porcine pepsinogen. *J. Biol. Chem. 264*, 4482–4489.

MANTAFOUNIS, D., and PITTS, J. 1990. Protein engineering of chymosin; Modification of the optimum pH of enzyme catalysis. *Protein Engineering 3*, 605–609.

McCAMAN, M. T., and CUMMINGS, D. B. 1986. A mutated bovine prochymosin zymogen can be activated without proteolytic processing at low pH. *J. Biol. Chem. 261*, 15345–15348.

NEWMAN, M.; SAFRO, M.; FRAZAO, C.; KHAN, G.; ZDANOV, A.; TICKLE, I. J.; BLUNDELL, T. L.; and ANDREEVA, N. 1991. X-ray analyses of aspartic proteinases IV. Structure and refinement at 2.2 A resolution of bovine chymosin. *J. Mol. Biol. 221*, 1295–1309.

NISHIMORI, K.; KAWAGUCHI, Y.; HIDAKA, M.; UOZUMI, T.; and BEPPU, T. 1982A. Expression of cloned calf prochymosin gene sequence in *Escherichia coli. Gene 19*, 337–344.

———. 1982B. Nucleotide sequence of calf prorennin cDNA cloned in *Escherichia coli. J. Biochem. 91*, 1085–1088.

PEARL, L., and BLUNDELL, T. 1984. The active site of aspartic proteinases. *FEBS Lett. 174*, 96–101.

PEDERSEN, V. B.; CHRISTENSEN, K. A.; and FOLTMANN, B. 1979. Investigations on the activation of bovine prochymosin. *Eur. J. Biochem. 94*, 573–580.

PEDERSEN, V. B., and FOLTMANN, B. 1975. Amino-acid sequence of the peptide segment liberated during activation of prochymosin (prorennin). *Eur. J. Biochem. 55*, 95–103.

RAAP, J.; KERLING, K. E. T.; VREEMAN, H. J.; and VISSER, S. 1983. Peptide substrate for chymosin (rennin): Conformational studies of κ-casein and some κ-casein-related oligopeptides by circular dichromism and secondary structure prediction. *Arch. Biochem. Biophys. 221*, 117–124.

SAFRO, M. G., and ANDREEVA, N. S. 1990. On the role of peripheral interactions in specificity of chymosin. *Biochemistry International 30*, 555–561.

SHIMADA, K., and CHEFTEL, J. C. 1989. Sulfhydryl group/disulfide bond interchange reaction during heat-induced gelation of whey protein isolate. *J. Agric. Food Chem. 37*, 161–168.

SIELECKI, A. R.; FEDOROV, A. A.; BOODHOO, A.; ANDREEVA, N. S.; and JAMES, M. N. G. 1990. Molecular and crystal structures of monoclinic porcine pepsin refined at 1.8 A resolution. *J. Mol. Biol. 214*, 143–170.

SIELECKI, A. R.; FUJINAGA, M.; READ, R. J.; and JAMES, M. N. G. 1991. Refined structure of porcine pepsinogen at 1.8A resolution. *J. Mol. Biol. 219*, 671–692.

STROP, P.; SEDLACEK, J.; KADERABKOVA, Z.; BLAKA, I.; PAVLICKOVA, L.; POHL, J.; FABRY, M.; KOSTKA, V.; NEWMAN, M.; FRAZAO, C.; SHEARER, A.; TICKLE, I.

J.; and BLUNDELL, T. L. 1990. Engineering enzyme subsite specificity: Preparation, kinetic characterization, and x-ray analysis at 2.0-A resolution of Val111 Phe site-mutated calf chymosin. *Biochemistry 29*, 9863–9871.

SUGUNA, K.; PADLAN, E. A.; SMITH, C. W.; CARLSON, W. D.; and DAVIES, D. R. 1987. Binding of a reduced peptide inhibitor to the aspartic proteinase from *Rhizopus chinensis*: Implications for a mechanism of action. *Proc. Natl. Acad. Sci. USA 84*, 7009–7013.

SUZUKI, J.; SASAKI, K.; SASAO, Y.; HAMU, A.; KAWASAKI, H.; NISHIYAMA, M.; HORINOUCHI, S.; and BEPPU, T. 1989. Alteration of catalytic properties of chymosin by site-directed mutagenesis. *Protein Engineering 2*, 563–569.

TONOUCHI, N.; SHOUN, H.; UOZUMI, T.; and BEPPU, T. 1986. Cloning and sequencing of a gene for *Mucor* rennin, an aspartate protease from *Mucor pusillus*. *Nucl. Acids Res. 14*, 7557–7568.

VISSER, S.; VAN ROOIJEN, P. J.; SCHATTENKERK, C.; and KERLING, K. E. T. 1976. Peptide substrates for chymosin (rennin). Kinetic studies with peptides of different chain length including parts of the sequence 101–112 of bovine κ-casein. *Biochim. Biophys. Acta 438*, 265–272.

————. 1977. Peptide substrates for chymosin (rennin). Kinetic studies with bovine κ-casein-(103–108)-hexapeptide analogues. *Biochim. Biophys. Acta 381*, 171–176.

VISSER, S.; VAN ROOIJEN, P. J.; and SLANGEN, C. J. 1980. Peptide substrates for chymosin (rennin). Isolation and substrate behavior of two tryptic fragments of bovine κ-casein. *Eur. J. Biochem. 108*, 415–421.

VISSER, S.; SLANGEN, C. J.; and VAN ROOIJEN, P. J. 1987. Peptide substrates for chymosin (rennin). Interaction sites in κ-casein-related sequences located outside the (103–108)-hexapeptide region that fits into the enzyme's active-site cleft. *Biochem. J. 244*, 553–558.

VOYNICK, I. M., and FRUTON, J. S. 1971. The comparative specificity of acid proteinases. *Proc. Natl. Acad. Sci. USA 68*, 257–259.

WATSON, F.; WOOD, S. P.; TICKLE, I. J.; SHEARER, A.; SIBANDA, B. L.; NEWMAN, M., KHAN, G.; FOUNDLING, S. I.; LOOPER, J.; VEERAPANDIAN, P.; and BLUNDELL, T. L. 1988. The evolution of three dimensional structure and specificity of aspartic proteinases: X-ray studies of endothiapepsin, mucorpepsin and chymosin. The 18th Linderstrom-Lang Conference, Elsinore, Denmark, 4–8 July 1988.

YAMASHITA, T.; TONOUCHI, N.; UOZYMI, T.; and BEPPU, T. 1987. Secretion of *Mucor* rennin, a fungal aspartic protease of *Mucor pusillus*, by recombinant yeast cells. *Mol. Gen. Genet. 210*, 462–467.

Chapter 6

Lipolytic Enzymes

Lipolytic enzymes consist of two major groups, the lipases, which are triacylglycerol acylhydrolase (EC 3.1.1.3), and the phospholipases A_1 (3.1.1.32) and A_2 (3.1.1.4), which are phosphoglyceride acyl hydrolases. Although phospholipases C (3.1.4.3) and D (3.1.4.4) are not acylhydrolases, they are nonetheless commonly included as lipolytic enzymes. The triacylglycerol lipases are found widely in animals, plants, and microorganisms. Animal lipases include pancreatic, gastric, and intestinal lipases, and also lipases found in milk. The present discussion covers the well-studied pancreatic lipase and phospholipase A_2. Microbial lipases will also be included because of their increasing importance in industrial applications.

ACTION AT LIPID/WATER INTERFACE

One of the unique characterisics of lipases is the phenomenon of interfacial catalysis. It was shown very early on that pancreatic lipase exhibits little activity when the water-soluble short fatty acid chain triglyceride, triacetin, is in the monomeric state. However, once the substrate concentration exceeds the solubility limit, the hydrolytic reaction increases dramatically with the same substrate in micelles or emulsion state (Fig. 6.1). The activity is independent of the molar substrate concentration, but controlled by the concentration of micellar substrates at interface. Other lipases are also strongly activated in a similar manner. In contrast, esterases act only on the water-soluble substrates and show a normal

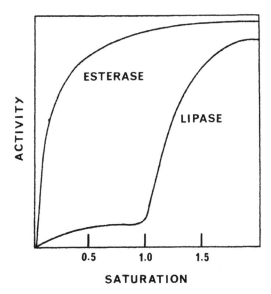

Figure 6.1.   Hydrolysis of triacetin by pancreatic lipase and esterase. Dotted line indicates saturation of substrate insolution (Sarda and Desnuelle 1958).

Michaelis-Menten-type dependence on substrate concentration. Lipase is often regarded as a distinct class of esterases that act specifically on water-insoluble esters at an oil-water interface.

A lipolytic reaction at the interface involves two equilibrium processes: (1) The enzyme must penetrate the interface in order to act. The adsorption of lipase to the interface induces a conformational change of the enzyme resulting in an increase in activity. This process is indicated by the equilibrium [E $\rightleftharpoons$ E*]. (2) Once the complex is formed, catalysis occurs, the products are formed, and the enzyme is regenerated as E*. Assuming enzyme inactivation at the interface is negligible, the overall reaction is represented by the following scheme (Verger 1976, 1980).

$$E \underset{k_d}{\overset{k_P}{\rightleftharpoons}} E^* \overset{S}{\underset{k_{-1}}{\overset{k_1}{\rightleftharpoons}}} E^*S \overset{k_{cat}}{\longrightarrow} P$$

It is also conceivable that desorption of the enzyme from the micelle occurs in each catalytic turnover. The products are formed with the enzyme released from the interface, and the enzyme E, now in the aqueous phase has to penetrate to the interface E* for substrate binding.

Several types of substrates are commonly employed in the study of interfacial catalysis, including emulsions, micelles, or liposomes, monolayers, and vesicles. Different interfaces exhibit differences in the kinetics of hydrolysis because the relative distribution of the two pathways through which E* is regenerated during the turnover is different. The kinetic aspects of interfacial catalysis are detailed in a number of reviews (Verger 1980; Jain and Berg 1989; Gelb et al. 1992).

TRIACYLGLYCERIDE LIPASES

PRIMARY AMINO ACID SEQUENCES

The complete amino acid sequence of porcine pancreatic lipase has been determined (De Caro et al. 1981). The primary structures of human, canine, and horse pancreatic lipases have been determined from their respective nucleotide sequences (Fig. 6.2) (Lowe et al. 1989; Mickel et al. 1989; Kerfelec et al. 1992). The porcine pancreatic enzyme is a single polypeptide of 449 amino acids, with a calculated MW 49,859. A carbohydrate chain composed of fucose, galactose, mannose, and N-acetylglucosamine is attached to Asn166, leading to a total molecular weight of 52 kD. Porcine pancreatic lipase exists in two forms, L_A and L_B, varying in carbohydrate content. Pancreatic lipases from horse, sheep, and bovine are not glycoproteins (Plummer and Sarda 1973).

The porcine enzyme contains six disulfides and two free sulfhydryl groups. The positions of the six disulfides are Cys4-Cys10, Cys237-Cys261, Cys433-Cys449, Cys285-Cys296, Cys299-Cys304, and Cys90-Cys103 (or Cys90-Cys101) (Benkouka et al. 1982). The disulfide bonds bridge the polypeptide chain into a succession of small loops, with considerable flexibility required for the formation of a productive enzyme-substrate complex at interfaces. The two sulfhydryl groups differ in reactivities; Cys181 reacts readily with a variety of sulfhydryl reagents, while the other sulfhydryl group (Cys101 or Cys103) is less reactive. Analysis of a refined structure of horse pancreatic lipase seems to suggest that Cys103 is the free Cys, and the disulfide is Cys90-Cys101 (Bourne et al. 1994). Human, canine, and horse pancreatic lipases show 86%, 67%, and 68% identity to porcine pancreatic lipase, respectively.

The genes coding for a wide variety of microbial lipases have been isolated, characterized, and their primary structures deduced from the corresponding nucleotide sequences. Included in this catagory are *Staphylococcus hyicus* (Gotz et al. 1985), *Pseudomonas fragi* (Kugimiya et al. 1986), *Staphylococcus aureus* (Lee and Iandolo 1986), *Geotrichum candidum* (Shimada et al. 1989), *Rizomucor miehei* (Boel et al. 1988), and *Candida antarctica* (Uppenberg et al. 1994). The

```
PPL          SEVCFPRLGCFSDDAPWAGIVQRP
CPL          K---YEQI-----AE--A-TAI--
EPL          N---YERL-----DS--A-IVE--
HPL          MLPLWTLSLLLGAVAGK---YERL-----DS--S-ITE--


PPL     25   LKILPP DKDVDTRFLLYTNQNQNNYQELVADPSTITN S
CPL     25   -KV--WSPERIG--------K-PN-F-TLLPSDPSTIEA-
EPL     25   -KI--WSPEKVN--------E-PD-F-EIVADPSTIQS -
HPL     25   -HI--WSPKDVN--------E-PN-F-EVAADSSSISG -


PPL     63   NFRMDRKTRFIIHGFIDKGEEDWLSNICKNLFKVESVNCI
CPL     65   --QTDK----T-----N----N--LDM-K-M----E----
EPL     64   --NTGR----I-----D----S--STM-Q-M----S----
HPL     64   --KTNR----I-----D----N--ANV-K-L----S----


PPL     103  CVDWKGGSRTGYTQASQNIRIVGAEVAYFVEVLKSSLGYS
CPL     105  -----K--Q-S-T--AN-V-V---Q--QMLSM-SANYS--
EPL     104  -----S--R-A-S--SQ-V-I---E--YLVGV-QSSFD--
HPL     104  -----G--R-G-T--SQ-I-I---E--YFVEF-QSAFG--


PPL     143  PSNVHVIGHSLGSHAAGEAGRRTNGTIERITGLDPAEPCF
CPL     145  --Q-QL---S--A-V-----S--P- LG-----D-V-AS-
EPL     144  --N-HI---S--S-A-----R--N-AVG-----D-A-PC-
HPL     144  --N-HV---S--A-A-----R--N-TIG-----D-A-PC-


PPL     183  QGTPELVRLDPSDAKFVDVIHTDAAPIIPNLGFGMSQTVG
CPL     184  ---P-E-----T--D--------A--LI-F----T--QM-
EPL     184  ---P-L-----S--Q--------I--FI-N----M--TA-
HPL     184  ---P-L-----S--K--------G--IV-N----M--VV-


PPL     223  HLDFFPNGGKQMPGCQKNILSQIVDIDGIWEGTRDFVACN
CPL     224  ---------EE----K--A-----NL----E-----V---
EPL     224  ---------KE----Q--V-----DI----Q-----A---
HPL     224  ---------VE----K--I-----DI----E-----A---


PPL     263  HLRSYKYYADSILNPDGFAGFPCDSYNVFTANKCFPCPSE
CPL     264  H-------SE--L------SYP-A--RA-ES------PDQ
EPL     264  H-------TD--L------GFS-A--SD-TA------SSG
HPL     264  H-------TD--V------GFP-A--NV-TA------PSG
```

Figure 6.2. Comparison of amino acid sequences of lipases: PPL (porcine pancreatic, De Caro et al. 1981), CPL (canine pancreatic, Mickel et al. 1989), EPL (equine pancreatic, Kerfelec et al. 1992), HPL (human pancreatic, 465 amino acids deduced from cDNA, 449 amino acids for the mature protein,16 amino acids for the signal peptide, Lowe et al. 1989). The N-terminal amino acids are double underlined. Catalytic residues are boldfaced.

```
PPL  303  GCPQMGHYADRFPGKTNGVSQVFYLNTGDASNFARWRYKV
CPL  304  ----------KFAVK-SDET-KYF-N---S--------GV
EPL  304  ----------RFPGR-KGVG-LFY-N---A--------RV
HPL  304  ----------RYPGK-NDVG-KFY-D---A--------KV


PPL  343  SVTLSGKKVTGHILVSLFGNEGNSRQYEIYKGTLQPDNTH
CPL  344  SI------RA--QAK-A---SK--TH-FN-FK-I-K-GS-H
EPL  344  DV------KV--HVL-S---NK--SR-YE-FQ-T-K-DN-Y
HPL  344  SV------KV--HIL-S---NK--SK-YE-FK-T-K-DS-H


PPL  383  SDEFDSDVEVGDLQKVKFIWYNNNVINPTLPRVGASKITV
CPL  484  -N---AKLD--TIEK---L-N--  -V-P-F-K---A--T-
EPL  484  -N---SDVE--DLEK---I-Y--  -I-L-L-K---S--T-
HPL  484  -N---SDVD--DLQM---I-Y--  -I-P-L-R---S--I-


PPL  423  ERNDGK VYDFCSQETVREEVLLTLNPC  449
CPL  423  QKGEEKTVHS---ES----D-----TP-  450
EPL  423  ERNDGS VFN---EE----D-----TA-  449
HPL  423  ETNVGK QFN---PE----E-----TP-  449
```

Figure 6.2. (Continued).

fungus *Geotrichum candidum* produces extracellular lipases in multiple iso-forms—lipases I and II, with pI of 4.56 and 4.46, respectively (Veeraragavan et al. 1990). The genes for both lipases I and II present in four different strains have been cloned and sequenced (Bertolini et al. 1994). Lipase I consists of 554 amino acid residues (19 residues in the signal peptide) with MW of 59,085. The enzyme contains five Cys, 7% carbohydrates, mostly mannose, at two potential N-gly-cosylation sites. Lipase II contains 544 amino acids (13 are in the signal peptide) with a calculated MW of 59,550. The enzyme has four Cys, and 6.5% carbohy-drates. The two isozymes show 84% similarity in the sequences. Four of the Cys residues are conserved in the two lipases. Except for the common sequence Gly-X-Ser-X-Gly, where the active Ser is located, there is little homology between the *Geotrichum candida* lipases and the other microbial lipases, except for the *Candida cylindracea* enzyme (~45% overall homology) (Shimada et al. 1989, 1990).

The *Rhizomucor miehei* enzyme also contains at least two isoforms, A and B. Lipase B could be a deglycosylated form of A (Huge-Jensen et al. 1987). A *Rhizomucor miehei* lipase cDNA gene has been sequenced and shown to encode for a zymogen of 363 amino acids with a polypeptide precursor 70 residues long located between a 24-amino acid signal peptide and the N-terminus of the 269-

residue mature enzyme (Boel et al. 1988). The zymogen and the mature protein have a predicted MW of 39,529 and 29,472, respectively.

THREE-DIMENSIONAL STRUCTURES

All the lipases whose structures have been elucidated contain a common α/β fold consisting of a central parallel β sheet where all the residues in the catalytic triad (His-Ser-Asp/Glu) are located (Fig. 6.3).

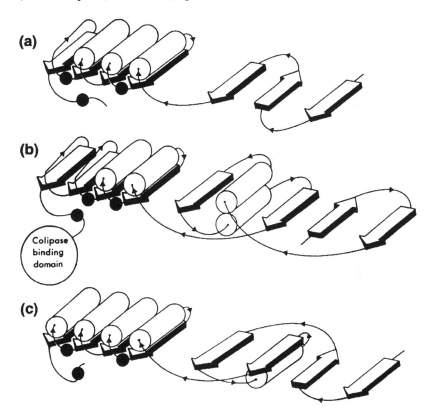

Figure 6.3. The common α/β fold of lipases: (a) *Rhizomucor miehei*, (b) *human pancreatic*, (c) *Geotrichum candidum*. All the molecules contain central, predominantly parallel β sheets. The core consists of four parallel strands, which often have helical connections; it also contains the catalytic triad and constitutes the most invariant structural motif of the fold. (From Derewenda and Sharp 1993. *TIBS 18*, 22, with permission. Copyright 1993 Elsevier Trends Journals, Cambridge.)

Pancreatic Lipases

Human pancreatic lipase is a polypeptide chain folded into two domains. The catalytic N-terminal (1–335 residues) domain has an α/β structure comprised of a central predominantly parallel β sheet of nine strands, and the C-terminal segment contains two layers of antiparallel 4-stranded β sheets. The N-terminal domain contains the catalytic residues, whereas the C-terminal domain is involved in colipase binding (Abousalham et al. 1992). This colipase binding domain is not found in microbial lipases. The β sheet structure is asymmetrical, with most of the connecting elements lining one side of the sheet. The distal side of the sheet is supported by an N-terminal helix. The essential Ser152 is located in the N-terminal domain at the edge of C-terminal β sheet layers. The active site containing the catalytic triad [His264-Ser153-Asp177] (His263-Ser152-Asp176 in porcine pancreatic lipase) is inaccessible to substrate, because it is concealed by a surface loop (residues 237–261) of a short α helix. A conformational change of this "flap" region (or the "lid") must occur to open a channel to the active site residues in catalysis.

A refined structure of horse pancreatic lipase is available (Fig. 6.4) and its comparison with that of the human enzyme shows very few differences (Bourne et al. 1994). As in the human enzyme, the horse lipase consists of two domains—

Figure 6.4. A stereo view of horse pancreatic lipase. Residues 4 of the catalytic triad and the disulfide bridges are represented in ball-and-stick form. The common core of eight β strands is represented in dark shading. (From Bourne et al. 1994. *J. Mol. Biol. 238,* 717, with permission. Copyright 1994 Academic Press, Ltd.)

the N-terminal domain (1–335) where the catalytic residues are found, and the C-terminal domain (336–446) where the colipase-binding site is located. The N-terminal domain has a typical α/β structure consisting of a central β sheet of nine strands interconnected by α helices. The catalytic residues are located between strands and helices. The Ser153-O_γ and Asp177-$O_{\delta 2}$ are hydrogen bonded to the $N_{\epsilon 2}$ and $N_{\delta 1}$ of His264, respectively. The active site, as in the human enzyme, is buried under a flexible "lid" that must be repositioned during interfacial activation to allow the substrate to enter the active Ser via a channel formed by hydrophobic side chains. A calcium binding site is found to be located between β strands 8 and 9. The functional role of this calcium ion is not known, although the human enzyme also contains a similar site.

Microbial Lipases

The *Rhizomucor miehei* lipase contains a single domain—a central β sheet of eight β strands connected by hairpins, loops, and helices in an α/β arrangement (Brady et al. 1990). Three disulfides are identified: Cys29-Cys268, Cys40-Cys43, and Cys235-Cys244. The active Ser144 is located in the catalytic site sequence Gly-His-Ser-Leu-Gly (Boel et al. 1988). The catalytic triad, consisting of His257, Ser144 and Asp203, is buried under a short surface helical lid (residues 85–91) that is held entirely by hydrophobic interactions (Derewenda et al. 1992A). The structure of lipase B from *Candida antarctica* has many features in common: a central α/β fold, an active site triad (His224-Ser105-Asp187), and "lid" movement (Uppenberg et al. 1994).

The *Geotrichum candidum* lipase I is also a single-domain protein that contains the basic structure of α/β arrangement with a mixed β sheet formed by 11 parallel strands, seven of which are parallel (Schrag et al. 1991) (Fig. 6.5). The helices and loops connecting many of the strands are found on both sides of the β sheet. There is a small three-stranded β sheet at the N-terminus. Two disulfides are present: Cys61-Cys105, and Cys276-Cys288. The enzyme is N-glycosylated at Asn283 and Asn364 or O-linked at Thr4. The active site consists of the catalytic triad His463-Ser217-Glu354, which is concealed by a "lid" of two parallel α helices (residues 66–76, 294–310). Conformational changes through the unwinding of these highly hydrophobic helices would allow the substrate access to the catalytic residues. The replacement of Asp with Glu in the catalytic triad is also found in the *Candida cylindracea* lipase, which shares ~45% homology of the sequence (Longhi et al. 1992). In the pancreatic lipases, the catalytic Asp is located in a loop at the C-terminal end of β strand 6. In microbial lipases, the acid residue (Asp/Glu) is found in the β turn connecting β strands 7 and 8.

The structure of a bacterial lipase from *Pseudomonas glumae* consists of three domains (Noble et al. 1993). The largest domain is an α/β arrangement with a central 6-stranded β sheet. The second domain consists of three helices, part of

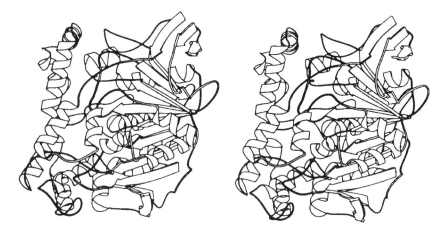

Figure 6.5. A stereo view of the lipase structure from *Geotrichum candidum* showing a side view of the 11-stranded mixed β sheet. (From Schrag et al. 1991. *Nature 351,* 762, with permission. Copyright 1991 *Macmillan Magazine,* Ltd.)

which may constitute the "lid." The third domain contains four helices and a β hairpin loop extending out from the main structure. A calcium loop is also present with ligands close to the catalytic His285. A three-domain structure is also observed in the *Candida rugosa* lipase (Grochulski et al. 1993).

CATALYTIC TRIAD IN THE ACTIVE SITE

All lipases consist of a catalytic triad of His-Ser-Asp-Glu, reminiscent of that present in serine proteinases. Specific site mutation of the triad residues in human pancreatic lipase demonstrates the essential role of these residues in catalysis. Substitution of Ser153 with Ala, Cys, Gly, Phe, Asp, Pro, Thr, and Val causes a drastic decrease in specific activity but has no effect on binding micelles (Lowe 1992). Crystallographic studies on human pancreatic lipase reveal a structure consistent with Ser153 as the nucleophile in a catalytic triad (Winkler et al. 1990). The essential serine residue is conserved in all lipases, including lipoprotein lipase and hepatic lipase (Fig. 6.6) (Yang et al. 1989; Mickel et al. 1989). The common sequence found in these lipases, Gly-X-Ser-X-Gly, agrees with the active site sequence in serine proteinases. The trypsin family has a characteristic sequence of Gly-Asp-Ser-Gly-Gly, and subtilisin Gly-X-Ser-X-Ala (Brenner 1988). The hydrophilic sequence surrounding the active site of serine proteinases is replaced by a highly hydrophobic sequence in the lipases. Porcine pancreatic, canine pancreatic, and human lipoprotein lipases show hydropathy indices of 8.7, 8.0, and 9.5, respectively, for the nine active site amino acid sequences. For chymotrypsin

PORCINE PANCREATIC	VHVIGHSLGSHAA	152	De Caro et al. 1981
CANINE PANCREATIC	VQLIGHSLGAHVA	154	Mickel et al. 1989
HUMAN PANCREATIC	VHVIGHSLGAHAA	153	Lowe et al. 1989
HORSE PANCREATIC	VHIIGHSLGSHAA	152	Kerfelec et al. 1992
Rizomucor miehei	VAVTGHSLGGATA	144	Boel et al. 1988
Geotrichum candidum	VMIFGESAGAMSV	217	Shimada et al. 1989
Pseudomonas fragi	VNLIGHSQGALTA	83	Kugimiya et al. 1986
Staphylococcus aureus	VHLVGHSMGGQTI	412	Lee and Iandolo 1986
BOVINE MILK LIPOPROTEIN	VHLLGYSLGAHAA	134	Yang et al. 1989
HUMAN LIOPROTEIN	VHLLGYSLGAHAA	159	Kirchgessner et al. 1989
MOUSE MACROPHAGE LIPOPROTEIN	VHLLGYSLGAHAA	132	Kirchgessner et al. 1987
RAT HEPATIC	VHLIGYSLGAHVS	147	Komaromy and Schotz 1987
HUMAN HEPATIC	VHLIGYSLGAHVS	152	Cai et al. 1989
HUMAN GASTRIC	LHYVGHSQGTTIG	153	Bodmer et al. 1987

Figure 6.6. Comparison of immediate environment of the active site serine in a number of lipases.

and trypsin, the indices are -2.7 and -10.8, respectively (Mickel et al. 1989). The hydrophobic environment of the active site serine may be an important factor for lipase-catalyzed hydrolysis at an oil-water interface. The catalytic triad in lipases is similar to that of the serine proteinases in certain respects. The serine is within hydrogen-bonding distance from $N_{\varepsilon 2}$ of the histidine. The other ring nitrogen ($N_{\delta 1}$) is hydrogen bonded to the carboxylate function of the Asp. The stereochemistry of the residues is comparable to that observed in the serine proteinases; the histidine imidazole ring and the hydrogen-bonded oxygens of the Asp and Ser are superimposable. But the other side chain atoms are very different in orientation (Winkler et al. 1990; Derewenda and Derewenda 1991; Derewenda et al. 1992A). There is little homology in other parts of the sequence and structure between the lipases and the serine proteinases. In addition, the lipase active site Ser is located in a β turn (type II) flanked by a β strand and a buried α helix. In the human pancreatic enzyme, the β strand is formed by residues 146–151, and the α helix by residues 153–164, with the Ser153 in an extended conformation. In the *Rhizomucor miehei* lipase, the catalytic Ser loop is located between β strand 4 (residues 137–143) on the C-terminal side, and the α helix of residues 145–159. This β-εSer-α structural motif, which is believed to be a common feature for lipases, constitutes the most rigid region of the molecule (Derewenda and Derewenda 1991). This unique motif is found also in a number of esterases, lending support to the contention that an evolutionary link may exist between lipases and esterases.

OXYANION HOLE

It is well established that oxyanion binding sites in trypsin (Gly193 and Ser195) and subtilisin (Ser221 and Asn155) form hydrogen bonds to the carbonyl oxygen of the scissile peptide bond. Such binding contributes to the stabilization of charge distribution and reduction of the ground state energy of the tetrahedral intermediate (Warshel et al. 1989). In the *Rhizomucor miehei* lipase, such an oxyanion hole is formed during the conformational change of the "lid" region during interface activation, and Ser82 is the likely candidate for the function of stabilization of the tetrahedral intermediate (Derewenda et al. 1992A). The oxyanion hole in *Geotrichum candidum* lipase may involve the main chain nitrogen of Ala218 and Ala132 as suggested by crystallography (Schrag and Cygler 1993). In human pancreatic lipase, the repositioning of the lid (which is stabilized by hydrogen bonding with colipase as discussed in the following section) induces a second conformational change involving the β5 loop in the active site (van Tilbeurgh et al. 1993). This change brings the Phe78 side chain into an ideal position for its main chain nitrogen to stabilize the negatively charged oxyanion (Ser153-O_γ) developed in ester hydrolysis. In the horse enzyme, the backbone NH of Phe78 and Leu154 may also be involved in oxyanion stabilization (Bourne et al. 1994). The *Candida rugosa* lipase has the oxyanion hole formed by the backbone NH atoms of Gly124 and Ala210 (Grochulski et al. 1993).

THE "LID"

A unique feature of all the lipases is that the catalytic triad is buried under a "lid" of a surface loop that must undergo a conformational change to open a channel for the active site for access of substrate. This finding substantiates the suggestion that interfacial activation of lipase is caused by a change of its conformation resulting from adsorption (Chapus et al. 1976; van Dam-Mieras et al. 1975; Chapus and Semeriva 1976). Hydrolysis of triacetin by pancreatic lipase is enhanced 100–500-fold by various interfaces. It is apparent from the molecular structure of lipases that the displacement of the "lid" represents conformational changes necessary for the active site to be accessible to the substrates. This repositioning of the "lid" is caused by interfacial activation and will be discussed further in conjunction with the phospholipase A_2 activation mechanism. It should be noted that guinea pig pancreatic lipase does not have a "lid" and is not interfacially activated (Hjorth et al. 1993), and that it exhibits both lipase and phospholipase activities.

MECHANISM OF CATALYSIS

The lipase-catalyzed reaction has long been recognized to be very similar to the hydrolytic action of the serine proteases in many respects (Brockerhoff 1973; Chapus and Semeriva 1976; Garner 1980). The formation of an acyl lipase intermediate has been detected in the hydrolysis of *p*-nitrophenyl acetate by porcine

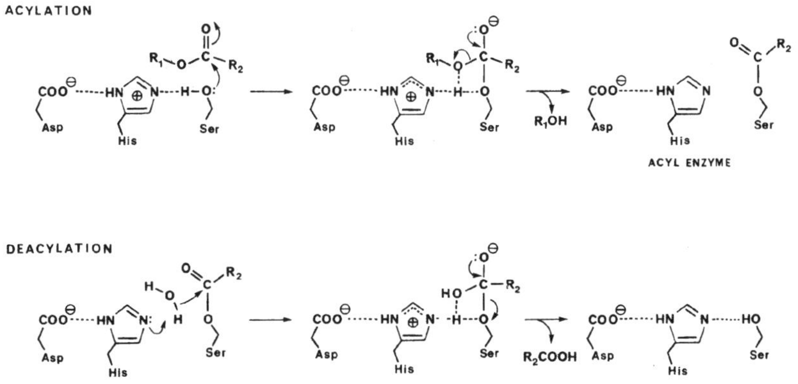

ACYLATION

ACYL ENZYME

DEACYLATION

Figure 6.7. The proposed reaction mechanism of lipase showing acylation and deacylation via the possible formation of stable tetrahedral intermediates.

pancreatic lipase (Semeriva et al. 1974). Kinetic studies on the hydrolysis of dissolved *p*-nitrophenyl acetate by pancreatic lipase also suggest that the reaction follows an acyl enzyme mechanism with $k_3 = k_{cat} = 0.11 \pm 0.02$ min^{-1}, $k_2 = 7 \pm 0.7$ min^{-1}, $K_{m,app} = 0.22 \pm 0.02$ mM, and $K_s = 15$ mM (Chapus et al. 1976). The relation $k_{cat}/K_m = k_2/K_s$ is consistent with acyl–enzyme formation. Recent evidence obtained from the crystal structures of two lipases on the presence of a catalytic triad Asp-His-Ser in the active center provides further support for the reaction mechanism.

If we accept the premise that lipase functions via acyl–enzyme formation and the Asp-His-Ser triad is involved, then the following mechanism can be envisioned to occur (Fig. 6.7). In the acylation step a covalent acyl-enzyme is formed by nucleophilic attack on the carbonyl carbon of the substrate by the essential Ser-OH. Deacylation of the acyl–enzyme occurs with H_2O being the nucleophile, giving the product and the enzyme. Both acylation and deacylation may possibly involve the formation of a tetrahedral intermediate.

The oxyanions developed in the tetrahedral intermediates in both acylation and deacylation are stabilized by hydrogen bonds with the residues in the oxyanion hole. The presence of a tetrahedral intermediate and its stabilization in the two pathways has been revealed by crystallographic studies of the *Candida rugosa* lipase complexed with transition state analog inhibitors (Grochulski et al. 1994). The catalytic His has its $N_{\varepsilon2}$ atom hydrogen bonded to both the reactive Ser-O_γ and the oxygen of the leaving alcohol of the substrate in the tetrahedral intermediate of the acylation step. Likewise, the His-$N_{\varepsilon2}$ forms hydrogen bonds with catalytic Ser-O_γ and the H_2O nucleophile in the deacylation step.

COLIPASE AND ITS FUNCTIONAL ROLE

Interfacial adsorption of an enzyme often results in the unfolding of the three-dimensional structure, and consequently inactivation. Lipases are no exception in this respect. Lipolytic hydrolysis slows down and eventually stops after an initial reaction period. In the physiological environment, the adsorption and hence the denaturation of pancreatic lipase is prevented by the presence of bile salts at these interfaces. At a low bile salt concentration, the enzyme is stabilized at the interface by a lowering of the surface energy. At an increasing concentration above a critical micelle concentration, bile salt inhibits lipase adsorption at the interface by physically excluding the enzyme from the interface or by a general detergent effect. Bile salts are principally cholic and chenodeoxycholic acids conjugated with glycine or taurine. In general, many amphipathic compounds are effective inhibitors.

Primary Structure of Colipase

For the pancreatic lipase to act on the substrate at the interface in the presence of bile salts, a protein cofactor, colipase, is necessary for enzyme-substrate binding. Porcine pancreatic colipase is excreted as a procolipase of 101 amino acid residues. It is activated by proteolytic removal of the N-terminal pentapeptide (Val-Pro-Asp-Pro-Arg) to yield the active colipase$_{96}$. Higher concentrations of trypsin further cleave a number of amino acids from the C-terminal to yield a shorter but active colipase$_{84}$ (Larsson and Erlanson-Albertsson 1981). Human and chicken procolipases have not been isolated (Sternby and Borgstrom 1979; Bosc-Bierne et al. 1981), although loss of the N-terminal pentapeptide during purification cannot be excluded. Multiple forms of colipase differing in electrophoretic properties also exist, possibly because of their content of Glu and Asp residues (Erlanson et al. 1973). Colipase I contains 13 (Asp + Glu) and 9 (Asn + Gln), whereas colipase II has 12 (Asp + Glu) and 10 (Asn + Gln) (Borgstrom et al. 1974). The two forms differ in only one acidic residue. Evidence for the existence of two colipases with a few different sequence substitutions has also been reported for horse pancreatic colipase (Julien et al. 1980).

The amino acid sequence of porcine pancreatic colipase has been determined (Erlanson et al. 1974A) and revised (Erlanson et al. 1974B) (Fig. 6.8). Complete sequences of horse colipase I and II (Bonicel et al. 1981; Pierrot et al. 1982) and human colipase$_{86}$ (Sternby et al. 1984B) and a partial sequence of chicken (Bosc-Bierne et al. 1981) and dogfish (Sternby et al. 1984A) colipases are also known. A high homology exists among these colipases.

Three-Dimensional Structure of Colipase

The porcine procolipase is a flattened molecule consisting of three finger-like regions (residues 21–34, 42–59, and 62–82) held by five disulfides (Cys12-

```
P-CL II    1    GIIINLDEGELCLNSAQCKSNCCQHDTILSLLRCALKARE
H-CL       1    -I----EN--L-M---Q---N--QHSSA-G-A--TSM-S-
E-CL A     1    -V----EA--I-L---E---E--HQEES-S-A--AAK-S-


P-CL II   41    NSECSAFTLYGVYYKCPCERGLTCEGDKSLVGSITNTNFG
H-CL      41    -----VK----I------------EG--TI----T-----
E-CL A    41    -----AW----V------------QV--TL----M-----


P-CL II   81    ICHNVG    86
H-CL      81    --HDAG    86
E-CL A    81    --FNAAKER    89
```

Figure 6.8. Amino acid sequences of colipases: P-CL II (porcine colipase II, Erlanson et al. 1974A, B), H-CL (humancolipase, Sternby et al. 1984A), E-CL (horse colipase A, Pierrot et al. 1982). Porcine and horse procolipase contain an N-terminal peptide VPDPR.

Cys23, Cys18-Cys34, Cys22-Cys56, Cys58-Cys82, and Cys44-Cys64 with sequence numbering starting at the pentapeptide) (van Tilbeurgh et al. 1992). The procolipases have an overall amphipathic character. The high hydrophobic finger-like region constitutes the lipid-binding site. Located opposite to this is the binding site for lipase consisting mostly of hydrophilic residues (van Tilbeurgh et al. 1993).

Lipid Binding Region.   The binding of colipase to the interface involves two regions—a short N-terminal segment and a three-finger region (Fig. 6.9). The N-terminal end contains three consecutive hydrophobic residues that are involved in lipid binding, Ile7-Ile8-Ile9 in both porcine and human procolipases, and Val-Ile-Ile and Leu-Ile-Phe in horse and chicken colipases, respectively. The finger regions are mobile, and contain mostly hydrophobic residues that interact with the bile salt-lipid substrate. In the center finger, three highly conserved Tyr residues (50, 53, and 54) are located in a strong hydrophobic loop, designated also as the tyrosine loop region. The three Tyr, together with His25, are recognized as part of the binding site for bile salt micelles (Wieloch and Falk 1978; Wieloch et al. 1979; De Caro et al. 1983; Granon 1986). Thus, colipase binds bile salt micelle in a 1:1 stoichiometry, via predominantly hydrophobic interactions (Borgstrom and Donner 1975; Charles et al. 1975). The dissociation constant K_d between porcine colipase and lipase in a 1:1 complex in buffer solution is $5 \times 10^{-7}\,M$ (Patton et al. 1978). In the presence of substrate and bile salt, K_d is lowered to $10^{-11}\,M$ (Larsson and Erlanson-Albertsson 1983).

The colipase binding site for the bile salt micelle also serves for the binding of the cofactor to the substrate at the interface (Chapus et al. 1975). The binding of colipase in the presence of bile salt to the substrate tributyrin has a K_d of $3.3 \times 10^{-7}\,M$ (Erlanson-Albertsson 1980). The binding is dependent on pH, salt

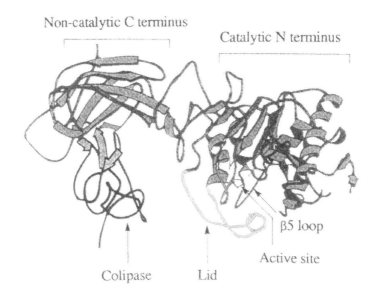

Non-catalytic C terminus

Catalytic N terminus

β5 loop

Active site

Colipase   Lid

Interfacial Activation

⇩

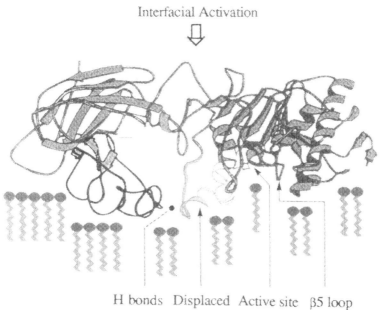

H bonds Displaced Active site β5 loop
          lid

Figure 6.9. The structures of a human pancreatic lipase-colipase complex and its conformational rearrangements in the presence of mixed phospholipid/bilesalt micelles. (From Riddihough 1993. *Nature 362*, 793, with permission. Copyright 1993 Macmillan Magazine, Ltd.)

184

concentration, and type and concentration of bile salt. At increasingly high concentration of bile salt, noncompetitive inhibition of colipase binding occurs.

Lipase Binding Region.   Lipase-colipase binding involves mostly hydrophilic residues located in the two hairpin loops and the C-terminal end of the colipase. Chemical modification studies suggest that two carboxyl groups essential for colipase function in binding lipase (Erlanson et al. 1977) are located in the end segments. Sequence comparison indicates that Asp7, Glu10, or Asp67 may be involved (Bosc-Bierne et al. 1985; Sternby et al. 1984B). The importance of ionic interaction between colipase and lipase is substantiated by the finding that the corresponding amino group of lipase involved in the binding is a Lys residue located at the C-terminal region of the enzyme (Mahe-Gouhier and Leger 1988; Chaillan et al. 1989). The colipase binding site in lipase is located entirely in the C-terminus, which is separated from the catalytic N-terminal domain by a short linking region (Abousalham et al. 1992). In addition, colipase also binds to the lipase lid in its open structure following activation. Three hydrogen bonds occur between the lid residues (Val247, Ser244, and Asn241) of the human pancreatic enzyme and the colipase residues (Arg33, Leu11, and Glu10) (van Tilbeurgh et al. 1993). Crosslinking of lipase and colipase using carbodiimides identifies Glu440 in the porcine pancreatic lipase as the residue in ionic interaction (Chaillian et al. 1992). Two salt bridges, and six hydrogen bonds account for these binding interactions (van Tilbeurgh et al. 1992).

Action of Colipase

The proposed mechanism of the colipase effect involves a system of four components: lipase, colipase, bile salt, and substrate. In the presence of bile salt, the lipase is excluded from interface adsorption. The function of colipase then is to bind to the substrate at the interface, and to anchor the lipase. The resulting complex stabilizes the lipase against surface denaturation and brings the lipase in proximity to the substrate, where interfacial activation and hydrolysis can occur (van Tilbeurgh et al. 1992). Because a lipid substrate interface covered by a monolayer of bile salts mimics a micelle, the binding between colipase and bile salt at the interface may also be of importance for the binding of colipase to the substrate molecule. A high concentration of bile salt can, by forming an increasing amount of free micelles, compete with the substrate for the binding of colipase (Chapus et al. 1975; Borgstrom et al. 1979). Microbial lipases from *Rhizopus arrhizus* and *Geotrichum candidum* are also inhibited by bile salts in the same manner as the pancreatic lipases. However the activity cannot be restored by the addition of pancreatic colipase (Canioni et al. 1977).

A detailed aspect of the action of colipase has been revealed by x-ray crystallography of the human pancreatic lipase-colipase (porcine) complex, cocrystallized with mixed micelles of phosphatidylcholine and bile salt (van Tilbeurgh 1993) (see Fig. 6.9). In the complex, colipase is seen to interact with the open lid of the lipase molecule, via three hydrogen bonds (Arg33, Leu11, and Glu10, and the lid residues Val247, Ser244, and Asn241). As described in a later section, the helical lid blocking the active site of the enzyme moves to expose the active site upon binding to the lipid-water interface. This rearrangement of the "lid" brings it closer toward the colipase for interaction. The conformation of this open structure of the lipase molecule represents a critical function of colipase. The open lid together with the colipase finger-like region form a continuous hydrophobic surface extending over 50 A. This surface allows enough extensive interactions with the lipid micelle at the interface to overcome the inhibitory effect of the bile salts.

SUBSTRATE BINDING AND SPECIFICITY

Substrate Binding Site

The substrate binding site of lipases, as revealed by the refined structure of the *Geotrichum candidum* enzyme (Schrag and Cygler 1993) and of the *Candida rugosa* enzyme-inhibitor complex (Grochulski et al. 1993, 1994A), consists of a hydrophobic tunnel extending from the catalytical Ser to the surface that can accommodate a C18 fatty acid chain. Longer fatty acid chains can fit in the tunnel with slight movement of some side chains. Movement of the "lid" in interfacial activation exposes the mouth of this tunnel to allow its accessibility to substrate molecules. Structural analysis of the *Candida rugosa* lipase-inhibitor complex indicates that at a high ratio of inhibitor to enzyme concentrations, a second molecule of inhibitor is covalently bound to a His residue at a site apart from the active site (Grochulski et al. 1994B). Modelling of triglyceride binding at the active site and the second site shows that one fatty acid acyl chain fits into the substrate binding tunnel where the active site is located, and a second fatty acid chain occupies the second binding site. The spatial arrangement of a triglyceride molecule assumes a turning fork configuration.

It is tempting to identify this second binding site with the so-called second "catalytic" site described in some literature. It has long been speculated that, in addition to a catalytic site similar to the serine proteinases, lipases also contain a second site that can hydrolyze soluble substrates. Treatment of porcine pancreatic lipase with a serine proteinase inhibitor, diethyl-*p*-nitrophenyl phosphate, results in specific modification of Ser152. The enzyme, however, does not react with another serine proteinase inhibitor, diisopropylfluorophosphate. The modified enzyme loses its ability for interface recognition and activity towards natural insoluble substrates (Chapus and Semeriva 1976; Guidoni et al. 1981) but retains hydrolytic action on monomeric (soluble) substrates, such as *p*-nitrophenylacetate. This residual activity has been related to a functional domain (336–449) in the C-terminal sequence of the porcine enzyme, indicative of a secondary active site involving His354 and Lys373 (De Caro et al. 1986). However, the occurrence of a second catalytic site is doubtful, or at least not a general feature, because His354 is not conserved in dog pancreatic lipase. Horse lipase, lacking the corresponding Lys residue, shows no residual activity (Kerfelec et al. 1992). It has been suggested that the residual activity may be the result of a migration of the acyl group from the reactive His in the triad to the Lys (Kaimal and Saroja 1989; Winkler et al. 1990).

Lipase Specificity

There are three known types of lipase specificity (Jensen et al. 1983).

(1) Substrate specificity: The enzyme hydrolyzes the various acylglycerols, tri-, di-, and monoglycerides (TG, DG, MG) or types of fatty acids at different rates.

(2) Positional specificity: The enzyme catalyzes the release of fatty acids at preferential positions (primary, secondary ester or random hydrolysis) on the acylglycerol molecule.

(3) Stereospecificity: The enzyme hydrolyzes the two primary esters (sn-1 or sn-3) at different rates. The stereochemistry of glycerol derivatives is expressed by the stereospecific numbering nomenclature, sn, which recognizes the fact that the two primary carbinol groups of the parent glycerol are not identical in their reaction with nonsymmetric structures.

Pancreatic lipase hydrolyzes only primary esters, and the sequence of hydrolysis follows TG > DG > MG. Substitution or unsaturation near the carboxyl end of the aliphatic ester chain leads to decreasing rate of hydrolysis. The rate of lipolytic hydrolysis of saturated acids generally increases with increasing chain length, usually with a preference for butyric acid (Brockerhoff 1970).

Microbial lipases generally show a similar mode of action on triglycerides as that exhibited by pancreatic lipase. Lipase from the mold *Geotrichum candidum* has a unique specificity for unsaturated fatty acids containing cis-9 or cis, cis-9,12 double bonds regardless of position in the TG (Jensen 1974). Trans-isomers and positional isomers, such as cis-6 or cis-11 are hydrolyzed at a very slow rate. *Staphylococcus aureus* lipase hydrolyzes TG, DG, and MG at similar rates (Rollof et al. 1987). Multiple forms of *Geotrichum candidum* lipase exist, depending on a particular strain, and also on the culture medium employed. These lipase species exhibit a different degree of substrate specificities toward oleate with respect to the other fatty esters (Baillargeon 1990; Jacobson et al. 1990). However, recent characterization of two *Geotrichum candidum* lipases I and II suggests that these are products of two separate lipase genes and show essentially the same substrate specifies (Sugihara et al. 1990). A lipase purified from *Geotrichum candidum* strain CMICC 335426 shows a marked specificity for unsaturated substrates with a cis-9 double bond (Sidebottom et al. 1991).

Pancreatic lipase hydrolyzes only the primary esters at the 1- and 3-positions of glycerides (Paltauf et al. 1974). Because 1,2- and 2,3-DG and 2-MG are unstable because of acyl migration to yield the 1,3-DG, prolonged reaction will cause complete hydrolysis of the triglycerides to glycerol. The enzyme shows no stereoselectivity for sn-1 or sn-3 positions in TG, although sn-3 stereoselectivity has been detected in DG analogs (Ransac et al. 1990; Rogalska et al. 1990).

Lipases from *Aspergillus flavus, Staphylococcus aureus, Geotrichum candidum, Penicillium cyclopium,* and *Candida cylindracea* lack positional specificity (Alford et al. 1964; Vadehra 1974; Rollof et al. 1987; Okumura et al. 1976; Benzonana and Esposito 1971). Lipases from *Aspergillus niger, Rhizopus delemar, Rhizopus arrhizus,* and *Mucor javanicus* exhibit specificity towards the 1,3-positions of TG (Okumura et al. 1976; Benzonana 1974; Linfield et al. 1984; Ishihara et al. 1975).

Pancreatic and microbial lipases do not show stereospecificity in the hydrolysis of triglycerides. Lingual and gastric (human and rat) lipases have specificity for the sn-3 position (Akesson et al. 1983; Rogalska et al. 1990). Lipoprotein lipases (milk, plasma, adipose tissue) and heparin-releasable hepatic lipase attack at the sn-1 primary ester of TG (Morley et al. 1975; Akesson et al. 1976). Phospholipase A$_1$ also acts on the sn-1 substituent of phosphoglyceride.

ACYL TRANSFER REACTION

Under conditions in which the reaction system consists of increasing proportions of nonaqueous solvent, the lipase-catalyzed hydrolysis reaction is reversed in favor of acyl transfer (for a discussion of low-water solvent media, see Halling 1989). In this type of reaction, the acyl group is transferred to the glycerol instead of water. Although proteinases such as chymotrypsin and subtilisin exhibit increased catalytic efficiency with solvent hydrophobicity, porcine pancreatic lipase shows very minor solvent effects (Kanerva et al. 1990). The catalytic activity of the enzyme depends on the layer of bound water, and hence on the water activity in the reaction mixture (Goldberg et al. 1990; Valivety et al. 1992A). In organic solvents, enzyme activity has been reported at a_w of about 0.1, and even as low as 0.0001 in some cases (Valivety et al. 1992B).

The acyl transfer reaction offers a novel way to synthesize acylglycerols (Jensen et al. 1978; Linfield et al. 1984; Hayes and Gulari 1990). Using lipase with desired specificity, it becomes possible to manipulate the esterification process to obtain a designated product. Lipases from *Geotrichum candidum* and *Penicillium cyclopium* catalyze esterification of long chain fatty acids at all three positions of the glycerol. *Aspergillus niger* and *Rhizopus delemar* lipases synthesize only 1(3)-MG and 1,3-DG (Tsujisaka et al. 1977).

The acyl transfer reaction, however, is most useful to the food industry in its application in the interesterification process (Tanaka et al. 1981; Stevenson et al. 1979). Hydrolysis and resynthesis cause acyl exchange between acylglycerol molecules to form interesterified product mixtures unusually unobtainable from chemical process (Macrae 1983). Exchange reactions between acylglycerol and free fatty acids produce new products enriched with the added fatty acid. *Geotrichum candidum* lipase, being specific for fatty acids with cis-9 or cis, cis-9,12 unsaturations, can be used in interesterification of olive oil with linoleate, generating triglycerides enriched in linoleate. Kinetic studies of interesterification reactions catalyzed by *Candida cylindracea* lipase in cyclohexane using trilaurin substrate suggest that k_{cat} values for esterification are three times those for hydrolysis ($k_{cat} = 3$ s^{-1} and 0.9 s^{-1}, respectively). In organic medium, it is likely that the hydrolysis step is the rate-limiting step (Miller et al. 1991).

Lipases also provide a novel method for synthesis of a variety of optically active compounds (Klibanov 1990). The following list includes a few examples of selected lipase-catalyzed reactions in nonaqueous medium (Fig. 6.10).

(1) Preparation of optically active esters from transesterification between tributyrin and a wide variety of chiral alcohols (*Candida cylindracea* lipase, Cambou and Klibanov 1984).

(2) Preparation of optically active alcohols, carboxylic acids, and their esters (*Candida cylindracea* lipase, Kirchner et al. 1985).

(3) Preparation of enantiomeric esters of hydroxy compounds (glycerol, serinol de-

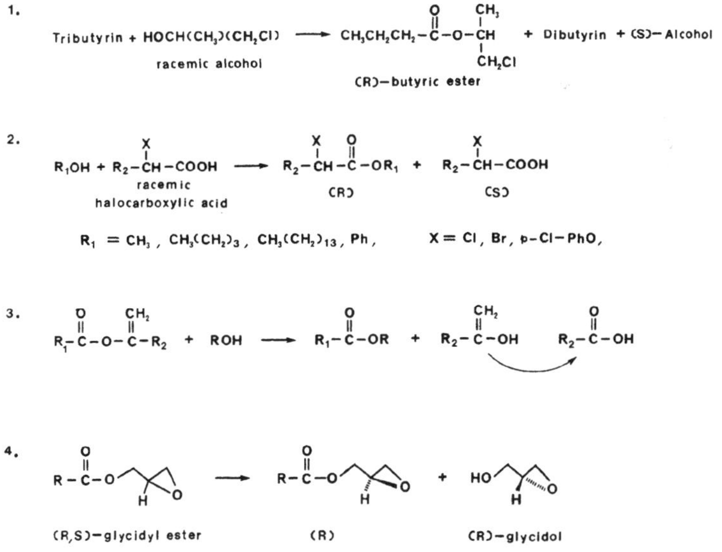

Figure 6.10. Examples of lipase-catalyzed reactions in nonaqueous solvent.

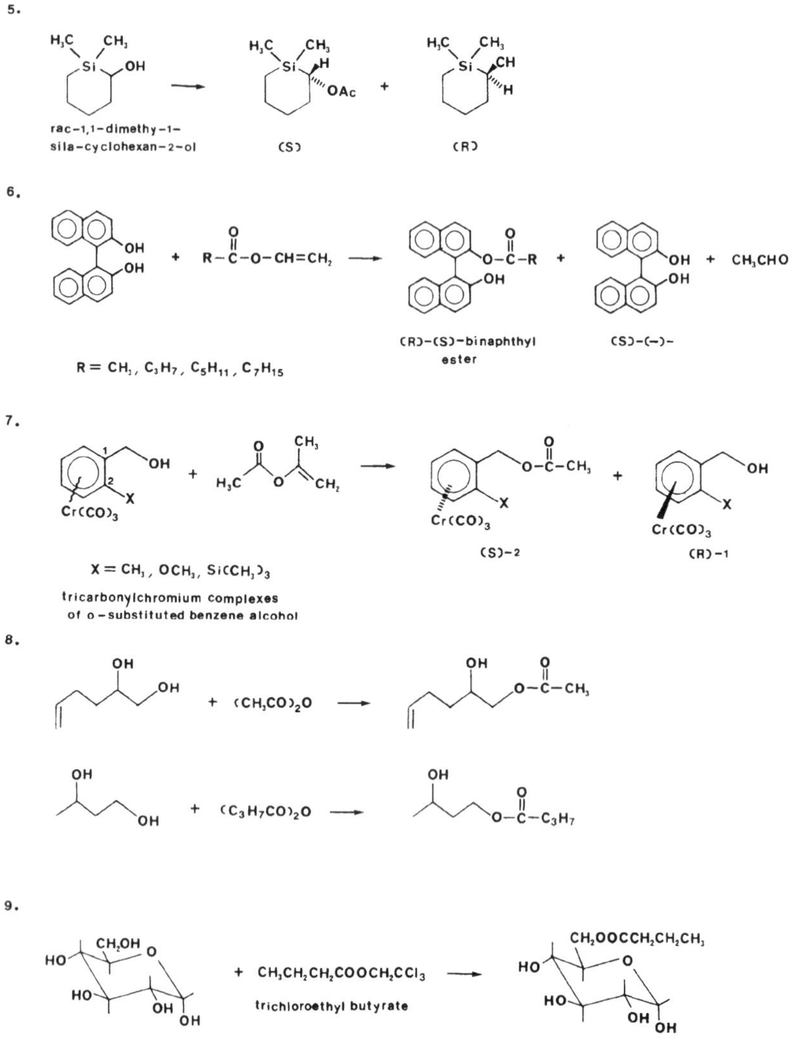
5.
rac-1,1-dimethy-1-
sila-cyclohexan-2-ol
(S)
(R)
6.
R = CH3, C3H7, C5H11, C7H15
(R)-(S)-binaphthyl
ester
(S)-(−)-
CH3CHO
7.
X = CH3, OCH3, Si(CH3)3
tricarbonylchromium complexes
of o-substituted benzene alcohol
(S)-2
(R)-1
8.
(CH3CO)2O
(C3H7CO)2O
9.
CH3CH2CH2COOCH2CCl3
trichloroethyl butyrate
CH2OOCCH2CH2CH3

Figure 6.10. (Continued).

rivatives, sugars, and other alcohols) using enol esters for acylating (*Pseudomonas* sp. lipase, Wang et al. 1988).

(4) Synthesis of β-adrenergic blocking agents (aryloxypropanolamines) in their active (*S*)-enantiomeric form (porcine pancreatic lipase, Kloosterman et al. 1988).

(5) Separation of enantiomers from a racemic mixture of dimethyl-1-sila-cyclohexanone (*Candida cylindracea* lipase, Fritsche et al. 1989).

(6) Acylation of binaphthol with enol esters to yield (R)-(+)-binaphthyl monoester (*Pseudomonas* sp. lipase, Inagaki et al. 1989).

(7) Asymmetric esterification of tricarbonylchromium complexes of *o*-substituted benzyl alcohol derivatives (*Pseudomonas* sp. lipase, Nakamura et al. 1990).

(8) Selective acylation of primary OH of hydroxy compounds (porcine pancreatic lipase, Ramaswamy et al. 1990).

(9) Selective esterification of carbohydrates (porcine pancreatic lipase, Therisod and Klibanov 1986; see also Seino et al. 1984; Sweers and Wong 1986; Hennen et al. 1988; Uemura et al. 1989).

It is obvious from all these reactions that lipases are quite tolerant with respect to the nature of the substrates in nonaqueous solvents. These results imply that lipase can also be used to catalyze hydrolysis of various substrates with regiospecificity in aqueous medium. For example, hydrolysis of $3',5'$-di-*O*-hexanoylpyrimidine nucleoside by the *Pseudomonas fluorescens* lipase yields the $5'$-*O*-acylnucleoside with the deacylation at the $3'$ position (Uemura et al. 1989). Asymmetric hydrolysis of α-acyloxy esters by immobilized microbial lipases produces chiral intermediates for the synthesis of (−)-indolmycin and diltiazem hydrochloride (Akita et al. 1989). Porcine pancreatic lipase has been shown to catalyze the hydrolysis of monosubstituted α-amino lactones in preference to the L-enantiomer. The enzyme also hydrolyzes disubstituted lactones with enantioselectivity (Gutman et al. 1990). Lipase has been used for the optical resolution of racemic esters of epoxy alcohols (Ladner and Whitesides 1984), and for the resolution of chiral acyclic amino alcohols (Foelsche et al. 1990). Recently, the preference for (1R)-enantiomers of secondary alchols in reactions catalyzed by many lipases and esterases has been related to the binding orientations of the substrate in the active site (Cygler et al. 1994). The lack of reactivity of the 1S-enantiomer may be the result of restrictive sterics imposed by the active site.

PANCREATIC PHOSPHOLIPASE A$_2$

Phospholipases catalyze the hydrolysis of phospholipids. The enzymes are classified according to their stereospecificity on the substrate. A phospholipase that hydrolyzes the sn-1 acyl ester is called phospholipase A$_1$ (Phosphatide 1-acylhydrolase, EC 3.1.1.32). A phospholipase that specifically acts on the sn-2 position is designated as phospholipase A_2 (Phosphatide 2-acylhydrolase, EC 3.1.1.4). Phospholipases C (Phosphatidylcholine cholinephosphohydrolase, EC 3.1.4.3) and D (Phosphatidylcholine phosphatidohydrolase, EC 3.1.4.4) are not acyl hydrolases, because these enzymes cleave the sn-3 phosphodiester bond. Phospholipase C hydrolyzes the phosphodiester bond near the glycerol side, whereas phospholipase D attacks the bond close to the polar side chain. Phospholipases secreted from the pancreas, as well as isolated from snake venoms, all belong to the A_2 type. Intracellular phospholipases from many mammalian tissues, including liver, spleen, lung, and heart, contain both A_1 and A_2 activities.

PRIMARY AMINO ACID SEQUENCES

The primary structures of a number of pancreatic phospholipases have been elucidated; these include enzymes from the porcine (Puijk et al. 1977), horse (Evenberg et al. 1977), and bovine pancreas (Fleer et al. 1978).

The porcine pancreatic phospholipase A_2 contains 124 amino acids, with seven disulfide bonds. The horse enzyme consists of a single polypeptide chain of 125 amino acids with a MW of 13,927. A similar structure is found for the bovine phospholipase A_2, which consists of 123 amino acids and seven disulfide bonds, and has a MW of 12,782. The sequences are highly homologous, with 71% of the amino acids conserved in the three species (Fig. 6.11).

The enzyme is synthesized as a zymogen which is activated by tryptic removal of seven amino acids from the N-terminus. The zymogen acts only on monomeric substrates, whereas the activated enzyme also shows high activity towards insoluble substrates, such as micelles.

```
PORCINE PLA₂   1    ALWQFRSMIKCAIPGSHPLMDFNNYGCYCGLGGSGTPVDE
BOVINE PLA₂    1    -L---N---K-K--S-E-LLD--N--------------D
HORSE PLA₂     1    -V---R---Q-T--N-K-YLE--D--------------E

PORCINE PLA₂   41   LDRCCETHDNCYRDAKNLDSCKFLVDNPYTESYSYSCSNT
BOVINE PLA₂    41   --R--QTH----KQ--K-D--KV-------NN-SY---NN
HORSE PLA₂     41   --A--QVH----TQ--E-S--RF-------ES-KF---GT

PORCINE PLA₂   81   EITCNSKNNACEAFICNCDRNAAICFSKAPYNKEHKNLDT
BOVINE PLA₂    81   -I--SSE----------D---------V-Y-K-H----K
HORSE PLA₂     81   -V--SDK----------D---------A-K-P-N----S

PORCINE PLA₂   121  -KYC   124
BOVINE PLA₂    121  -NC    123
HORSE PLA₂     121  -RKAC  125
```

Figure 6.11. Comparison of amino acid sequences of porcine pancreatic (Puijk et al. 1977), bovine pancreatic (Fleer et al. 1978), and horse pancreatic (Evenberg et al. 1977) phospholipase A_2.

THREE-DIMENSIONAL STRUCTURES

The tertiary structures of both the bovine and porcine pancreatic phospholipase A_2 are known (Fig. 6.12) (Dijkstra et al. 1978, 1981B, 1983A). Bovine pancreatic phospholipase A_2 has a compact molecular structure consisting of two long antiparallel α helices (helix C = 40–58, helix E = 90–108), connected by two disulfide bonds (Cys44-Cys105, Cys51-Cys98). The active site residues are located in these two helices. Two short α helices, A and B (residues 1–13, and 19–22, respectively) are in the N-terminal region. The polypeptide segment lying between helices B and C is characterized by two disulfide bridges—one linking helix C (Cys29-Cys45), and the other linking the C-terminal cysteine residue (Cys27-Cys123). The long C-terminal segment (109–123) therefore forms a surface loop over the center core (helices C and E). The remaining helix D (59–64) together with two antiparallel β strands (74–78 and 80–85) connect the remaining ends of helices C and E. Helix D forms a disulfide bridge (Cys61-Cys91) with helix E, while the β strands are linked to helix A and helix E via the disulfide bonds, Cys77-Cys11 and Cys84-Cys96, respectively. Extensive hydrogen bonding occurs between main chain (peptide) and side chain, and also among side chain atoms (Dijkstra et al. 1981B). The high number of disulfide bonds provides considerable stability for the enzyme against denaturation at the interface.

The porcine enzyme has a very similar structure, except for the segment covering residues 59–70. Val63 in the bovine enzyme is replaced by a Phe, with

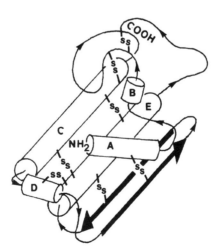

Figure 6.12. Schematic drawing showing the helices and β structure of the bovine phospholipase A_2 molecule. (From Dijkstra et al. 1978. *J. Mol. Biol. 124,* 57, with permission. Copyright 1978 Academic Press, Ltd.)

its side chain oriented towards the hydrophobic interior. Consequently, helix D in bovine phospholipase A_2 structure is replaced by a conformation of a 3_{10} helix (Dijkstra et al. 1983A, B).

Active Site

The two highly conserved catalytic residues, His48 and Asp99, are located in a depression surrounded by hydrophobic residues: Phe5, Ile9, Phe22, Ala102, Ala103, Phe106, and the Cys29-Cys45 disulfide (Dijkstra et al. 1981A). This His-Asp pair act together with the H_2O nucleophile as a catalytic triad analogous to that of serine proteinases. The His48 $N_{\epsilon 2}$ is in close proximity to one of the carboxylate oxygens of Asp99, and a water molecule in close contact with $N_{\delta 1}$ acts as nucleophile in place of the Ser-OH in serine proteinases. The active site of the porcine enzyme has structures similar to those of the bovine enzyme. The hydrophobic channel formed around the active site is also found in other phospholipases and is well characterized in the cobra-venom enzyme (White et al. 1990). The structure of this channel and its functional importance in interfacial catalysis will be discussed in a later section.

Two highly conserved residues, Tyr52 and Tyr73, which form a hydrogen bond network with the carboxylate oxygen of Asp99, have been suggested to be a critical part of the catalytic step. Substitution of either one or both of the Tyr residues by site-directed mutation demonstrates that the phenolic OH groups involved in hydrogen bonding are not essential for catalysis, but are related to the conformational stability of the active site (Kuipers et al. 1990; Dupureur et al. 1992). The Tyr73 $\cdots$ Asp99 hydrogen bond contributes a stabilization energy of 4–5 kcal/mol as the only linkage between the β segment (74–78) and the helix E (where Asp99 is located) (Dupureur et al. 1992). The phenolic hydroxyl group of Tyr52 is less critical because the Tyr52 $\cdots$ Asp99 hydrogen bond in this case connecting helices C and E are also held together by two disulfide bonds.

Calcium Binding Site

The catalytic activity of phospholipases requires calcium ion as a cofactor, which is located in the active site and coordinated by seven oxygen ligands. The ligands involved in the calcium coordination include the following: the main chain oxygens of Tyr28, Gly30, and Gly32, two water molecules, and the two side chain oxygens of Asp49. Coordination of these closely located residues tends to fold the peptide chain in the region (25–42) into a loop surrounding the calcium ion. The coordination together with the interactions among a number of conserved amino acid residues in this loop serve to stabilize the conformation of the calcium-binding site in a precise orientation with respect to the active site.

The porcine enzyme shows identical coordination, although substitutions with different amino acids occur in the loop, which may affect the binding affinity of the site for calcium ion. The bovine and porcine enzymes have K_d values for calcium of 8.0 and 2.5 mM, respectively (Meyer et al. 1979A)

Interface Recognition Site

The N-terminal sequence has been shown to be involved in the interaction of phospholipase A$_2$ with micellar interfaces (van Dam-Mieras et al. 1975). The interface recognition site of the bovine enzyme consists of a number of N-terminal residues: Ala1, Leu2, Trp3, Asn6, Glu17, Leu19, Leu20, Asn23, Asn24, Leu31, Lys56, Val65, Asn67, Tyr69, Thr70, Asn72, Lys116, Asn117, Asp119, Lys120, and Lys123 (Fig. 6.13) (Dijkstra et al. 1981A, 1984; Meyer et al. 1979A, B). All the residues lie in a plane around the entrance to the active site except residues 64–66, which bulge out forming a protuberance. Although these amino acid residues are not entirely conserved in other phospholipases (Fig. 6.11), there is similarity in folding, and in the hydrophilic/hydrophobic patterns. The Tyr69 side chain may interact with the phosphate of the substrate (Renetseder et al. 1988).

At neutral or acidic pH, the terminal α-amino group of Ala1 is protonated, and hydrogen bonded to the main chain oxygen of Asn-71, and the side chain oxygen of Gln4. A third hydrogen bond forms between the Ala-αNH$_3$ and an internal H$_2$O molecule, which in turn links to Tyr52, Asp99 (side chain oxygen) and Pro68 (main chain oxygen). In the porcine enzyme, the basic arrangement is

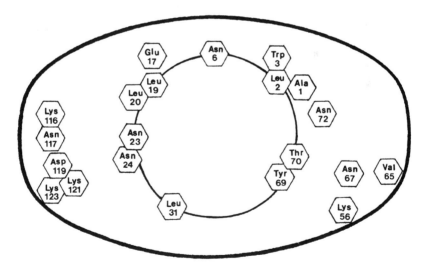

Figure 6.13. Schematic representation of the face of the molecule, which is directed towards the phospholipid layer. (From Dijkstra et al. 1981A. *Nature 289*, 605, with permission. Copyright 1981 Macmillan Magazine, Ltd.)

similar, except that the H_2O molecule is hydrogen bonded to Tyr52 and Asp99. In addition, Leu2 is hydrogen bonded to Tyr69.

In contrast to the mature enzymes, the zymogens consist of N-terminal residues that are highly disordered, and the loop (62–72) exhibits no defined structure (Dijkstra et al. 1982). The difference in this N-terminal structure may account for the fact that the zymogens act on monomeric substrates, and exhibit no activity towards insoluble micelles. Deletion of part of the surface loop (62–66) from porcine pancreatic phospholipase A_2 by site-directed mutagenesis results in a 16-fold increase in catalytic activity on micellar substrates (Kuipers et al. 1989). It is conceivable that the deletions allow the binding site to form a smooth plane that interacts more effectively with aggregated substrate.

The bovine enzyme exhibits a lower affinity for various interfaces than the porcine enzyme. Bovine phospholipase A_2 has a dissociation constant $K_d > 13$ mM, whereas the porcine enzyme has a K_d of 1.8 mM (Meyer et al. 1979A) in the interaction with n-hexadecylphosphocholine at pH 6.0 and 25°C.

Second Calcium Binding Site

In addition to the calcium binding site at the active site, porcine pancreatic phospholipase A_2 has a second calcium binding site located close to the N-terminal region. The second calcium is coordinated with the side chains of Glu71, Glu92, and Ser72 (Donne-Op den Kelder et al. 1983; Dijkstra et al. 1983A). The second binding site has a low affinity for calcium ($K_d = 20$ mM). It functions to promote enzyme-interface interaction at a high pH range.

At alkaline pH, the Ala1 α-amino group is deprotonated, resulting in an interruption of the conformation of the interface recognition site. The binding of calcium at the N-terminal region tends to retain the conformation by shifting the pK of the α-NH$_3^+$ group from 8.4 to 9.3 (Slotboom et al. 1978). A similar site is not found in the bovine enzyme.

MECHANISM OF CATALYSIS

In the reaction catalyzed by phospholipase A_2, the Ca^{++} serves to bind the sn-2 carbonyl oxygen and sn-3 phosphate, positioning the phospholipid substrate in a correct orientation. The binding of the Ca^{++} also contributes to polarization of the ester carbonyl bond, making it more susceptible to nucleophilic attack by the H_2O molecule assisted by the His48 imidazole (Fig. 6.14) (Verheij et al. 1980; Slaich et al. 1992).

The tetrahedral oxyanion thus formed is stabilized by the main chain NH group of Gly30 and the calcium ion. Because H_2O is the nucleophile, phospholipase A_2-catalyzed hydrolysis does not proceed via an acyl enzyme intermediate (Wells 1971). In the breakdown of the tetrahedral intermediate, the proton pulled

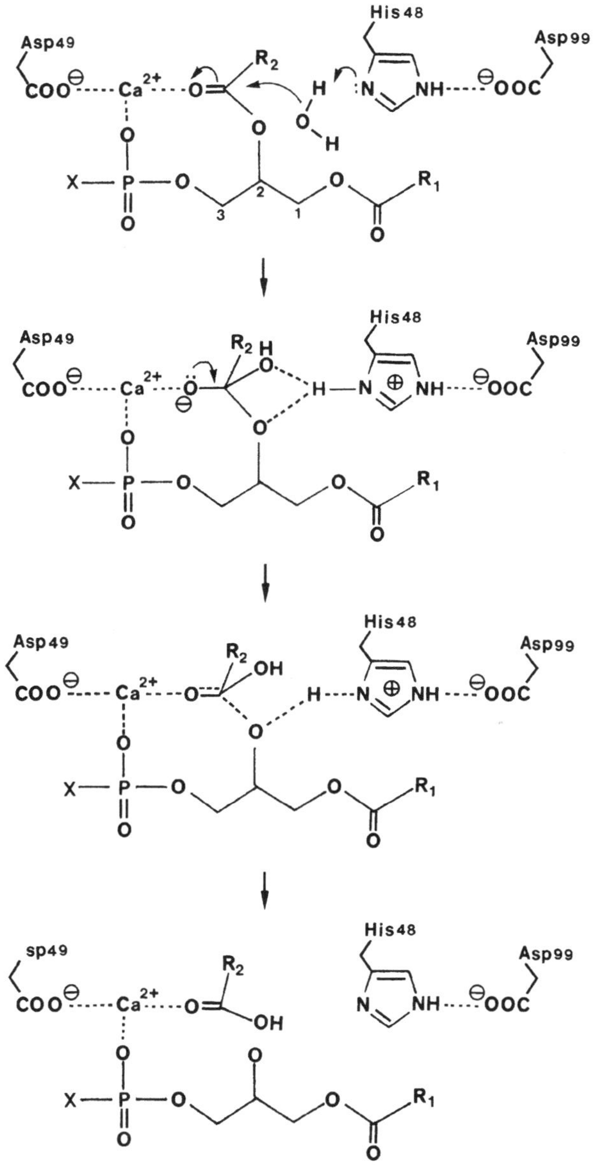

Figure 6.14. The proposed mechanism of phospholipase A$_2$. Details described in text. (From Scott et al. 1990).

off from the H_2O to His48 is transferred to the C2 oxygen, via a doubly hydrogen bonded structure (Waszkowycz and Hillier 1990).

In the catalytic mechanism originally proposed (Verheij et al. 1980), the functional role of Asp99 and His48 involves a concerted transfer of protons (from H_2O to His, and from His to Asp) in a charge-relay type mechanism. Recent studies on the serine proteinases suggest that the Asp residue in the catalytic triad stabilizes the tetrahedral intermediate by electrostatic interactions, rather than an actual transfer of protons (Kossiakoff and Spencer 1981; Warshel et al. 1989) (see Chapter 5). The same conclusion may apply also to the role of Asp99 in phospholipase A_2 (Dennis 1983; Waszkowycz and Hillier 1989, 1990).

MOLECULAR BASIS FOR INTERFACIAL CATALYSIS

Both lipases and phospholipases act at the lipid-water interface, and various hypotheses have been suggested in the past to explain this phenomenon of interfacial activation. Crystallographic investigations on the structure of enzyme-inhibitor complexes offer an explanation on a molecular basis with regard to how the enzyme binds to the lipid substrate and initiates hydrolysis of the ester bond at a micelle surface.

The crystal structure of *Rhizomucor miehei* lipase reveals the active site buried underneath a helical lid of 15 amino acids (Leu85–Asp91) with two hinge regions (residues 82–84 and 92–95) (Brzozowski et al. 1991; Derewenda et al. 1992A). The lid and the hinges undergo conformational changes equivalent to a hinge-type rigid-body motion by a translation of over 12 A. The movement exposes the hydrophobic side of the lid that originally shields the active site (Fig. 6.15B). The resulting increase in the hydrophobic area around the active site provides a hydrophobic seal at the interface with the lipid phase. There are 12 hydrophobic residues clustered around the entrance to the active site, with a hydrophobic surface area of $\sim$750 A^2. The energy of stabilization contributed by the formation of this hydrophobic seal is estimated to be $\sim$4 kcal/mol (Derewenda et al. 1992A). In response to the formation of the hydrophobic contact, the enzyme catalytic residues are now accessible for action on the triglyceride molecules on the micelle surface. That interfacial activation involving conformational rearrangement of a "lid" was also evident in the *Candida rugosa* lipase when both the inactive (closed) and active (open) forms were studied (Grochulski et al. 1993; 1994A, B).

Pancreatic lipase undergoes similar movement of the "lid" segment, caused by a reorganization of the hinge region (residues 238–247 and 256–260). The open lid moves its position close to the colipase bound at the enzyme noncatalytic C domain (Fig. 6.9). Three hydrogen bonds hence established between the open lid and colipase stabilize the conformation of the open structure (van Tilbeurgh

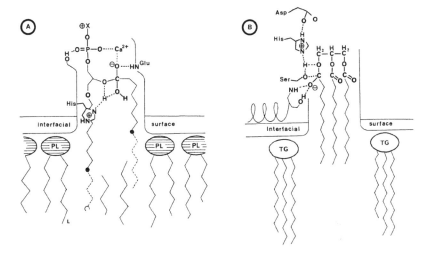

Figure 6.15. Diagrams of (A) phospholipase A₂, and (B) lipase as interfacial substrate complexes. PL phospholipid, TG triacyglyceride. (From (A) Scott 1990. *Science 250,* 1545, with permission. Copyright 1990 American Association for the Advancement of Science, (B) Blow 1991. *Nature 351,* 444, with permission. Copyright 1990 Macmillan Magazine, Ltd.)

et al. 1992, 1993). The open lid joins the finger regions (lipid binding site) of the colipase in forming a continuous hydrophobic surface that strongly holds the lipid-colipase complex to the interface against the bile salt effect. As already discussed, the opening of the lid also causes a second conformational change in the β5 loop in the active site, which positions and stabilizes the oxyanion hole for catalysis.

Phospholipase A₂ also exhibits interfacial activation, but there is no significant movement of the protein backbone as shown in the case of lipase (Thunnissen et al. 1990). A hydrophobic channel, where the active site is located deep inside, is formed by the anchoring of a flexible "flap" on one side of the wall upon binding to the substrate (Fig. 6.15A). Interfacial binding involves the enzyme surface around the entrance of the channel (mostly the N-terminal helix and neighboring residues) and the micellar surface. The formation of the hydrophobic channel at the interface facilitates the diffusion of phospholipid molecules from the micelle into the active site. The phospholipid molecule drawn into the channel maintains its conformation similar to those in the crystalline and micellar structures. The sn-1 and sn-2 substituents are parallel inside the hydrophobic channel, and the sn-3 projects away from the sn-1 and sn-2 chains to form interactions with the calcium ion. The conformation of the phospholipid molecule in micelle aggregate satisfies the optimum condition for enzyme binding and action. Although this mechanism of interfacial catalysis is based primarily on the studies of crystal structures of the Chinese cobra and bee venom enzyme complexes with a phosphonate transition state analog, it is believed that this model may represent a general feature for this class of enzyme (White et al. 1990; Scott et al. 1990).

REFERENCES

ABOUSALHAM, A.; CHAILLAN, C.; KERFELEC, B.; FOGLIZZO, E.; and CHAPUS, C. 1992. Uncoupling of catalysis and colipase binding in pancreatic lipase by limited proteolysis. *Protein Engineering 5*, 105–111.

AKESSON, B.; GRONOWITZ, S.; and HERSLOF, B. 1976. Stereospecificity of hepatic lipases. *FEBS Lett. 71*, 241–244.

———. 1983. Stereospecificity of different lipases. *Lipids 18*, 313–318.

AKITA, H.; ENOKI, Y.; YAMADA, H.; and OISHI, T. 1989. Enzymatic hydrolysis in organic solvents for kinetic resolution of water-insoluble α-acyloxy esters with immobilized lipases. *Chem. Pharm. Bull. 37*, 2876–2878.

ALFORD, J. A.; PIERCE, D. A.; and SUGGS, F. G. 1964. Activity of microbial lipases on natural fats and synthetic triglycerides. *J. Lipid Res. 5*, 390–394.

BABA, T.; DOWNS, D.; JACKSON, K. W.; TANG, J.; and WANG, C.-S. 1991. Structure of human milk bile salt activated lipase. *Biochemistry 30*, 500–510.

BAILLARGEON, M. W. 1990. Purification and specificity of lipases from *Geotrichum candidum*. *Lipids 25*, 841–848.

BENKOUKA, F.; GUIDONI, A. A.; DECARO, J. D.; BONICEL, J. J.; DESNUELLE, P. A.; and ROVERY, M. 1982. Porcine pancreatic lipase. The disulfide bridges and the sulfhydryl groups. *Eur. J. Biochem. 128*, 331–341.

BENZONANA, G. 1974. Some properties of an exocellular lipase from *Rhizopus arrhizus*. *Lipids 9*, 166–172.

BENZONANA, G., and ESPOSITO, S. 1971. The positional and chain specificities of *Candida cylindracea* lipase. *Biochim. Biophys. Acta 231*, 15–22.

BERTOLINI, M. C.; LARAMEE, L.; THOMAS, D. Y.; CYGLER, M.; SCHRAG, J. D.; and VERNET, T. 1994. Polymorphism in the lipase genes of *Geotrichum candidum* strains. *Eur. J. Biochem. 219*, 119–125.

BLOW, D. 1991. Lipases research the surface. *Nature 351*, 444–445.

BODMER, M. W.; ANGAL, S.; YARRANTON, G. T.; HARRIS, T. J. R.; LYONS, A.; KING, D. J.; PIERONI, G.; RIVIERE, C.,; VERGER, R.; and LOWE, P. A. 1987. Molecular cloning of a human gastric lipase and expression of the enzyme in yeast. *Biochim. Biophys. Acta 909*, 237–244.

BOEL, E.; HUGE-JENSEN, B.; CHRISTENSEN, M.; THIM, L.; and FILL, N. P. 1988. *Rhizomucor miehei* triglyceride lipase is synthesized as a precursor. *Lipids 23*, 701–706.

BONICEL, J.; COUCHOUD, P.; FOGLIZZO, E.; DESNUELLE, P.; and CHAPUS, C. 1981. Amino acid sequence of horse colipase B. *Biochim. Biophys. Acta 669*, 39–45.

BORGSTROM, B., and DONNER, J. 1975. Binding of bile salts to pancreatic colipase and lipase. *J. Lipid Res. 16*, 287–292.

BORGSTROM, B.; ERLANSON, C.; and STERNBY, B. 1974. Further characterization of two co-lipases from porcine pancreas. *Biochem. Biophys. Res. Comm. 59*, 902–906.

BORGSTROM, B.; ERLANSON-ALBERTSSON, C.; and WIELOCH, T. 1979. Pancreatic colipase: Chemistry and physiology. *J. Lipid Res. 20*, 805–816.

BOSC-BIERNE, I.; PERROT, C.; SARDA, L.; and RATHELOT, J. 1985. Inhibition of pancreatic colipase by antibodies and F_{ab} fragments. Selective effects of two fractions of antibodies on the functional sites of the cofactor. *Biochim. Biophys. Acta 827*, 109–118.

BOSC-BIERNE, I.; ROTHELOT, J.; CANIONI, P.; JULIEN, R.; BECHIS, G.; GREGOIRE, J.; ROCHAT, H.; and SARDA, L. 1981. Isolation and partial structural characterization of chicken pancreatic colipase. *Biochim. Biophys. Acta. 667*, 225–232.

BOURNE, Y.; MARTINEZ, C.; KERFELEC, B.; LOMBARDO, D.; CHAPUS, C.; and CAMBILLAU, C. 1994. Horse pancreatic lipase. The crystal structure refined at 2.3 A resolution. *J. Mol. Biol. 238*, 709–732.

BRADY, L.; BRZOZOWSKI, A. M.; DEREWENDA, Z. S.; DODSON, E.; DODSON, G.; TOLLEY, S.; TURKENBURG, J. P.; CHRISTIANSEN, L.; HUGE-JESEN, B.; NORSKOV, L.; THIM, L.; and MENGE, U. 1990. A serine protease triad forms the catalytic centre of a triacylglycerol lipase. *Nature 343*, 767–770.

BRENNER, S. 1988. The molecular evolution of genes and proteins: A tale of two serines. *Nature 334*, 528–530.

BROCKERHOFF, H. 1970. Substrate specificity of pancreatic lipase. Influence of the structure of fatty acids on the reactivity of esters. *Biochim. Biophys. Acta 212*, 92–101.

————. 1973. A model of pancreatic lipase and the orientation of enzymes at interfaces. *Chem. Phys. Lipids 10*, 215–222.

BRZOZOWSKI, A. M.; DEREWENDA, U.; DEREWENDA, Z. S.; DODSON, G. G.; LAWSON, D. W.; TURKENBURG, J. P.; BJORKLING, F.; HUGE-JENSEN, B.; PATKAR, S. A.; and THIM, L. 1991. A model for interfacial activation in lipases from the structure of a fungal lipase-inhibitor complex. *Nature 351*, 491–494.

CAI, S.-J.; WONG, D. M.; CHEN, S.-H.; and CHAN, L. 1989. Structure of the human hepatic triglyceride lipase gene. *Biochemistry 28*, 8966–8971.

CAMBOU, B., and KLIBANOV, A. M. 1984. Preparative production of optically active esters and alcohols using esterase-catalyzed stereospecific transesterification in organic media. *J. Am. Chem. Soc. 106*, 2687–2692.

CANIONI, P.; JULIEN, R.; RATHELOT, J.; and SARDA, L. 1977. Pancreatic and microbial lipase: A comparison of the interaction of pancreatic colipase with lipases of various origins. *Lipids 12*, 393–397.

CHAILLAN, C.; KERFELEC, B.; FOGLIZZO, E.; and CHAPUS, C. 1992. Direct involvement of the C-terminal extremity of pancreatic lipase (403–449) in colipase binding. *Biochem. Biophys. Res. Comm. 184*, 206–211.

CHAILLAN, C.; ROGALSKA, E.; CHAPUS, C.; and LOMBARDO, D. 1989. A cross-linked complex between horse pancreatic lipase and colipase. *FEBS Lett. 257*, 443–446.

CHAPUS, C.; SARI, H.; SEMERIVA, M.; and DESNUELLE, P. 1975. Role of colipase in the interfacial adsorption of pancreatic lipase at hydrophilic interfaces. *FEB Lett.* *58*, 155–158.

CHAPUS, C., and SEMERIVA, M. 1976. Mechanism of pancreatic lipase action. 2. Catalytic properties of modified lipases. *Biochemistry 15*, 4988–4991.

CHAPUS, C.; SEMERIVA, M.; BOVIER-LAPIERRE, C.; and DESNUELLE, P. 1976. Mechanism of pancreatic lipase action. 1. Interfacial activation of pancreatic lipase. *Biochemistry 15*, 4980–4987.

CHARLES, M.; ASTIER, M.; SAUVE, P.; and DESNUELLE, P. 1975. Interactions of colipase with bile salt micelles. I. Ultracentrifugation studies. *Eur. J. Biochem. 58*, 555–559.

CYGLER, M.; GROCHULSKI, P.; KAZLAUSKAS, R. J.; SCHRAG, J. D.; BOUTHILLIER, F.; RUBIN, B.; SERREQI, A. N.; and GUPTA, A. K. 1994. A structural basis for the chiral preferences of lipases. *J. Am. Chem. Soc. 116*, 3180–3186.

DATCHEVA, V. K.; KISS, K.; SOLOMON, L.; and KYLER, K. S. 1991. Asymmetric hydroxylation with lipoxygenase: The role of group hydrophobicity on regioselectivity. *J. Am. Chem. Soc. 13*, 270–274.

DE CARO, J. D.; Behnke, W. K.; Bonicel, J. J.; Desnuelle, P. A.; and Rovery, M. 1983. Nitration of the tyrosine residues of porcine pancreatic colipase with tetranitromethane and properties of the nitrated derivatives. *Biochim. Biophys. Acta 747*, 253–262.

DE CARO, J.; BOUDOUARD, M.; BONICEL, J.; GUIDONI, A.; DESNUELLE, P.; and ROVERY, M. 1981. Porcine pancreatic lipase: Completion of the primary structure. *Biochim. Biophys. Acta 671*, 129–138.

DE CARO, J. D.; ROUIMI, P.; and ROVERY, M. 1986. Hydrolysis of *p*-nitrophenyl acetate by the peptide chain fragment (336–449) of porcine pancreatic lipase. *Eur. J. Biochem. 158*, 601–607.

DENNIS, E. A. 1983. Phospholipases. *The Enzymes 16*, 307–353.

DEREWENDA, U.; BRZOZOWSKI, A. M.; LAWSON, D. W.; and DEREWENDA, Z. S. 1992B. Catalysis at the interface: The anatomy of a conformation change in a triglyceride lipase. *Biochemistry 31*, 1532–1541.

DEREWENDA, Z. S., and DEREWENDA, U. 1991. Relationships among serine hydrolases: Evidence for a common structural motif in triacylglyceride lipases and esterases. *Biochem. Cell Biochem. 69*, 842–851.

DEREWENDA, Z. S.; DEREWENDA, U.; and DODSON, G. G. 1992A. The crystal and molecular structure of the *Rhizomucor miehei* triacylglyceride lipase at 1.9 A resolution. *J. Mol. Biol. 227*, 818–839.

DEREWENDA, Z. S., and SHARP, A. M. 1993. News from the interface: The molecular structures of triacylglyceride lipases. *TIBS 18*, 20–25.

DIJKSTRA, B. W.; DRENTH, J.; KALK, K. H.; and VANDERMAELEN, P. J. 1978. Three-dimensional structure and disulfide bond corrections in bovine pancreatic phospholipase A$_2$. *J. Mol. Biol. 124*, 53–60.

DIJKSTRA, B. W.; DRENTH, J.; and KALK, K. H. 1981A. Active site and catalytic mechanism of phospholipase A$_2$. *Nature 289*, 604–606.

DIJKSTRA, B. W.; KALK, K. H.; HOL, W. G. J.; and DRENTH, J. 1981B. Structure of bovine pancreatic phospholipase A$_2$ at 1.7 A resolution. *J. Mol. Biol. 147*, 97–123.

DIJKSTRA, B. W.; KALK, K. H.; DRENTH, J.; DE HAAS, G. H.; EGMOND, M. R.; and SLOTBOOM, A. J. 1984. Role of the N-terminus in the interaction of pancreatic phospholipase A$_2$ with aggregated substrates. Properties and crystal structure of transaminated phospholipase A$_2$. *Biochemistry 23*, 2759–2766.

DIJKSTRA, B. W.; RENETSEDER, R.; KALK, K. H.; HOL, W. G. J.; and DRENTH, J. 1983A. Structure of porcine pancreatic phospholipase A$_2$ at 2.6 A resolution and comparison with bovine phospholipase A$_2$. *J. Mol. Biol. 168*, 163–179.

DIJKSTRA, B. W.; VAN NES, G. J. H.; KALK, K. H.; BRANDENBURG, N. P.; HOL, W. G. J.; and DRENTH, J. 1982. The structure of bovine pancreatic prophospholipase A$_2$ at 3.0 A resolution. *Acta. Cryst. B38*, 793–799.

DIJKSTRA, B. W.; WEIJER, W. J.; and WIERENGA, R. K. 1983B. Polypeptide chains with similar amino acid sequences but a distinctly different conformation. *FEBS Lett. 164*, 25–27.

DONNE-OPDEN KELDER, G. M.; DE HAAS, G. H.; and EGMOND, M. R. 1983. Localization of the second calcium ion binding site in porcine and equine phospholipase A$_2$. *Biochemistry 22*, 2470–2478.

DUPUREUR, C. M.; YU, B.-Z.; JAIN, M. K.; NOEL, J.-P.; DENG, T.; LI, Y.; BYEON, I.-J. L.; and TSAI, M.-D. 1992. Phospholipase A$_2$ engineering. Structural and functional roles of highly conserved active site residues tyrosine-52 and tyrosine-73. *Biochemistry 31*, 6402–6413.

ERLANSON-ALBERTSSON, C. 1980. Measurement of the binding of colipase to a triglycerol substrate. *Biochim. Biophys. Acta 617*, 371–382.

ERLANSON, C.; BARROWMAN, J. A.; and BORGSTROM, B. 1977. Chemical modifications of pancreatic colipase. *Biochim. Biophys. Acta 489*, 150–162.

ERLANSON, M. C.; BIANCHETTA, J.; JOFFRE, J.; GUIDONI, A.; and ROVERY, M. 1974A. The primary structure of porcine colipase II. 1. The amino acid sequence. *Biochim. Biophys. Acta 359*, 186–197.

ERLANSON, C.; CHARLES, M.; ASTIER, M.; and DESNUELLE, P. 1974B. The primary structure of porcine colipase II. II. The disulfide bridges. *Biochim. Biophys. Acta 359*, 198–203.

ERLANSON, C.; FERNLUND, P.; and BORGSTROM, B. 1973. Purification and characterization of two proteins with co-lipase activity from porcine pancreas. *Biochim. Biophys. Acta 310*, 437–445.

EVENBERG, A.; MEYER, H.; GAASTRA, W.; VERHEIJ, H. M.; and DE HAAS, G. H. 1977. Amino acid sequence of phospholipase A$_2$ from horse pancreas. *J. Biol. Chem. 252*, 1189–1196.

FLEER, E. A. M.; VERHEIJ, H. M.; and DE HAAS, G. H. 1978. The primary structure of bovine pancreatic phospholipase A$_2$. *Eur. J. Biochem. 82*, 261–269.

FOELSCHE, E.; HICKEL, A.; HONIG, H.; and SEUFER-WASSERTHAL, P. 1990. Lipase-catalyzed resolution of acylic amino alcohol precursors. *J. Org. Chem. 55*, 1749–1753.

FRITSCHE, K.; SYLDATK, C.; WAGNER, F.; HENGELSBERG, H.; and TACKE, R. 1989. Enzymatic resolution of *rac*-1,1-dimethyl-1-sila-cyclohexan-2-ol by ester hydrolysis or transesterification using a crude lipase preparation of *Candida cylindracea. Appl. Microbiol. Biotechnol. 31*, 107–111.

FUJII, T.; TATARA, T.; and MINAGAWA, M. 1986. Studies on applications of lipolytic enzyme in detergent. I. Effect of lipase from *Candida cylindracea* on removal of olive oil from cotton fabric. *JAOCS 63*, 796–799.

GARDNER, C. W. 1980. Boronic acid inhibitors of porcine pancreatic lipase. *J. Biol. Chem. 255*, 5064–5068.

GELB, M. H.; JAIN, M. K.; and BERG, O. 1992. Interfacial enzymology of phospholipase A₂. *Bioorganic and Medicinal Chemistry Letters 2*, 1335–1342.

GOLDBERG, M.; THOMAS, D.; and LEGOY, M.-D. 1990. Water activity as a key parameter of synthesis reactions: The example of lipase in biphasic (liquid/solid) media. *Enzyme Microb. Technol. 12*, 976–981.

GOTZ, F.; POPP, F.; KORN, E.; and SCHLEIFER, K. H. 1985. Complete nucleotide sequence of the lipase from *Staphylococcus hyicus* cloned in *Staphylococcus carnosus. Nucl. Acids Res. 13*, 5895–5906.

GRANON, S. 1986. Spectrofluorimetric study of the bile salt micelle binding site of pig and horse colipases. *Biochim. Biophys. Acta 874*, 54–60.

GROCHULSKI, P.; BOUTHILLIER, F.; KAZLAUSKAS, R. J.; SERREQUI, A. N.; SCHRAG, J. D.; ZIOMEK, E.; and CYGLER, M. 1994A. Analogs of reaction intermediates identify a unique substrate binding site in *Candida rugosa* lipase. *Biochemistry 33*, 3494–3500.

GROCHULSKI, P.; LI, Y.; SCHRAG, J. D.; BOUTHILLIER, F.; SMITH, P.; HARRISON, D.; RUBIN, B.; and CYGLER, M. 1993. Insights into interfacial activation from an open structure of *Candida rugosa* lipase. *J. Biol. Chem. 268*, 12843–12847.

GROCHULSKI, P.; LI, Y.; SCHRAG, J. D.; and CYGLER, M. 1994B. Two conformational states of *Candida rugosa* lipase. *Protein Science 3*, 82–91.

GUIDONI, A.; BENKOUKA, F.; DE CARO, J.; and ROVERY, M. 1981. Characterization of the serine reacting with diethyl *p*-nitrophenyl phosphate in porcine pancreatic lipase. *Biochim. Biophys. Acta 660*, 148–150.

GUTMAN, A. L.; ZUOBI, K.; and GUIBE-JAMPEL, E. 1990. Lipase catalyzed hydrolysis of γ-substituted α-aminobutyrolactones. *Tetrahedron Letters 31*, 2037–2038.

HALLING, P. J. 1989. Lipase-catalyzed reactions in low-water organic media: Effects of water activity and chemical modification. *Biochem. Soc. Trans. 17*, 1142–1145.

HAYES, D. G., and GULARI, E. 1990. Esterification reactions of lipase in reverse micelles. *Biotechnol. Bioengineer. 35*, 793–801.

HENNEN, W. J.; SWEERS, H. M.; WANG, Y.-F.; and WONG, C.-H. 1988. Enzymes in carbohydrate synthesis: Lipase-catalyzed selective acylation and deacylation of furanose and pyranose derivatives. *J. Org. Chem. 53*, 4939–4945.

HJORTH, A.; CARRIERE, F.; CUDREY, C.; WOLDIKE, H.; BOEL, E.; LAWSON, D. M.; FERRATO, F.; CAMBILLAU, C.; DODSON, G. G.; THIM, L.; and VERGER, R. 1993. A structural domain (the lid) found in pancreatic lipases is absent in the guinea pig (phospho)lipase. *Biochemistry 32*, 4702–4707.

HUGE-JENSEN, B.; GALLUZZO, D. R.; and JENSEN, R. G. 1987. Partial purification and characterization of free and immobilized lipases from *Mucor miehei*. *Lipids 22*, 559–565.

INAGAKI, M.; HIRATAKE, J.; HISHIOKA, T.; and ODA, J. 1989. Lipase-catalyzed stereoselective acylation of [1,1'-Binaphthyl]-2,2'-diol and deacylation of its esters in an organic solvent. *Agric. Biol. Chem. 53*, 1879–1884.

ISHIHARA, H.; OKUYAMA, H.; IKEZAWA, H.; and TEJIMA, S. 1975. Studies on lipase from *Mucor javanicus*. *Biochim. Biophys. Acta 388*, 413–422.

JACOBSEN, T.; OLSEN, J.; and ALLERMANN, K. 1990. Substrate specificity of *Geotrichum candidum* lipase preparations. *Biotechnol. Lett. 12*, 121–126.

JAIN, M. K., and BERG, O. G. 1989. The kinetics of interfacial catalysis by phospholipase A_2 and regulation of interfacial activation: Hopping versus scooting. *Biochim. Biophys. Acta 1002*, 127–156.

JENSEN, R. G. 1974. Characteristics of the lipase from the mold, *Geotrichum candidum*: A review. *Lipids 9*, 149–157.

JENSEN, R. G., DEJONG, F. A., and CLARKS, R. H. 1983. Determination of lipase specificity. *Lipids 18*, 239–252.

JENSEN, R. G.; GERRIOR, S. A.; HAGERTY, M. M.; and MCMAHON, K. E. 1978. Preparation of acylglycerols and phospholipids with the aid of lipolytic enzymes. *JAOCS 55*, 422–427.

JULIEN, R.; BECHIS, G.; GREGOIRE, J.; RATHELOT, J.; ROCHAT, H.; and SARDA, L. 1980. Evidence for the existence of two isocolipases in horse pancreas. *Biochim. Biophys. Res. Comm. 95*, 1245–1252.

KAIMAL, T. N. B., and SAROJA, M. 1989. The active site composition of porcine pancreatic lipase: Possible involvement of lysine. *Biochim. Biophys. Acta 999*, 331–334.

KANERVA, L. T.; VIHANTO, J.; HALME, M. H.; LOPONEN, J. M.; and EURANTO, E. K. 1990. Solvent effects in lipase-catalyzed transesterification reactions. *Acta Chem. Scand. 44*, 1032–1035.

KERFELEC, B.; FOGLIZZO, E.; BONICEL, J.; BOUGIS, P. E.; and CHAPUS, C. 1992. Sequence of horse pancreatic lipase as determined by protein and cDNA sequencing. Implication for *p*-nitrophenyl acetate hydrolysis by pancreatic lipases. *Eur. J. Biochem. 206*, 279–287.

KIRCHGESSNER, T. G.; CHUAT, J.-C.; HEINZMANN, C.; ETIENNE, J.; GUILHOT, S.; SVENSON, K.; AMEIS, D.; PILON, C.; D'AURIOL, L.; ANDALIBI, A.; SCHOTZ, M. C.; GALIBERT, F.; and LUSIS, A. J. 1989. Organization of the human lipoprotein lipase gene and evolution of the lipase gene family. *Proc. Natl. Acad. Sci. 86*, 9647–9651.

KIRCHGESSNER, T. G.; SVENSON, K. L.; LUSIS, A. J.; and SCHOTZ, M. C. 1987. The sequence of cDNA encoding lipoprotein lipase. *J. Biol. Chem. 262*, 8463–8466.

KIRCHNER, G.; SCOLLAR, M. P.; and KLIBANOV, A. M. 1985. Resolution of racemic mixture via lipase catalysis in organic solvents. *J. Am. Chem. Soc. 107*, 7072–7076.

KLIBANOV, A. M. 1990. Asymmetric transformations catalyzed by enzymes in organic solvents. *Acc. Chem. Res. 23*, 114–120.

KLOOSTERMAN, M.; ELFERINK, V. H. M.; VAN LERSEL, J.; ROSKAM, J.-H.; MEIJER, E.

M.; HULSHOF, L. A.; and SHELDON, R. A. 1988. Lipases in the preparation of β-blockers. *TIBTECH 6*, 251–256.

KOMAROMY, M., and SCHOTZ, M. C. 1987. Cloning of rat hepatic lipase cDNA: Evidence for a lipase gene family. *Proc. Natl. Acad. Sci. USA 84*, 1526–1530.

KOSSIAKOFF, A. A., and SPENCER, S. A. 1981. Direct determination of the protonation states of aspartic acid-102 and histidine-57 in the tetrahedral intermediate of the serine protease: Neutron structure of trypsin. *Biochemistry 20*, 6462–6474.

KUGIMIYA, W.; OTANI, Y.; HASHIMOTO, Y.; and TAKAGI, Y. 1986. Molecular cloning and nucelotide sequence of the lipase gene from *Pseudomonas fragi. Biochem. Biophys. Res. Comm. 141*, 185–190.

KUIPERS, O. P.; FRANKEN, P. A.; HENDRIKS, R.; VERHEIJ, H. M.; and DE HAAS, G. H. 1990. Function of the fully conserved residues Asp99, Tyr52 and Tyr73 in phospholipase A_2. *Protein Engineering 4*, 199–204.

KUIPERS, O. P.; THUNNISSEN, M. M. G. M.; DE GEUS, P.; DIJKSTRA, B. W.; DRENTH, J.; VERHEIJ, H. M.; and DE HAAS, G. H. 1989. Enhanced activity and altered specificity of phospholipase A_2 by deletion of a surface loop. *Science 244*, 82–85.

LADNER, W. E., and WHITESIDES, G. M. 1984. Lipase catalyzed hydrolysis as a route to esters of chiral epoxy alcohol. *J. Am. Chem. Soc. 106*, 7251–7252.

LARSSON, A., and ERLANSON-ALBERTSSON, C. 1981. The identity and properties of two forms of activated colipase from porcine pancreas. *Biochim. Biophys. Acta 664*, 538–548.

———. 1983. The importance of bile salt for the reactivation of pancreatic lipase by colipase. *Biochim. Biophys. Acta 750*, 171–177.

LEE, C. Y., and IANDOLO, J. J. 1986. Lysogenic conversion of Staphylococcal lipase is caused by insertion of the bacteriophage L54a genome into the lipase structural gene. *J. Bacteriol. 166*, 385–391.

LINFIELD, W. M.; BARAUSKAS, R. A.; SIVIERI, L.; SEROTA, S.; and STEVENSON, R. W., SR. 1984. Enzymatic fat hydrolysis and synthesis. *JAOCS 61*, 191–195.

LONGHI, S.; LOTTI, M.; FUSETTI, F.; PIZZI, E.; TRAMONTANO, A.; and ALBERGHINA, L. 1992. Homology-derived three-dimensional structure prediction of *Candida cylindracea* lipase. *Biochim. Biophys. Acta 1165*, 129–133.

LOWE, M. F.; ROSENBLUM, J. L.; and STRAUSS, A. W. 1989. Cloning and characterization of human pancreatic lipase cDNA. *J. Biol. Chem. 264*, 20042–20048.

LOWE, M. F. 1992. The catalytic site residues and interfacial binding of human pancreatic lipase. *J. Biol. Chem. 267*, 17069–17073.

MACRAE, A. R. 1983. Lipase-catalyzed interesterifaction of oils and fats. *JAOCS 60*, 291–294.

MAHE-GOUHIER, N., and LEGER, C. L. 1988. Immobilized colipase affinities for lipases B, A, C and their terminal peptide (336–449): The lipase recognition site lysine residues are located in the C-terminal region. *Biochim. Biophys. Acta 962*, 91–97.

MARAGANORE, J. M., and HEINRIKSON, R. L. 1986. Which class of serine is involved in the lipase mechanism. *TIBS 11*, 497–498.

MARGOLIN, A. L.; TAI, D.-F.; and KLIBANOV, A. M. 1987. Incorporation of D-amino acids into peptides via enzymatic condensation in organic solvents. *J. Am. Chem. Soc. 109*, 7885–7887.

MEYER, H.; PUIJK, W. C.; DIJKMAN, R.; FODA-VAN DER HOORN; M. M. E. L., PATTUS, F.; SLOTBOOM, A. J.; and DE HAAS G. H. 1979A. Comparative studies of tyrosine modification in pancreatic phospholipases. 2. Properties of the nitrotyrosyl, aminotyrosyl, and dansylaminotyrosyl derivatives of pig, horse, and ox phospholipases A$_2$ and their zymogens. *Biochemistry 18*, 3589–3597.

MEYER, H.; VERHOEF, H.; HENDRIKS, F. F. A.; SLOTBOOM, A. J.; and DE HAAS, G. H. 1979B. Comparative studies of tyrosine modification in pancreatic phospholipases. 1. Reaction of tetranitromethane with pig, horse, and ox phospholipase A$_2$ and their zymogens. *Biochemistry 18*, 3582–3588.

MICKEL, F. S.; WEIDENBACH, F.; SWAROVSKY, B.; LAFORGE, K. S.; and SCHEELE, G. A. 1989. Structure of the canine pancreatic lipase gene. *J. Biol. Chem. 264*, 12895–12901.

MILLER, D. A.; PRAUSNITZ, J. M.; and BLANCH, H. W. 1991. Kinetics of lipase-catalyzed interesterification of triglycerides in cyclohexane. *Enzyme Microb. Technol. 13*, 98–103.

MORLEY, N. H.; KUKSIS, A.; BUCHNEA, D.; and MYHER, J. J. 1975. Hydrolysis of diacylglycerols by lipoprotein lipase. *J. Biol. Chem. 250*, 3414–3418.

NAKAMURA, K.; ISHIHARA, K.; OHNO, A.; UEMURA, M.; NISHIMURA, H.; and HAYASHI, Y. 1990. Kinetic resolution of (η^6-arene)chromium complexes by a lipase. *Tetrahedron Letters 31*, 3603–3604.

NELSON, J. H. 1972. Enzymatically produced flavors for fatty systems. *JAOCS 49*, 559–562.

NOBLE, M. E. M.; CLEASBY, A.; JOHNSON, L. N.; EGMOND, M. R.; and FRENKEN, L. G. J. 1993. The crystal structure of triacylglycerol lipase from *Pseudomonas glumae* reveals a partially redundant catalytic aspartate. *FEBS Lett. 331*, 123–128.

OKUMURA, S.; IWAI, M.; and TSUJISAKA, Y. 1976. Positional specificities of four kinds of microbial lipases. *Agric. Biol. Chem. 40*, 655–660.

OTA, T.; TAKANO, S.; and HASEGAWA, T. 1990. Synthesis of C$_{18}$-fatty acid esters in organic solvent by lipase from *Candida cylindracea*. *Agric. Biol. Chem. 54*, 1571–1572.

PALTAUF, F.; ESFANDI, F.; and HOLASEK, A. 1974. Stereospecificity of lipases. Enzymic hydrolysis of enantiomeric alkyl diacylglycerols by lipoprotein lipase, lingual lipase and pancreatic lipase. *FEBS Lett. 40*, 119–123.

PATTON, J. S.; ALBERTSSON, P.-A.; ERLANSON, C.; and BORGSTROM, B. 1978. Binding of porcine pancreatic lipase and colipase in the absence of substrate studied by two-phase partition and affinity chromatography. *J. Biol. Chem. 253*, 4195–4202.

PIERROT, M.; ASTIER, J.-P.; ASTIER, M.; CHARLES, M.; and DRENTH, J. 1982. Pancreatic colipase: Crystallographic and biochemical aspects. *Eur. J. Biochem. 123*, 347–354.

PLUMMER, T. H., JR., and SARDA, L. 1973. Isolation and characterization of the glycoptides of porcine pancreatic lipases L$_A$ and L$_B$. *J. Biol. Chem. 248*, 7865–7869.

PUIJK, W. C.; VERHEIJ, H. M.; and DE HAAS, G. H. 1977. The primary structure of phospholipase A$_2$ from porcine pancreas. *Biochim. Biophys. Acta 492*, 254–259.

RAMASWAMY, S.; MORGAN, B.; and OEHLSCHLAGER, A. C. 1990. Porcine pancreatic lipase mediated selective acylation of primary alcohols in organic solvents. *Tetrahedron Letters 31*, 3405–3408.

RANSAC, S.; ROGALSKA, E.; GARGOURI, Y.; DEVEER, A. M. T. J.; PALTAUF, F.; DE HAAS, G. H.; and VERGER, R. 1990. Stereoselectivity of lipases. I. Hydrolysis of enantiomeric glyceride analogues by gastric and pancreatic lipases, a kinetic study using the monomolecular film technique. *J. Biol. Chem. 265*, 20263–20270.

RATHELOT, J.; BOSC-BIERNE, I.; GUY-CROTTE, O.; DELORI, P.; ROCHAT, H.; and SARDA, L. 1983. Isolation and characterization of colipase from porcine and human pancreatic juice by immunoaffinity chromatography. *Biochim. Biophys. Acta. 744*, 115–118.

RENETSEDER, R.; DIJKSTRA, B. W.; HUIZINGA, K.; KALK, K. H.; and DRENTH, J. 1988. Crystal structure of bovine pancreatic phospholipase A$_2$ covalently inhibited by *p*-bromo-phenacyl-bromide. *J. Mol. Biol. 200*, 181–188.

RIDDIHOUGH, G. 1993. Picture an enzyme at work. *Nature 362*, 793.

ROGALSKA, E.; RANSAC, S.; and VERGER, R. 1990. Stereoselectivity of lipases. II. Stereoselective hydrolysis of triglycerides by gastric and pancreatic lipases. *J. Biol. Chem. 265*, 20271–20276.

ROLLOF, J.; HEDSTROM, S. A.; and NILSSON-EHLE, P. 1987. Positional specificity and substrate preference of purified *Staphylococcus aureus* lipase. *Biochim. Biophys. Acta 921*, 370–377.

SARDA, L., and DESNUELLE, P. 1958. Action of pancreatic lipase on emulsified esters. *Biochim. Biophys. Acta 30*, 513–521.

SCHRAG, J. D., and CYGLER, M. 1993. 1.8 A refined structure of the lipase from *Geotrichum candidum*. *J. Mol. Biol. 230*, 575–591.

SCHRAG, J. D.; LI, Y.; WU, S.; and CYGLER, M. 1991. Ser-His-Glu triad forms the catalytic site of the lipase from *Geotrichum candidum*. *Nature 351*, 761–764.

SCHRAG, J. D.; WINDLER, F. K.; and CYGLER, M. 1992. Pancreatic lipases: Evolutionary intermediates in a positional change of catalytic carboxylates? *J. Biol. Chem. 267*, 4300–4303.

SCOTT, D. L.; WHITE, S. P.; OTWINOWSKI, Z.; YUAN, W.; GELB, M. H.; and SIGLER, P. B. 1990. Interfacial catalysis: The mechanism of phospholipase A$_2$. *Science 250*, 1541–1546.

SEINO, H.; UCHIBORI, T.; NISHITANI, T.; and INAMASU, S. 1984. Enzymatic synthesis of carbohydrate esters of fatty acid (1) Esterification of sucrose, glucose, fructose, and sorbitol. *JAOCS 61*, 1761–1765.

SEMERIVA, M.; CHAPUS, C.; BOVIER-LAPIERRE, C.; and DESNUELLE, P. 1974. On the transient formation of an acyl enzyme intermediate during the hydrolysis of *p*-nitrophenyl acetate by pancreatic lipase. *Biochem. Biophys. Res. Comm. 58*, 808–813.

SHIMADA, Y.; SUGIHARA, A.; IIZUMI, T.; and TOMINAGA, Y. 1990. cDNA cloning and characterization of *Geotrichum candidum* lipase II. *J. Biochem. 107*, 703–707.

SHIMADA, Y.; SUGIHARA, A.; TOMINAGA, Y.; IIZUMI, T.; and TSUNASAWA, S. 1989. cDNA molecular cloning of *Geotrichum candidum* lipase. *J. Biochem. 106*, 383–388.

SIDEBOTTOM, C. M.; CHARTON, E.; DUNN, P. P.; MYCOCK, G.; DAVIES, C.; SUTTON, J. L.; MACRAE, A. R.; and SLABAS, A. R. 1991. *Geotrichum candidum* produces several lipases with markedly different substrate specificities. *Eur. J. Biochem. 202*, 485–491.

SIMS, H. F., and LOWE, M. E. 1992. The human colipase gene: Isolation, chromosomal location, and tissue-specific expression. *Biochemistry 31*, 7120–7125.

SLAICH, P. K.; PRIMROSE, W. U.; ROBINSON, D. H.; WHARTON, C. W.; WHITE, A. J.; DRABBLE, K.; and ROBERTS, G. C. K. 1992. The binding of amide substrate analogues to phospholipase A_2. Studies by ^{13}C-nuclear-magnetic-resonance and infrared spectroscopy. *Biochem. J. 288*, 167–173.

SLOTBOOM, A. J.; JANSEN, E. H. J. M.; VLIJM, IT.; and PATTUS, F. 1978. Ca^{++} Binding to porcine pancreatic phospholipase A_2 and its function in enzyme-lipid interaction. *Biochemistry 17*, 4593–4600.

STERNBY, B., and BORGSTROM, B. 1979. Purification and characterization of human pancreatic colipase. *Biochim. Biophys. Acta 572*, 235–243.

STERNBY, B.; ENGSTROM, A.; and HELLMAN, U. 1984A. Purification and characterization of pancreatic colipase from the dogfish (*Squalus acanthius*). *Biochim. Biophys. Acta 789*, 159–163.

STERNBY, B.; ENGSTROM, A.; HELLMAN, U.; VIHERT, A. M.; STERNBY, N.-H.; and BORGSTROM, B. 1984B. The primary sequence of human pancreatic colipase. *Biochim. Biophys. Acta 784*, 75–80.

STEVENSON, R. W.; LUDDY, F. E.; and ROTHBART, H. L. 1979. Enzymatic acyl exchange to vary saturation in di- and triglycerides. *JAOCS 56*, 676–680.

SUGIHARA, A.; SHIMADA, Y.; and TOMINAGA, Y. 1990. Separation and characterization of two molecular forms of *Geotrichum candidum* lipase. *J. Biochem. 107*, 426–430.

SWEERS, H. M., and WONG, C.-H. 1986. Enzyme-catalyzed regioselective deacylation of protected sugars in carbohydrate synthesis. *J. Am. Chem. Soc. 108*, 6421–6422.

TANAKA, T.; ONO, E.; ISHIHARA, M.; YAMANAKA, S.; and TAKINAMI, K. 1981. Enzymatic acyl exchange of triglyceride in n-hexane. *Agric. Biol. Chem. 45*, 2387–2389.

THERISOD, M., and KLIBANOV, A. M. 1986. Facile enzymatic preparation of monoacylated sugars in pyridine. *J. Am. Chem. Soc. 108*, 5638–5640.

THUNNISSEN, M. M. G. M.; AB, E.; KALK, K. H.; DRENTH, J.; KIJKSTRA, B. W.; KUIPERS, O. P.; DIJKMAN, R.; DE HAAS, G. H.; and VERHEIJ, H. M. 1990. X-Ray structure of phospholipase A_2 complexed with a substrate-derived inhibitor. *Nature 347*, 689–691.

TRIANTAPHYLIDES, C.; LANGRAND, G.; ILLET, H.; RANGHEARD, M. S.; BUONO, G.; and BARATTI, J. 1988. On the use of lipase specificity. Application to flavour chemistry. In: *Bioflavour '87*, ed. P. Schreier, Walter de Gruyter, Berlin and Hawthorne, New York.

TSUJISAKA, Y.; OKUMURA, S.; and IWAI, M. 1977. Glyceride synthesis by four kinds of microbial lipase. *Biochim. Biophys. Acta 489*, 415–422.

UEMURA, A.; NOZAKI, K.; YAMASHITA, J.-I.; and YASUMOTO, M. 1989. Regioselective deprotection of 3', 5'-O-acylated pyrimidine nucleosides by lipase and esterase. *Tetrahedron Letters 30*, 3819–3820.

UPPENBERG, J.; HANSEN, M. T.; PATKAR, S.; and JONES, T. A. 1994. The sequence, crystal structure determination and refinement of two crystal forms of lipase B from *Candida antarctica*. *Structure 2*, 293–308.

VADEHRA, D. V. 1974. *Staphylococcal* lipases. *Lipids 9*, 158–165.

VALIVETY, R. H.; HALLING, P. J.; and MACRAE, A. R. 1992A. Reaction rate with suspended lipase catalyst shows similar dependence on water activity in different organic solvents. *Biochim. Biophys. Acta 1118*, 218–222.

———. 1992B. *Rhizomucor miehei* lipase remains highly active at water activity below 0.0001. *FEBS Lett. 301*, 258–260.

VAN DAM-MIERAS, M. C. E.; SLOTBOOM, A. J.; PIETERSON, W. A.; and DE HAAS, G. H. 1975. The interaction of phospholipase A₂ with micellar interfaces. The role of the N-terminal region. *Biochemistry 14*, 5387–5394.

VAN TILBEURGH, H.; EGLOFF, M.-P.; MARTINEZ, C.; RUGANI, N.; VERGER, R.; and CAMBILLAU, C. 1993. Interfacial activation of the lipase-prolipase complex by mixed micelles revealed by x-ray crystallography. *Nature 362*, 814–820.

VAN TILBEURGH, H.; SARDA, L.; VERGER, R.; and CAMBILLAU, C. 1992. Structure of the pancreatic lipase-procolipase complex. *Nature 359*, 159–162.

VEERARAGAVAN, K.; COLPITTS, T.; and GIBBS, B. F. 1990. Purification and characterization of two distinct lipases from *Geotrichum candidum*. *Biochim. Biophys. Acta 1044*, 26–33.

VERGER, R. 1976. Interfacial enzyme kinetics of lipolysis. *Ann. Rev. Biophys. Bioeng. 5*, 77–177.

———. 1980. Enzyme kinetics of lipolysis. *Methods in Enzymology 64*, 340–392.

VERHAGEN, J.; VELDINK, G. A.; EGMOND, M. R.; VLIEGENTHART, J. F. G.; BOLDINGH, J.; and VAN DER STAR, J. 1978. Steady-state kinetics of the anaerobic reaction of soybean lipoxygenase-1 with linoleic acid and 13-L-hydroperoxylinoleic acid. *Biochim. Biophys. Acta 529*, 369–379.

VERHEIJ, H. M.; VOLWERK, J. J.; JANSEN, E. H. J. M.; PUYK, W. C.; DIJKSTRA, B. W.; DRENTH, J.; and DE HAAS, G. H. 1980. Methylation of histidine-48 in pancreatic phospholipase A₂. Role of histidine and calcium ion in a catalytic mechanism. *Biochemistry 19*, 743–750.

WANG, Y.-F.; LALONDE, J. J.; MOMONGAN, M.; BERGBREITER, D. E.; and WONG, C.-H. 1988. Lipase-catalyzed irreversible transesterifications using enol esters as acylating reagents: Preparative enantio- and regioselective synthesis of alcohols, glycerol derivatives, sugars, and organometallics. *J. Am. Chem. Soc. 110*, 7200–7205.

WARSHEL, A.; NARAY-SZABO, G.; SUSSMAN, F.; and HWANG, J.-K. 1989. How do serine proteases really work? *Biochemistry 28*, 3629–3637.

WASZKOWYCZ, B., and HILLIER, I. H. 1989. Aspects of the mechanism of catalysis in phospholipase A₂. A combined ab initio molecular orbital and molecular mechanics study. *J. Chem. Soc. Perkin Trans. II 1989*, 1795–1800.

———. 1990. A theoretical study of hydrolysis by phospholipase A₂: The catalytic role of the active site and substrate specificity. *J. Chem. Soc. Perkin Trans. II 1990*, 1259–1268.

WELLS, M. A. 1971. Evidence for O-acyl cleavage during hydrolysis of 1, 2-diacyl-*sn*-glycero-3-phosphorylcholine by the phospholipase A₂ of *Crotalus adamanteus* venom. *Biochim. Biophys. Acta 248*, 80–85.

WHITE, S. P.; SCOTT, D. L.; OTWINOWSKI, Z.; GELB, M. H.; and SIGLER, P. B. 1990. Crystal structure of cobra-venom phospholipase A₂ in a complex with a transition-state analogue. *Science 250*, 1560–1563.

WIELOCH, T.; BORGSTROM, B.; FALK, K.-E.; and FORSEN, S. 1979. High-resolution proton magnetic resonance study of porcine colipase and its interactions with taurodeoxycholate. *Biochemistry 18*, 1622–1628.

WIELOCH, T., and FALK, K.-E. 1978. An NMR study of a tyrosine and two histidine residues in the structure of porcine pancreatic colipase. *FEBS Lett. 85*, 271–274.

WINKLER, F. K.; D'ARCY, A.; and HUNZIKER, W. 1990. Structure of human pancreatic lipase. *Nature 343*, 771–774.

XIE, Z.-F., and SAKAI, K. 1989. Preparation of a chiral building block based on 1,3-*syn*-diol using *Pseudomonas fluorescens* lipase and its application to the synthesis of a hunger modulator. *Chem. Pharm. Bull. 37*, 1650–1752.

YANG, C.-Y.; GU, Z.-W.; YANG, H.-X.; ROHDE, M. F.; GOTTO, A. M., JR.; and POWNALL, H. J. 1989. Structure of bovine milk lipoprotein lipase. *J. Biol. Chem. 264*, 16822–16827.

YOKOZEKI, K.; YAMANAKA, S.; TAKINAMI, K.; HIROSE, Y.; TANAKA, A.; SONOMOTO, K.; and FUKUI, S. 1982. Application of immobilized lipase to regio-specific interesterification of triglyceride in organic solvent. *Eur. J. Appl. Microbiol. Biotechnol. 14*, 1–5.

Chapter 7

Pectic Enzymes

Pectic enzymes constitute a unique group of enzymes that catalyze the degradation of pectic polymers in plant cell walls. Depolymerization of pectin is generally associated with the process of fruit ripening. These enzymes therefore play a significant role in the changes occurring in postharvest storage of fruits and vegetables. The control of these enzymes in transgenic tomato fruit has become one of the successful examples in the application of antisense RNA to manipulate gene expression (Kramer et al. 1990).

Pectic enzymes are also well known in food processing as the cause of cloud loss in citrus juices (Pilnik and Rombouts 1979; Pilnik and Voragen 1993). The cold-break and hot-break processes used in the tomato industry are based on the controlling of the activities of pectic enzymes for desirable viscosity and consistency of the product. Pectic enzymes are used in clarification of apple and grape juice. They are commonly used for treatment of pulp in facilitating juice extraction (Voragen et al. 1986). Enzyme-peeled grapefruit are produced by vacuum infusion of the fruit peel with pectic enzymes. These are commercially marketed as minimally processed products in the form of whole peeled fruit or intact segments.

STRUCTURE OF PECTIC SUBSTANCES

Pectic enzymes act on pectic substances that are heterogeneous in chemical structure and molecular size. The basic chemical structure of pectin is α-D-galacturonan or α-D-galacturonoglycan—a linear chain of 1,4-α-D-galactopyranosyluronic acid units. The polymer contains varying degrees of esterification, with certain carboxyl groups esterified with methoxyl groups. In some cases, such as in sugar beet pectin, esterification with acetyl groups occurs. The backbone chain is generally interrupted by L-rhamnose linked α-1,2 and β-1,4 to the preceding and succeeding galacturonic acid, respectively. Branching also occurs with neutral sugars, forming side chains of α-1,5-arabinan and β-1,4-galactan.

Pectic substances can be classified according to the modification of the backbone chain (BeMiller 1986).

(1) Pectic acids: Galacturonans that contain negligible amounts of methoxyl groups. Pectates are salts of pectic acids.

(2) Pectinic acids: Galacturonans with various amounts of methoxyl groups. Pectinates are salts of pectinic acids.

(3) Pectins: A generic name for mixtures of widely differing composition containing pectinic acid as the major component. Pectins in native form as located in the cell wall may be interlinked with other structural polysaccharides and proteins to form insoluble protopectin. Solubility of pectins can be achieved by degradation such as heating in acid medium. The soluble pectin thus obtained is partially degraded and heterogeneous.

CLASSIFICATION OF PECTIC ENZYMES

Pectic enzymes can be classified according to their mode of action.

(1) Polygalacturonase (PG): Catalyzes the hydrolytic cleavage of α-1,4-glycosidic bonds. The exo-PG (exo-poly(1,4-α-D-galacturonide)galacturonohydrolase, EC 3.2.1.67) cleaves from the nonreducing end, and the endo-PG (endo-poly(1,4-α-D-galacturonide)glycanohydrolase, EC 3.2.1.15) attacks the substrate randomly.

(2) Pectinesterase (PE) (Pectin pectylhydrolase, EC 3.1.1.11): Catalyzes the hydrolysis of the methyl ester groups, resulting in the deesterification of pectin. The enzyme acts preferentially on a methyl ester group of a galacturonate unit next to a nonesterified galacturonate unit.

(3) Pectate lyase (PEL): Catalyzes the cleavage of nonesterified galacturonate units via β-elimination. Both exo-PEL (exo-poly(1,4-α-D-galacturonide)lyase, EC 4.2.2.9) and endo-PEL (endo-poly(1,4-α-D-galacturonide)lyase, EC 4.2.2.2) enzymes exist. Pectate and low-methoxyl pectin are the preferred substrates for the enzymes. Both the exo- and the endo-enzymes, in general, have an alkaline pH optimum from 8.0 to 11, and require Ca^{++} for activity. Pectate lyases are not found in plants, but in bacteria and fungi. These microbial extracellular enzymes play an important part in plant pathogenesis, causing tissue degradation of cell walls and softening, and rotting of plant tissues (Collmer and Keen 1986).

(4)   Pectin lyase (PNL): Catalyzes cleavage of esterified galacturonate units by β-elimination. All PNLs studied are endo-enzymes.

POLYGALACTURONASE

Earlier studies of PG largely come from microbial sources. PG is commonly found in the extracellular secretions of pathogenic species of fungi and bacteria. Some examples of these studies include *Saccharomyces fragilis* (Demain and Phaff 1954), *Aspergillus niger* (Koller and Neukom 1969; Ruttkowski et al. 1990; Bussink et al. 1990), *Lactobacillus plantarum* (Sakellaris et al. 1989), *Cochliobolus carbonum* (Scott-Craig et al. 1990), *Neurospora crassa* (Polizeli et al. 1991), *Ascomycete* species (Keon and Waksman 1990), *Rhizopus arrhizus* (Liu and Luh 1978), and *Fusarium oxysporum* (Martinez et al. 1991). However, it is the tomato PG that has received overwhelming attention in recent years in relation to genetic engineering of vine-ripe fruit. In fact, PG in higher plants was first studied in ripe tomato fruit.

TOMATO ENZYMES

Tomato PG exists in two forms; both are endo-enzymes (Pressey and Avants 1973). PG1 has a MW of 84 kD and is 50% inactivated at 78°C. PG2 has a MW of 44 kD and is 50% inactivated at 57°C. PG1 has an optimum stability at pH 4.3, whereas PG2 is most stable at pH 5.6. Analysis using SDS-PAGE seems to suggest that PG1 is a dimer of PG2 (Tucker et al. 1980), but later studies suggest that PG1 is produced by the combination of PG2 with a β subunit (fomerly known as converter) (Tucker et al. 1981; Pressey 1986; Knegt et al. 1988). PG polypeptide chains are glycosylated. PG2 contains 4.6% neutral sugars (D-mannose, L-fucose, D-xylose) linked via an N-acetylglucosaminylasparaginyl bond (Moshrefi and Luh 1983). Slight variations on the MW have been reported (Pressey and Avants 1973; Moshrefi and Luh 1983; Tucker et al. 1980), and two PG2 isozymes, PG2A and PG2B were isolated in fully ripe fruits with MW 43 kD and 46 kD, respectively (Ali and Brady 1982). These results may be attributed to post-translational glycosylation or modification. Later investigations have demonstrated that PG2A and PG2B are indeed two glycosylated forms of the same polypeptides. The two isozymes consist of a single catalytic PG polypeptide different only in the degree of glycosylation (DellaPenna and Bennett 1988).

AMINO ACID SEQUENCES

PG amino acid sequences have been deduced from cDNA nucleotides isolated from *Pseudomonas solanacearum* (Huang and Schell 1990), *Oenothera orga-*

nensis pollen (Brown and Crouch 1990), maize pollen (Niogret et al. 1991), and avocado fruit (Dopico et al. 1993), in addition to tomato fruit (Grierson et al. 1986; Sheehy et al. 1987). The tomato PG2 gene encodes a mature protein of 373 amino acids, with a calculated MW of 40,279 (Fig. 7.1). The enzyme is synthesized as a precursor, which undergoes post-translational modification including removal of the signal peptide, glycosylation at four potential glycosylation sites, and modification of the C-terminus (Sheehy et al. 1987). A single PG gene codes for each of the recognized isozymes, consistent with the conclusion that PG1 and PG2 consist of the same catalytic polypeptide, and that PG1 is formed by the association of PG2 with a β subunit.

β-SUBUNIT

The conversion of PG2 to PG1 in tomatoes during ripening has received considerable attention. A heat-stable nondialyzable factor (converter, now known as β subunit) was first isolated from green tomatoes capable of in vitro conversion of PG2 to PG1 by Tucker et al. (1981). This β subunit was later confirmed as a glycoprotein immunologically distinct from the PG polypeptide (Pressey 1986; Knegt et al. 1988; Pogson et al. 1991). PG1 is made up of one molecule of β subunit and one molecule of PG2 (Knegt et al. 1991). The β subunit is noncatalytic; only the PG polypeptide has enzyme activity.

The cDNA encoding the β subunit of tomato PG1 has been characterized (Fig. 7.2) (Zheng et al. 1992). The predicted structure is a large precursor (69 kD, 630 amino acids) consisting of a hydrophobic signal sequence (30 amino acids), an N-terminal propeptide (78 amino acids), and a large C-terminal propeptide (233 amino acids), in addition to the mature protein domain. The unglycosylated mature protein consists of 289 amino acids of 31.5 kD with a calculated pI of 4.9. The protein has a high amount of Gly, Tyr, and Phe, and a repeating motif with a consensus structure of FTNYGxxGNGGxxx where "x" is mostly polar amino acids. The functional significance of this unique motif in the protein is not clear.

SPECIFIC ROLES IN RIPENING

There have been considerable efforts to elucidate the specific roles of PG1, PG2, and β subunit in the ripening process of tomato fruits. The enzyme activity in the early stages of ripening is mainly due to PG1. PG1 continues to increase during ripening, and is detected at all stages of ripening. The lower-MW and less heat-stable PG2 appears late in the process, and becomes the major enzyme at the ripe stage (Brady et al. 1982; Pressey 1986). This apparent sequential appearance of the two enzymes is the result of the regulatory action of the β subunit, which is

```
Tomato PG                   MVIQRNSILLLIITFASSISTCRSNVID
Avocado PG          MALTRLLLPISILWFCFY-SHTILQKDPLLICVNGDPGF-

Tomato PG                   DNLFKQVYDNILEQEFAHDFQAYLSYLSKNIESNNNIDKV
Avocado PG    41    QRAYPTYFGP--D --S SIMGFEPSI LSL-RF-PVGGP

Tomato PG                   DKNGIKVINVLSFGAKGDGKTYDNIAFEQAWNEACSSRTP
Avocado PG    78    ETSPDTD-S-DD---R---TD -TK---K--KD----GSV

Tomato PG     38    VQFVVPKNKNYLLKQITFSGPCRSSISVKIFGSLEASSKI
Avocado PG    117     LI--E--------------K-DLR---R-TI----DQ

Tomato PG     78    SDY  KDRRLWIAFDSVQNLVVGGGGTINGNGQVWWPSSC
Avocado PG    155   --WVGHN-KR--E-EDIS--TLE---------ET--D---

Tomato PG     116   KINKSLPCRDAPTALTFWNCKNLKVNNLKSKNAQQIHIKF
Avocado PG    195   -RK-----KS-------RS----I-SD-SI-DS-KM-LS-

Tomato PG     156   ESCTNVVASNLMINASAKSPNTDGVHVSNTQYIQISDTII
Avocado PG    235   DKCQD-I-----VT-PEH------I-ITG--R-HVMNSV-

Tomato PG     196   GTGDDCISIVSGSQNVQATNITCGPGHGISIGSLGSGNSE
Avocado PG    275   ---------E---K--I----------------DR---

Tomato PG     236   AYVSNVTVNEAKIIGAENGVRIKTWQGGSGQASNIKFLNV
Avocado PG    315   -H--G-L-DGGNLFDTT--L---------SAK----Q-I

Tomato PG     276   EMQDVKYPIIIDQNYCDRVEPCIQQFSAVQVKNVVYENIK
Avocado PG    355   V-HN-TN------Y---SKD--PE-E---KVS--A-M--R

Tomato PG     316   GTSATKVAIKFDCSTNFPCEGIIMENINLVGESGKPSEAT
Avocado PG    395   ----SE--V-----KSS--QGY-VG------NG--ETTMS

Tomato PG     356   CKNVHFNNAEHVTPHCTSLEISEDEALLYNY   373
Avocado PG    435   -S-IVQGLLREGLSTFLFMKRRVH-CSY    462
```

Figure 7.1. Amino acid sequences of polygalacturonases from tomato PG2 (Sheehy et al. 1987) and avocado (Dopico et al. 1993). Both sequences are deduced from cDNA sequencing. The N-terminal amino acids are double underlined, and the C-terminal amino acids are underscored with gray lines.

β-subunit		MHTKIHLPPCILLLLLFSLPSFNVVVGGDGESGNPFTPKG
	41	YLIRYWKKQISNDLPKPWFLLNKASPLNAAQYATYTKLVA
	81	DQNALTTQLHTFCSSANLMCAPDLSPSLEKHSGDIHFATY
	121	SDKNFTNYGTNEPGIGVNTFKNYSEGENIPVNSFRRYGRG
	161	SPRDNKFDNYASDGNVIDQSFNSYSTSTAGGSGKFTNYAA
	201	NANDPNLHFTSYSDQGTGGVQKFTIYSQEANAGDQYFKSY
	241	GKNGNGANGEFVSYGNDTNVIGSTFTNYGQTANGGDQKFT
	281	SYGFNGNVPENHFTNYGAGGNGPSETFNSYRDQSNVGDDT
	321	FTTYVKDANGGEANFTNYGQSFNEGTDVFTTYGKGGNDPH
	361	INFKTYGVNNTFKDYVKDTATFSNYHNKTSQVLASLMEVN
	401	GGKKVNNRWVEPGKFFREKMLKSGTIMPMPDIKDKMPKRS
	441	FLPRVIASKLPFSTSKIAELKKIFHAGDESQVEKMIGDAL
	481	SECERAPSAGETKRCVNSAEDMIDFATSVLGRNVVVRTTE
	521	DTKGSNGNIMIGSVKGINGGKVTKSVSCHQTLYPYLLYYC
	561	HSVPKVRVYEADILDPNSKVKINHGVAICHVDTSSWGPSH
	601	GAFVALGSGPGKIEVCHWIFENDMTWAIAD 630

Figure 7.2. Amino acid sequence of β subunit of tomato fruit polygalacturonase isozyme 1 (deduced from cDNA, Zheng et al. 1992). The N- and C-terminal amino acids are indicated by double lines and gray lines, respectively.

found at low levels in green tomatoes, and in increasing amounts with ripening (Knegt et al. 1988; Pressey 1986).

It is now generally agreed that PG2 is the only endogenous PG in tomatoes, and that PG1 is formed during ripening as β subunit is produced to react with PG2. This has been conclusively demonstrated by monitoring the changes in PGs using differential extraction under varying NaCl concentrations and buffer pHs (Pressey 1988). The fact that a single gene encodes for the catalytic PG polypeptide as described earlier provides supporting evidence for this conclusion. Recent studies on the expression patterns of β subunit and PG genes indicate that β subunit mRNA appears early in fruit developing (10 days after pollination) and increases to a maximum at 30 days, a time coinciding with the onset of ripening. Thereafter, the amount of β subunit mRNA decreases sharply, whereas the amount of PG mRNA increases significantly (Zheng et al. 1992).

Taking all these results into account, it is reasonable to suggest that the endogenous PG2 produced during all the early stages of ripening must be converted to PG1 because of the presence of β subunit. The accumulation of PG2 at later stages is due to the exhaustion of β subunit. The functional role of this conversion process is to regulate PG activity at a precise phase of fruit development. The β subunit acts to anchor PG2 onto the cell wall, forming PG1, which is the active form in depolymerizing pectin in the middle lamella (Knegt et al. 1988, 1991). Studies on the PG isozymes and pectin depolymerization demonstrate that PG1, and not PG2, is responsible for polyuronide degradation and solubilization (DellaPenna et al. 1990; Huber 1983).

Although convincing evidence has accumulated supporting the action of PG enzyme as degradation of polyuronide, the implication of this reaction as the primary cause of tomato fruit softening is a subject of continuous debate, which will not be described in this discussion. One recent study worth mentioning involves the insertion and expression of a chimeric PG gene in transgenic *rin* (ripening inhibition) tomato fruits that results in polyuronide degradation but does not cause fruit softening (Giovannoni et al. 1989).

MECHANISMS AND ACTION PATTERNS

Exo-PG hydrolyzes the nonreducing end unit of a polygalacturonic chain, giving galacturonic acid as a major reaction product (Fig. 7.3). Depolymerization is interrupted by branching that occurs in the substrate. The rate of hydrolysis increases with substrate size, and reaches a maximum with a degree of polymerization of 20 for carrot and peach exo-PGs (Pressey and Avants 1975). Exo-PG action causes a large increase in the formation of reducing groups and a slow increase in viscosity. Polyuronide degradation in ripening does not result in the accumulation of galacturonic acid; only endo-PG is involved.

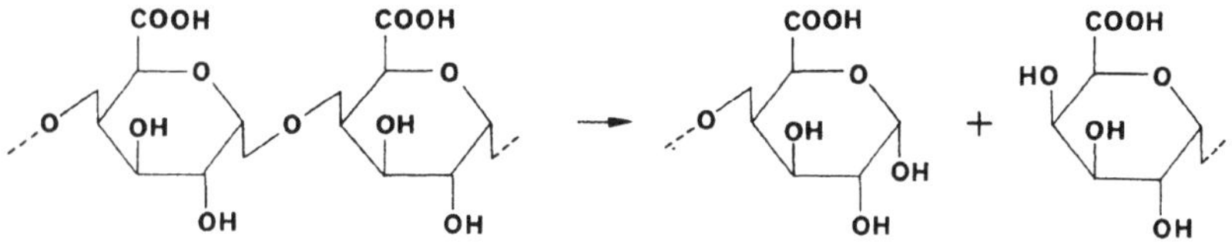

Figure 7.3.   Depolymerization action of polygalacturonase.

Endo-PGs depolymerize pectic acid randomly, resulting in a rapid decrease in the viscosity of the substrate solution. The specificity and action pattern of endo-PG are determined by the nature of the active site. The rate of hydrolysis by yeast endo-PG decreases with a shortening of the substrate (Patel and Phaff 1959). The binding site of *Aspergillus niger* endo-PG is comprised of four subsites and cleavage occurs between subsites 1 and 2 (Fig. 7.4). A tetramer substrate undergoes (3 + 1) cleavage into trigalacturonic and galacturonic acid. A pentamer gives two productive complexes, both satisfying the complete occupancy of all four subsites, yielding a (4 + 1) and (3 + 2) cleavage. Likewise, for a hexagalac-

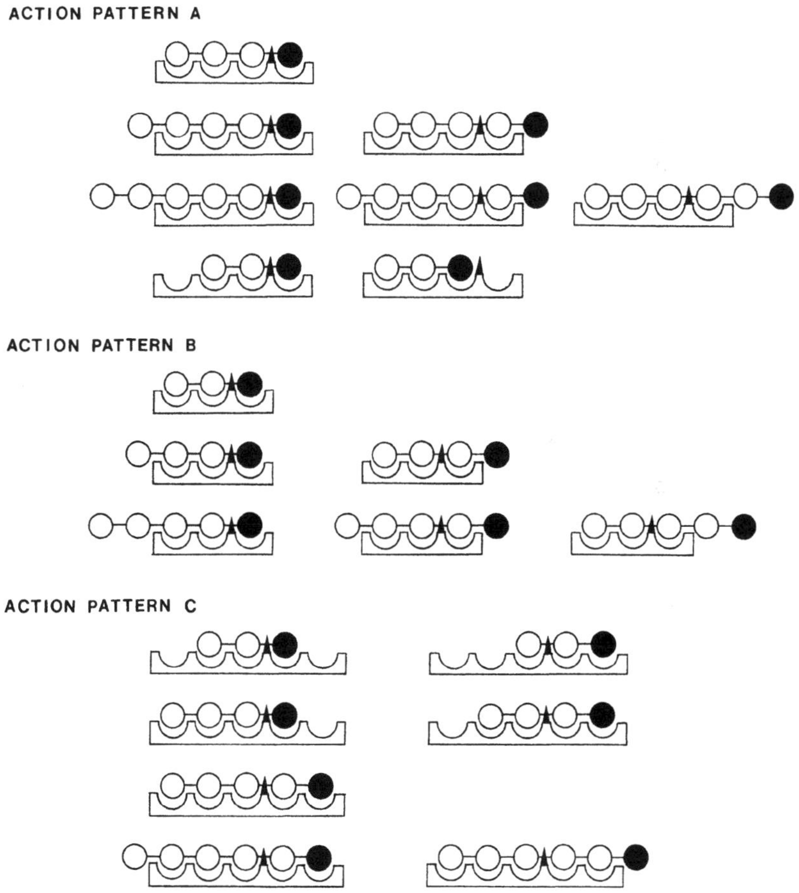

ACTION PATTERN A

ACTION PATTERN B

ACTION PATTERN C

Figure 7.4. Action patterns of endo-polygalacturonases from (A) *Aspergillus niger,* (B) tomato, and (C) *Erwinia carotovora.*

turonic acid substrate, three productive complexes can be envisioned: $(5+1)$, $(4+2)$, and $(3+3)$. Trigalacturonic acid binds to the active site to form either a productive or nonproductive complex. A productive binding in this case yields a $(2+1)$ cleavage. A nonproductive binding is formed when the three monomeric units bind to subsites 2, 3, and 4, without interacting with the catalytic groups located between subsites 1 and 2. The substrate in this case acts as a competitive inhibitor ($K_i = 0.67$ mM).

The action pattern exhibited by *Aspergillus niger* endo-PG also occurs in the enzymes secreted by *Saccharomyces fragilis* and *Acrocylindrium* (Rexova-Benkova and Markovic 1976). Tomato endo-PG shows an alternative action pattern. Trimers are cleaved at a relatively rapid rate. Tetramers gives both $(3+1)$ and $(2+2)$ cleavage, whereas pentamers undergo $(4+1)$, $(3+2)$, and $(2+3)$ cleavage. This is more consistent with an active site comprising three subsites.

Extracellular endo-PG produced by *Erwinia carotovora* hydrolyzes pentamers only in a $(3+2)$ pattern. However, hexamers yield $(3+3)$ and $(4+2)$ cleavages. Pentamers are cleaved five times faster than tetramers. The active site of this enzyme likely contains five subsites.

PECTINESTERASE

Pectinesterase (PE) is widely present in various fruit plants. The enzyme usually exists in multiple forms, and is found in cell wall fractions. Fungal PEs have also been isolated as extracellular enzymes (Kohn et al. 1983; Markovic and Kohn 1984; Forster and Rasched 1985). In general, plant PEs show maximum activity at a slightly alkaline pH range. Low concentrations of metal cations, such as Ca^{++}, tend to enhance the enzyme activity.

CHARACTERISTICS OF PLANT PECTINESTERASES

Tomato fruit contains at least two PE isozymes (Lee and Macmillan 1968; Tucker et al. 1982). Both isozymes PE1 and PE2 increase during the initial phase of ripening. After the "breaker" stage, there is a decrease in PE1, but an increasing accumulation of PE2 until the full color stage of development. PE2 has a MW of 23 kD and a pH optimum of 7.6. The enzyme shows 50% inactivation after heating at 67°C for 5 min. In other studies, five isozymes can be detected, and the major isozyme has a MW of 27.5 kD, and a pH optimum of 7.5–8.5 (Kohn et al. 1983). The enzyme is irreversibly inhibited by iodine in a noncompetitive manner (Markovic and Patocka 1977) in that K_m is unchanged (1.33×10^6 M methyl D-galactopyranosyl residues), but V_{max} is affected by the inhibitor. Tomato PE also exhibits end product (polygalacturonic acid) inhibition with $K_i = 7 \times$

10^{-3} M, although the isozyme used in the investigation is unidentified (Lee and Macmillan 1968). Moderate concentrations of $CaCl_2$ (0.005 M) or NaCl (0.05 M) are required for maximum activity.

Soybean PE is a 33-kD protein with maximum activity at pH close to 8 (Moustacas et al. 1986). Polygalacturonic acid, the product from pectin demethylation, is a competitive inhibitor.

Two PE isozymes are found in banana pulp; both having the same MW of 30 kD, but different pIs of 8.8 and 9.3 (Hultin et al. 1966; Brady 1976). The PEs show optimum activity at pH 7.5. The activity is increased by the addition of 0.2 M NaCl, but the pH optimum is shifted to pH 6.0. The enzyme is also inhibited by a variety of low-molecular weight polyols, such as glycerol, sucrose, glucose, maltose, and galactose.

Orange PE is another well-studied enzyme. Two isozymes, PE1 and PE2, have been purified; both are 36.2 kD proteins, but with different pIs of 10.05 and ≥ 11.0, respectively (Versteeg et al. 1978). PE1 has a pH optimal of 7.6, whereas that of PE2 is 8.0. Kinetic studies indicate K_m = 0.083 and 0.004 mg pectin ml^{-1}, and K_i = 0.42 and 0.0016 mg polygalacturonate ml^{-1}, for PE1 and PE2, respectively. Orange PE also exhibits end product inhibition (Termote et al. 1977).

Grapefruit (March white) pulp contains two isozymes—one more thermally stable and less susceptible to proteolysis (Seymour et al. 1991A). The degree of stability may be related to the extent of glycosylation of the enzyme molecules (Seymour et al. 1991B). The thermally stable, and the thermally labile PE have MW of 51 kD and 36 kD, and K_m values of 1.02 and 0.27 mg pectin ml^{-1}, respectively.

Apple (Castaldo et al. 1989) and kiwi (Giovane et al. 1990) both contain two PE isozymes. The kiwi enzymes possess the same MW of 57 kD and same pI of 7.3, but differ in thermal stability with PE1 being more stable. In addition, PE1 has a K_m of 1.82 mg pectin ml^{-1}, compared with 0.76 mg ml^{-1} for PE2.

It is interesting to note that microbial PEs are not all alkaline proteins. *Trichoderma reesei* PE has a pI of 8.3–9.5 and a pH optimum of 7.6 (Markovic and Kohn 1984). The *Aspergillus* enzymes, however, have a pI and pH optimum in the acidic region (Kohn et al. 1983). Maximum activity for the *Aspergillus niger* enzyme is observed at pH 4.5 at 40°C (Baron et al. 1980). Alkaline and acidic PEs apparently demethylate pectic substrate in a different pattern. The former gives deesterified pectin that forms weak gels with calcium; the latter yields deesterified pectin that is capable of forming stronger gels with calcium ion (Ishii et al. 1979).

AMINO ACID SEQUENCES

The primary structure of tomato PE is known (Fig. 7.5). Protein sequencing identifies 305 amino acid residues with a MW of 33,239 (Markovic and Jornvall

```
Tomato PE1        MINGTNLDELISRAKVALAMLASVTTPNDEVLRPGLGKMP
A. niger                                         MVKSILA

Tomato PE1        SWVSSRDRKLMESSGKDIGANAVVAKDGTGKYRTLAEAVA
A. niger          -VLFAATALAASRMTAPS--I ----S- -D-D-ISA--D

Tomato PE1   24   AAPDKSKTRYVIYVKRGTYKENVEVSSRKMNLMIIGDGMY
A. niger     29   -LSTT-TETQT-FIEE-S-DEQ-YIPALSGK-IVY-QTED

Tomato PE1   64   ATIITGSLNVVDGSTTFHSATLAAVGKGFILQDICIQNTA
A. niger     69   T-TY-SN-VNITHAIALADVDNDDETATLRNYAEGSAIYN

Tomato PE1  104   GPAK        HQAVALRVGADKSVINRCRIDAYQDTLYA
A. niger    109   LNIANTCGQAC---L-VSAY-SEQGYYA-QFTG-----L-

Tomato PE1  137   HSQRQFYQSSYVTGTIDFIFGNAAVVFQKCQLVARKPGKY
A. niger    149   ETGY-V-AGT-IE-AV-----QH-RAWFHECDIRVLE-PS

Tomato PE1  177   QQNMVTAQGRTDPNQATGTSIQFCDIIASPDLKPVVKEFP
A. niger    189   SASI --N--SSESDDSYYV-HKSTVA-A -GND -SSGT

Tomato PE1  217   TYLGRPWKKYSRTVVMESSLGGLIDPSGWAEWHGDFALKT
A. niger    226   Y------SQ-A-VCFQKT-MTDV-NHL--T--STSTPNTE

Tomato PE1  257   LYYGEFMNNGPGAGTSKRVKWPGYHVITDPAEAMSFTVAK
A. niger    266   NVTFVEYG-TGTGAEGP-ANFSSELTEPITISWLLGSDWE

Tomato PE1  297   LIQGGSWLRSTDVAYVDGLYDYSDIKLLFVYVTRHL 317
A. niger    306   DWVDT-YIN 314
```

Figure 7.5. Amino acid sequences of pectinesterases: tomato PE1 (Ray et al. 1988), *Aspergillus niger* (Khanh et al. 1990). Both sequences are deduced from DNA sequences. The N- and C-terminal amino acids are indicated by double lines and gray lines, respectively.

1986). The enzyme contains two disulfide bridges (Cys98-Cys125, and Cys166-Cys200). Cys166 is conserved in all pectinesterases with known sequences (Markovic and Jornvall 1992). The amino acid sequence deduced from cDNA nucleotides shows discrepancies—18 of the 27 variations may result in charge alteration in the protein (Ray et al. 1988). Another cDNA gives a partial sequence that exhibits 94% similarity with the complete cDNA sequence (Harriman et al. 1991). The discrepancies may reflect the existence of isoforms, although only two different isoforms are known in tomato. Cloning and characterization of PE genes from *Erwinia chrysanthemi* (Plastow 1988), *Aspergillus niger* (Khanh et al. 1990), and *Pseudomonas solanacearum* (Spok et al. 1991) have also been reported. The *Pseudomonas* PE clone encodes a protein of 396 amino acids with a calculated MW of 41,004. The amino acid sequences of the four known PEs share about 30% similarity. Six conserved regions can be assigned, although their functional significance awaits further investigation.

MECHANISM OF DEMETHYLATION

PE removes methoxyl groups in pectin by a nucleophilic attack of the enzyme on the ester, resulting in the formation of an acyl-enzyme intermediate with the release of a methanol. This is then followed by deacylation, which is the hydrolysis of the intermediate to regenerate the enzyme and a carboxylic acid (Fig. 7.6).

PEs of plant origins exhibit an action pattern that results in the formation of blocks of carboxylate groups along the pectin chain. The *Trichoderma reesei* enzyme, being a basic protein, shows a similar pattern. Enzymes of the *Aspergillus* species with pH optimum in the acidic region, catalyze random attack (Kohn et al. 1983; Markovic and Kohn 1984).

The effect of metal ions in the activation of pectinesterase may be related to its interaction with the substrate (Nari et al. 1991). Polygalacturonic acid has been shown to be a competitive inhibitor in PE-catalyzed hydrolysis reaction. Pectin consisting of a blockwise arrangement of carboxylate groups may act in a similar way. Binding of metal ions to carboxylate groups in this case tend to neutralize the inhibition effect of the pectin substrate on the enzyme. Excess metal

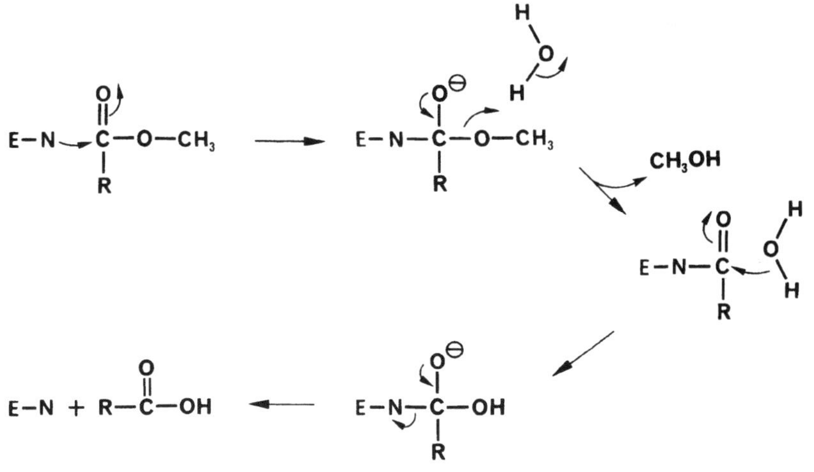

Figure 7.6. Reaction mechanism of pectinesterase. Details described in text.

ions, however, actually cause inactivation of PE because metal ions are bound to carboxylate groups neighboring to the ester bonds that are required for hydrolysis to occur.

PECTATE LYASE

Pectate lyases (PELs) are microbial extracellular enzymes. Enzymes of the genera *Erwina* and *Bacillus* are among the best known to produce soft-rot symptoms in plants (Perombelon and Kelman 1980; Collmer and Keen 1986), although the enzyme is also found in other microorganisms, including *Aeromonas* (Hsu and Vaughn 1969), *Pseudomonas* (Liao et al. 1988; Liao 1991), *Xanthomonas* (Nasuno and Starr 1967), *Aspergillus* (Dean and Timberlake 1989), and *Fusarium* (Crawford and Kolattukudy 1987).

ENZYME CHARACTERISTICS

Erwinia chrysanthemi produces pectate lyases in isoforms that can be grouped according to their pIs: acidic (pH 4–5), neutral (pH 7–8.5), and alkaline (pH 9–10) (Garibaldi and Bateman 1971; Pupillo et al. 1976). The number of isozymes in each group may vary depending on the strains. Most strains studied produce five or at least four isozymes—one acidic (PELA), two neutral (PELB and C), and two alkaline (PELD and E) (Bertheau et al. 1984; van Gijsegem 1986). All isozymes require Ca^{++} for maximum activity, and have a pH optimum of 8.0–10. Studies on the strain EC16 indicate that all the four PELs produced are endo-enzymes (PELA (pI 4.2), PELB (pI 8.8), PELC (pI 9.0), and PELE (pI 10.0) (Barras et al. 1987). Alkaline PELs are effective in causing maceration of plant tissues followed by the neutral isozymes, whereas the acidic PEL shows no effect (Garibaldi and Bateman 1971; Barras et al. 1987).

In *Erwinia carotovora* strains, at least three PELs are secreted, all with alkaline pIs—PELI (pI 9.7), PELII (pI 10.2), and PELIII (pI 10.35). All the three isozymes have a pH optimum of 9. The temperature optima for activity of PELII and PELIII are 50°C and 60°C, respectively. PELI has a low heat stability (Quantick et al. 1983; Sugiura et al. 1984; Yoshida et al. 1991A,B). In one strain (GIR726), four kinds of endo-PEL isozymes are observed with very high alkaline pIs (10.0, 10.6, 10.3, 10.9) and a MW range of 28 kD to 33 kD (Tanabe et al. 1984). The pH optima for activity are 9.3 for PELII, 9.5 for PELIV, and 9.7 for PELI and III, respectively. Similar to PELs from various other sources, these isozymes are activated by Ca^{++}. The enzyme activity increases by 50–70% with 0.5 mM Ca^{++} concentration. The isozyme with pI > 10 consists of 2.5–4.8% neutral sugar residues (Sugiura et al. 1984). An exo-PEL with MW of 76 kD has also been isolated from *Erwinia carotovora* subsp. carotovora (Kegoya et al. 1984).

PELs isolated from a number of *Bacillus* species also possess maceration properties (Dave and Vaughn 1971; Hasegawa and Nagel 1966). All the *Bacillus* PELs exhibit endo-action and require Ca^{++} for activation. The *Bacillus macerans* PEL shows maximum activity at pH 9.0 and 60°C. The enzyme has a MW of 35 kD, and a pI of 10.3 (Miyazaki 1991). Enzyme activity is enhanced by Ca^{++} or Mg^{++}. The PEL attacks pectate randomly, and also degrades pectin with 4.3% esterification. Certain strains of *Bacillus subtilis* produce PEL that causes rapid maceration of vegetable tissues in carrot and potato (Chesson and Codner 1978). A purified PEL excreted from *Bacillus subtilus* strain S0113 has a pI of 9.6 and a MW of 42 kD. The enzyme has optimum activity at pH 8.4 and 42°C. As in all other PELs, the enzyme is activated by Ca^{++}. The K_m and V_{max} of the purified enzyme for polygalacturonic acid are 0.862 gm $\cdot$ l^{-1} and 1.48 μmol $\cdot$ min^{-1}mg^{-1}, respectively (Nasser et al. 1990).

AMINO ACID SEQUENCES

The amino acid sequences of all the four PELs secreted by *Erwinia chrysanthemi* EC16 are deduced from the encoding genes (Fig. 7.7). The *pel* genes occur in two loosely linked chromosomal clusters, with two structurally homologous genes (*pelA/pelE* and *pelB/pelC*), although the genes are all independently regulated. The 3' end of the remains of a deleted *pelD* gene is located between the *pelA* and *pelE* genes (Tamaki et al. 1988). PELA (pI 4.2) contains 361 amino acids (including signal peptide) with a calculated MW of 38,756. PELB (pI 8.8) consists of 392 amino acids in the mature protein, with a calculated MW of 37,922. The amino acid sequence predicted for PELC contains 375 amino acid residues (including 22 in the signal peptide). The MW estimates to be 37,676. The PELE mature protein has 355 amino acids and a calculated MW of 38,037 (Keen and Tamaki 1986; Tamaki et al. 1988). PELA and PELE sequences are 62% identical. PELB and PELC have 84% homology in the amino acid sequences. However, PELB and PELE show little similarity in the sequences.

In the *Erwinia chrysanthemi* strain 3937, there are five *pel* genes (*A* to *E*) distributed over five different regions of the chromosome (Reverchon et al. 1989; Hugouvieux-Cotte-Pattat and Robert-Baudouy 1987). The *pelA* gene has been isolated and sequenced. It codes the acidic PELA (pI 5.2) containing 361 amino acids with a calculated MW of 38,412. The mature protein shows 90% similarity in sequence with the PELA of strain EC16 (Favey et al. 1992). The *pelIII* gene of *Erwinia carotovara* encodes a protein of 352 amino acids with a calculated MW of 38,253 (Yoshida et al. 1991).

```
EC16 PLa              MMNKASGRSFTRSSKYLLATLIAGMMASGVSA
EC-3937 PLe           MNNSRMSSVSTQKTTGRSALGT-SA--AI--TT-MVS-AS


         33    AELVSDKALESAPTVGWASQNGFTTGGAAATSDNIYIVTN
         41    -ASLQTTKATE-AST---T-SGG-----K-SSSK--A-KS


         73    ISEFTSAL SAGAEAKIIQIKGTIDISGGTPYTDFADQKA
         81    ----KA--NGTDSSP----VT-A------KA----D----


        112    RSQINIPANTTVIGLGTDAKFINGSLIIDGTDGTNNVIIR
        121    ----S--S---I--I-NKG--T----VVK-VS   ---L-


        152    NVYIQTPIDVEPHYEKGDGWNAEWDAMNITNGAHHVWIDH
        158    -L--E--V--A----E----------VVID STD---V--


        192    VTISDGNFTDDMYTTKDGETYVQHDGALDIKRGSDYVTIS
        197    ------SL---K----N--K------S-----------V-


        232    NSLIDQHDKTMLIGHNDTNSAQDKGKLHVTLFNNVFNRVT
        237    --RFEL----I----S-N-GS--A---R--FH--L-DR-G


        272    ERAPRVRYGSIHSFNNVFKGDAKDPVYRYQYSFGIGTSGS
        277    --T----F--V-AY---YV--VNHKA-------------


        312    VLSEGNSFTIANLSASKA   CKVVKKFNGSIFSDNGSV
        317    L---S-A---D-MKKISGRDKE-S---A---K----K--I


        348    LNGSAVDLS  GCGFSAYTSKIPYIYDVQPMTTELAQSIT
        357    I--ASYNLNGC-F-----SA----K-SA-TI--S-AN--S


        386    DNAGSGKL   393
        397    S---Y---   404


ECA PLb               MKYLLPTAAAGLLLLAAQPAMA ANTGGYATTDGGDVSGA
EC16 PLc              --S-ITPIT-----ALS--LL-ATD-----A-A--N-T--


         40    VKKTARSLQEIVDIIEAAKVDSKGKKVKGGAYPLIITYNG
         41    -S---T-M-D--N--D--RL-AN----------V---T-
```

Figure 7.7.   Comparison of amino acid sequences of pectate lyases: EC16 PLa (*Erwinia chrysanthemi* EC16 PEL A, Tamaki et al. 1988), EC-3937 PLe (*Erwinia chrysanthemi* 3937 PEL E, Reverchon et al. 1989), ECA PLb (*Erwinia carotovora* PEL B, Lei et al. 1987), EC16 PLc (*Erwinia chrysanthemi* EC16 PEL C, Tamaki et al. 1988). All sequences are deduced from cDNA sequencing.

```
 80    NEDSLIKAAEKNICGQWSKDARGVQIKEFTKGITIQGTNG
 81    ------N--AA--------DP---E----------I-A--

120    SSANFGVWIVNSSNVVVRNMRFGYMPGGAQDGDAIRVDNS
121    ------I--KK--D---Q---I--L----KD--M----D-

160    PNVWIDHNKIFAQSFECKGTPDNDTTFESAVDIKKGSTNV
161    ----V---E---ANH--D--------------GA-NT-

200    TVSYNYIHGIKKVGLSGASNTDTGRNLTYHHNIYRDVNSR
201    ---------V-----D-SSSS-----I-----Y-N---A-

240    LPLQRGGLVHAYTNLYDGITGSGFNVRQKGIALIESNWFE
241    ------------N---TN---S-L----N-Q----N----

280    IALNPVTARNDSSNFGTWELRNNNITKPADFSKYNITWGR
281    K------S-Y-GK-----V-KG----------T-S---TA

320    PSTPHVNADDWKSTGKFPSISYKYTPVSAQCVKDKLANYA
321    DTK-Y----S-T---T--TVAYN-S----------PG--

360    GVGKNLAVLTAANCK    374
361    ------T--STA--    375
```

Figure 7.7   (Continued)

THREE-DIMENSIONAL STRUCTURE

The crystal structure of PELC from *Erwinia chrysanthemi* EC16 has been determined at 2.2-A resolution (Fig. 7.8) (Yoder et al. 1990; 1993). The protein molecule assumes a parallel β helix structure consisting of three parallel β sheets each composed of 7 to 10 parallel β strands all coiled into a large right-handed helical cylinder. The N-terminal end and the C-terminal end of the helical cylinder are covered by a 3-turn helix, and an extended loop, respectively. PELE, which shares a 31% similarity in the amino acid sequence, also has a very similar structure. The core regions of PELC and PELE superimpose well (Yoder et al. 1994). Within the core, a series of amino acid side chains pack into stacks. Four stacking motifs are observed in both enzymes: (1) An Asn ladder with Asn residues located in consecutive turns of the parallel β helix forming a prominent stack via a network of hydrogen bonds; (2) A Ser stack formed by hydrogen bonding; (3) A number of aliphatic stacks involving side chains of either Val, Ile, Leu, or Ala; (4) An aromatic stack involving hydrophobic interactions among Phe and Tyr residues. All the stacking motifs have the amino acid side chains arranged parallel to the axis of the parallel β helical cylinder. These novel side chains stacks tend to stabilize the overall structure of the enzyme.

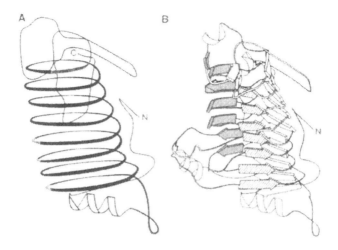

Figure 7.8. (A) Schematic representation of the PELC backbone illustrating the parallel β helix topology. The gray regions indicate the positions where several external loops are deleted for clarity. (B) Ribbon diagram of the PELC polypeptide backbone. Disulfide bonds are indicated by thick lines. (From Yoder et al. 1993. *Science 260*, 1504, with permission. Copyright 1993 American Association for the Advancement of Science. Figure courtesy of Professor Francis Jurnak, University of California, Riverside.)

The Ca^{++} binding site is likely located in a narrow groove consisting of several charged (basic and acidic) side chains on the exterior surface of the parallel β helix. The shape of the groove accommodates a dodecamer substrate. The active site may also be located in the vicinity.

REACTION MECHANISM

Evidence of PEL-catalyzed β-elimination of pectic acid was first obtained by the identification of a digalacturonic acid in the product as 4-*O*-α-D-(1,5-dehydro-galacturonosyl)-D-galacturonic acid (Fig. 7.9) (Hasegawa and Nagel 1962). Details of the enzyme reaction are not available. It is known that in alkaline medium, pectin undergoes deesterification, accompanied by degradation via β-elimination. Similar splitting of glycosidic bonds also occurs in neutral pH solution at elevated temperature. Pectate, which consists mostly of carboxyl groups, is relatively stable. These observations are likely to be related to the electron-withdrawing methyl ester that renders the C5-H increasingly acidic. Abstraction of the proton at C5

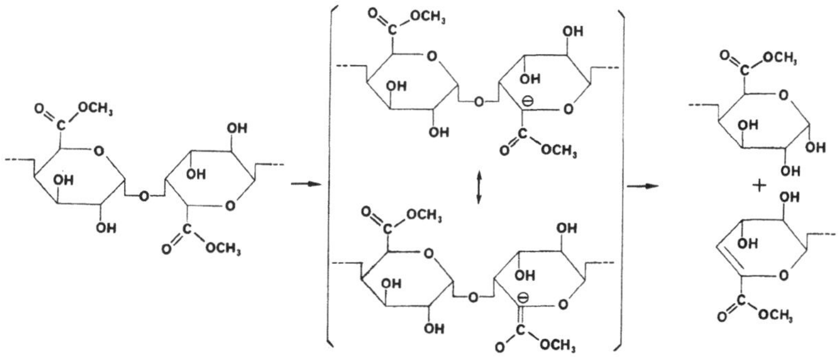

Figure 7.9. The proposed reaction mechanism of pectate lyase. Details described in text.

by hydroxyl ions results in the formation of a carbanion that is stabilized by the C6 carboxyl group. This step is then followed by elimination at C4 with the formation of a double bond between C4 and C5 (BeMiller and Kumari 1972). The rate of elimination of pectin is enhanced by the addition of Ca^{++} and K^+ (Keijbets and Pilnik 1974). The effect of cations may suggest a neutralization of the anionic groups in the pectin chain, thereby increasing the accessibility of hydroxyl ions. Elimination reactions catalyzed by enzymes often occur via carbanion intermediates.

For PEL-catalyzed reaction, the rate of cleavage depends on the chain length of the substrate, similar to the PG enzymes. The *Bacillus polymyxa* PEL attacks tetramers preferentially at the nonreducing end to yield monomers and unsaturated trimers. However, for unsatured tetramers, the cleavage is shifted to the middle, giving unsaturated dimers as predominant products (Anderson and Nagel 1964). The four extracellular PELs of *Erwinia chrysanthemi* EC16 expressed as recombinant enzymes in *Escherichia coli* show some unique depolymerization patterns (Preston et al. 1991; 1992). PELA exhibits a random endoaction. Compared with other PELs, depolymerization catalyzed by PELA is relatively limited. No single predominant product is formed in the process. PELB and PELC, however, catalyze the formation of trimer as the major (68–72%) product. In PELE-catalyzed depolymerization, 76% of the final product consists of dimers with only 2% trimers. The action of these enzymes shows an endo pattern, but with selective cleavage at some preferential bonds as determined by the size of the binding site and the location of catalytic residues in the active site. The profiles of these products obtained from these four enzymes seem to correlate well with the fact that PELE is a strong tissue-macerating enzyme, whereas the acidic isozyme usually has little activity.

REFERENCES

ALI, Z. M., and BRADY, C. J. 1982. Purification and characterization of the polygalacturonases of tomato fruits. *Aust. J. Plant Physiol. 9*, 155–169.

ANDERSON M. M., and NAGEL, C. W. 1964. Effect of the unsaturated bond on the degradation of the tetragalacturonic acids by a transeliminase. *Nature 203*, 649.

BARON, A.; ROMBOUTS, F.; DRILLEAU, J. F.; and PILNIK, W. 1980. Purification et propriétés de la pectinestérase produite par *Aspergillus niger*. *Lebensm-Wiss. u.-Technol. 13*, 330–333.

BARRAS, F.; THURN, K. K.; and CHATTERJEE, A. K. 1987. Resolution of four pectate lyase structural genes of *Erwinia chrysanthemi* (EC16) and characterization of the enzymes produced in *Escherichia coli*. *Mol. Gen. Genet. 209*, 319–325.

BEMILLER, J. N. 1986. An introduction to pectins: Structure and properties. In: Chemistry and Function of Pectins, M. L. Fishman and J. J. Jen, eds., *ACS Sym. Ser. 310*, American Chemical Society, Washington, D.C.

BEMILLER, J. N., and KUMARI, G. V. 1972. Beta-elimination in uronic acids: Evidence for an ElcB mechanism. *Carbohydr. Res. 25*, 419–428.

BERTHEAU, Y.; MADGIDI-HERVAN, E.; KOTOUJANSKY, A.; NGUYEN, The, C.; ANDRO, T.; and COLENO, A. 1984. Detection of depolymerase isoenzymes after electrophoresis or electrofocusing, or in titration curves. *Anal. Biochem. 139*, 383–389.

BRADY, C. J. 1976. The pectinesterase of the pulp of the banana fruit. *Aust. J. Plant Physiol. 3*, 163–172.

BRADY, C. J.; MACALPINE, G.; MCGLASSON, W. B.; and UEDA, Y. 1982. Polygalacturonase in tomato fruits and the induction of ripening. *Aust. J. Plant Physiol. 9*, 171–178.

BROWN, S. M., and CROUCH, M. L. 1990. Characterization of a gene family abundantly expressed in *Oenothera organensis* pollen that shows sequence similarity to polygalacturonase. *The Plant Cell 2*, 263–274.

BUSSINK, H. J. D.; KESTER, H. C. M.; and VISSER, J. 1990. Molecular cloning; nucleotide sequence and expression of the gene encoding prepro-polygalacturonase II of *Aspergillus niger*. *FEBS Lett. 273*, 127–130.

CASTALDO, D.; QUAGLIUOLO, L.; SERVILLO, L.; BALESTRIERI, C., and GIOVANE, A. 1989. Isolation and characterization of pectin methylesterase from apple fruit. *J. Food Sci. 54*, 653–655.

CHESSON, A., and CODNER, R. C. 1978. The maceration of vegetable tissue by a strain of *Bacillus subtilus*. *J. Appl. Bacteriol. 44*, 347–364.

COLLMER, A., and KEEN, N. T. 1986. The role of pectic enzymes in plant pathogenesis. *Ann. Rev. Phytopathol. 24*, 383–409.

CRAWFORD, M. S., and KOLATTUKUDY, P. E. 1987. Pectate lyase from *Fusarium solani* f. sp. *pisi*: Purification, characterization, *in vitro* translation of the mRNA, and involvement in pathogenicity. *Arch. Biochem. Biophys. 258*, 196–205.

DAVE, B. A., and VAUGHN, R. H. 1971. Purification and properties of a polygalacturonase acid trans-eliminase produced by *Bacillus pumilus*. *J. Bacteriol. 108*, 166–174.

DEAN, R. A., and TIMBERLAKE, W. E. 1989. Regulation of the *Aspergillus nidulans* pectate lyase gene (*pelA*). *The Plant Cell 1*, 275–284.

DELLAPENNA, D., and BENNETT, A. B. 1988. *In vitro* synthesis and processing of tomato fruit polygalacturonase. *Plant Physiol. 86*, 1057–1063.

DELLAPENNA, D.; LASHBROOK, C. C.; TOENJES, K.; GIOVANNORII, J. J.; FISCHER, R. L.; and BENNETT, A. B. 1990. Polygalacturonase isozymes and pectin depolymerization in transgenic *rin* tomato fruit. *Plant Physiol. 94*, 1882–1886.

DEMAIN, A. L., and PHAFF, H. J. 1954. Hydrolysis of the oligogalacturonides and pectic acid by yeast. *J. Biol. Chem. 210*, 381–393.

DOPICO, B.; LOWE, A. L.; WILSON, I. D.; MERODIO, C.; and GRIERSON, D. 1993. Cloning and characterization of avocado fruit mRNAs and their expression during ripening and low-temperature storage. *Plant Mol. Biol. 21*, 437–449.

FAVEY, S.; BOURSON, C.; BERTHEAU, Y.; KOTOUJANSKY, A.; and BOCCARA, M. 1992. Purification of the acidic pectate lyase and nucleotide sequence of the corresponding gene (*pelA*) of *Erwinia chrysanthemi* strain 3937. *J. Gen. Microbiol. 138*, 499–508.

FORSTER, H., and RASCHED, I. 1985. Purification and characterization of extracellular pectinesterases from *Phytophthora infestans*. *Plant Physiol. 77*, 109–112.

GARIBALDI, A., and BATEMAN, D. F. 1971. Pectic enzymes produced by *Erwinia chrysanthemi* and their effects on plant tissue. *Physiol. Plant Pathol. 1*, 25–40.

GIOVANE, A.; QUAGLIUOLO, L.; CASTALDO, D.; SERVILLO, L.; and BALESTRIERI, C. 1990. Pectin methyl esterase from *Actinidia chinensis* fruits. *Phytochemistry 29*, 2821–2823.

GIOVANNONI, J. J.; DELLAPENNA, D.; BENNETT, A. B.; and FISCHER, R. L. 1989. Expression of a chimeric polygalacturonase gene in transgenic *rin* (ripening inhibitor) tomato fruit results in polyuronide degradation but not fruit softening. *The Plant Cell 1*, 53–63.

GRIERSON, D.; TUCKER, G. A.; KEEN, J.; RAY, J.; BIRD, C. R.; and SCHUCH, W. 1986. Sequencing and identification of a cDNA clone for tomato polygalacturonase. *Nucl. Acids Res. 14*, 8595–8603.

HARRIMAN, R. W.; TIEMAN, D. M.; and HANDA, A. K. 1991. Molecular cloning of tomato pectin methylesterase gene and its expression in Rutgers, ripening inhibitor, nonripening, and never ripe tomato fruits. *Plant Physiol. 97*, 80–87.

HASEGAWA, S., and NAGEL, C. W. 1962. The characterization of an α,β-unsaturated digalacturonic acid. *J. Biol. Chem. 237*, 619–621.

———. 1966. A new pectic acid transeliminase produced exocellularly by a *Bacillus*. *J. Food Sci. 31*, 838–845.

HSU, E. J., and VAUGHN, R. H. 1969. Production and catabolite repression of the constitutive polygalacturonic acid trans-eliminase of *Aeromonas liquefaciens*. *J. Bacteriol. 98*, 172.

HUANG, J., and SCHELL, M. A. 1990. DNA sequence analysis of *pglA* and mechanisms of export of its polygalacturonase product from *Pseudomonas solanacearum*. *J. Bacteriol. 172*, 3879–3887.

HUBER, D. J. 1983. Polyuronide degradation and hemicellulose modifications in ripening tomato fruit. *J. Am. Soc. Hort. Sci. 108*, 405–409.

HUGOUVIEUX-COTTE-PATTAT, N., and ROBERT-BAUDOUY, J. 1987. Hexuronate catabolism in *Erwinia chrysanthemi*. *J. Bacteriol. 169*, 1223–1231.

HULTIN, H. O.; SUN, B.; and BULGER, J. 1966. Pectin methyl esterases of the banana. Purification and properties. *J. Food Sci. 31*, 320–327.

ISHII, S.; KIHO, K.; SUGIYAMA, S.; and SUGIMOTO, H. 1979. Low-methoxyl pectin prepared by pectinesterase from *Aspergillus japonicus*. *J. Food Sci. 44*, 611–614.

KEEN, N. T., and TAMAKI, S. 1986. Structure of two pectate lyase genes from *Erwinia chrysanthemi* EC16 and their high-level expression in *Escherichia coli*. *J. Bacteriol. 168*, 595–606.

KEIJBETS, M. J. H., and PILNIK, W. 1974. β-Elimination of pectin in the presence of anions and cations. *Carbohydr. Res. 33*, 359–362.

KEON, J. P. R., and WAKSMAN, G. 1990. Common amino acid domain among endopolygalacturonases of *Ascomycete fungi*. *Appl. Environ. Microbiol. 56*, 2522–2528.

KHANH, N. Q.; ALBRECHT, H.; RUTTHOWSKI, E.; LOFFLER, F.; GOTTSCHALK, M.; and JANY K.-D. 1990. Nucleotide and derived amino acid sequence of a pectinesterase cDNA isolated from *Aspergillus niger* strain RH5344. *Nucl. Acids Res. 18*, 4262.

KNEGT, E.; VERMEER, E.; and BRUINSMA, J. 1988. Conversion of the polygalacturonase isoenzymes from ripening tomato fruits. *Physiologia Plantarum 72*, 108–114.

KNEGT, E.; VERMEER, E.; PAK, C.; and BRUINSMA, J. 1991. Function of the polygalacturonase converter in ripening tomato fruit. *Physiologia Plantarum 82*, 237–242.

KOGOYA, Y.; SETOGUCHI, M.; YOKOHIKI, K.; and HATANAKA, C. 1984. Affinity chromatography of exopolygalacturonate lyase from *Erwinia carotovora* subsp. *carotovora*. *Agric. Biol. Chem. 48*, 1055–1060.

KOHN, R.; MARKOVIC, O.; and MACHOVA, E. 1983. Deesterification mode of pectin by pectin esterases of *Aspergillus foetidus*, tomatoes and alfalfa. *Collection Czechoslovak Chem. Comm. 48*, 790–797.

KOLLER, A., and NEUKOM, H. 1969. Untersuchungen über den Abbaumechanisms. einer gereinigten polygalakturonase aus *Aspergillus niger*. *Eur. J. Biochem. 7*, 485–489.

KRAMER, M.; SANDERS, R. A.; SHEEHY, R. E.; MELIS, M.; KUEHN, M.; and HIATT, W. R. 1990. Field evaluation of tomatoes with reduced polygalacturonase of antisense RNA. *Plant Biol. 11* (Horticultural Biotechnology) 347–355.

LEE, M., and MACMILLAN, J. D. 1968. Mode of action of pectic enzymes. I. Purification and certain properties of tomato pectinesterase. *Biochemistry 7*, 4005–4010.

LEI, S.-P.; LIN, H.-C.; WANG, S.-S.; CALLAWAY, J.; and WILCOX, G. 1987. Characterization of the *Erwinia carotovora pelB* gene and its product pectate lyase. *J. Bacteriol. 169*, 4379–4383.

LIAO, C.-H. 1991. Cloning of pectate lyase gene pel from *Pseudomonas fluorescens* and detection of sequences homologous to *pel* in *Pseudomonas viridiflava* and *Pseudomonas putida*. *J. Bacteriol. 173*, 4386–4393.

LIAO, C.-H.; HUNG, H. Y.; and CHATTERJEE, A. K. 1988. An extracellular pectate lyase is the pathogenicity factor of the soft-rotting bacterium *Pseudomonas viridiflava*. *Mol. Plant-Microbe Interact. 1*, 199–206.

LIU, Y. K., and LUH, B. S. 1978. Purification and characterization of endo-polygalacturonase from *Rhizopus arrhizus*. *J. Food Sci. 43*, 721–726.

MARKOVIC, O., and JORNVALL, H. 1986. Pectinesterase. The primary structure of the tomato enzyme. *Eur. J. Biochem. 158*, 455–462.

———. 1992. Disulfide bridges in tomato pectinesterase: Variations from pectinesterases of other species; conservation of possible active site segments. *Protein Science 1*, 1288–1292.

MARKOVIC, O., and KOHN, R. 1984. Mode of pectin deesterification by *Trichoderma reesei* pectinesterase. *Experientia 40*, 842–843.

MARKOVIC, O., and PATOCKA, J. 1977. Action of iodine on the tomato pectinesterase. *Experientia 33*, 711–713.

MARTINEZ, M. J.; ALCONADA, M. T.; GUILLEN, F.; VAZQUEZ, C.; and REYES, F. 1991. Pectic activities from *Fusarium oxysporum* f. sp. *melons*: Purification and characterization of an exopolygalacturonase. *FEMS Microbiol. Lett. 81*, 145–150.

MATSUI, I.; ISHIKAWA, K.; MIYAIRI, S.; FUKUI, S.; and HONDA, K. 1992A. A mutant α-amylase with enhanced activity specific for short substrates. *Biochemistry 31*, 5232–5236.

———. 1992B. Alteration of bond-cleavage pattern in the hydrolysis catalyzed by *Saccharomycopsis* α-amylase altered by site-directed mutagenesis. *Biochemistry 31*, 5232–5236.

MIYAZAKI, Y. 1991. Purification and characterization of endo-pectate lyase from *Bacillus macerans*. *Agric. Biol. Chem. 55*, 25–30.

MOSHREFI, M., and LUH, B. S. 1983. Carbohydrate composition and electrophoretic properties of tomato polygalacturonase isoenzymes. *Eur. J. Biochem. 135*, 511–514.

MOUSTACAS, A.-M.; NARI, J.; DIAMANTIDIS, G.; NOAT, G.; and CRASNIER, M. 1986. Electrostatic effects and the dynamics of enzyme reactions at the surface of plant cells. 2. The role of pectin methyl esterase in the modulation of electrostatic effects in soybean cell walls. *Eur. J. Biochem. 155*, 191–197.

NARI, J.; NOAT, G.; and RICARD, J. 1991. Pectin methylesterase, metal ions and plant cell-wall extension. *Biochem. J. 279*, 343–350.

NASSER, W.; CHALET, F.; and ROBERT-BAUDOUY, J. 1990. Purification and characterization of extracellular pectate lyase from *Bacillus subtilis*. *Biochimie 72*, 689–695.

NASUNO, S., and STARR, M. P. 1967. Polygalacturonic acid *trans*-eliminase of *Xanthomonas campestris*. *Biochem. J. 104*, 178.

NIOGRET, M.-F.; DUBALD, M.; MANDARON, P.; and MACHE, R. 1991. Characterization of pollen polygalacturonase encoded by several cDNA clones in maize. *Plant Mol. Biol. 17*, 1155–1164.

PATEL, D. S., and PHAFF, H. J. 1959. On the mechanism of action of yeast endopolygalacturonase on oligogalacturonides and their reduced and oxidized derivatives. *J. Biol. Chem. 234*, 237–241.

PEROMBELON, M. C. M., and KELMAN, A. 1980. Ecology of the soft rot *Erwinias*. *Ann. Rev. Phytopathol. 18*, 361–387.

PILNIK, W., and ROMBOUTS, F. M. 1979. Ultilization of pectic enzymes in food production. In: *Proceedings of the Fifth Internal Congress of Food Science and Technology*, Hideochiba, ed., Kodansha, Ltd., and Elsevier Scientific Publ. Co.

PILNIK, W., and VORAGEN, A. G. J. 1993. Pectic enzymes in fruit and vegetable juice manufacture. In: *Enzymes in Food Processing*. 3d ed., T. Nagodawithana and G. Reed, eds., Academic Press, New York and San Diego.

PLASTOW, G. S. 1988. Molecular cloning and nucleotide sequence of the pectin methyl esterase gene of *Erwinia chrysanthemi* B374. *Mol. Microbiol. 2*, 247–254.

POGSON, B. J.; BRADY, C. J.; and ORR, G. R. 1991. On the occurrence and structure of subunits of endopolygalacturonase isoforms in mature-green and ripening tomato fruits. *Aust. J. Plant Physiol. 18*, 65–79.

POLIZELI, M. L. T. M.; JORGE, J. A.; and TERENZI, H. F. 1991. Pectinase production by *Neurospora crassa*: Purification and biochemical characterization of extracellular polygalacturonase activity. *J. Gen. Microbiol. 137*, 1815–1823.

PRESSEY, R. 1986. Changes in polygalacturonase isoenzymes and converter in tomatoes during ripening. *HortScience 21*, 1183–1185.

———. 1988. Reevaluation of the changes in polygalacturonases in tomatoes during ripening. *Planta 174*, 39–43.

PRESSEY, R., and AVANTS, J. K. 1973. Two forms of polygalacturonase in tomatoes. *Biochim. Biophys. Acta 309*, 363–369.

———. 1975. Modes of action of carrot and peach exopolygalacturonases. *Phytochemistry 14*, 957–961.

PRESTON III, J. F.; RICE, J. D.; CHOW, M. C.; and BROWN, B. J. 1991. Kinetic comparisons of trimer-generating pectate and alginate lyases by reversed-phase ion-pair liquid chromatography. *Carbohydr. Res. 215*, 147–157.

PRESTON III, J. F.; RICE, J. D.; INGRAM, L. O.; and KEEN, N. T. 1992. Differential depolymerization mechanisms of pectate lyases secreted by *Erwinia chrysanthemi* EC16. *J. Bacteriol. 174*, 2039–2042.

PUPILLO, P.; MAZZUCCHI, U.; and PIERINI, G. 1976. Pectic lyase isozymes produced by *Erwinia chrysanthemi* Burkh. et al. in polypectate broth or in *Dieffenbachia* leaves. *Physiol. Plant Pathol. 9*, 113–120.

QUANTICK, P.; CERVONE, F.; and WOOD, R. K. S. 1983. Isoenzymes of a polygalacturonate trans-eliminase produced by *Erwinia atroseptica* in potato tissue and in liquid culture. Physiol. *Plant Pathol. 22*, 77–86.

RAY, J.; KNAPP, J.; GRIERSON, D.; BIRD, C.; and SCHUCH, W. 1988. Identification and sequence determination of a cDNA clone for tomato pectin esterase. *Eur. J. Biochem. 174*, 119–124.

REVERCHON, S.; HUANG, Y.; BOURSON, C.; and ROBERT-BAUDOUY, J. 1989. Nucleotide sequences of *Erwinia chrysanthemi ogl* and *pelE* genes negatively regulated by the *kdgR* gene product. *Gene 85*, 125–134.

REXOVA-BENKOVA, L., and MARKOVIC, O. 1976. Pectic enzymes. *Adv. Carbohydr. Chem. Biochem. 33*, 323–385.

RUTTKOWSKI, E.; LABITZKE, R.; KHANH, N. W.; LOFFLER, F.; GOTTSCHALK, M.; and

JANY, K.-D. 1990. Cloning and DNA sequence analysis of a polygalacturonase cDNA from *Aspergillus niger* RH5344. *Biochim. Biophys. Acta 1087*, 104–106.

SAKELLARIS, G.; NIKOLAROPOULOS, S.; and EVANGELOPOULOS, A. E. 1989. Purification and characterization of an extracellular polygalacturonase from *Lactobacillus plantarum* strain BA11. *J. Appl. Bacteriol. 67*, 77–85.

SCOTT-CRAIG, J. S.; PANACCIONE, D. G.; CERVONE, F.; and WALTON, J. D. 1990. Endopolygalacturonase is not required for pathogenicity of *Cochliobolus carbonum* on maize. *The Plant Cell 2*, 1191–1200.

SHEEHY, R. E.; PEARSON, J.; BARDY, C. J.; and HIATT, W. R. 1987. Molecular characterization of tomato fruit polygalacturonase. *Mol. Gen. Genet. 208*, 30–36.

SPOK, A.; STUBENRAUCH, G.; SCHORGENDORFER, K.; and SCHWAB, H. 1991. Molecular cloning and sequencing of a pectinesterase gene from *Pseudomonas solanacearum*. *J. Gen. Microbiol. 137*, 131–140.

SUGIURA, J.; YASUDA, M.; KAMIMIYA, S.; IZAKI, K., and TAKAHASHI, H. 1984. Purification and properties of two pectate lyases produced by *Erwinia carotovora*. *J. Gen. Appl. Microbiol. 30*, 167–175.

TAMAKI, S. J.; ROBESON, G. M.; MANULIS, S.; and KEEN, N. T. 1988. Structure and organization of the *pel* genes from *Erwinia chrysanthemi* EC16. *J. Bacteriol. 170*, 3468–3478.

TANABE, H.; KOBAYASHI, Y.; MATUO, Y.; NISHI, N., and WADA, F. 1984. Isolation and fundamental properties of endo-pectate lyase pI − isozymes from *Erwinia carotovara*. *Agric. Biol. Chem. 48*, 2113–2120.

TERMOTE, F.; ROMBOUTS, F. M.; and PILNIKI, W. 1977. Stabilization of cloud in pectinesterase active orange juice by pectic acid hydrolysates. *J. Food Biochem. 1*, 15–34.

TUCKER, G. A.; ROBERTSON, N. G.; and GRIERSON, D. 1980. Changes in polygalacturonase isoenzymes during the "ripening" of normal and mutant tomato fruit. *Eur. J. Biochem. 112*, 119–124.

———. 1981. The conversion of tomato-fruit polygalacturonase isoenzyme 2 into isoenzyme 1 *in vitro*. *Eur. J. Biochem. 115*, 87–90.

———. 1982. Purification and changes in activities of tomato pectinesterase isoenzymes. *J. Sci. Food Agric. 33*, 396–400.

VAN GIJSEGEM, F. 1986. Analysis of the pectin-degrading enzymes secreted by three strains of *Erwinia chrysanthemi*. *J. Gen. Microbiol. 132*, 617–624.

VERSTEEG, C.; ROMBOUTS, F. M.; and PILNIK, W. 1978. Purification and some characteristics of two pectinesterase isoenzymes from orange. *Lebensm-Wiss. u.-Technol. 11*, 267–274.

VORAGEN, A. G. J.; SCHOLS, H. A.; SILIHA, H. A. I.; and PILNIK, W. 1986. Enzymic lysis of pectic substances in cell walls: Some implications for fruit juice technology. In: Chemistry and Function of Pectins, M. L. Fishman, and J. J. Jen, eds., *ACS Sym. Ser. 310*, American Chemical Society, Washington, D.C.

YODER, M. D.; DECHAINE, D. A.; and JURNAK, F. 1990. Preliminary crystallographic analysis of the plant pathogenic factor, pectate lyase C from *Erwinia chrysanthemi*. *J. Biol. Chem. 265*, 11429–11431.

YODER, M. D.; KEEN, N. T.; and JURNAK, F. 1993. New domain motif: The structure of pectate lyase C, a secreted plant virulence factor. *Science 260*, 1503–1507.

YODER, M. D.; LIETZKE, S. E.; and JURNAK, F. 1994. Unusual structural features in the parallel β-helix in pectate lyases. *Structure 1*, 241–251.

YOSHIDA, A.; ITO, K.; KAMIO, Y.; and IZAKI, K. 1991A. Purification and properties of pectate lyase III of *Erwinia carotovora* Er. *Agric. Biol. Chem. 55*, 601–602.

YOSHIDA, A.; IZUTA, M.; ITO, K.; KAMIO, Y.; and IZAKI, K. 1991B. Cloning and characterization of the pectate lyase III gene of *Erwinia carotovora* Er. *Agric. Biol. Chem. 55*, 933–940.

ZHENG, L.; HEUPEL, R. C.; and DELLAPENNA, D. 1992. The β subunit of tomato fruit polygalacturonase isoenzyme 1: Isolation, characterization, and identification of unique structural features. *The Plant Cell 4*, 1147–1156.

Chapter 8

Lipoxygenase

Lipoxygenase (linoleate:oxygen oxidoreductase, EC 1.3.11.12) is a dioxygenase that catalyzes the oxygenation of polyunsaturated fatty acids (LH) containing a *cis*, *cis*-1,4-pentadiene system to hydroperoxides (LOOH). The enzyme exists in multiple forms, three in soybeans and peas, and two in corn. Lipoxygenases from different sources, as well as their isozymes, may differ in substrate specificity, pH optimum, and oxidation activity as indicated in Table 8.1. Lipoxygenase activity is widespread in plants. Most of our present information regarding lipoxygenase comes from studies on soybean lipoxygenase 1 (LOX-1), the isozyme first isolated and crystallized from soybean (Theorell et al. 1947). Only in recent years have mammalian lipoxygenases been identified as key enzymes in the leukotriene pathways related to inflammatory or immunological reactions (Needleman et al. 1986; Borgeat 1989). The present discussion focuses on the well-studied soybean LOX-1.

AMINO ACID SEQUENCES

The primary structures of soybean LOX-1 (Shibata et al. 1987), LOX-2 (Shibata 1988), and LOX-3 (Yenofsky et al. 1988) have been deduced from their respective cDNA sequences. The sequences of the following lipoxygenases have also been described: pea (Ealing and Casey 1988, 1989), human 5-lipoxygenase (Dixon et al. 1988; Matsumoto et al. 1988; Funk et al. 1989), human 15-lipoxygenase (Sigal

Table 8.1. Characteristics of Lipoxygenases from Soybean, Pea, and Potato[a]

	Soybean			Pea		Potato	
	LOX-1	LOX-2	LOX-3	LOX-1	LOX-2	LOX-1	LOX-2
MW	98,000	98,000	98,000	95,000	95,000	92,000	92,000
pH optimum	9.0	6.5	4.5–9.0	7.0	6.0	5.5	5.5
pI	5.68	6.25	6.15	6.25	5.82	4.94	4.99
9/13 LOOH	10:90	50:50	—	50:50	—	95:5	85:15
Cooxidation activity	low	high	—	high	low	—	—

[a]Potato (Mulliez et al. 1987); Pea (Yoon and Klein 1979); Soybean (Christopher 1972; Galliard and Chan 1980 Steczko et al. 1990).

et al. 1988), porcine leukocyte 12-lipoxygenase (Yoshimoto et al. 1990), and rabbit reticulocyte 15-lipoxygenase (Fleming et al. 1989).

Soybean LOX-1 is composed of a single polypeptide chain of 838 amino acids, with MW 94,038, as deduced from the nucleotide sequence. The enzyme contains four Cys but no disulfides. The majority of the Trp residues (12 of 13) are located in a hydrophobic region between residues 617–783 (Shibata et al. 1987). LOX-2 has a MW of 97,036 with 865 amino acids, and contains 7 Cys and 15 Trp (Shibata et al. 1988). The deduced amino acid sequence of LOX-3 consists of 857 amino acids with a MW of 96,663. There are 5 Cys and 14 Trp, but no disulfides in the enzyme molecule (Yenofsky et al. 1988). The amino acid sequences of LOX-1 and LOX-2 are 81% identical, whereas those of LOX-1 and LOX-3 show 70% similarity (Fig. 8.1). When compared with known sequences of mammalian lipoxygenases, the regions in the middle and C-terminal exhibit significant similarities.

```
LOX1    1     MFS AG           HKIKGTVVLMPKN    E        L
LOX2    1     -FSVP-VSGILNRGGG----------R--VLDFNSVADLT
LOX3    1     -  L -  GLLHRG   ----------R--VLD   VNSVT


LOX1   21     E V     NP D  GSAVDNLNAFLGRSVSLQLISATKA
LOX2   41     KGN-GGLIGTGLNVV--TL-N-T-----S-A--------P
LOX3   31     S -GGIIGQGLDLV--TL-T-T-----P-S--------A


LOX1   51     DAHGKGKVGKDTFLEGINTSLPTLGAGESAFNIHFEWDGS
LOX2   81     L-N----V--D------IV--------E---N-Q----ES
LOX3   69     D-N----L--A------IT--------Q---K-N----DG


LOX1   91     MGIPGAFYIKNYMQVEFFLKSLTLEAISNQGTIRFVCNSW
LOX2  121     M--P-------Y--V--Y-K-----DVP-Q-T-R------
LOX3  109     S--L-------F--T--F-V-----DIP-H-S-H------


LOX1  131     VYNTKLYKSVRIFFANHTYVPSETPAPLVSYREEELKSLR
LOX2  161     V--T---Y--V------H--V------A--G------KN--
LOX3  149     I--A---F--D------Q--L------P--K------HN--


LOX1  171     GNGTGERKEYDRIYDYDVYNDLGNPDKSEKLARPVLGGSS
LOX2  201     -D-K-----HD-I----------N--HG-NF---I---SS
LOX3  189     -D-T-----WE-V---------D--KG-NH---V---ND
```

Figure 8.1. Comparison of amino acid sequences of soybean lipoxygenases: LOX1 (Shibata et al. 1987), LOX2 (Shibata et al. 1988), and LOX3 (Yenofsky et al. 1988). All three sequences are deduced from cDNA sequencing.

```
LOX1   211    TFPYPRRGRTGRGPTVTDPNTE KQGEVFYVPRDENLGHL
LOX2   241    -H----------Y--RK-Q-S- KPGE- -V----NF---
LOX3   229    -F----------K--RK-P-S-SRSND- -L----AF---

LOX1   250    KSKDALEIGTKSLSQIVQPAFESAFDLKSTPIEFHSFQDV
LOX2   279    --S-F-AY-I--L--Y-L-AFE-V---NF--N--D--QD-
LOX3   268    --S-F-TY-L--V--N-L-LLQ-A---NF--R--D--DE-

LOX1   290    HDLYEGGIKLPRDVISTIIPLPVIKELYRTDGQHILKFPQ
LOX2   319    RD-HE------TEV--T-M----V--LF----EQV----P
LOX3   308    HG-YS------TDI--K-S----L--IF----EQA----P

LOX1   330    PHVVQVSQSAWMTDEEFAREMIAGVNPCVIRGLEEFPPKS
LOX2   359    -H-I---K-------------V-----CV--G-Q----K-
LOX3   348    -K-I---K-------------L-----NL--C-K----R-

LOX1   370    NLDPAIYGDQSSKITADSLD  LDGYTMDEALGSRRLFML
LOX2   399    N--PTI--EQT-K--ADA-D  -D-Y-V---LASR---M-
LOX3   388    K--SQV--DHT-Q--KEH-EPN-E-L-V---IQNK---L-

LOX1   408    DYHDIFMPYVRQINQLNSAKTYATRTILFLREDGTLKPVA
LOX2   437    DY--VF---I-R--QTY A-A---------REN---K-V-
LOX3   428    GH--PI---L-R--AT ST-A---------KND---R-L-

LOX1   448    IELSLPHSAGDLSAAVSQVVLPAKEGVESTIATS KAYVI
LOX2   476    -------PA--L-G-V---I---K-----T-WLLA----V
LOX3   467    -------PQ--Q-G-F---F---D-----S-WLLA----V

LOX1   487    VNDSCYHQLMSHWLNTHAAMEPFVIATHRHLSVLHPIYKL
LOX2   516    ---------M--------VI---I---N----AL------
LOX3   507    ---------V--------VV---I---N----VV------

LOX1   527    LTPHYRNNMNINALARQSLINANGIIETTFLPSKYSVEMS
LOX2   556    -T----DT----A---Q--I-AD-I--KS--PSKH-----
LOX3   547    -H----DT----G---L--V-DG-V--QT--WGRY-----

LOX1   567    SAV YKNWVFTDQALPADLIKRGVAIKDPSTPHGVRLLIE
LOX2   596    S-- --N---------------V--K---A---L--L--
LOX3   587    --V--D---------------M--E---C---I--V--
```

Figure 8.1.   (Continued).

```
LOX1   606   DYPYAADGLEIWAAIKTWVQEYVPLYYARDDDVKNDSELQ
LOX2   635   -----V------A------Q---S---AR--DVKP-S---
LOX3   626   -----V------D------H---F---KS--TLRE-P---

LOX1   646   HWWKEAVEKGHGDLKDKPWWPKLQTLEDLVEVCLIIIWIA
LOX2   675   QW---A--K----L-DK-----L--I-E---I-T----T-
LOX3   666   AC---L--V----K-NE-----M--R-E---A-A----I-

LOX1   686   SALHAAVNFGQYPYGGLIMNRPTASRRLLPEKGTPEYEEM
LOX2   715   ----------------F-L----S---LL----TP----M
LOX3   706   ----------------L-L----L---FM----SA----L

LOX1   726   INNHEKAYLRTITSKLPTLISLSVIEILSTHASDEVYLGQ
LOX2   755   VKSHQ----R---S-FQ--VD--------R--------Q
LOX3   746   RKNPQ----K---P-FQ--ID--------R--------E

LOX1   766   RDNPHWTSDSKALQAFQKFGNKLKEIEEKLVRRNNDPSLQ
LOX2   795   ----H----SK--Q--QK-----KE--E--ARK---QS-S
LOX3   786   ----N----TR--E--KR-----AQ--N--SER---EK-R

LOX1   806   GNRLGPVQLPYTLLYPSSEEGLTFRGIPNSISI   838
LOX2   835   --L----L-----H-N- ----C---------   865
LOX3   826   --C----M-----L-S-K----F---------   857
```

Figure 8.1.   (Continued).

Only two of the cysteines, Cys127 and Cys492 (LOX-1 numbering), are conserved in all three isozymes. Two Trp residues are conserved in a hydrophobic region homologous in the three izozymes. In addition, there is a histidine-rich region in a highly conserved segment between residues 489–532 (Fig. 8.2). This cluster of five conserved His, together with one His in the C-terminal conserved region, may play a role in the coordination of the iron (refer to the section later in this chapter, The Role of Iron in LOX-Catalyzed Reactions). In soybean LOX-1, His499, 504, and 690 may be ligands for the heme iron.

THREE-DIMENSIONAL STRUCTURE

Soybean LOX-1 has been crystallized (Theorell et al. 1947), and preliminary x-ray characterization of the enzyme presented (Boyington et al. 1990; Steczko et al. 1990; Stallings et al. 1990; Minor et al. 1993). A high-resolution analysis of the soybean LOX-1 structure is available (Boyington et al. 1993A,B). The

```
SB-LOX1    482    KAYVIVNDSCYHQLMSHWLNTHAAMEPFVIATHRHLSVLHPIYKLLTPHY
SB-LOX2    511    -AY-VVN-SCYH--M-HW-N-HAVI-P-I---N-H-SALH--Y---T-HY
SB-LOX3    502    -AY-VVN-SCYH--V-HW-N-HAVV-P-I---N-H-SVVH--Y---H-HY
P-LOX      506    -AY-IVN-SCYH--V-HW-N-HAVV-P-V---N-H-SCLH--Y---Y-HY
5-LOX      351    -IW-RSS-FHVH-TITHL-R-HLVS-V-G--MY-Q-PAVH--F---VAHV
12-LOX     345    -CW-RSS-FQLHE-H-HL-RGHLMA-VIAV--M-C-PSIH--F---I-HF
15-LOX     344    -CW-RSS-FQLHE-Q-HL-RGHLMA-VIVV--M-C-PSIH--F--II-HL


SB-LOX1    681    IIWIASALHAAVNFGQYPYGGLIMNRP
SB-LOX2    710    I-WT---LH--------PYGGF-L-R-
SB-LOX3    701    I-WT---LH--------PYGGL-L-R-
P-LOX      705    V-WT---LH--------SYGGL-L-R-
5-LOX      543    V-FT---QH--------DWCSW-P-A-
12-LOX     533    C-FTCTGQHSSNHL--LDWYTWVP-A-
15-LOX     532    C-FTCTGQHASVHL--LDWYSWVP-A-
```

Figure 8.2.   Comparison of His-rich regions and C-terminal conserved region of seven lipoxygenases: SB-1 (soybean lipoxygenase-1, Shibata et al. 1987), SB-2 (soybean lipoxygenase-2, Shibata et al. 1988), SB-3 (soybean lipoxygenase-3, Yenofsky et al. 1988), P-LOX (pea seed lipoxygenase (Ealing and Casey 1988), 5-LOX (human 5-lipoxygenase, Dixon et al. 1988), 12-LOX (porcine leukocyte 12-lipoxygenase (Yoshimoto et al. 1990), 15-LOX (human 15-lipoxygenase (Sigal et al. 1988).

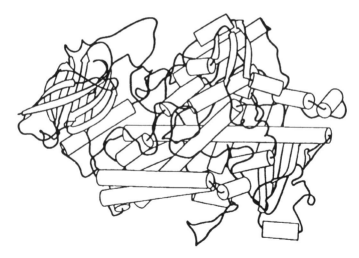

Figure 8.3.   Ribbon diagram of soybean lipoxygenase-1. (From Boyington et al. 1993A. *Biochem. Soc. Trans. 21*, 746, V with permission. Copyright 1993 The Biochemical Society and Portland Press.)

enzyme molecule has a dimension of 90 × 65 × 60 A) and consists of two domains (Fig. 8.3). Domain I, which is not found in mammalian LOX, is an 8-stranded antiparallel barrel. The interior core of the barrel is densely packed with hydrophobic side chains. Domain II comprises the C-terminal 693 residue segment, which folds into a bundle of 23 helices with two antiparallel β sheets on the surface. The longest one, helix 9, extends along the entire central part of the domain. Helices 9 and 18 both contain a segment with the uncommon structure of a π helix. All the His ligands for the iron coordination are located in these two regions.

The nonheme iron is coordinated to the $N_{\epsilon 2}$ atoms of His499, His504, and His690, and the carboxylate oxygen of the C-terminal Ile839, forming a distorted octahedron structure with the 5 and 6 positions vacant. The orientation of these His ligands is directed by extensive hydrogen-bonding networks with side chains in the vicinity. The use of x-ray absorption and other studies suggests that the iron is liganded via four nitrogen and two oxygen linkages (Navaratnam et al. 1988). Site-directed substitution of the three His residues by site-directed mutagenesis results in complete inactivation, but has no effect on substrate binding (Steczko et al. 1992).

Two cavities located on each side of the iron center connect to the surface of the enzyme molecule. Both are heavily lined with highly conserved hydrophobic residues. Cavity I, facing His504-$N_{\epsilon 2}$, forms an 18-A-long tunnel that could allow O_2 to pass through. Cavity II, which faces His499, His504, and Ile839, also forms a channel but with two bents for positioning the substrate in close proximity to the reaction center. His499-$N_{\epsilon 2}$ and Ile839 carboxylate oxygen are the possible bases involved in H abstraction.

SPECIFICITY OF LIPOXYGENASE-CATALYZED REACTION

The lipoxygenase-catalyzed reaction exhibits substrate specificity, positional specificity, and stereospecificity. In the general scheme of lipoxygenase reaction, the orientation of the *cis, cis*-pentadiene system of the substrate molecule is considered to be planar on the enzyme, and the hydrogen abstraction occurs either above or below the plane, H(L_S) or H(D_R). The reaction is accompanied by a double bond shift with a *cis-trans* isomerization. Oxygen insertion takes place on the opposite side of the plane from which the hydrogen is abstracted (Fig. 8.4) (Egmond et al. 1972). The oxygen molecule originates from O_2, and not from H_2O (Dolev et al. 1967).

A new system of nomenclature has been proposed for lipoxygenases to better characterize the stereochemistry of the enzymes (Kuhn et al. 1986). The following formulation takes into consideration the substrate, and the positional and stereospecificity of a particular lipoxygenase:

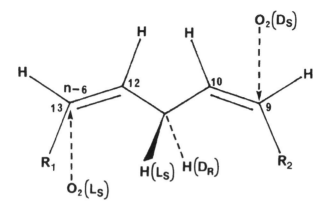

Figure 8.4. Oxygenation of linoleic acid by lipoxygenase. The pentadiene is planar. The L_S (n − 8) hydrogen atom and the D_R (n − 8) hydrogen atom point outward and downward from the plane, respectively. Oxygenation from above gives hydroperoxides with D-configurations; from below it yields hydroperoxides with L-configurations. $R_1 = CH_3(CH_2)_4^-$ and $R_2 = COOH(CH_2)_7^-$. The symbols L and D refer to Fischer convention; R and S refer to Cahn, Ingold, and Prelog convention. (From Egmond et al. 1972. *Biochem. Biophys. Res. Comm.* **48**, 1060, with permission. Copyright 1972 Academic Press, Inc.)

$$[C_{n-x}(C_x)]\text{-}L(D)_{\text{pro-S(R)}} : [\pm 2(4) \text{ - } L(D)_{S(R)}]$$

Polyenoic fatty acid/oxygen oxidoreductase

where n = total number of carbons in the substrate.

x = position of the methylene carbon of H abstraction.

n − x = indication of H abstraction determined by the methyl end (= ω numbering system).

x = indication of H abstraction determined by the nonmethyl end.

$L(D)_{\text{pro-S(R)}}$ = steric configuration of the abstracted H according to the Fischer convention, and the Cahn, Ingold, and Prelog convention.

$L(D)_{S(R)}$ = steric configuration of the chiral carbon after oxygen insertion, according to the Fischer convention, and the Cahn, Ingold, and Prelog convention.

$\pm 2(4)$ = double bond shift: the radical moves 2 or 4 carbons in the direction of the carboxylic (−) or methyl (+) end of the substrate.

For example, soybean LOX has a nomenclature of $[C_{n-8}]\text{-}L_{\text{pro-S}} : [+2]\text{-}L_S$ with the following characteristics:

(1)   The H abstraction occurs at C8 from the methyl end.

(2)   The L_S-hydrogen at C8 is removed in the abstraction step.

(3) The conjugation of the double bonds results in the radical shifting 2 carbons in the direction of the methyl end.

(4) The oxygenated carbon has the L_S configuration.

A commonly used nomenclature for lipoxygenases, especially those from mammalian sources, is to describe the enzyme according to the position of oxygenation on the substrate, arachidonic acid (all-*cis*,5,8,11,14-eicosatetraenoic acid). For example, soybean LOX-1 is a 15- lipoxygenase. Another system occasionally used in the literature designates the position of peroxidation from the methyl (ω) end. In this case, soybean LOX-1 is an (n-6)-lipoxygenase.

Soybean LOX-1 catalyzes the oxygenation reaction at a pH optimum of 9.0 with high regio- and stereospecificity. The substrates that exhibit high reaction rates contain *cis* double bonds at the n − 6 and n − 9 positions (such as in linoleic acid). Chain length and number of double bonds seem to have little effect. Isomers such as 8,11- and 10,13-18:2 are oxidized to a limited extent. It is apparent that the n-6 carbon must have a double bond, and the carboxyl group must be unhindered for the enzyme to act (Holman et al. 1969). The enzyme converts linoleic acid predominantly (95%) to 13S-hydroperoxy-9Z,11E-octadienoic acid—abstraction of H(L_S) from C11, followed by oxygen addition to the n-6 position. In contrast, LOX-2 shows no preference for either the C9 or C13 position as the site of oxygenation. Recent work suggests that soybean LOX-1 oxygenates a keto fatty acid, 12-keto-9Z-octadecenoic acid methyl ester not containing a double allylic methylene. Oxygenation occurs at C9, yielding 9,12-diketo-10E-octadecenoic acid methyl ester as the product (Kuhn et al. 1991).

Lipoxygenase from corn germ shows positional specificity in the oxygenation of linoleic acid different from that observed for the soybean enzyme. At pH 9.0, the product is predominantly (85%) 12S-hydroperoxy-9Z,11E-octadecadienoic acid, whereas at pH 6.0, 9S-hydroperoxy-10E,12Z-octadecadienoic acid (80%) is obtained (Gardner and Weisleder 1970; Veldink et al. 1972).

Lipoxygenase from wheat flour forms mainly 9-hydroperoxy-10E,12Z-octadecadienoic acid from linoleic acid. The alkyl radical formed by the H abstraction from C11 shifts two carbons to the carboxylic end in contrast to the reaction catalyzed by soybean LOX-1 (Graveland 1973; Kuhn et al. 1985). Potato tuber contains three lipoxygenase isozymes; LOX-1 yields 9S-hydroperoxy-10E,12Z-octadecadienoic acid. Lipoxygenases from ripe tomato fruits convert linoleic acid to 9S-hydroperoxy-10E,12Z-octadecadienoic acid with high regio- and stereospecificity (Regdel et al. 1994).

Lipoxygenase-catalyzed reactions are complicated by increasing evidence of reaction products that may not be adequately described by the stereochemistry described above. For example, soybean LOX-1 products also include 5% 9-LOOH containing a mixture of *S* (73%) and *R* (27%). For the potato enzyme, the 9-LOOH is in the *S* configuration, but the 13-LOOH consists of both *R* and

S configurations (61% and 39%). Furthermore, both 13-LOOH and 9-LOOH products consist of small amounts of trans, trans-isomers (Funk et al. 1987; Nikolaev et al. 1990).

To account for the formation of some of these products, the substrate can be assumed to orient in both directions at the active site, so that H abstraction and oxygenation in the formation of 9-LOOH and 13-LOOH occur without changing the spatial alignment (Fig. 8.5) (Funk et al. 1987; Gardner 1989). Both the methyl and the carboxylic end chains are implicated as the binding site to a hydrophobic pocket of the active site. This directional mode of binding is supported by data indicating that the difference in the hydrophobic content between the proximal and distal ends of the substrate affects the regiospecificity of oxygenation (Datcheva et al. 1991).

In the general scheme, a *cis, cis*-pentadiene system is a substrate require-

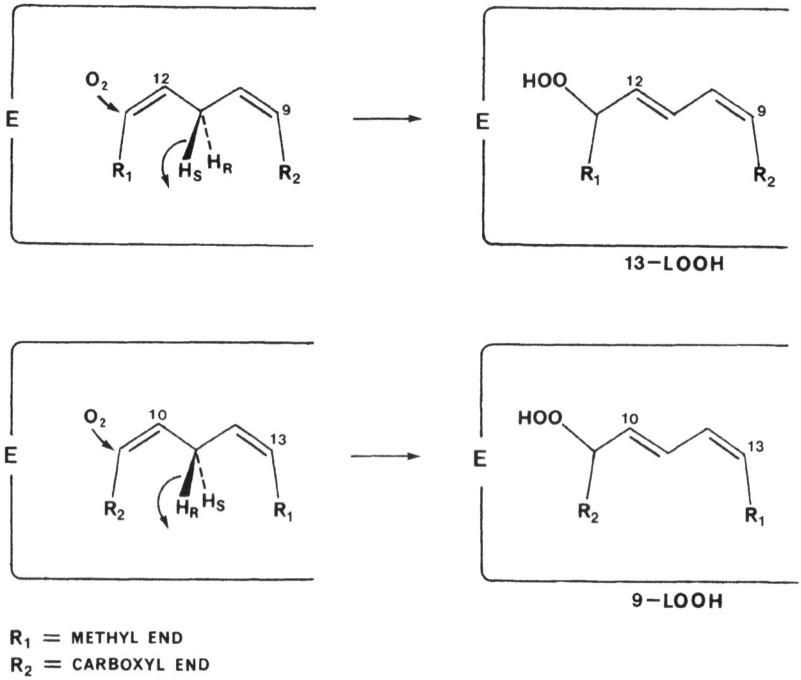

R$_1$ = METHYL END
R$_2$ = CARBOXYL END

Figure 8.5. Lipoxygenase-catalyzed reactions in two substrate orientations. Oxidation of linoleic acid to either the 9*S*-LOOH or 13*S*-LOOH is spatially identical when the substrate is arranged head-to-tail in opposite orientations. (From Gardner 1989. *Biochim. Biophys. Acta 1001,* 279, with permission. Copyright 1989 Elsevier Science Publisher B.V.).

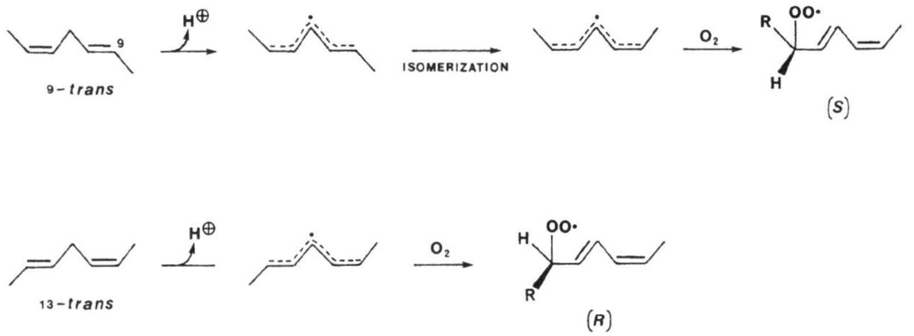

Figure 8.6. Isomerization of C9-C10 bond in the oxygenation of trans fatty acids by lipoxygenase. (From Funk et al. 1987. *Biochemistry 26*, 6883, with permission. Copyright 1987 American Chemical Society.)

ment. Recent studies indicate that soybean LOX-1 utilizes the geometric isomers 9*E*,12*Z*-, and 9*Z*,12*E*-C18:2 as substrates (Funk et al. 1987). Both substrates are converted predominantly to 13-hydroperoxy-9*Z*,11*E*-octadecadienoic acid. It is apparent in the formation of 13-LOOH that there is a *trans* ↔ *cis* isomerization at the C9-C10 bond (Fig. 8.6). For the formation of the minor product, 9-LOOH, the C12-C13 bond in the 9*Z*,12*E*-isomers shows no isomerization. The *trans-cis* isomerization of the C9-C10 bond is assumed to take place at the stage of formation of the pentadienyl radical, resulting from the steric effect created by the enzyme, and possibly catalyzed by the iron. However, the reason for the lack of isomerization at the C12-C13 bond is not clear.

Multiple Positional Specificity

In the oxidation of arachidonic acid, soybean LOX-1 produces 15*S*-hydroperoxy-5*Z*,8*Z*,11*Z*,13*E*-eicosatetraenoic acid. A second oxygenation step produces two dioxygenated products—8*S*,15*S*-dihydroperoxy-5*Z*,9*E*,11*Z*,13*E*-eicosatetraenoic acid, and 5*S*,15*S*-dihydroperoxy-6*E*,8*Z*,11*Z*,13*E*-eicosatetraenoic acid (van Os et al. 1981). These products are formed by H abstraction from three separate methylene carbons (C13, C10, C7) respectively. Soybean LOX-2 also catalyzes a double oxygenation of arachidonic acid, but the product upon reduction yields a prostaglandin (F$_{2\alpha}$), 9,11,15-trihydroxyprosta-5,13-dienoic acid (Bild et al. 1978).

Potato tuber LOX-1 oxidizes arachidonic acid to 5*S*-hydroperoxy-6*E*-8*Z*,11*Z*,14*Z*-eicosatetraenoic acid, by elimination of the D-hydrogen at C7. The enzyme also possesses 8-lipoxygenase activity, in that a minor product, leukotriene$_{A4}$ is formed (Shimizu et al. 1984; Mulliez et al. 1987).

This dual function of lipoxygenase has been extensively studied in mammalian reticulocyte lipoxygenase. In the conversion of arachidonic acid, both 15- and 12-hydroperoxyeicosatetraenoic acids (9:1 ratio) are formed via hydrogen abstraction from C13 and C10 (Kuhn et al. 1983). The reaction rates of soybean LOX-1 and the reticulocyte enzyme on a series of C20:4 fatty acids have been studied. It is estimated that the reaction rate depends on the alignment of the allylic group and the hydrogen acceptor in the active site of the enzyme. Close alignment favors a fast rate of reaction, and hence predominantly a single product. Improper alignment means the neighboring allylic methylene group is getting closer to the site for H abstraction, resulting in possibly two products. Repositioning of the substrate into the proper alignment favorable for the reaction also leads to a decrease in the reaction rate (Kuhn et al. 1990).

The reticulocyte enzyme prefers substrates with an $n-9$ methylene. Because arachidonic acid contains allylic methylene groups at $n-8$ and $n-11$, realignment of the substrate with respect to the active site receptor leads to decreased reaction rate, and mixed products, 12- and 15-hydroperoxyeicosatetraenoic acids (12- and 15-HPETE). In the case of soybean LOX-1, the optimal substrate contains an $n-8$ doubly allylic methylene group. Hence, oxygenation of the arachidonic acid results in a single product, 15-HPETE, with fast reaction rate. The second oxygenation step has 15-HPETE as substrate; the reaction is slow, because it is difficult to attain the proper alignment for H abstraction from the methylene C10 or C7.

Lipoxins

A class of trihydroxytetraenes, lipoxins, has been found to be produced by the combined actions of 5- and 15-lipoxygenases in the metabolic pathway of arachidonic acid in mammalian systems (Fitzsimmons et al. 1985). Soybean lipoxygenase is also known to catalyze the same reaction in the conversion of arachidonic acid to lipoxin A and B (Fig. 8.7) (Sok et al. 1988). The intermediates are 5S-, and 15S-dihydroperoxy-6E,13E, 8Z,11Z-eicosatetraenoic acid (5S,15S-diHPETE) and 15S-hydroxy-5,6-epoxy-7E,9E,13E,11Z-eicosatetraenoic acid. The latter compound is an unstable tetraene epoxide which, upon hydrolysis, forms the lipoxins.

THE ROLE OF IRON IN LIPOXYGENASE-CATALYZED REACTIONS

Soybean LOX-1 contains one atom of iron per molecule. The iron is tightly bound, and can be removed only by the Fe^{++} chelator, 4,5-dihydroxy-*m*-benzenedisulfonic acid, and not by the Fe^{+++} chelator such as *o*-phenanthroline (unless a reducing agent is included) (Pistorius and Axelrod 1974). Even with a Fe^{++} chelator, the removal of the iron is slow. Removal of the iron by complexation is complete and rapid only when the enzyme is denatured.

Lipoxygenase in its native (resting) form is colorless, and EPR silent. The

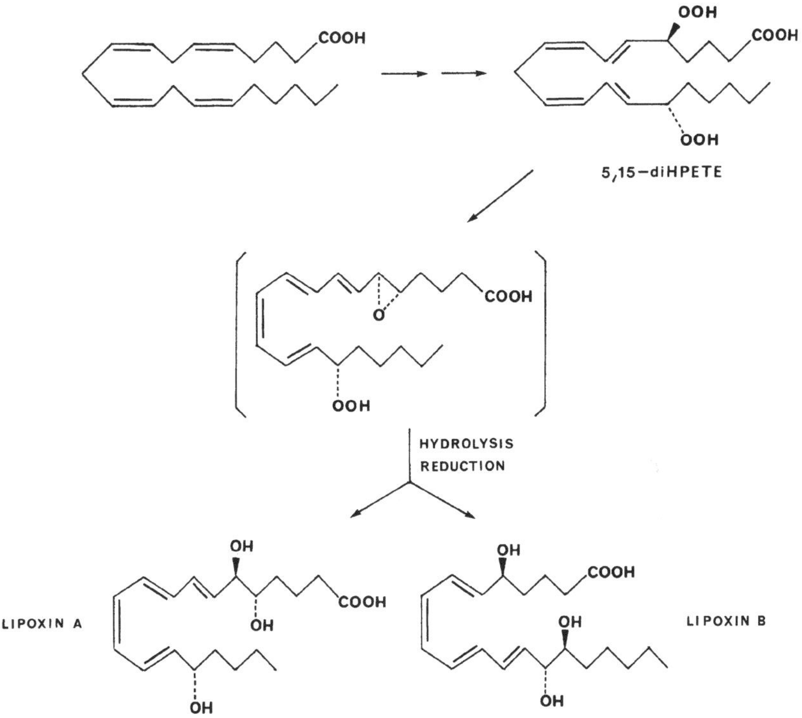

Figure 8.7. Pathway of soybean lipoxygenase-catalyzed formation of lipoxin A and B isomers. (From Sok et al. 1988. *Biochem. Biophys. Res. Comm. 153*, 846, with permission. Copyright 1988 Academic Press, Inc.)

iron is paramagnetic, high-spin Fe(II) (S = 4/2) (Slappendel et al. 1982) and coordinated to four imidazole nitrogens and/or two carboxylate oxygen ligands (Navaratnam et al. 1988). An octahedral symmetry has been suggested based on studies by magnetic CD and Mössbauer spectroscopy (Whittaker and Solomon 1986; Dunham et al. 1990; Funk and Carroll 1990). The ferrous form is converted by the addition of an equimolar concentration of the product, 13-LOOH, to a yellow ferric enzyme (de Groot et al. 1975B). The iron in this yellow enzyme form is high-spin Fe(III) (S = 5/2) (Cheesebrough and Axelrod 1983). In the Fe(III) active site, at least three His ligands coordinate in an equatorial plane, and the hydroperoxide binds to the Fe(III) as the fourth exogenous ligand (Zhang et al. 1991). The replacement of one imidazole ligand in the Fe(III) enzyme has been confirmed by another study (Heijdt et al. 1992). The dissociation of one His ligand in the activation process [E-Fe(II) → E-Fe(III)] suggests the possible involvement of His as proton acceptor in initial reaction leading to the formation of pentadienyl radical (see following section, Mechanism in Lipoxygenase-Catalyzed Reactions).

When an excess (3 to 5 molar) of 13-LOOH is added to the native enzyme or the active yellow form, an unstable purple form is obtained (de Groot et al. 1975B; Pistorus and Axelrod 1976; Slappendel et al. 1981). These two enzyme forms differ in symmetry—the yellow form having largely axial ligands, and the latter rhombic symmetry (Slappendel et al. 1983; Feiters et al. 1986). The interconversion of purple ↔ yellow enzymes is accompanied by a concomitant conversion of the hydroperoxy to a hydroxyepoxy product. In the case of 13-LOOH linoleate, the product of conversion is 11-hydroxy-12:13-epoxy-9Z-octadecenoic acid (Garssen et al. 1976). The pseudo-first-order rate constant k_2 for the conversion of the native enzyme to the yellow and purple forms by 13-LOOH linoleate is 50 s^{-1} at 4.4°C. The rate constant k_{cat} for the 13-LOOH formation from linoleic acid and O$_2$ catalyzed by soybean LOX-1 is 232 s^{-1} at 4.4°C (Egmond et al. 1977).

MECHANISM OF LIPOXYGENASE-CATALYZED REACTIONS

For the lipoxygenase-catalyzed dioxygenation reaction, two mechanisms have been proposed: (1) Formation of a pentadienyl radical and peroxy radical intermediate involving enzyme complexes of Fe(II) and Fe(III) species (the radical mechanism), (2) ferric ion-assisted deprotonation of the pentadiene and binding to the carbanion followed by insertion of dioxygen into the iron-carbon bond (the organoiron-mediated pathway).

Radical Mechanism

The radical reaction scheme under both aerobic and anaerobic conditions has been described for soybean LOX-1. In the reaction scheme outlined in Fig. 8.8 and in the following discussion, the substrates are linoleic acid and oxygen.

The Aerobic Cycle.   In the aerobic cycle, the Fe(III) in the active enzyme [E-Fe(III)] oxidizes the 1,4-diene of the substrate via a one-electron transfer that leads to the stereospecific H abstraction and subsequent formation of a pentadienyl radical [E-Fe(II)-L·]. The abstraction of C11-H from linoleic acid by soybean LOX-1 has been shown to be the rate-limiting step in the overall reaction (Egmond et al. 1973).

Oxygen then combines stereospecifically with the enzyme-free radical to form the enzyme-peroxy radical [E-Fe(II)-LOO·]. This oxygenation step is followed by a one-electron transfer in the reoxidation of the Fe(II), and formation of the enzyme-peroxy anion [E-Fe(III)-LOO⁻]. Protonation of the enzyme-peroxy anion releases the LOOH and regenerates the active enzyme [E-Fe(III)].

The Anaerobic Cycle.   Under anaerobic conditions, the enzyme-free rad-

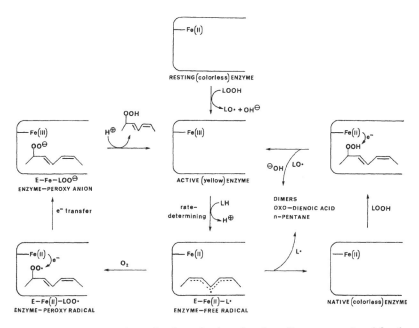

Figure 8.8. Reaction scheme for the activation of soybean lipoxygenase-1 and for the catalytic mechanism under aerobic and anaerobic conditions. (From de Groot et al. 1975B. *Biochim. Biophys. Acta 377,* 77, with permission. Copyright 1975 Elsevier Science Publisher B.V.)

ical complex dissociates into the ferrous enzyme [E-Fe(II)] and the fatty acid radical [L·]. The breakdown of LOOH is also catalyzed by an initial one-electron transfer from Fe(II). The alkoxy radical [LO·] formed can undergo a number of reactions, including reaction with more LOOH to form the peroxy radical [LOO·]. Rearrangement and decomposition of these fatty acid radicals yields oxo-dienoic acid and pentane (Garssen et al. 1971). Monooxygenated dimeric acids and dimers containing oxygen are formed from reactions among these radicals (Fig. 8.9) (Garssen et al. 1972; Vergagen et al. 1977).

For the aerobic reaction, the formation of both the pentadienyl radical [L·] and the peroxy radical [LOO·] has been confirmed by EPR spectroscopy (Chamulitrat and Mason 1990; Nelson and Cowling 1990; Nelson et al. 1990). Both radicals are bound to the enzyme, and in proximity to the active site iron. In the anaerobic reaction, the involvement of pentadienyl radical has been demonstrated (deGroot et al. 1975A). Also detected is the formation of alkoxy and acyl radicals in the breakdown of the hydroperoxide products (Davies and Slater 1987). A recent study confirms the formation of an alkyl free radical from the β-scission

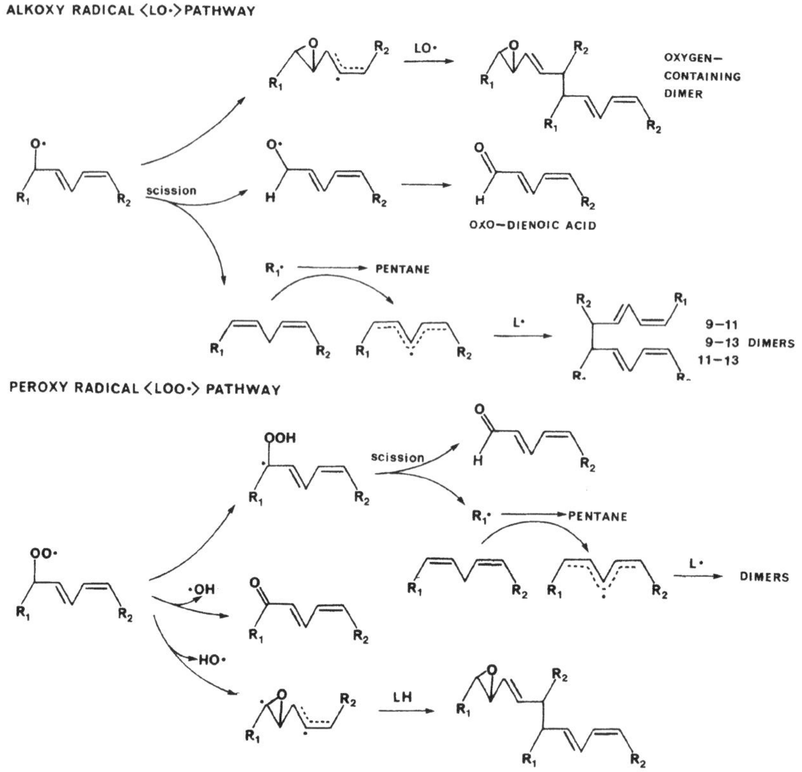

ALKOXY RADICAL ⟨LO·⟩ PATHWAY

PEROXY RADICAL ⟨LOO·⟩ PATHWAY

Figure 8.9.   Two possible pathways (alkoxy radical and peroxy radical) for the anaerobic reaction between soybean lipoxygenase, linoleic acid, and 13-LOOH. (From Garssen et al. 1972. *Biochem. J. 130*, 441, with permission. Copyright 1972 The Biochemical Society and Portland Press.)

of the fatty alkoxy radical (such as pentyl radical from linoleic alkoxy radical) (Chamulitrat and Mason 1990).

A similar mechanism that also involves radical intermediates, but differs in the role of the iron, has been proposed by Gibian and Galaway (1977). In this mechanism, the Fe(III) in the active enzyme [E-Fe(III)] oxidizes the olefin via a one-electron transfer followed by a rate-limiting E1 elimination assisted by base catalysis to form the pentadienyl radical (Fig. 8.10). The direct oxidation of olefins by metal ions via a one-electron transfer, observed and confirmed recently (Bill et al. 1990), requires no coordination of the iron to the olefin, but is facilitated

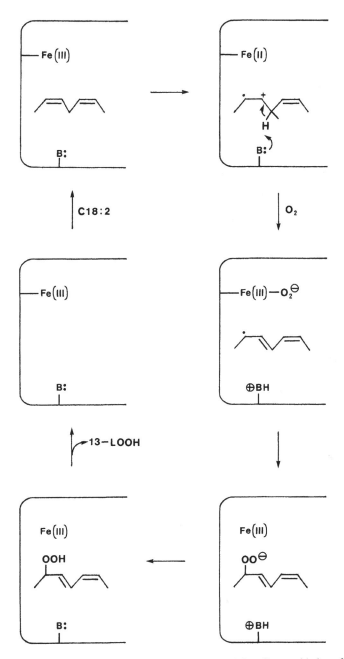

Figure 8.10. Reaction mechanism of lipoxygenase involving direct oxidation of olefins by metal ions. (From Gibian and Galaway 1977. *Bioorganic Chem. 1*, 133, with permission, Copyright 1977 Academic Press, Inc.)

by π interaction between the double bonds in the 1,4-diene system (homoconjugation).

In the oxygenation step, the binding of O_2 to the [E-Fe(II)] produces [E-Fe(II)-O_2 $\leftrightarrow$ E-Fe(III)-O_2^-. Dissociation of the complex yields O_2^-, which reacts readily with the pentadienyl radical to give the peroxy anion. Protonation of the peroxy anion produces the 13-LOOH. For the oxygenation step, the metal-bound O^- is more reactive than a dioxygen.

Organoiron-Mediated Pathway

The rate-limiting step in this scheme involves a general base-catalyzed abstraction of the proton at the methylene carbon, facilitated by a concomitant electrophilic attack (S_E2) of Fe(III) to the carbon. The resulting organoiron intermediate is carbanion in character, which reacts with dioxygen by σ-bond insertion assisted by an electron transfer from the olefin to the iron. The LOOH is formed by protonation of the peroxy anion (Fig. 8.11) (Corey and Nagata 1987; Corey et al. 1989).

A chemical model of organoiron-mediated oxygenation of allylic organotin compounds has been described (Corey and Walker 1987). Indirect evidence in support of this pathway include the following: (1) The catalytic constant for 16,17-dehydroarachidonic acid is $\sim$25 times higher than that obtained for arachidonic acid. (2) For 16,17,18,19-(E,E)-bis-dehydroarachidonic acid, the formation of hydroperoxides is not obtained. (3) The turnover number values for arachidonic acid, and 16,17-dehydroarachidonic acid increase with increasing O_2 pressure. All these results are more in accord with organoiron rather than free-radical intermediates. There is, however, no direct evidence for the critical organoiron intermediate required for this mechanism. Furthermore, in a study of a series of keto analogs and corresponding alcohols of arachidonic acid, the reactivity is shown to decrease (Wiseman and Nichols 1988). Electron-withdrawing substituents would be expected to increase reactivity if the reaction involves anionic intermediates. The result therefore is inconsistent with deprotonation as the rate-determining step and formation of organoiron. Similar results have also been obtained in the enzymatic oxygenation of difluoroarachidonic acid (Kwok et al. 1987).

Activation of The Native Enzyme

The lipoxygenase-catalyzed reactions as presented in either mechanism imply the requirement of a initial step in that the native enzyme [E-Fe(II)] must first be converted to the activated yellow form [E-Fe(III)]. It has been reported that under aerobic conditions the native enzyme coordinates with O_2, and a one-electron transfer from Fe(II) to O_2 yield a diamagnetic complex [E-Fe(II)-O_2 $\leftrightarrow$ E-Fe(III)-O_2^-], which reacts with substrate to yield hydroperoxide products (Calpin et al. 1978; Vliegenthart et al. 1979). The LOOH formed then oxidizes the Fe(II)

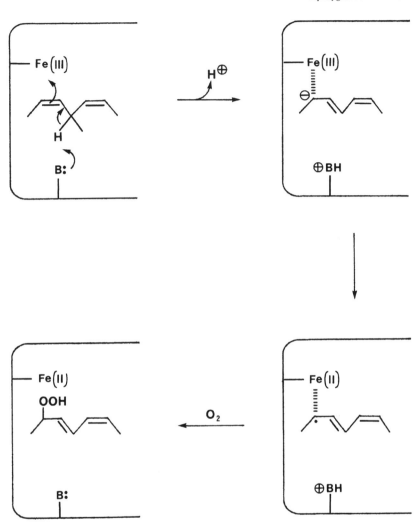

Figure 8.11. Organoiron-mediated oxygenation as a possible pathway for lipoxygenase-catalyzed reaction. (From Corey and Walker 1987. *J. Am. Chem. Soc. 109*, 8108, with permission. Copyright 1987 American Chemical Society.)

enzyme to E-Fe(III). EPR and magnetic susceptibility studies, however, show no coordination between dioxygen and the iron of the native enzyme (Petersson et al. 1985).

In the radical mechanism presented in Fig. 8.8, activation of the catalytically inactive native enzyme [E-Fe(II)] is achieved by oxidation via a one-electron transfer from Fe(II) to LOOH, similar to the Haber-Weiss reaction [E-Fe(II) + LOOH $\rightleftharpoons$ E-Fe(II)-LOOH $\rightarrow$ E-Fe(III) + LO· + OH$^-$] (Egmond et al. 1975). This scheme of activation by reaction product (LOOH), which may be related to the aspect of lag period characteristic to the LOX-catalyzed reaction (discussed below), does not provide adequate explanation of the initial process of activation. The first LOOH molecule has to come from a source other than from enzyme conversion of the substrate.

Characteristic of The Lag Period

Lipoxygenase-catalyzed dioxygenation of polyunsaturated fatty acids exhibits a characteristic lag period before the reaction reaches its steady-state kinetics. This lag period observed in the reaction started with the native enzyme [E-Fe(II)] can be eliminated by the addition of hydroperoxide product LOOH (Smith and Lands 1972). Glutathione peroxidase in the presence of glutathione causes an increase in the lag time because of reduction of LOOH. The origin of this lag period is due to the initial slow oxidative conversion of the inactive native form of the enzyme [E-Fe(II)] to the fully active [E-Fe(III)] promoted by the autocatalytic action of LOOH (Schilstra et al. 1992, 1994) as described above. The interaction between the inactive native soybean enzyme and 13-LOOH shows kinetic parameters consistent with a fast bimolecule association ($K_d = 3 \times 10^{-5}$ M) followed by a slow unimolecular process. The substrate linoleic acid is a competitive inhibitor with $K_i = 10^{-5} M$ (Egmond et al. 1975).

An alternative explanation relates the lag period to the action of substrate inhibition (Egmond et al. 1976; Lagocki et al. 1976; Wang et al. 1993). In this model, both [E-Fe(II)] and [E-Fe(III)] are catalytically active, albeit with differences in the degree of activity. The enzyme activity is allosterically controlled by a regulatory site. Binding of substrate to the regulatory site inhibits the enzyme activity, whereas the displacement of bound substrate by the reaction product restores the enzyme activity. The lag period in this model originates from the differences in the affinity of the substrate and the product at the allosteric site of the enzyme. However, the regulatory site proposed in this scheme has not been identified.

THE FATE OF HYDROPEROXIDES

The hydroperoxides formed by lipoxygenase-catalyzed reactions do not accumulate in the cell tissue, but undergo further enzymatic rearrangements or break down to secondary products.

Hydroperoxide Isomerase

This enzyme catalyzes the conversion of hydroperoxide to α-ketol fatty acids. The enzyme activity has been identified in corn germ (Gardner 1970, 1979; Gerritsen et al. 1976), flax seed (Veldink et al. 1970), alfalfa seed (Esselman and Clagett 1974), cotton seedlings (Vick and Zimmerman 1981), eggplant (Grossman et al. 1983), and French beans (Kermasha et al. 1986). The corn enzyme uses both 9- and 13-LOOH as substrates. The flax seed enzyme prefers the 13-LOOH. The enzyme from alfalfa catalyzes the conversion of both 9- and 13-LOOH to predominantly γ-ketols.

The mechanism for the action of linoleic acid hydroperoxide isomerase on 9-LOOH may involve the transfer of one of the hydroperoxide oxygens to the vicinal olefinic carbon. The resulting epoxy cation intermediate is susceptible to nucleophilic substitution (S_N2) at C9 or C13 to yield 10-oxo-9-hydroxy-12Z-octadecenoic acid (an α-ketol), or 10-oxo-13-hydroxy-11E-octadecenoic acid (a γ-ketol), respectively (Gardner 1979). A similar mechanism applies to the 13-LOOH. In the reaction, the hydroperoxide oxygen is incorporated in the oxo group, whereas the hydroxyl group originates from H_2O acting as the nucleophile. The configuration at C9 is inverted from S to R. Recent evidence suggests that the epoxy cation undergoes elimination of a proton to yield a reactive allene oxide intermediate (Fig. 8.12) (Hamberg 1987; Baertschi et al. 1988). The formation of allene oxide intermediate has been established by UV, CD, NMR, and labelling experiments (Brash et al. 1988; Hamberg and Hughes 1988; Crombie and Morgan 1991). The hydrolysis in the following step is nonenzymatic. The enzyme-catalyzed reaction is essentially a dehydration step, converting the LOOH to allene oxide. For this reason, it has been suggested that the correct name for the enzyme is "hydroperoxide dehydrase" (Hamberg 1987). In the presence of lipoxygenase, α-ketol is a substrate for oxidation to ketol hydroperoxide (Grechkin et al. 1991). In this case, lipoxygenase catalyzes oxidation of a substrate containing a 3(Z)-buten-1-onyl function instead of the common 1(Z),4(Z)-pentadienyl system.

Allene Oxide Cyclase

This enzyme catalyzes the cyclization of allene oxide to oxo-phytodienoic acid (Fig. 8.12) (Hamberg and Fahlstadius 1990). The enzyme, originally named hydroperoxide cyclase (Vick et al. 1980), has been found in flax seed (Vick and Zimmerman 1979; Vick et al. 1980), cotton seedling (Vick and Zimmerman 1981), and *Vicia faba* pericarp (Vick and Zimmerman 1983).

In the case of corn, the 13S-hydroperoxy-9Z,11E,15Z-octadecatrienoic acid is converted to 12,13S-oxido-9Z,11E,15Z-octadecadecenoic acid (allene oxide intermediate) by the enzyme hydroperoxide dehydrase. Spontaneous hydrolysis (nonenzymatic) of the latter leads to 12-oxo-hydroxy-9Z,15Z-octadecadienoic acid (an α-ketol, predominantly 13R isomers). In the presence of the allene oxide cyclase, the allene oxide intermediate undergoes formation of a delocalized car-

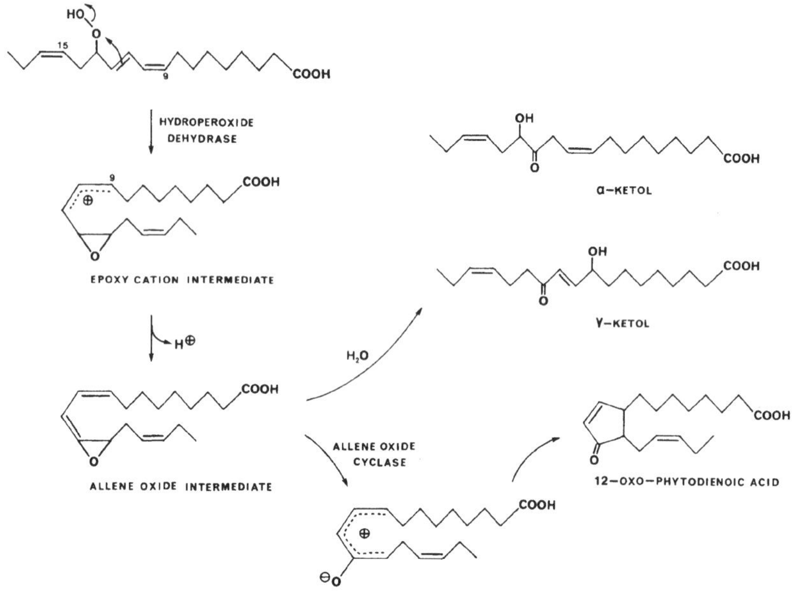

Figure 8.12. Sequential actions of hydroperoxide dehydrase and allene oxide cyclase on 13S-hydroperoxy-9Z,11E,15Z-octadecatrienoic acid. (From Hamberg and Hughes 1988. *Lipids 23, 474*, with permission. Copyright 1988 American Oil Chemists' Society, and from Hamberg and Fahlstadius 1990. *Arch. Biochem. Biophys. 276*, 525, with permission. Copyright 1990 Academic Press, Inc.)

bocation and is converted into 12-oxo-10,15Z-phytodienoic acid (9S,13S-isomer) (Fig. 8.12) (Hamberg 1988). The same reaction can occur nonenzymatically, although the product is a racemic mixture (9R,13R- and 9S,13S-isomers) (Brash et al. 1987). The final product, 12-oxo-phytodienoic acid is a precursor of jasmonic acid, a plant growth regulator related to the promotion of senescence.

Potato tuber contains an enzyme that converts 9-LOOH linoleic acid to a divinyl ether derivative (colneleic acid). However, a similar reaction does not occur with the 13-LOOH (Galliard and Matthew 1975). The mechanism likely involves the formation of an epoxide carbocation, followed by elimination of the pro-R hydrogen at C8, and cleavage of the C9-C10 bond (Fig. 8.13) (Fahlstadius and Hamberg 1990).

Hydroperoxide Lyase

This enzyme cleaves hydroperoxide to oxoacids and carbonyls. The enzyme has been identified and partially purified from tomato (Galliard and Matthew 1977), cucumber (Galliard et al. 1976; Phillips and Galliard 1978), watermelon

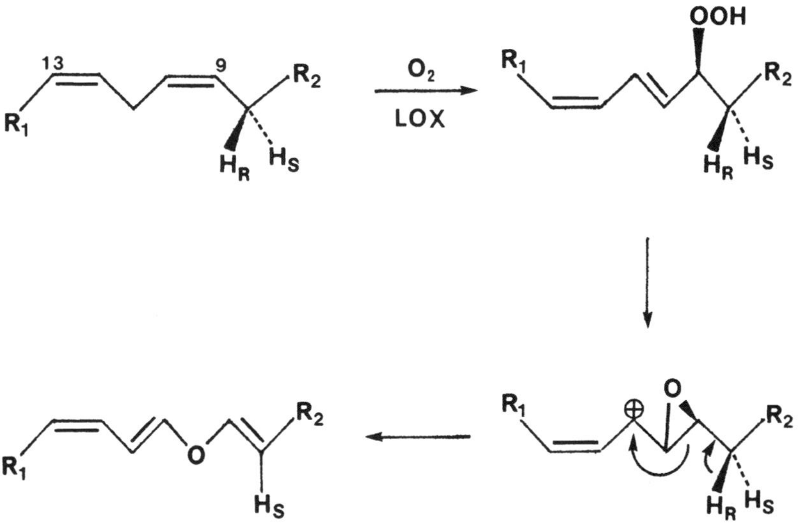

Figure 8.13. Biosynthesis of colneleic acid from 9*S*-hydroperoxyoctadeca-10*E*,12*Z*-dienoic acid. (From Fahlstadius and Hamberg 1990. *J. Chem. Soc. Perkin Trans. I 1990*, 2027, with permission. Copyright 1990 Royal Society of Chemistry.)

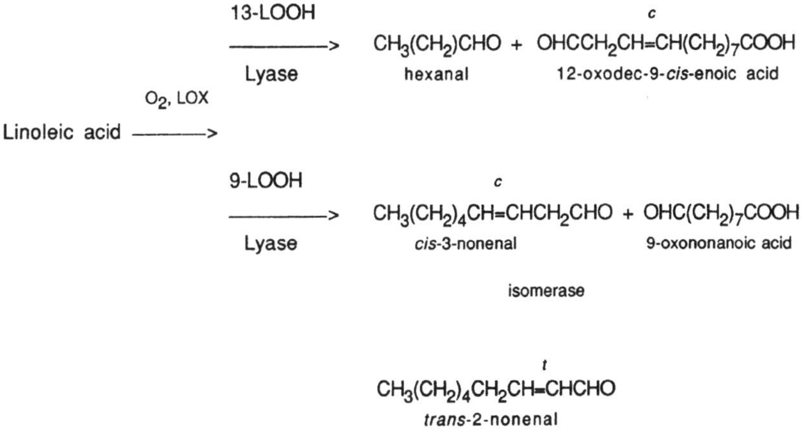

Figure 8.14. Enzymatic sequences for the formation of carbonyl fragments from linoleic acid in extracts of cucumber fruits. (From Galliard et al. 1976. *Biochim. Biophys. Acta.* *441*, 191, with permission. Copyright 1976 Elsevier Science Publisher B.V.)

seedlings (Vick and Zimmerman 1976), pear (Kim and Grosch 1981), French bean (Kermasha et al. 1986), soybean seedlings (Olias et al. 1990), and leaf tissue (Gardner 1991).

The enzymatic sequences for the formation of carbonyl fragments from linoleic acid in cucumber fruits is presented in Fig. 8.14. The major volatile fragments from the 9-LOOH are *cis*-3-nonenal and *trans*-2-nonenal. Hexanal is the major volatile compound formed from the 13-LOOH. The tomato fruit enzyme is specific for the 13-LOOH. The lyase in watermelon catalyzes the conversion of 13-LOOH to hexanal and 12-oxo-10*E*-dodecenoic acid. The latter compound is a precursor of traumatin, a wound hormone found in many plants (Zimmerman and Coudron 1979). The lyase from soybean seedling acts on 13-LOOH, but not 9-LOOH. The cleavage of 13-LOOH of linoleic and linolenic acids forms the volatile aldehydes, hexanal and *cis*-3-hexenal respectively, and 12-oxo-9Z-dodecenoic acid (Olis et al. 1990). The enzyme from pears is optimally active at pH 6.5 and cleaves exclusively the 9-LOOH of linoleic and linolenic acids. The former substrate gives *cis*-3-nonenal and 9-oxononanoic acid (Kim and Grosch 1981).

Cooxidation

Soybean lipoxygenase shows cooxidizing activity on polyenes, especially β-carotene. This ability to bleach pigments is applied to bread making. Commercially, soybean flour is added to wheat flour. Soybean LOX-1 is less effective than LOX-2 or LOX-3 in cooxidation. The oxidation of pigments by the enzyme is believed to be mediated by the peroxy radical. LOX-1 converts most peroxy radicals into hydroperoxide, whereas LOX-2 and LOX-3 release the radicals more readily (Schieberle et al. 1981). It has also been observed that LOX-1 catalyzes the exchange of dioxygen with the hydroperoxide oxygen. Radical deoxygenation of the hydroperoxide (LOOH $\rightleftharpoons$ LOO· $\rightleftharpoons$ L· + O_2) may also be involved in cooxidation (Matthew and Chan 1983).

The production of singlet oxygen has been detected in reactions catalyzed by soybean lipoxygenase isozymes, with LOX-3 being the most active. The singlet oxygen forms from recombinations of peroxy radicals [2RR'CHOO· $\rightarrow$ RR'CHOH + RR'CO + O_2*] (Kanofsky and Axelrod 1986). The singlet oxygen generated can further undergo "ene" reaction, cycloaddition, or "dioxetane" reaction with polyenes.

FUNCTIONAL ROLES OF LIPOXYGENASE IN PHYSIOLOGICAL SYSTEMS

The physiological role of lipoxygenase in plants remains inconclusive in spite of recent advances in understanding the structure and mechanism of this enzyme. It has been postulated that the hydroperoxide products are involved in plant senescence, wound resistance, insect repelling, and enzyme regulatory functions (Gardner 1991; Siedow 1991).

Lipoxygenase activity is known to increase during senescence and is likely to be involved in the initiation of membrane peroxidation and deteriorative changes. The level of mRNA coding for all three soybean isozymes is high in the embryo and cotyledons (Altschuler et al. 1989). Lipoxygenase activity increases 3-fold in senescing cotyledons of common bean (*Phaseolus vulgaris*) seeds (Pauls and Thompson 1984). A similar increase is also observed in the peanut cotyledons during early stages of germination, and is then followed by a continuous decline (Prakash et al. 1990). However, biochemical and immunological studies indicate a constant decrease of all three isozymes in soybean cotyledons (Peterman and Siedow 1985). Investigation on mutant seeds of soybean lacking LOX-1, LOX-2, or LOX-3 also suggests that the enzymes are not related to senescening changes (Wang et al. 1990).

Certain products of lipoxygenase-catalyzed reactions are growth regulator, or wound hormone precursor. The product of hydroperoxide dehydrase and allene oxide cyclase, 12-oxo-phytodienoic acid, is further metabolized by plants into jasmonic acid, a plant hormone associated with growth, abscission, and other physiological effects. The enzyme, 12-oxo-phytodienoic acid reductase, that catalyzes the conversion process has been characterized from the kernel and seedling of corn (Vick and Zimmerman 1986). Methyl jasmonate can act as a chemical signal for the induction of synthesis of proteinase inhibitors as a defense mechanism (Farmer and Ryan 1990).

Hydroxy and epoxy unsaturated fatty acids from the lipoxygenase metabolic pathways of hydroperoxides contain antifungal activities. Mono- and tri-hydroxy products such as 9-hydroxyoctadecadienoic acid, and 9,12,13-trihydroxyoctadec-10-enoic acid are identified as the active compounds from rice against blast fungus (Ohta et al. 1990A,B). 13-Hydroperoxy-9Z,11E-octadecadienoic acid obtained from LOX-3-catalyzed conversion of linoleic acid, as well as the corresponding monohydroxy product from isomerization, exhibit high inhibition of fungal growth (Ohta et al. 1991). Hydroperoxides are known to have a repelling effect against pests on soybeans (Mohri et al. 1990).

REFERENCES

ALTSCHULER, M.; GRAYBURN, W. S.; COLLINS, G. B.; and HILDEBRAND, D. F. 1989. Developmental expression of lipoxygenases in soybeans. *Plant Science 63*, 151–158.

AOSHIMA, H.; KAJIWARA, T.; HATANAKA, A.; NAKATANI, H.; and HIROMI, K. 1977. Kinetic study of lipoxygenase-hydroperoxylinoleic acid interaction. *Biochim. Biophys. Acta 486*, 121–126.

BAERTSCHI, S. W.; INGRAM, C. D.; HARRIS, T. M.; and BRASH, A. R. 1988. Absolute configuration of *cis*-12-oxophytodienoic acid of flaxseed: Implications for the mechanism of biosynthesis from the 13(*S*)-hydroperoxide of linolenic acid. *Biochemistry 27*, 18–24.

BILD, G. S.; BHAT, S. G.; RAMADOSS, C. S.; and AXELROD, B. 1978. Biosynthesis of a prostaglandin by a plant enzyme. *J. Biol. Chem. 253*, 21–23.

BILL, T. J.; CHEN, S.; PASCAL, R. A., Jr.; SCHWARTZ, J. 1990. "Outer-sphere" oxidation of nonconjugated dienes by simple iron (III) complexes: A new mechanistic consideration for oxidation of arachidonic acid by lipoxygenase. *J. Am. Chem. Soc. 112*, 9019–9020.

BORGEAT, P. 1989. Biochemistry of the lipoxygenase pathways in neutrophils. *Can. J. Physiol. Pharmacol. 67*, 936–942.

BOYINGTON, J. C.; GAFFNEY, B. J.; and AMZEL, L. M. 1990. Crystallization and preliminary x-ray analysis of soybean lipoxygenase-1, a non-heme iron-containing dioxygenase. *J. Biol. Chem. 265*, 12771–12773.

———. 1993A. Structure of soybean lipoxygenase-1. *Biochem. Soc. Trans. 21*, 744–748.

———. 1993B. The three-dimensional structure of an arachidonic acid 15-lipoxygenase. *Science 260*, 1482–1486.

BRASH, A. R.; BAERTSCHI, S. W.; INGRAM, C. D.; and HARRIS, T. M. 1987. On noncyclooxygenase prostaglandin synthesis in the sea whip coral, *Plexaura homomalla*: An 8(*R*)-lipoxygenase pathway leads to formation of an α-ketol and a racemic prostanoid. *J. Biol. Chem. 262*, 15829–15839.

———. 1988. Isolation and characterization of natural allene oxides: Unstable intermediates in the metabolism of lipid hydroperoxides. *Proc. Natl. Acad. Sci. USA 85*, 3382–3386.

CHAMULITRAT, W., and MASON, R. P. 1989. Lipid peroxyl radical intermediates in the peroxidation of polyunsaturated fatty acids by lipoxygenase. *J. Biol. Chem. 264*, 20968–20973.

———. 1990. Alkyl free radicals from the β-scission of fatty acid alkoxyl radicals as detected by spin trapping in a lipoxygenase system. *Arch. Biochem. Biophys. 282*, 65–69.

CHEESBROUGH, T. M., and AXELROD, B. 1983. Determination of the spin state of iron in native and activated soybean lipoxygenase 1 by paramagnetic susceptibility. *Biochemistry 22*, 3837–3840.

CHRISTOPHER, J. P.; PISTORIUS, E. K.; and AXELROD, B. 1972. Isolation of a third isoenzyme of soybean lipoxygenase. *Biochim. Biophys. Acta 284*, 54–62.

COREY, E. J., and NAGATA, R. 1987. Evidence in favor of an organoiron-mediated pathway for lipoxygenation of fatty acids by soybean lipoxygenase. *J. Am. Chem. Soc. 109*, 8107–8108.

COREY, E. J., and WALKER, J. C. 1987. Organoiron-mediated oxygenation of allylic organotin compounds. A possible model for enzymatic lipoxygenation. *J. Am. Chem. Soc. 109*, 8108–8109.

COREY, E. J.; WRIGHT, S. W.; and MATSUDA, S. P. T. 1989. Stereochemistry and mechanism of the biosynthesis of leukotriene A4 from 5(*S*)-hydroperoxy-6(*E*),

8,11,14(*Z*)-eicosatetraenoic acid. Evidence for an organoiron intermediate. *J. Am. Chem. Sco. 111*, 1452–1455.

CROMBIE, L., and MORGAN, D. O. 1991. The enzymic conversion of 13-hydroperoxylinoleic acid into 13-hydroxy-12-oxooctadec-9(*Z*)-enoic acid and 9-hydroxy-12-oxooctadec-10(*E*)-enoic acid: Isotopic evidence for an allene epoxide intermediate. *J. Chem. Soc. Perkin Trans. 1, 1991*, 577–580.

CUCUROU, C.; BATTIONI, J. P.; THANG, D. C.; NAM, N. H.; and MANSUY, D. 1991. Mechanisms of inactivation of lipoxygenases by phenidone and BW755C. *Biochemistry 30*, 8964–8969.

DAVIES, M. J., and SLATER, T. F. 1987. Studies on the metal-ion and lipoxygenase-catalyzed breakdown of hydroperoxides using electron-spin-resonance spectroscopy. *Biochem. J. 245*, 167–173.

DATCHEVA, V. K.; KISS, K.; SOLOMON, L.; and KYLER, K. S. 1991. Asymmetric hydroxylation with lipoxygenase: The role of group hydrophobicity on regioselectivity. *J. Am. Chem. Soc. 113*, 270–274.

DE GROOT, J. J. M. C.; GARSSEN, G. J.; VELDINK, G. A.; VLIEGENTHART, J. F. G.; and BOLDINGH, J. 1975A. On the interaction of soybean lipoxygenase-1 and 13-L-hydroperoxylinoleic acid, involving yellow and purple coloured enzyme species. *FEBS Lett. 56*, 50–54.

DE GROOT, J. J. M. C.; VELDINK, G. A.; VLIEGENTHART, J. F. G.; BOLDINGH, J.; WEVER, R.; and VAN GELDER, B. F. 1975B. Demonstration by EPR spectroscopy of the functional role of iron in soybean lipoxygenase-1. *Biochim. Biophys. Acta 377*, 71–79.

DIXON, R. A. F.; JONES, R. E.; DIEHL, R. E.; BENNETT, C. D.; KARGMAN, S.; and ROUZER, C. A. 1988. Cloning of the cDNA for human 5-lipoxygenase. *Proc. Natl. Acad. Sci. USA 85*, 416–420.

DOLEV, A.; ROHWEDDER, W. K.; MOUNTS, T. L.; and DUTTON, H. J. 1967. Mechanism of lipoxidase reaction. II. Origin of the oxygen incorporated into linoleate hydroperoxide. *Lipids 2*, 33–36.

DUNHAM, W. R.; CARROLL, R. T.; THOMPSON, J. F.; SANDS, R. H.; and FUNK, M. O. 1990. The initial characterization of the iron environment in lipoxygenase by Mossbauer spectroscopy. *Eur. J. Biochem. 190*, 611–617.

EALING, P. M., and CASEY, R. 1988. The complete amino acid sequence of a pea (*Pisum sativum*) seed lipoxygenase predicted from a near full-length cDNA. *Biochem. J. 253*, 915–918.

———. 1989. The cDNA cloning of a pea (*Pisum sativum*) seed lipoxygenase. *Biochem. J. 264*, 929–932.

EGMOND, M. R.; BRUNORI, M.; and FASELLA, P. M. 1976. The steady-state kinetics of the oxygenation of linoleic acid catalyzed by soybean lipoxygenase. *Eur. J. Biochem. 61*, 93–100.

EGMOND, M. R.; FASELLA, P. M.; VELDINK, G. A.; VLIEGENTHART, F. G.; and BOLDINGH, J. 1977. On the mechanism of action of soybean lipoxygenase-1. A stopped-flow kinetic study of the formation and conversion of yellow and purple enzyme species. *Eur. J. Biochem. 76*, 469–479.

EGMOND, M. R.; FINAZZI-AGRO, A.; FASELLA, P. M.; VELDINK, G. A.; VLIEGENTHART,

F. G. 1975. Changes in the fluorescence and absorbance of lipoxygenase-1 induced by 13-Ls-hydroperoxylinoleic acid and linoleic acid. *Biochim. Biophys. Acta 397*, 43–49.

EGMOND, M. R.; VELDINK, G. A.; VLIEGENTHART, J. F. G.; and BOLDINGH, J. 1973. C-11 H-abstraction from linoleic acid, the rate-limiting step in lipoxygenase catalysis. *Biochem. Biophys. Res. Comm. 54*, 1178–1184.

EGMOND, M. R.; VLIEGENTHART, J. F. G.; and BOLDINGH, J. 1972. Stereospecificity of the hydrogen abstraction at carbon atom n-8 in the oxygenation of linoleic acid by lipoxygenases from corn germs and soya beans. *Biochem. Biophys. Res. Comm. 48*, 1055–1060.

ESSELMAN, W. J., and CLAGETT, C. O. 1974. Products of a linoleic hydroperoxide-decomposing enzyme of alfalfa seed. *J. Lipid Res. 15*, 173–178.

FAHLSTADIUS, P., and HAMBERG, M. 1990. Stereospecific removal of the pro-*R* hydrogen at C-8 of (9S)-hydroperoxyoctadecadienoic acid in the biosynthesis of colneleic acid. *J. Chem. Soc. Perkin Trans I 1990*, 2027–2030.

FARMER, E. E., and RYAN, C. A. 1990. Interplant communication: Airborne methyl jasmonate induces synthesis of proteinase inhibitors in plant leaves. *Proc. Natl. Acad. Sci. USA 87*, 7713–7716.

FEITERS, M. C.; AASA, R.; MALMSTROM, B. G.; SLAPPENDEL, S.; VELDINK, G. A.; and VLIEGENTHART, J. F. G. 1985. Substrate fatty acid activation in soybean lipoxygenase-1 catalysis. *Biochim. Biophys. Acta 831*, 302–305.

FEITERS, M. C.; AASA, R.; MALMSTROM, B. G.; VELDINK, G. A.; and VLIEGENTHART, J. F. G. 1986. Spectroscopic studies on the interactions between lipoxygenase-2 and its product hydroperoxides. *Biochim. Biophys. Acta 873*, 182–189.

FITZSIMMONS, B. J.; ADAMS, J.; EVANS, J. F.; LEBLANC, Y.; and ROKACH, J. 1985. The lipoxins. *J. Biol. Chem. 260*, 13008–13012.

FLEMING, J.; THIELE, B. J.; CHESTER, J.; O'PREY, J.; JANETZKI, S.; AITKEN, A.; ANTON, I. A.; RAPOPORT, S. M.; and HARRISON, P. R. 1989. The complete sequence of rabbit erythroid cell-specific 15-lipoxygenase mRNA: Comparison of the predicted amino acid sequence of the erythrocyte lipoxygenase with other lipoxygenases. *Gene 79*, 181–188.

FUNK, C. D.; HOSHIKO, S.; MATSUMOTO, T.; RADMARK, O.; and SAMUELSSON, B. 1989. Characterization of the human 5-lipoxygenase gene. *Proc. Natl. Acad. Sci. USA 86*, 2587–2591.

FUNK, M. O.; ANDRE, J. C.; and OTSUKI, T. 1987. Oxygenation of trans polyunsaturated fatty acids by lipoxygenase reveals steric features of the catalytic mechanism. *Biochemistry 26*, 6880–6884.

FUNK, M. O., JR., and CARROLL, R. T. 1990. Role of iron in lipoxygenase catalysis. *J. Am. Chem. Soc. 112*, 5375–5376.

GALLIARD, T., and CHAN, H. W.-S. 1980. Lipoxygenase. In: *The Biochemistry of Plants, A Comprehensive Treatise.* vol. 4, P. K. Stumpf and E. E. Conn, eds., Academic Press, New York.

GALLIARD, T., and MATTHEW, J. A. 1975. Enzymatic reactions of fatty acid hydroperoxides in extracts of potato tuber. *Biochim. Biophys. Acta 398*, 1–9.

————. 1977. Lipoxygenase-mediated cleavage of fatty acids to carbonyl fragments in tomato fruits. *Phytochemistry 16*, 339–343.

GALLIARD, T.; PHILLIPS, D. R.; and REYNOLDS, J. 1976. The formation of *cis*-3-nonenal, *trans*-2-nonenal and hexanal from linoleic acid hydroperoxide isomers by a hydroperoxide cleavage enzyme system in cucumber (*Cucumis sativus*) fruits. *Biochim. Biophys. Acta 441*, 181–192.

GALPIN, J. R.; VELDINK, G. A.; VLIEGENTHART, E. G.; and BOLDINGH, J. 1978. The interaction of nitric oxide with soybean lipoxygenase-1. *Biochim. Biophys. Acta 536*, 356–362.

GARDNER, H. W. 1970. Sequential enzymes of linoleic acid oxidation in corn germ: Lipoxygenase and linoleate hydroperoxide isomerase. *J. Lipid. Res. 11*, 311–321.

————. 1979. Stereospecificity of linoleic acid hydroperoxide isomerase from corn germ. *Lipids 14*, 209–211.

————. 1989. Soybean lipoxygenase-1 enzymatically forms both (9*S*)- and (13*S*)-hydroperoxides from linoleic acid by a pH-dependent mechanism. *Biochim. Biophys. Acta 1001*, 274–281.

————. 1991. Recent investigations into the lipoxygenase pathway of plants. *Biochim. Biophys. Acta 1084*, 221–239.

GARDNER, H. W., and WEISLEDER, D. 1970. Lipoxygenase from *Zea-mays*: 9-D-hydroperoxy-*trans*-10, *cis*-12-octadecadienoic acid from linoleic acid. *Lipids 5*, 678–683.

GARSSEN, G. J.; VELDINK, G. A.; VLIEGENTHART, J. F. G.; and BOLDINGH, J. 1976. The formation of threo-11-hydroxy-*trans*-12:3-epoxy-9-*cis*-octadecenoic acid by enzymic isomerization of 13-L-hydroperoxy-9-cis, 11-*trans*-octadecadienoic acid by soybean lipoxygenase-1. *Eur. J. Biochem. 62*, 33–36.

GARSSEN, G. J.; VLIEGENTHART, J. F. G.; and BOLDINGH, J. 1971. An anaerobic reaction between lipoxygenase, linoleic acid and its hydroperoxides. *Biochem. J. 122*, 327–332.

————. 1972. The origin and structures of dimeric fatty acids from the anaerobic reaction between soya-bean lipoxygenase, linoleic acid and its hydroperoxide. *Biochem. J. 130*, 435–442.

GERRITSEN, M.; VELDINK, G. A.; VLIEGENTHART, J. F. G.; and BOLDINGH, J. 1976. Formation of α- and γ-ketols from ¹⁸O-labelled linoleic acid hydroperoxides by corn germ hydroperoxide isomerase. *FEBS Letts. 67*, 149–152.

GIBIAN, M. J., and GALAWAY, R. A. 1977. Chemical aspects of lipoxygenase reactions. *Bioorganic Chemistry 1*, 117–136.

GRAVELAND, A. 1973. Enzymatic oxidation of linolenic acid in aqueous wheat flour suspensions. *Lipids 8*, 606–611.

GRECHKIN, A. N.; KURAMSHIN, R. A.; LATYPOV, S. K.; SAFONOVA, Y. Y.; GAFAROVA, T. E.; and ILYASOV, A. V. 1991. Hydroperoxides of α-ketols. Novel products of the plant lipoxygenase pathways. *Eur. J. Biochem. 199*, 451–457.

GROSSMAN, S.; BERGMAN, M.; and SOFER, Y. 1983. Purification and partial characterization of eggplant linoleate hydroperoxide isomerase. *Biochim. Biophys. Acta 752*, 65–72.

HAMBERG, M. 1987. Mechanism of corn hydroperoxide isomerase: Detection of 12,13(*S*)-oxide-9(*Z*),11-octadecadienoic acid. *Biochim. Biophys. Acta 920*, 76–84.

———. 1988. Biosynthesis of 12-oxo-10,15(*Z*)-phytodienoic acid: Identification of an allene oxide cyclase. *Biochim. Biophys. Res. Comm. 156*, 543–550.

HAMBERG, M., and FAHLSTADIUS, P. 1990. Allene oxide cyclase: A new enzyme in plant lipid metabolism. *Arch. Biochem. Biophys. 276*, 518–526.

HAMBERG, M., and HUGHES, M. A. 1988. Fatty acid allene oxides. III. Albumin-induced cyclization of 12,13(*S*)-epoxy-9(*Z*),11-octadecadienoic acid. *Lipids 23*, 469–475.

HEIJDT, L. M.; FEITERS, M. C.; NAVARATHAM, S.; NOLTING, N.-F.; HERMES, C.; VELDINK, G. A.; and VLIEGENTHART, F. G. 1992. X-ray absorption spectroscopy of soybean lipoxygenase-1. Influence of lipid hydroperoxide activation and lyophilization on the structure of the non-heme iron active site. *Eur. J. Biochem. 207*, 793–802.

HOLMAN, R. T.; EGWIM, P. O.; and CHRISTIE, W. W. 1969. Substrate specificity of soybean lipoxidase. *J. Biol. Chem. 244*, 1149–1151.

KANOFSKY, J. R., and AXELROD, B. 1986. Singlet oxygen production by soybean lipoxygenase isozymes. *J. Biol. Chem. 261*, 1099–1104.

KERMASHA, S.; VAN DE VOORT, F. R.; and METCHE, M. 1986. Conversion of linoleic acid hydroperoxide by French bean hydroperoxide isomerase. *J. Food Biochem. 10*, 285–303.

KIM, I.-S., and GROSCH, W. 1981. Partial purification and properties of a hydroperoxide lyase from fruits of pear. *J. Agric. Food Chem. 29*, 1220–1225.

KUHN, H.; EGGERT, L.; ZABOLOTSKY, O. A.; MYAGKOVA, G. I.; and SCHEWE, T. 1991. Keto fatty acids not containing doubly allylic methylenes are lipoxygenase substrates. *Biochemistry 30*, 10269–10273.

KUHN, H.; HEYDECK, D.; WIESNER, R.; and SCHEWE, T. 1985. The positional specificity of wheat lipoxygenase; the carboxylic groups as signal for the recognition of the site of the hydrogen removal. *Biochim. Biophys. Acta 830*, 25–29.

KUHN, H.; SCHEWE, T.; and RAPOPORT, S. M. 1986. The stereochemistry of the reactions of lipoxygenases and their metabolites. Proposed nomenclature of lipoxygenases and related enzymes. *Adv. Enzymol. 58*, 273–311.

KUHN, H.; SPRECHER, H.; and BRASH, A. R. 1990. On singular or dual positional specificity of lipoxygenases. *J. Biol. Chem. 265*, 16300–16305.

KUHN, H.; WIESNER, R.; SCHEWE, T.; and RAPOPORT, S. M. 1983. Reticulocyte lipoxygenase exhibits both n-6 and n-9 activities. *FEBS Lett. 153*, 353–356.

KWOK, P.-Y.; MUELLNER, F. W.; and FRIED, J. 1987. Enzymatic conversion of 10,10-difluoroarachidonic acid with PGH synthase and soybean lipoxygenase. *J. Am. Chem. Soc. 109*, 3692–3698.

LAGOCKI, J. W.; EMKEN, E. A.; LAW, J. H.; and KEZDY, F. J. 1976. Kinetic analysis of the action of soybean lipoxygenase on linoleic acid. *J. Biol. Chem. 251*, 6001–6006.

MATSUMOTO, T.; FUNK, C. D.; RADMARK, O.; HOOG, J.-O.; JORNVALL, H.; and SA-

MUELSSON, B. 1988. Molecular cloning and amino acid sequence of human 5-lipoxygenase. *Proc. Natl. Acad. Sci. USA 85*, 26–30.

MATTHEW, J. A., and CHAN, H. W.-S. 1983. Soybean lipoxygenase-1 catalyzed exchange of molecular oxygen in 13-*S*-hydroperoxy-9-*cis*,11-*trans*-octadecadienoic acid. *J. Food Biochem. 7*, 1–6.

MINOR, W.; STECZKO, J.; BOLIN, J. T.; OTWINOWSKI, Z.; and AXELROD, B. 1993. Crystallographic determination of the active site iron and its ligands in soybean lipoxygenase L-1. *Biochemistry 32*, 6320–6323.

MOHRI, S.; ENDO, Y.; MATSUDA, K.; KITAMURA, K.; and FUJIMOTO, K. 1990. Physiological effects of soybean seed lipoxygenases on insects. *Agric. Biol. Chem. 54*, 2265–2270.

MULLIEZ, E.; LEBLANC, J.-P.; GIRERD, J.-J.; RIGAUD, M.; and CHOTTARD, J.-C. 1987. 5-Lipoxygenase from potato tubers. Improved purification and physiochemical characteristics. *Biochim. Biophys. Acta 916*, 13–23.

NAVARATNAM, S.; FEITERS, M. C.; AL-HAKIM, M.; ALLEN, J. C.; VELDINK, G. A.; and VLIEGENTHART, J. F. G. 1988. Iron environment in soybean lipoxygenase-1. *Biochim. Biophys. Acta 956*, 70–76.

NEEDLEMAN, P.; TURK, J.; JAKSCHIK, B. A.; MORRISON, A. R.; and LEFKOWITH, J. B. 1986. Arachidonic acid metabolism. *Ann. Rev. Biochem. 55*, 69–102.

NELSON, M. J.; BATT, D. G.; THOMPSON, J. S.; and WRIGHT, S. W. 1991. Reduction of the active-site iron by potent inhibitors of lipoxygenases. *J. Biol. Chem. 266*, 8225–8229.

NELSON, M. J., and COWLING, R. A. 1990. Observation of a peroxyl radical in samples of "purple" lipoxygenase. *J. Am. Chem. Soc. 112*, 2820–2821.

Nelson, M. J.; SEITZ, S. P.; and COWLING, R. A. 1990. Enzyme-bound pentadienyl and peroxyl radicals in purple lipoxygenase. *Biochemistry 29*, 6887–6903.

NIKOLAEV, V.; REDDANNA, P.; WHELAN, J.; HILDENBRANDT, G.; and REDDY, C. C. 1990. Stereochemical nature of the products of linoleic acid oxidation catalyzed by lipoxygenases from potato and soybean. *Biochem. Biophys. Res. Comm. 170*, 491–496.

OHTA, H.; SHIDA, K.; PENG, Y.-L.; FURUSAWA, I.; SHISHIYAMA, J.; AIBARA, S.; and MORITA, Y. 1990A. The occurrence of lipid hydroperoxide-decomposing activities in rice and the relationship of such activities to the formation of antifungal substances. *Plant Cell Physiol. 31*, 1117–1122.

————. 1990B. A lioxygenase pathway is activated in rice after infection with the rice blast fungus *Magnaporthe grisea. Plant Physiol. 97*, 94–98.

OLIAS, J. M.; RIOS, J. J.; VALLE, M.; ZAMORA, R.; SANZ, L. C.; and AXELROD, B. 1990. Fatty acid hydroperoxide lyase in germinating soybean seedlings. *J. Agric. Food Chem. 38*, 624–630.

PAULS, K. P., and THOMPSON, J. E. 1984. Evidence for the accumulation of peroxidized lipids in membranes of senescing cotyledons. *Plant Physiol. 75*, 1152–1157.

PETERMAN, T. K., and SIEDOW, J. N. 1985. Behavior of lipoxygenase during establishment, senescence, and rejuvenation of soybean cotyledons. *Plant Physiol. 78*, 690–695.

PETERSSON, L.; SLAPPENDEL, S.; and VLIEGENTHART, J. F. G. 1985. The magnetic susceptibility of native soybean lipoxygenase-1. Implications for the symmetry of the iron environment and the possible coordination of dioxygen to Fe(II). *Biochim. Biophys. Acta 828*, 81–85.

PETERSSON, L.; SLAPPENDEL, S.; FEITERS, M. C.; and VLIEGENTHART, J. F. G. 1987. Magnetic susceptibility studies on yellow and anaerobically substrate-treated yellow soybean lipoxygenase-1. *Biochim. Biophys. Acta 913*, 228–237.

PHILLIPS, D. R., and GALLIARD, T. 1978. Flavour biogenesis. Partial purification and properties of a fatty acid hydroperoxide cleaving enzyme from fruits of cucumber. *Phytochemistry 17*, 355–358.

PHILLIPS, D. R.; MATTHEW, J. A.; REYNOLDS, J.; and FENWICK, G. R. 1979. Partial purification and properties of a cis-3: trans-2-enal isomerase from cucumber fruit. *Phytochemistry 18*, 401–404.

PISTORIUS, E. K., and AXELROD, B. 1974. Iron, an essential component of lipoxgenase. *J. Biol. Chem. 249*, 3183–3186.

———. 1976. Evidence for participation of iron in lipoxygenase reaction from optical and electron spin resonance studies. *J. Biol. Chem. 251*, 7144–7148.

PRAKASH, T. R.; SWAMY, P. M.; SUGUNA, P.; and REDDANNA, P. 1990. Characterization and behavior of 15-lipoxygenase during peanut cotyledonary senescence. *Biochem. Biophys. Res. Comm. 172*, 462–470.

QUAX, W. J.; MRABET, N. T.; LUITEN, R. G. M.; SCHUURHUIZEN, P. W.; STANSSENS, P.; and LASTERS, I. 1991. Enhancing the thermostability of glucose isomerase by protein engineering. *Bio/Technology 9*, 738–742.

REGDEL, D.; KUHN, H.; and SCHEWE, T. 1994. On the reaction specificity of the lipoxygenase from tomato fruits. *Biochim. Biophys. Acta 1210*, 297–302.

SCHIEBERLE, P.; GROSCH, W.; KEXEL, H.; and SCHMIDT, H.-L. 1981. A study of oxygen isotope scrambling in the enzymatic and non-enzymatic oxidation of linoleic acid. *Biochim. Biophys. Acta 666*, 322–326.

SCHILSTRA, M. J.; VELDINK, G. A.; VERHAGEN, J.; and VLIEGENTHART, J. F. G. 1992. Effect of lipid hydroperoxide on lipoxygenase kinetics. *Biochemistry 31*, 7692–7699.

SCHILSTRA, M. J.; VELDINK, G. A.; and VLIEGENTHART, F. G. 1994. The dioxygenation rate in lipoxygenase catalysis is determined by the amount of iron (III) lipoxygenase in solution. *Biochemistry 33*, 3974–3979.

SHIBATA, D.; STECZKO, J.; DIXON, J. E.; HERMODSON, M.; YAZDANPARAST, R.; and AXELROD, B. 1987. Primary structure of soybean lipoxygenase-1. *J. Biol. Chem. 262*, 10080–10085.

SHIBATA, D.; STECZKO, J.; DIXON, J. E.; ANDREWS, P. C.; HERMODSON, M.; and AXELROD, B. 1988. Primary structure of soybean lipoxygenase L-2. *J. Biol. Chem. 263*, 6816–6821.

SHIMIZU, T.; RADMARK, O.; and SAMUELSSON, B. 1984. Enzyme with dual lipoxygenase activities catalyzes leukotriene A_4 synthesis from arachidonic acid. *Proc. Natl. Acad. Sci. USA 81*, 689–693.

SIEDOW, J. N. 1991. Plant lipoxygenase: Structure and function. *Ann. Rev. Plant Physiol. Plant Biol. 42*, 145–188.

SIGAL, E.; CRAIK, C. S.; HIGHLAND, E.; GRUNBERGER, D.; COSTELLO, L. L.; DIXON,

R. A. F.; and NADEL, J. A. 1988. Molecular cloning and primary structure of human 15-lipoxygenase. *Biochem. Biophys. Res. Comm. 157*, 457–464.

SLAPPENDEL, S.; MALMSTROM, B. G.; PETERSSON, L.; EHRENBERG, A.; VELDINK, G. A.; and Vliegenthart, J. F. G. 1982. On the spin and valence state of iron in native soybean lipoxygenase-1. *Biochem. Biophys. Res. Comm. 108*, 673–677.

SLAPPENDEL, S.; VELDINK, G. A.; VLIEGENTHART, J. F. G.; AASA, R.; and MALMSTROM, B. G. 1981. EPR spectroscopy of soybean lipoxygenase-1. Description and quantification of the high-spin Fe(III) signals. *Biochim. Biophys. Acta 667*, 77–86.

———. 1983. A quantitative optical and EPR study on the interaction between soybean lipoxygenase-1 and 13-*L*-hydroperoxylinoleic acid. *Biochim. Biophys. Acta 747*, 32–36.

SMITH, W. L., and LANDS, W. E. 1972. Oxygenation of unsaturated fatty acids by soybean lipoxygenase. *J. Biol. Chem. 247*, 1038–1047.

SOK, D.-E.; PHI, T. S.; JUNG, C. H.; CHUNG, Y. S.; and KANG, J. B. 1988. Soybean lipoxygenase-catalyzed formation of lipoxin A and lipoxin B isomers from arachidonic acid via 5,15-dihydroperoxyeicosatetraenoic acid. *Biochem. Biophys. Res. Comm. 153*, 840–847.

STALLINGS, W. C.; KROA, B. A.; CARROLL, R. T.; METZGER, A. L.; and FUNK, M. O. 1990. Crystallization and preliminary x-ray characterization of a soybean seed lipoxygenase. *J. Mol. Biol. 211*, 685–687.

STECZKO, J.; DONOHO, G. P.; CLEMENTS, J. C.; DIXON, J. E.; and AXELROD, B. 1992. Conserved histidine residues in soybean lipoxygenase: Functional consequences of their replacement. *Biochemistry 31*, 4053–4057.

STECZKO, J.; MUCHMORE, C. R.; SMITH, J. L.; and AXELROD, B. 1990. Crystallization and preliminary x-ray investigation of lipoxygenase 1 from soybeans. *J. Biol. Chem. 265*, 11352–11354.

THEORELL, H.; HOLMAN, R. T.; and AKESON, A. 1947. Crystalline lipoxidase. *Acta Chem. Scand. 1*, 571–576.

VAN OS, C. P. A.; RIJKE-SCHILDER, G. P. M.; VAN HALBEEK, H.; VERHAGEN, J.; and VLIEGENTHART, J. F. G. 1981. Double dioxygenation of arachidonic acid by soybean lipoxygenase-1. Kinetics and regio-stereo specificities of the reaction steps. *Biochim. Biophys. Acta 663*, 177–193.

VELDINK, G. A.; GARSSEN, G. J.; VLIEGENTHART, J. F. G.; and BOLDINGH, J. 1972. Positional specificity of corn germ lipoxygenase as a function of pH. *Biochem. Biophys. Res. Comm. 47*, 22–26.

VELDINK, G. A.; VLIEGENTHART, J. F. G.; and BOLDINGH, J. 1970. The enzymic conversion of linoleic acid hydroperoxide by flax-seed hydroperoxide isomerase. *Biochem. J. 120*, 55–60.

VERGAGEN, J.; BOUMAN, A. A.; VLIEGENTHART, J. F. G.; and BOLDINGH, J. 1977. Conversion of 9-D- and 13-L-hydroperoxylinoleic acids by soybean lipoxygenase-1 under anaerobic conditions. *Biochim. Biophys. Acta 486*, 114–120.

VICK, B. A.; FENG, P.; and ZIMMERMAN, D. C. 1980. Formation of 12-[^{18}O]oxo-*cis*-10, *cis*-15-phytodienoic acid from 13-[^{18}O]hydroperoxylinolenic acid by hydroperoxide cyclase. *Lipids 15*, 468–471.

VICK, B. A., and ZIMMERMAN, D. C. 1976. Lipoxygenase and hydroperoxide lyase in germinating watermelon seedlings. *Plant Physiol. 57,* 780–788.

————. 1979. Substrate specificity for the synthesis of cyclic fatty acids by a flaxseed extract. *Plant Physiol. 63,* 490–494.

————. 1981. Lipoxygenase, hydroperoxide isomerase, and hydroperoxide cyclase in young cotton seedlings. *Plant Physiol. 67,* 92–97.

————. 1983. The biosynthesis for jasmonic acid: A physiological role for plant lipoxygenase. *Biochem. Biophys. Res. Comm. 111,* 470–477.

————. 1986. Characterization of 12-oxo-phytodienoic acid reductase in corn. *Plant Physiol. 80,* 202–206.

VICK, B. A., and ZIMMERMAN, D. C. 1986. Characterization of 12-oxo-phytodienoic acid reductase in corn. *Plant Physiol. 80,* 202–206.

VLIEGENTHART, J. F. G.; VELDINK, G. A.; and BOLDINGH, J. 1979. Recent progress in the study on the mechanism of action of soybean lipoxygenase. *J. Agric. Food Chem. 27,* 623–626.

WANG, J.; FUJIMOTO, K.; MIYAZAWA, T.; ENDO, Y.; and KITAMURA, K. 1990. Sensitivities of lipoxygenase-lacking soybean seeds to accelerated aging and their chemiluminescence levels. *Phytochemistry 29,* 3739–3742.

WANG, Z.-X.; KILLILEA, S. D.; and SRIVASTAVA, D. K. 1993. Kinetic evaluation of substrate-dependent origin of the lag phase in soybean lipoxygenase-1 catalyzed reactions. *Biochemistry 32,* 1500–1509.

WHITTAKER, J. W., and SOLOMON, E. I. 1986. Spectroscopic studies on ferrous nonheme iron active sites: Variable-temperature MCD probe of ground- and excited-state splitting in iron superoxide dismutase and lipoxygenase. *J. Am. Chem. Soc. 108,* 835–836.

WISEMAN, J. S. 1989. α-Secondary isotope effects in the lipoxygenase reaction. *Biochemistry 28,* 2106–2111.

WISEMAN, J. S., and NICHOLS, J. S. 1988. Ketones as electrophilic substrates of lipoxygenase. *Biochem. Biophys. Res. Comm. 54,* 544–549.

YENOFSKY, R. L.; FINE, M.; and LIU, C. 1988. Isolation and characterization of a soybean (Glycine max) lipoxygenase-3-gene. *Mol. Gen. Genet. 211,* 215–222.

YOON, S., and KLEIN, B. P. 1979. Some properties of pea lipoxygenase isoenzymes. *J. Agric. Food Chem. 27,* 955–962.

YOSHIMOTO, T.; SUZUKI, H.; YAMAMOTO, S.; TAKAI, T.; YOKOYAMA, C.; and TANABE, T. 1990. Cloning and sequence analysis of the cDNA for arachidonic 12-lipoxygenase of porcine leukocytes. *Proc. Natl. Acad. Sci. USA 87,* 2142–2146.

ZHANG, Y.; GEBHARD, M. S.; and SOLOMON, E. I. 1991. Spectroscopic studies of the non-heme ferric active site in soybean lipoxygenase: Magnetic circular dichroism as a probe of electronic and geometric structure. Ligand-field origin of zero-field splitting. *Biochemistry 113,* 5162–5175.

ZIMMERMAN, D. C., and COUDRON, C. A. 1979. Identification of traumatin, a wound hormone, as 12-oxo-*trans*-10-dodecenoic acid. *Plant Physiol. 63,* 536–541.

Chapter 9

Polyphenol Oxidase

John R. Whitaker

Polyphenol oxidase (1,2-benzenediol:oxygen oxidoreductase; EC 1.10.3.1), also known as tyrosinase, phenolase, catechol oxidase, monophenol oxidase, cresolase, and catecholase, was first discovered in 1856 by Schoenbein (1856) in mushrooms. He noted that something in mushrooms catalyzed the aerobic oxidation of certain compounds in plants. The enzyme is found in many plant tissues (Sherman et al. 1991), in some fungi (especially those that produce brown filaments, Osuga et al. 1994), and in some higher animals, including insects (Sugumaran 1988) and humans (Witkop 1985). In higher plants, the enzyme protects the plant against insects and microorganisms and, when wounded, it forms an impervious scab of melanin against further attack by microorganisms and desiccation (Szent-Györgyi and Vietorisz 1931).

It has been suggested by some researchers that the primary function of polyphenol oxidase is modulation of photosystem I or photosystem II reduction of molecular oxygen (Mehler reaction) under low oxygen tension (Sherman et al. 1991), a hypothesis not yet accepted. In insects, it is involved in sclerotization of the exoskeleton (Sugumaran 1988) and in protection against other organisms by encapsulating them in melanin (Sugumaran 1990). In humans, polyphenol oxidase is responsible for pigmentation of the skin, hair, and eye. Racial variation in pigmentation of the skin depends upon the size, number, and degree of pigmentation of the melanosomes and their state of aggregation (Witkop 1985). Genetic regulation of the mammalian pigment system is very complex. At least 147 genes at 53 loci affect coat color in mice (Silvers 1979). The genes affecting pigmentation in humans are numerous and complex (Quevedo 1971; Witkop 1979). There are at least 40 clinical manifestations of hypopigmentation and 27 of hyperpigmentation (Witkop 1985).

In mice and humans, biosynthesis of melanin occurs in melanocytes, which are specialized dendritic, pigment-forming cells. The melanocytes form pigments in specialized granules, melanosomes, which are transported via dendrites to keratinocytes in skin and hair, which are incapable of producing melanin themselves (Witkop 1985). In plants, polyphenol oxidase is localized exclusively in the chloroplasts on the thylakoid membrane with the cytochemical reaction product accumulating in the lumen of the thylakoid (Olah and Mueller 1981; Vaughn and Duke 1981; Vaughn 1987).

Polyphenol oxidase is a very important enzyme in determining the quality and economics of fruit and vegetable storage and processing (Osuga et al. 1994). Up to one-half of some fruits are lost because of browning. Bruises, cuts, and other mechanical damage that allow O_2 penetration lead to rapid browning in many fruits and vegetables because of melanin formation. The color of damaged products, the off-taste and the loss of nutritional quality are unacceptable to the consumer. Peaches, apricots, apples, grapes, bananas, strawberries, and several tropical fruits and juices brown, as do the vegetables Irish potatoes, some lettuce, and other leafy vegetables. Black spot development in shrimp is a major economic problem.

The action of polyphenol oxidase is desirable in tea, coffee, cocoa, prunes, black raisins, black figs (i.e., Black Mission), and zapote. The melanins produced by polyphenol oxidase are excellent sun blockers when applied to the skin, and some of the soluble pigments have potential as food colorants (V. Kahn and J. R. Whitaker 1994, unpublished data).

The purpose of this chapter is to examine: (1) the molecular properties; (2) the reactions catalyzed; and (3) the mechanism of action of polyphenol oxidase.

MOLECULAR PROPERTIES

Molecular Weights

The overall diversity of molecular weights of polyphenol oxidase is not known. Sherman et al. (1991) reported that the molecular weights of polyphenol oxidases from four species of higher plants ranged from 33.0 to >200.0 kD. The generally accepted molecular weight of mushroom polyphenol oxidase is 128.0 kD (Mason 1965). The primary amino acid sequences (from cDNA sequence of the gene) for two potato, a tomato, broad bean, and grape polyphenol oxidases have been determined (Table 9.1). The molecular weights of higher plant polyphenol oxidases range from 40.680 to 58.082 kD, although there is some evidence that the broad bean polyphenol oxidase may be further processed to 40.0–42.0 kD (Cary et al. 1992; Robinson and Dry 1992). The polyphenol oxidases from the fungi *Streptomyces glaucescens* and *S. antibioticus* have molecular weights of 30.900 and 30.736 kD, respectively, while the polyphenol oxidase from fungal

Table 9.1. Comparison of Number of Amino Acids, Molecular Weights, and Histidine and Half-Cystine Content of Eleven Polyphenol Oxidases

PPO	Amino acids		MW (Mature) (kD)	Histidines		Half-Cystine	
	Proprotein	Mature		Pro-	Mature	Pro-	Mature
POT 1[a]	588	500	56.500	20	18	11	9
POT 2[a]	584	501	56.613	20	18	10	8
Tomato[b]	587	500	56.500	19	18	10	8
Broad bean[c]	606	514[d]	58.082[d]	17	14	8	7
GPO1[e]	607	360	40.680	10	9	6	5
S.g.[f]	—	273	30.900	—	11	—	0
S.a.[g]	302	272	30.736	11	11	0	0
N.c.[h]		407	46.000	—	10	—	1
Mushroom[i]		~1130	128.000	—	7	—	2
H.s.[j]	560	548	62.610	14	14	18	17
M.m.[k]	537	513	57.872	17	17	16	16

[a]Potato cDNA clones 1 and 2 (Hunt et al. 1993).

[b]Tomato cDNA (Shahar et al. 1992).

[c]Broad bean cDNA (Cary et al. 1992).

[d]This protein may be processed at a second site to give mature PPO of 40.0–42.0 kD.

[e]Grape cDNA (Robinson, S.P., personal communication, 1994).

[f]Streptomyces glaucescens polyphenol oxidase protein sequence determined by a combination of amino acid and cDNA sequence analyses (Huber et al. 1985).

[g]Streptomyces antibioticus polyphenol oxidase protein determined from cDNA (Bernan et al. 1985).

[h]Neurospora crassa polyphenol oxidase amino acid sequence (Lerch 1978).

[i]Based on 31.0 kD weight (Jolley et al. 1969B).

[j]Homo sapiens (human) polyphenol oxidase protein determined from cDNA clone (Kwon et al. 1987).

[k]Mus musculus (mouse) polyphenol oxidase protein determined from cDNA clone (Shibahara et al. 1986).

Neurospora crassa has a molecular weight of 46.000 kD (Table 9.1). The molecular weight of mushroom polyphenol oxidase is 128.0 kD. The higher animal polyphenol oxidases from human and mouse have molecular weights of 62.610 and 57.872 kD, respectively.

Signal Peptides and Processing of Proenzymes

The signal peptides appear to differ in length and protease specificity among the different polyphenol oxidase families (Table 9.2). In the higher plants, processing of the proenzyme gives signal peptides of 88, 83, 87, and 92 amino acid residues for potato 1, potato 2, tomato, and broad bean propolyphenol oxidases, respectively. In grapes, the signal peptide is 247 amino acids long, whereas in *S. antibioticus*, human, and mouse proenzymes the signal peptides are 30, 12, and 24 amino acid residues, respectively (Table 9.2).

The proteases that process the nine proenzymes have similar specificities (Table 9.2). On the carbonyl side Ala is the amino acid in 6 of the 10 proenzymes, whereas Met and Ser appear to be recognized in *S. glaucescens* and human propolyphenol oxidases, respectively. There appears to be conserved specificity on the amino side of the scissile bond, although three of the four higher plant proenzymes have Ser, as does *N. crassa* proenzyme. It is interesting that the human

Table 9.2. Length of Signal Peptides and Scissile Peptide Bonds in Polyphenol Oxidases Proenzyme Processing[a]

Polyphenol oxidase	Length of signal peptide	Scissile peptide bond[b]
Potato-1	88	88 89 -Ala-Ser-
Potato-2	83	83 84 -Ala-Ser-
Tomato	87	87 88 -Ala-Ala-
Broad bean	92	92 93 -Ala-Ser-
S.g.	—	-Met-Thr-
S.a.	30	30 31 -Ala-Asp
N.c.	—	? -Ser
H.s.	12	12 13 14 -Ser-Phe-Gln
M.m.	24	24 25 26 -Ala-Gln-Phe

[a]See Table 9.1 for references and abbreviations used.

[b]Numbering from N-terminal end of proenzyme, except for S.g. and N.c.

protease recognizes Phe-Gln, whereas the mouse protease recognizes Gln-Phe on the amino side of the scissile bond.

Histidine and Half-Cystine Residues

There are many histidine residues in the mature enzymes (Table 9.1). The significance of this will be important when the amino acid sequences and the mechanism of action of the enzymes are discussed. The higher plant and animal polyphenol oxidases have a substantial number of half-cystine residues (Table 9.1; 1.4–1.8% and 3.1%, respectively), whereas the fungal enzymes have zero or one in the case of *N. crassa* polyphenol oxidase.

Inactive Forms of Polyphenol Oxidase

Several reseachers have noted that some of the polyphenol oxidases in tissues appear to be in an inactive form. Several explanations for these observations have been put forward. They include: (1) a proenzyme; (2) membrane-bound enzyme; (3) a need for activators; and (4) inactivity of the catalytic mechanism.

The cDNA sequencing of polyphenol oxidase genes showed that proenzymes exist for the human (Kwon et al. 1987), mouse (Shibahara et al. 1986), *S. glaucescens* (Huber et al. 1985), *S. antibioticus* (Bernan et al. 1985), potato (Hunt et al. 1993), tomato (Shahar et al. 1992), and broad bean (Cary et al. 1992) polyphenol oxidases. One might expect that processing of the proenzymes would occur rapidly following biosynthesis and transport and that the processed enzyme would be the only protein found. However, polyphenol oxidase was found to be present in a proenzyme form in the hemolymph larvae of the insects *Manduca sexta* and *Sarcophaga bullata* (Sugumaran 1990). A serine-type protease was identified that activates the proenzymes. The protease is present as a proprotease in the hemolymph and is processed in vitro by chymotrypsin, subtilisin, or trypsin. The hemolymph also contains at least three protease inhibitors that apparently regulate the activity of the propolyphenol oxidase processing protease (Sugumaran 1990). The system appears to be delicately balanced to control the level of active polyphenol oxidase in the hemolymph, much like the blood-clotting and pancreatic systems in animals.

Evidence for membrane-bound polyphenol oxidase in plants is based on evidence that detergent extraction of the tissues substantially increases the level of polyphenol oxidase activity (Rodriquez and Flurkey 1992). Whether this is a mechanism for controlling the level of polyphenol oxidase in the tissue and whether the detergent treatment could increase the activation of a tissue-bound proenzyme by the protease is not clear.

Flurkey (1994) has shown that soluble higher plant polyphenol oxidases are activated by sodium dodecylsulfate and Sugumaran (1994) reported that phospholipids activate insect polyphenol oxidases. As discussed under the mechanism of polyphenol oxidase catalysis, the "resting" polyphenol oxidase can be in the *met* form, requiring a reducing compound, BH_2 (Eq. 1) to bring it into an active form.

Primary Amino Acid Sequences

The primary amino acid sequences of nine polyphenol oxidases have been published (Figs. 9.1–9.3). Most have been done by sequencing the cDNAs. The sequence of *N. crassa* polyphenol oxidase is the only primary sequence determined solely by amino acid sequencing (Lerch 1978). The sequencing of *S. glaucescens* was largely determined by amino acid sequencing, with some help from cDNA sequencing (Huber et al. 1985).

Higher plant enzymes.   The proenzymes for polyphenol oxidases from Irish potato (Pot 1 and Pot 2; Hunt et al. 1993), tomato (Shahar et al. 1992), and broad bean (*Vica faba* L; Cary et al. 1992) have been sequenced using the cDNA method (Fig. 9.1). Pot 2, tomato, and broad bean proenzymes have 96.6, 92.2, and 38.1% strict homology, respectively, when compared to Pot 1 proenzyme (Table 9.3). Tomato proenzyme has 91.4 and 40.0% strict homology to Pot 2 and broad bean proenzymes, respectively. All of the proenzymes are processed at a common Ala carbonyl of a peptide bond (Fig. 9.1; Table 9.2) with the amino group contributed by Ser in the case of Pot 1, Pot 2, and BB (broad bean) proenzymes, and Ala in the case of Tom proenzyme.

The mature polyphenol oxidases of Irish potato (Pot 1 and Pot 2), tomato, and broad bean increase in strict homology about 1 or 2% over that of the proenzymes (Table 9.3). This is largely due to the increase in overall contribution of the rather conserved regions of the active site (see below). Therefore, the homologies among the four plant polyphenyl oxidases appear to be about the same for both the proenzymes and the enzymes.

The Pot 1, Pot 2, Tom, and BB polyphenol oxidases have nine, eight, eight, and seven half-cystine residues, respectively (Fig. 9.1). Eight of the half-cystine residues are conserved among the Pot 1, Pot 2, and Tom enzymes, whereas five half-cystine residues are conserved among the four enzymes. It is anticipated that the secondary and tertiary structures will be similar, when elucidated. There are 18, 18, 18, and 14 histidine residues in Pot 1, Pot 2, Tom, and BB polyphenol oxidases, respectively. All 18 histidine residues are conserved among the Pot 1, Pot 2, and Tom polyphenol oxidases, whereas 10 histidine residues are conserved among all four enzymes. Six of the 10 conserved histidine residues are presumed to be in the active site (see below).

Fungal enzymes.   The primary amino acid sequences of the polyphenol oxidases of *N. crassa, S. glaucescens,* and *S. antibioticus* are shown in Fig. 9.2. The polyphenol oxidases of *S. glaucescens* and *S. antibioticus* have 17.0 and 16.5% strict homology, respectively, with *N. crassa* polyphenol oxidase (Table 9.4). The *N. crassa* polyphenol oxidase is 1.5 times larger than the *S. glaucescens*

AMINO ACID SEQUENCE RELATIONS AMONG POTATO, TOMATO AND BROAD
BEAN POLYPHENOL OXIDASES

```
POT1            SSSSTTTIPLCTNK SLSSSFTTNNSSFLSKPSQ
POT2            --L---N-- ------------------
TOM       MSS S --I---L------ --------T---L------
BB        MTSISAL-FIS-INVSSNS-I-H--VYPFLQKQHQ-SKLR


POT1            LFLHGRRNQSFKVSCNANNNVGEHDKNLDTVD RRNVLLG
POT2            ---------------------------A-- -------
TOM             ------------------V    ---P-A-- -------
BB        KPKRQVTCS-NNNQN-PKEE    QELSNI-GH-----I-
                                 1
POT1     1 LGGLYGAANLAPLASASPIPPPDLKSCGVAHVTEGVDVTY
POT2     1 --------------------------------K-----S-
TOM      1 --------------T-A-----------T---K-----I-
BB       1 ---I--TLATN-S-L----S----SK-VPPSDLPSGTTPP


POT1    25 S  CCPPVPDDIDSVPYYKFPPMTKLRIRPPAHAADEEYV
POT2    25 -  -----------------S------------------
TOM     25 -  -----------------S------------------
BB      25 NIN----YSTK-TDF   ---SNQP--V-QA--LV-N-FL


POT1    63 AKYQLATSRMRELDKDSFDPLGFKQQANIHCAYCNGAYKV
POT2    63 ----------------------------------------
TOM     63 --------------ND-PF--------------------
BB      62 E--KK--EL-KA-PSNDPRNFT  ----------D---SQ


POT1   103 GGKE    LQVHFSWLFFPFHRWYLYFYERILGSLINDPTF
POT2   103 ----    --------------------------------
TOM    103 ----    --------------------------------
BB     100 I-FPDLK----G--------------------------


POT1   140 ALPYWNWDHPKGMRIPPMFDREGSSLYDDKRNQNHRNGTI
POT2   140 ----------------------------------------
TOM    140 ------------------------------E----------
BB     140 ---F--Y-A-D--QL-SIYTDKA-P---EL--AS-QPP-L


POT1   180 IDLGHFGQEVDTPQLQIMTNNLTLMYRQMVTNAPCPSQFF
POT2   180 -------K--------------------------------
TOM    180 -------K--------------------------------
BB     180 ---NFCDSDS-IHGDELIKT--SI----VYS-GKTSRL-L


POT1   220 GAAYPLGTEPSPGMGTIENIPHTPVHIWTGDSPRQKNGEN
POT2   220 -------------Q----------------K--------
TOM    220 ----LWVLN----Q----------------K------D
BB     220 -NP-RA-DAEPQ-A-----V--A---T----NT -T-I-D


POT1   260 MGNFYSAGLDPIFYCHHANVDRMWDEWKLIGGKRRDLSNK
POT2   260 ----------------------------------------
TOM    260 ----------------------N-----------TD-
BB     259 --I----AR-----S--S----L-YI--TL---KH-FTD-
```

Figure 9.1 Amino acid sequence relationships among potato (Pot 1 and Pot 2; Hunt et al. 1993), tomato (Shahar et al. 1992), and broad bean (BB; Cary et al. 1992) polyphenol oxidase cDNAs. The numbering system, based on mature potato 1 polyphenol oxidase, is from the cDNA sequences. The numeral 1 marks the proteolytic processing of the pro-enzymes to give the mature enzymes. Amino acid homologies are shown by dashes (−). A space is left for alignment purposes.

```
POT1  300  DWLNSEFFFYDENRNPYRVKVRDCLDSKKMGFSYAPMPTP
POT2  300  ---------------------S----------------
TOM   300  ---------------------V--------D-------
BB    299  ---E-G-L-----K-LV--N---S--ID-L-YA-QDV-I-

POT1  340  WRNFKP IRKTTAGK VNTASIAPVTKVFPLAKLDRAIS
POT2  340  ------ ------- I----------------------
TOM   340  ------ ---SSS-- ---------S-------------
BB    339  -EKA--VPPP-KVQ-L-EVEVNDGNLRKS-TIF-V-QQ-P

POT1  377   FSITRP ASSRTTQEKNEQEEILTFNK VAYDDTKYV R
POT2  377   ------ --------------------KI-----Q-- -
TOM   377   ------ -------------------- IS---RN-- -
BB    379  RKYV-F-LVLNNKVSAIVKRPKK-RSK-EKEEEEEVL-IE

POT1  413  FDVF LN VDK TV NADELDK A   EFAGSYTSLPHV
POT2  414  ---- -- --- -- ------- -   ------------
TOM   413  ---- -- --- -- ------- -   ------------
BB    419  GIE-YM-IAI-FD-YIN--D--VG-GNT-----FVNI--S

POT1  444  HGNNTNHVTSVTFKLAITELLEDNGLEDEDTIAVTLVPKV
POT2  445  -------------TV--------I----------------
TOM   444  --S---------K---------I------------I--A
BB    459  AHGHK-KKIITSLR-G--D----LHV-GD-N-V------C

POT1  484  GGEGVSIESVEIKLEDC 501
POT2  485  ---------------- 502
TOM   484  ---------------- 501
BB    499  -SGQ-K-NN---VF-- 515
```

Figure 9.1. (Continued)

and *S. antibiotics* polyphenol oxidases. As shown in Fig. 9.2, most of the ho-mology is in the regions of the active site (residues 95–122 and 277–331; *N. crassa* PPO numbering). There is 87.5% strict homology between *S. glaucescens* and *S. antibioticus* polyphenol oxidases. Unlike the higher plant and animal poly-phenol oxidases, only the *N. crassa* polyphenol oxidase contains a single half-

Table 9.3. Amino Acid Sequence Homologies Among the Plant Polyphenol Oxidases[a]

Enzymes \ Enzymes	Pot 1 (%)	Pot 2 (%)	Tom (%)
Pot 1(%)	—		
Pot 2 (%)	96.6[b] (97.4)[c]	—	
Tom (%)	92.2 (93.8)	91.4 (93.6)	—
BB (%)	38.1 (42.1)	38.0 (42.1)	40.0 (42.2)

[a]Strict amino acid homologies for potato 1, potato 2, tomato, and broad bean (BB) propolyphenol oxidases and polyphenol oxidases. See Table 9.1 for references.

[b]Strict homologies for the proenzymes.

[c]Strict homologies for the enzymes.

NEUROSPORA CRASSA, STREPTOMYCES GLAUCESCENS, AND S. ANTIBIOTICUS
POLYPHENOL OXIDASES

```
N.c.     1 Ac-STDIKFAITGVPTTPSSNGAVPLRRELRDLQQNYPEQFNL
S.g.     1                                   TVRK-Q
S.a.     1                                   TVRK-Q

N.c.    41 YLLGLRDFQGLDEAKLDSYYQVAGIHGMPFKPWAGVPSDT
S.g.     7 AT-TADEKRRFVA-V-ELKRSGRYDEFVTTHNAFIIG --
S.a.     7 AS-TAEEKRRFVA-L-ELKRTGRYDAFVTTHNAFILG --

N.c.    81 DWSQPGSSGFGGYCTHSSILFITWHRPYLALYEQALYASV
S.g.    46 -          A-ERTG-R-PS-LP---R--LEF-R--  Q--
S.a.    46 -          N-ERTG-R-PS-LP---RF-LEF-R--  Q--

N.c.   121 QAVAQKFPVEGGLRAKYVAAAKDFRAPYFDWASQPPKGTL
S.g.    77 D-                     SVAL--W--
S.a.    77 D-                     SVAL--W--

N.c.   161 AFPESLSSRTIQVVDVDGKTKSINNPLH RFTFHPVNPSP
S.g.    88             SADRTARASLWAPDFLGGTG-SLDGR-MDG-
S.a.    88             SADRSTRSSLWAPDFLGGTG-SRDGQ-MDG-

N.c.   200 GNFSAAWSRYPSTVRYPNRLTGASRDERIAPILADELASL
S.g.   119   -A-SAGNW-IN--VDG-AYLRRSLGTAVR     --PT
S.a.   119   -A-SAGNW-IN--VDG-TFLRRALGAGVS     --PT

N.c.   240 RNNVSLLLLSYKDFDAFSYNRWDPNTNPGDFGSLEAVHNE
S.g.   152 -AE-GSV-GMATYDT-PWNSAS-GFR-HLEGWRGVNL--R
S.a.   152 -AE-DSV-AMATYDM-PWNSGS-GFR-HLEGWRGVNL--R

N.c.   280 IHDRTGGNGHMSSLEVSAFDPLFWLHHVNVDRLWSIWQDL
S.g.   192 V-VWV--  R- ATGM-PN--V-----AYV-K--AE--RR
S.a.   192 V-VWV--  Q- ATG--PN--V-----AYI-K--AE--RR

N.c.   320 NPNSFMTPRPA PYSTFVAQEGESQSKSTPLEPFWDKSAA
S.g.   229 H-G-GYL-AAGT-DVVDLNDRMKPWNDTS-ADLLDHTAHY
S.a.   229 H-S-PYL-GGGT-NVVDLNETMKPWNDT--AALLDHTRHY

N.c.   359 NFWISEQVKDSITFGYAYPETQKWKYSSVKEYQAAIRKSV
S.g.   269 T-DTD 274
S.a.   269 T-DV 273

N.c.   399 TALYGSNVF 408
```

Figure 9.2. Amino acid sequence relationships among *Neurospora crassa* (*N.c.*; Lerch 1978), *Streptomyces glaucescens* (*S.g.*; Huber et al. 1985), and *S. antibioticus* (*S.a.*; Bernan et al. 1985) processed polyphenol oxidases. Numbering is according to the *N. crassa* polyphenol oxidase. See Fig. 9.1 legend for explanations of other symbols used.

Table 9.4.   Amino Acid Sequence Homologies Among the Fungal Polyphenol
Oxidases[a]

Enzymes \ Enzymes	S.g.	S.a.
S. glaucescens	—	
S. antibioticus	87.5	—
N. crassa	17.0	16.5

[a]See Table 9.1 for references. The data are for the processed enzymes.

cystine residue. Therefore, half-cystine residues are not universally necessary for polyphenol oxidases. There are 10, 11, and 11 histidine residues in *N. crassa, S. glaucescens,* and *S. antibioticus* polyphenol oxidases, respectively. Six histidine residues, common to all three enzymes, are presumed to be in the active sites. All 11 histidine residues are common to *S. glaucescens* and *S. antibioticus* polyphenol oxidases.

Animal enzymes.   The primary amino acid sequences of *Homo sapiens* (human) and *Mus musculus* (mouse) polyphenol oxidases are shown in Fig. 9.3. The strict homology between the two is 41.0%. This appears low. However, the homology extends throughout the total sequence, not just the active regions, clearly indicating that these two polyphenol oxidases are more closely related than they are to the higher plant and fungal polyphenol oxidases. Of the 16 half-cystine residues of mouse polyphenol oxidase, 13 align with 13 of the 17 half-cystine residues of human polyphenol oxidase. Although the secondary and tertiary structures of these polyphenol oxidases are not known, it is likely that these structures are very similar for the human and mouse enzymes, given the overall homology and the 81% homology in locations of the half-cystine residues. There are 14 and 17 histidine residues in human and mouse polyphenol oxidase, respectively. Seven of the 14 histidine residues of human polyphenol oxidase are homologous with 7 of the 17 histidine residues of mouse polyphenol oxidase (Fig. 9.3). Of more significance, six of the seven common histidines are in the postulated active sites of the two enzymes (see below). Therefore, conserved half-cystine residues appear to be a function of the total protein, whereas the conserved histidine residues appear to be a function of the active site.

Homologies among all polyphenol oxidases.   There is little overall homology among the three groups of polyphenol oxidases. Computer alignment using algorithms indicates 19.4, 18.8, 13.0, 10.4, and 10.9% homology of *S. glaucescens, S. antibioticus, N. crassa,* human, and mouse polyphenol oxidases, respectively, with Pot 1 polyphenol oxidase (data not shown). Numerous gaps (–) must be used in the alignments leading to 765 spacings to accommodate all nine polyphenol oxidases. Note that the largest polyphenol oxidase, human, contains only 549 amino acids. Visual inspection of the computer alignment indicates

HUMAN AND MOUSE POLYPHENOL OXIDASES

```
H.s.    1 FQTSAGHFPRACVSSKNLMEKECCP    PWSG  TGVCGQ
M.m.    1        Q---E-ANIEA-RRGV---DLL-S--PG-DP--S

H.s.   36 LSGRGSCQN ILLSNAPLGPQFPFTGVDDRESWPSVFYNR
M.m.   34 S----R-VAV-AD-R -HSRHY-HD-K----A--LR-F--

H.s.   75 TCQCSGNFMGFNCGNCKFGFW GPNCTERRLLVRRNIFDL
M.m.   73 ----ND--S-H---T-RP- -R-AA-NQKI-T----LL--

H.s.  114 SAPEKDK  FF AYLTLAKHTISSDYVIPIGTYGQMKN
M.m.  112 - -- E-SH-VR- -DM--R-THPQF--ATRRLEDILGPD

H.s.  149 GSTPMF NDINIYDLFVWMHYYVSMDAL LG GYEIWRDI
M.m.  149 -N--Q-E- -SV-NY---T--- -VKKTF--T-Q-SFG-V

H.s.  186 DFAHEAPAFLPWHRLFLLRWEQEIQKLTGDQNFTIPYWDW
M.m.  187 --S--G----T---YH--QL-RDM-EMLQEPS-SL---NF

H.s.  226 RDA EKCDICTDEYMGGQHPTNPNLLSPASFFSSWQIVC
M.m.  227 ATGKNV--V---DL--SRSNFDST----N-V--Q-RV--E

H.s.  264 SRLEEYNSHQSLCNGTPEG PLRRNP GNHDKSTTPRLPS
M.m.  267 - ----DTLGT---S- --G------A--VGRPAVQ---E

H.s.  302 SADVEFCLSLTQYESGSMDKAANFSFRNTLEGF ASPLTG
M.m.  305 PQ--TQ--EVRVFDTPPFYSNSTD-----V--YS- - --

H.s.  341 IAD ASQSSMHNAL HIYMNGHVPG TG SANDRIF LLT
M.m.  343 KY-P-VR- L-- -A-LFL--T G-Q-HL-P--P--V--

H.s.  376 MHLLT  VFLRQWLQRHRPLQEVYP EANAPIGHNRESY
M.m.  380 -TF-DA--D E---R-YNADISTF-L- -------Q -N

H.s.  412 MVPFIPLYRNGDFFISSKD LGYDYSYLQDSDPDSFQDYI
M.m.  417 ----W-P-T-TEM-VTAP-N---A-EV -WPGQEFTVSE-

H.s.  451 KSYLEQASRIWSWLLGAAMVGAVLTALLAGPVSLLCRHKR
M.m.  456        -TIAVV --LL LVA-IF -VA-C-I-SRS

H.s.  491 KQLPEEKQPLLMEKEGLPQLVSEPFIKGLGNRVGPKSPDL
M.m.  483 TK N-AN----TDHYQRYAEDY-ELPNPNHSM- 515

H.s.  531 TLTQSNVQVPENICWYFL 549
```

Figure 9.3. Amino acid sequence relationships between *Homo sapiens* (*H.s.,* human; Kwon et al. 1987) and *Mus musculus* (*M.m.,* mouse; Shibahara et al. 1986) for the processed enzymes from cDNA sequencing. The numbering system is that for the human enzyme. See Fig. 9.1 for explanations of other symbols used.

questionable homologous alignments in several areas. Therefore, it is difficult to do more than roughly estimate overall homology among the nine polyphenol oxidases.

Homologies among the active sites.   Above, it was noted that there are two regions, A and B, of the primary amino acid sequences that appear to have good homologies among the nine polyphenol oxidases. These regions are shown in Fig. 9.4. Region A contains 26 amino acid residues of which six amino acid residues are strictly conserved among the 10 sequences available. Two histidine

```
ACTIVE SITE REGION A

POT1   110   HFSWLFFPFHRWYLYFYERILGSLIN      135
POT2   110   -------------------------      135
TOM    110   -------------------------      135
BB     110   -G-----------------------      135
S.g.    54   -R-PS-L----YL-E- --A-Q-V        76
S.a.    54   -R-PS-L-W--RF-LEF--A-Q-V        77
N.c.    96   -S-I--ITW--P--AL--QA-YASVQ     121
H.s.   189   -EAPA-L-W--LF-LRW-QEIQK-TG     214
M.m.   190   -EAPG-L-W--LF-LLW-QEIRE-TG     215
pMT4   191   -EGPA-LTW--YH-LQL--DMQEMLQ     216


ACTIVE SITE REGION B

POT1   241   HTPVHIWTGDSPRQKNGENMGNFY SAGLDPIFYCHHANVDRMWDEWKLIGGKRRD    296
POT2   241   ----------K------------- ------L------------------------    296
TOM B  241   ----------K------------- --------------------N----------    296
TOM Aa 328   ----------K---G---D----- ------L-----------N-----------    382
BB     241   -A---T----NT -T-I-D--I-- --AR-----S--S----L-YI--TL---KH-    295
S.g.   189   -NR--V-V-GR        -ATGM -P N--V-WL---Y--KL-A--Q    --H    229
S.a.   189   -NR--V-V-GQ        -ATGV -P N--V-WL---YI-LL-A--Q    --H    229
N.c.   277   -NEI-DR--G         --H -SSLEV-- F--L-WL--V----L-SI-QDLN    320
H.s.   350   -NAL--Y        M--HVP-TG  -- N----LL---F--SIFEQ- -   Q-H   390
M.m.   352   -NAL--F        M--T -SQVQG-- N----LL---F--SIFEQ- -   --H   393
pMT4   353   -NLA-LF        L--TG -QTHL-P N----VLL-TFT-AVF--- -    --Y   394
h.E    318   -NWG-VMKMARLQDPDHGV-SDTST-L R-----RY-RFI-NIF QKYIATL PHY    373
h.D    320   -NWG-VM IARIHDAD-GV-DDTST-L R-----RY-RWM-NIF QEY- H-L    374
             Cu  Cu                                          Cu
```

* IN h.E. RESIDUES 335 RFNENP 340 AND IN h.D. RESIDUES 336 RYRTNP 341 LEFT OUT FOR ALIGNMENT PURPOSES.
a Proenzyme

Figure 9.4.   Amino acid sequence relationships in and near the active sites of potato (Pot 1 and Pot 2), tomato (Tom A and Tom B), broad bean (BB), *S. glaucescens* (*S.g.*), *S. antibioticus* (*S.a.*), *Neurospora crassa* (*N.c.*), *Homo sapiens* (*H.s.*), *Mus musculus* (*M.m.*), putative mouse transcript 4 (pMT 4), hemocyanin E (spider, *Eurypelma californicum,* Schartau et al. 1983), and hemocyanin D (spider, *E. californicum;* Schneider et al. 1983) polyphenol oxidases. Shown are regions A and B containing the presumed copper-binding histidine residues. The Cu indicates histidines that ligand to copper in *N. crassa* polyphenol oxidase, and presumably in the other polyphenol oxidases and hemocyanins.

residues presumed to be in the active site are strictly conserved, as well as a phenylalanine, an arginine, a leucine, and a glutamic acid residue. Five other amino acid positions show conservative replacements.

Region B amino acid sequences range from 41 to 62 residues in length, including the addition of Tom A and two hemocyanin sequences to the list. Eight residues, strictly conserved, are three histidines, two aspartic acids, one serine, one proline, and one phenylalanine residue; there are at least five residues that show conservative homologies. The three histidines are presumed to be in the active site as ligands of copper (Fig. 9.9), based on their similar location in *N. crassa* polyphenol oxidase and in the hemocyanins. Of course, the extent of homologies are based on the reference enzyme, which is potato 1 polyphenol oxidase in Fig. 9.4. Other frames of reference have been published.

Secondary, Tertiary, and Quaternary Structures

The secondary and tertiary structures of a polyphenol oxidase have not been determined. The nine polyphenol oxidases for which the primary structures are discussed in the previous section are single polypeptide enzymes. Mushroom polyphenol oxidase, of 128.0 kD, with four subunits would have a quaternary structure. The system is known to undergo association forming monomers (~ 32.5 kD) to octamers, with the tetrameric form predominating under most conditions (Jolley et al. 1969A,B). But the lowest unit with activity is not known.

The partial primary amino acid sequence and secondary and tertiary structures of *Panulirus interruptus* (spiny lobster; Gaykema et al. 1984) hemocyanin subunit *a* has been determined. The subunit folds into three domains. Domain 2 contains the binuclear copper site responsible for binding and transporting O_2 in the lobster (Gaykema et al. 1984). Each of the two coppers is liganded to three histidine residues (copper A, His 196, 200, and 226; copper B, His 346, 350, and 386). All of these histidine residues are located on four α helices that fold in proximity to each other. There are striking similarities between the amino acid sequence around the binuclear copper sites of the lobster hemocyanin and that of region B of the polyphenol oxidases (Fig. 9.4). The spacing of the active site histidines along the polyphenol oxidase peptide chain is similar but not the same as that of lobster hemocyanin. In *N. crassa* polyphenol oxidase, Huber et al. (1985) determined that Cu(A) is coordinated to His 187, 193, and 281, and Cu(B) to His 95, 104, and 306. The three-dimensional structure of domain 2 of subunit *a* of *P. interruptus* hemocyanin is shown in Fig. 9.5. Is it possible that the polyphenol oxidases fold in a similar pattern?

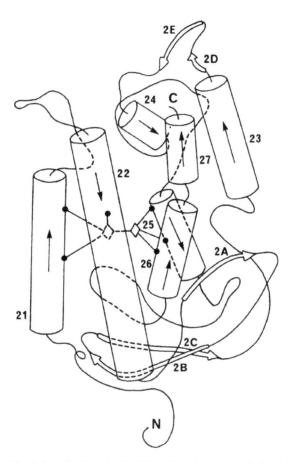

Figure 9.5.   The *b* domain (domain 2) of *Panulirus interruptus* (spiny lobster) hemocyanin showing the binding sites for the two coppers (indicated as diamonds) on the α helices. All six ligands of the binuclear copper binding site on the α helices are shown. From Gaykema et al. 1984. Nature 309, p. 28. With permission. Copyright 1984 MacMillan Magazines Ltd.

REACTIONS CATALYZED

In theory, polyphenol oxidases catalyze two quite different types of reactions as shown in Eqs. 1 and 2.

p-CRESOL   4-METHYLCATECHOL

$$\text{(p-cresol)} + O_2 + BH_2 \longrightarrow \text{(4-methylcatechol)} + B + H_2O \qquad (1)$$

CATECHOL   o-BENZOQUINONE

$$2\,\text{(catechol)} + O_2 \longrightarrow 2\,\text{(o-benzoquinone)} + 2\,H_2O \qquad (2)$$

In Eq. 1, BH_2 is an o-diphenol required as a cofactor. The polyphenol oxidases that perform the monohydroxylation reaction (Eq. 1) show hysteresis (lag phase) if BH_2 is not added initially to the reaction. The enzymes can slowly produce BH_2 by o-hydroxylation of the monophenol; the source of the initial reducing equivalents required is not known. As little as $1 \times 10^{-7} M$ catechol added to the reaction, as BH_2, gives normal kinetics.

There is disagreement as to whether all polyphenol oxidases can perform the monohydroxylation reaction (Eq. 1). Purified (homogeneous) peach (Wong et al. 1971A) and pear (Rivas and Whitaker 1973) polyphenol oxidases did not oxidize monophenols, such as p-cresol, within several hours reaction time. *S. glaucescens* polyphenol oxidase hydroxylates a large number of monophenols (Lerch and Ettlinger 1972; Table 9.5), as do mushroom and human skin polyphenol oxidases (Bouchilloux et al. 1963). However, the monohydroxylation reaction can be up to 200 times slower for some substrates in comparison to the rate of oxidation of o-diphenols (Tables 9.5–9.7).

Substrate Specificity

The hydroxylation reaction (Eq. 1) is difficult to study independently of oxidation of the o-diphenol (Eq. 2), except by monitoring the reaction using o-tritiated monophenol and measuring the rate of release of tritium (as THO). Sometimes, a change in absorbance at 280 nm can be used to distinguish between

Table 9.5. Michaelis-Menten (K_m), Catalytic (k_{cat}), and k_{cat}/K_m Constants and Lag-Time (t) of Different Monophenol Substrates[a]

Substrate	K_m (mM)	k_{cat} (s^{-1})	t (min)	k_{cat}/K_m (mM^{-1} s^{-1})
L-Tyrosine	0.410 (7.9)[b]	13.2 (0.084)[b]	5.00	32.2
L-Tyrosine methyl ester	1.667 (32)	235 (1.5)	0.85	141
L-Tyrosine ethyl ester	0.945 (18)	223 (1.4)	0.55	236
L-Tyrosinamide	1.540 (30)	172 (1.1)	0.55	112
L-Tyrosine hydrazide	2.960 (57)	151 (0.96)	1.75	51.0
L-Tyrosyl-L-leucine	1.815 (35)	357 (2.3)	0.95	197
L-Tyrosyl-glycyl-glycine	3.058 (59)	310 (2.0)	1.30	101
N-Formyl-L-tyrosine	0.689 (13)	104 (0.66)	5.00	151
N-Acetyl-L-tyrosine	0.553 (11)	182 (1.2)	3.30	329
N-Chloroacetyl-L-tyrosine	0.287 (5.5)	249 (1.6)	2.20	868
N-Trifluoroacetyl-L-tyrosine	0.467 (9.0)	330 (2.1)	2.80	707
Glycyl-L-tyrosine	5.341 (103)	412 (2.6)	4.20	77.1
L-Leucyl-L-tyrosine	2.225 (43)	152 (0.96)	9.70	68.3
N-Acetyl-L-tyrosine ethyl ester	0.268 (5.2)	243 (1.5)	1.15	907
N-Chloroacetyl-L-tyrosine ethyl ester	0.052 (1.0)	158 (1.0)	1.10	3038
N-Trifluoroacetyl-L-tyrosine ethyl ester	0.045 (0.87)	104 (0.66)	1.05	2311
N-Acetyl-L-tyrosinamide	0.225 (4.3)	227 (1.4)	0.70	1009
N-Acetyl-L-tyrosine hydrazide	0.247 (4.8)	376 (2.4)	0.32	1522
p-Cresol	0.242 (4.7)	11.6 (0.073)	3.76	47.9
p-Hydroxyphenylacetic acid	1.164 (22)	48.8 (0.31)	16.00	41.9
p-Hydroxyphenylpropionic acid	0.488 (9.4)	134 (0.85)	4.30	275
L-Tyrosine methyl ester (α)[c]	0.462 (8.9)	148 (0.94)	0.70	320
L-Tyrosine methyl ester (γ)[c]	0.262 (5.0)	129 (0.82)	0.70	492

[a]Adapted from Lerch and Ettlinger (1972). Activity was measured polarographically at 30°C in 0.1 M phosphate buffer, pH 7.0; [O$_2$] was 0.24 mM (air-saturated buffer). *Streptomyces glaucescens* polyphenol oxidase was used.

[b]Number in parentheses is relative to the value for N-chloroacetyl-L-tyrosine ethyl ester.

[c]Kinetic constants for α and γ isozymes of mushroom tyrosinase (kinetic constants based on the subunit molecular weight of 32.400 kD) (Jolley et al. 1969B).

Table 9.6. Michaelis-Menten (K_m), Catalytic (k_{cat}), and k_{cat}/K_m Constants for Different o-Diphenol Substrates[a]

Substrate	K_m (mM)	k_{cat} (s^{-1})	k_{cat}/K_m ($mM^{-1} s^{-1}$)
3,4-Dihydroxy-L-phenylalanine	5.774 (6.3)	1445 (0.44)	250
3,4-Dihydroxy-DL-phenylalanine	12.558 (14)	1542 (0.46)	123
3,4-Dihydroxy-D-phenylalanine	15.020 (16)	184 (0.055)	12.3
Catechol	4.508 (4.9)	395 (0.12)	87.6
Homocatechol	1.906 (2.1)	2340 (0.70)	1230
t-Butylcatechol	0.915 (1.0)	3320 (1.0)	3630
3,4-Dihydroxybenzoic acid	1.023 (1.1)	8.2 (0.0025)	8.01
3,4-Dihydroxyphenylacetic acid	1.169 (1.3)	142 (0.043)	121
3,4-Dihydroxyphenylpropionic acid	1.055 (1.2)	1830 (0.55)	1734
Caffeic acid	0.372 (0.41)	476 (0.14)	1280
Chlorogenic acid	0.748 (0.82)	785 (0.24)	1050
3-Hydroxytyramine hydrochloride	11.900 (13)	875 (0.26)	73.5
3,4-Dihydroxy-L-phenylalanine, α[b]	0.375 (0.41)	414 (0.12)	1100
3,4-Dihydroxyl-L-phenylalanine, γ[b]	0.400 (0.44)	670 (0.20)	1680

[a]Adapted from Lerch and Ettlinger (1972). Activity was measured polarographically at 30°C in 0.1 M phosphate buffer, pH 7.0, using *Streptomyces glaucescens* polyphenol oxidase; [O_2] was 0.24 mM (air-saturated buffer).

[b]Kinetic constants for α and γ isozymes of mushroom tyrosinase (kinetic constants based on the subunit molecular weight of 32.400 kD) (Jolley et al. 1969B).

287

Table 9.7. Ratios of k_{cat}/K_m and k_{cat}/k_{cat}, in Parentheses, for some Monophenol/
o-Diphenol Substrate Pairs[a]

Substrate pairs			
Monophenol	Diphenol	(1)[b]	(2)[c]
L-Tyrosine	3,4-Dihydroxy-L-phenylalanine	7.81	(110)
p-Cresol	Homocatechol	25.7	(222)
p-Hydroxyphenylpropionic acid	3,4-Dihydroxyphenylpropionic acid	6.31	(13.6)
p-Hydroxyphenylacetic acid	3,4-Dihydroxyphenylacetic acid	2.89	(2.89)
L-Tyrosine methyl ester	3,4-Dihydroxy-L-phenylalanine	1.77	(6.17)
L-Tyrosine methyl ester (α)[d]	3,4-Dihydroxyl-L-phenylalanine (α)[b]	3.44	(2.62)
L-Tyrosine methyl ester (γ)[d]	3,4-Dihydroxy-L-phenylalanine (γ)[b]	3.41	(5.20)

[a]Calculated from data in Tables 9.5 and 9.6.

[b](k_{cat}/K_m)diphenol/(k_{cat}/K_m) monophenol from Tables 9.5 and 9.6.

[c]k_{cat}, diphenol/k_{cat}, monophenol from Tables 9.5 and 9.6.

[d]Based on kinetic constants for α and γ isozymes of mushroom polyphenol oxidase, using a subunit molecular weight of 32.400 kD (Jolly et al. 1969B).

hydroxylation and further oxidation of the *o*-diphenol early on in the reaction. O_2 uptake cannot be used, as it measures both reactions, because oxidation of the *o*-diphenol is the faster of the two reactions.

The polyphenol oxidases have broad substrate specificity, as shown in Tables 9.5 and 9.6 for *S. glaucescens* polyphenol oxidase. Using k_{cat}/K_m as the specificity coefficient, the best monophenol substrate for *S. glaucescens* polyphenol oxidase is *N*-chloroacetyl-L-tyrosine ethyl ester (Table 9.5), with a specificity coefficient of 3,038 mM^{-1} s^{-1}. Replacement of the *N*-chloroacetyl group with an *N*-acetyl group reduces the specificity coefficient to 907 mM^{-1} s^{-1}, whereas elimination of the *N*-chloroacetyl group (to give L-tyrosine ethyl ester) results in a specificity coefficient of 236 mM^{-1} s^{-1}. Elimination of the ethyl ester group (to give *N*-chloroacetyl-L-tyrosine) results in a specificity coefficient of 868 mM^{-1} s^{-1}. Elimination of both the *N*-chloroacetyl- and ethyl ester groupings (to give L-tyrosine) results in a specificity coefficient of 32.2 mM^{-1} s^{-1}; L-tyrosine is the worst substrate among the list of 21 substrates for the enzyme. Thus, *S. glaucescens* polyphenol oxidase prefers to have both the amino and carboxyl groups of L-tyrosine derivatized. The differences in specificity coefficients for monophenols are largely a result of changes in K_m. The K_m values increase up to 103 times (glycyl-L-tyrosine relative to that of *N*-chloroacetyl-L-tyrosine ethyl ester), whereas k_{cat} varies by a factor of 3 (for *p*-hydroxyphenylacetic acid relative to *N*-chloroacetyl-L-tyrosine ethyl ester; ignoring the values for L-tyrosine and *p*-cresol). The longest lag period under the conditions used was 16.0 min for *p*-hydroxyphenylacetic acid; the shortest was 0.32 min for *N*-acetyl-L-tyrosine hydrazide. D-Tyrosine and derivatives are not substrates.

The best o-diphenol substrate for *S. glaucescens* polyphenol oxidase is 4-t-butylcatechol with a specificity coefficient of 3,630 mM^{-1} s^{-1}. Elimination of the t-butyl group to give catechol results in a specificity coefficient of 87.6 mM^{-1} s^{-1} (a 41.4 decrease in specificity). The worse substrate is 3,4-dihydroxybenzoic acid with a specificity coefficient of 8.01 mM^{-1} s^{-1} (a 453 times worse substrate than t-butylcatechol). K_m is similar for both t-butylcatechol and 3,4-dihydroxybenzoic acid, whereas k_{cat} differs by a factor of 405. Therefore, the difference between the two substrates in specificity coefficient is due to differences in k_{cat}. As shown in Table 9.6, the primary contributor to differences in the specificity coefficients for the o-diphenols is k_{cat}. There is a 400 times difference in k_{cat} for the substrates, whereas K_m varies only 16 times (except for three substrates, K_m varies only 6 times). As shown in Table 9.7, the ratio of specificity coefficients for o-diphenols/monophenols varied 76.8 times between homocatechol/p-cresol and 3,4-dihydroxyphenylacetic acid/p-hydroxyphenylacetic acid.

The relative specificities of polyphenol oxidases on substrates vary from one source to another (Table 9.8). For example, peach polyphenol oxidase has only 52% the activity on 4-methylcatechol as on catechol, whereas broad bean leaf polyphenol oxidase has 200–225% the activity on 4-methylcatechol as on

Table 9.8. Relative Substrate Specificities of Three Polyphenol Oxidases

Substrate	Activity Relative to Catechol			
	Potato[a]	Peach[b]	Broad Bean Leaf[c]	Pear[d]
Di- or triphenolic compounds				
Catechol	100	100	100	100
4-Methylcatechol		51.5	200–225	72.3
d-Catechin		31.8		7.79
Chlorogenic acid	140	22.2	8	71.8
Caffeic acid	76.5	0	12.5	4.41
Protocatechuic acid		16.3	0.11	
3,4-Dihydroxy-L-phenylalanine	54.3	40.5	50	
Dopamine		45.6		15.6
Gallic acid		25.7	0.22	
Pyrogallol			85–95	
Monophenolic compounds				
p-Cresol	5.5	0	4	
p-Coumaric acid	nil	0	0.05	0

[a]Macrae and Duggleby (1968); pH 7.0.
[b]Wong et al. (1971A). For isozyme A of clingstone peach at pH 6.8 and 30°C.
[c]Robb et al. (1966).
[d]Rivas and Whitaker (1973); 35°C and pH 6.2; isozyme B.

catechol. Potato, peach, broad bean leaf, and pear polyphenol oxidases have 140, 22.2, 8, and 71.8%, respectively, the activity on chlorogenic acid as on catechol.

Kinetics of Monophenol and *o*-Diphenol Oxidations

As discussed above, at least some polyphenol oxidases can catalyze both the *o*-hydroxylation of monophenols (Eq. 1) and the further oxidation of *o*-diphenols to *o*-benzoquinones (Eq. 2).

Monophenol oxidation. Most investigators of monophenol oxidation have measured the rate of formation of *o*-benzoquinone, with the assumption that the rate-determining step is the *o*-hydroxylation of the monophenol and that the *o*-diphenol formed is immediately further oxidized to the *o*-benzoquinone. The data of Table 9.7 indicate that this assumption is approximately correct for *p*-cresol/homocatechol, where the rate of hydroxylation of the monophenol is quite slow (Table 9.5) and (k_{cat}/K_m)diphenol/(k_{cat}/K_m)monophenol is 25.7. But for the other four *o*-diphenol/monophenol pairs, the ratios range from 7.81 to 1.77. Therefore, if the results for *S. glaucescens* polyphenol oxidase are indicative of results for other polyphenol oxidases, we must accept the published reports on the kinetics for monophenol oxidation cautiously.

The kinetic results obtained solely by polarographic methods (O_2 uptake) must be viewed with a great deal of caution, as O_2 uptake by both monophenol hydroxylation and *o*-diphenol oxidation are measured (compounded by further uptake of O_2 in oxidation of *o*-benzoquinone to melanin; Table 9.9). The only kinetic method that can give correct results is to measure the rate of tritium

Table 9.9. 4-Methyl-2,3-benzoquinone Formation and 4-Methylcatechol Disappearance in the Polyphenol Oxidase-Catalyzed Reaction[a]

Time (sec)	(1)[b] Quinone Formed (μmol)	(2)[c] O_2 Consumption (μmol)	Ratio[d] (2)/(1)
18	0.256	0.261	1.02
54	0.512	0.653	1.28
93.6	0.768	1.133	1.48
198	1.024	1.568	1.53
480	1.408	2.464	1.75

[a]Adapted from Mayer et al. (1966).

[b]Measured spectrophotometrically.

[c]Measured polarographically.

[d]*o*-Benzoquinone formation from 4-methylcatechol consumes 1 gram-atom of O_2 per mole of catechol, whereas conversion of 4-methylcatechol to melanin consumes 2.4 gram-atoms O_2 per mole of catechol oxidized.

removal from the *o*-position of the monophenol, as it is hydroxylated. Good kinetic studies have not been done yet using this method.

As shown by Eq. 1, monophenol hydroxylation involves three reactants, the monophenol, O_2, and BH_2. The concentrations of all three affect the observed initial velocity, which is not constant, but increases to a maximum when sufficient BH_2 is produced to saturate the enzyme (Fig. 9.6). As shown in Table 9.5, the lag period (Fig. 9.6) is a function of the nature of the monophenol, ranging from 0.32 min for *N*-acetyl-L-tyrosine hydrazide to 16.0 min for *p*-hydroxyphenylacetic acid. The lag period is also a function of the monophenol, enzyme, and O_2 concentrations, as well as pH and temperature. The velocity of the reaction is measured tangent to the activity curve *after* maximum velocity is reached (Fig. 9.6). By adding various amounts of BH_2 to the reaction initially, it is possible to determine approximately the K_d value for BH_2. With 4-methylcatechol, the value is about 0.1 μM for mushroom polyphenol oxidase (J. R. Whitaker, unpublished data).

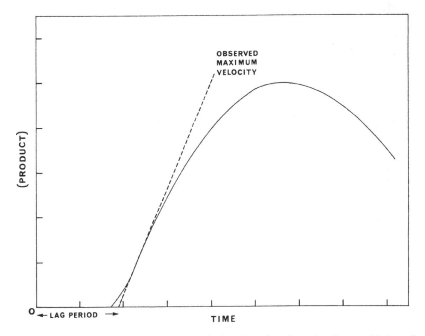

Figure 9.6. Theoretical graph for the velocity of product formation from oxidation of monophenols by polyphenol oxidases in the absence of BH_2. There is an initial lag period followed by maximum velocity and then slowing of the velocity as the enzyme undergoes reaction inactivation. The negative velocity phase is due to melanin formation, which absorbs at a different wavelength. See text for more detailed explanation.

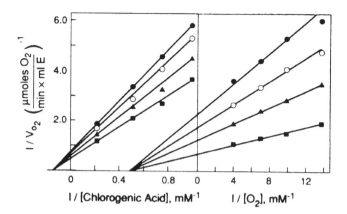

Figure 9.7.   Effect of O_2 and chlorogenic acid concentrations on initial velocity, v_0, of pear polyphenol oxidase-B-catalyzed reactions. The reactions were followed with an oxygen electrode at pH 4.0 and 30.0°C, permitting initial v_0 to be determined. Rivas and Whitaker. 1973. Plant Physiol. 52, p. 505. With permission. Copyright 1973 American Society of Plant Physiologists.

If sufficient BH_2 is added initially to saturate the enzyme, then the monophenol hydroxylation rate is kinetically a two-substrate reaction, depending on both the concentration of monophenol (MP) and O_2. Only when [MP] *and* [O_2] $\geq 100\ K_m$ for each substrate will the true values for $K_m(O_2)$ and $K_m(MP)$ and k_{cat} ($= V_{max}/[E]_0$) be obtained (see Fig. 9.7). The results reported in Table 9.5 were done by varying the monophenol concentration but keeping [O_2] at 0.24 mM (solubility of O_2 at the temperature used). Rivas and Whitaker (1973) reported that $K_m(O_2)$ for pear polyphenol oxidase was 0.11 mM; at 0.24 mM [O_2] the enzyme would be 69% saturated with O_2. Therefore, the k_{cat} and $K_m(MP)$ values for *S. glaucescens* polyphenyl oxidase (Table 9.5) are likely to be in error by a factor $K_m(O_2)_{satd}/K_m(O_2)_{0.24}$ mM

o-Diphenol oxidation.   Kinetically, the rate of oxidation of *o*-diphenols to *o*-benzoquinones is a second-order reaction depending on the concentrations of both *o*-diphenol and O_2 (Lerch and Ettlinger 1972; Rivas and Whitaker 1973) (Eq. 2). No BH_2 needs to be added because the *o*-diphenol meets the need (if any). It has been shown that both *S. glaucescens* (Lerch and Ettlinger 1972) and pear (Rivas and Whitaker 1973) polyphenol oxidases follow an Ordered Sequential Bi Bi Mechanism (Figs. 9.7 and 9.8; Cleland 1963A–C). The order in which the two products are released from the enzyme is unknown. The initial velocities of the reactions can be followed spectrophotometrically at the λ_{max} for the respective *o*-benzoquinone or polarimetrically (O_2 uptake; one equivalent of oxygen/equivalent *o*-diphenol maximum). The spectrophotometric and polarographic

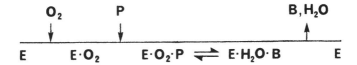

Figure 9.8. Kinetic mechanism for pear polyphenol oxidase-B catalysis, diagrammed according to the Cleland nomenclature (Cleland 1967A,B,C) for an Ordered Sequential Bi Bi Mechanism. P is an *o*-diphenol; B is an *o*-benzoquinone; $E \cdot O_2$ is a binary enzyme $\cdot O_2$ complex; $E \cdot O_2 \cdot P$ is a ternary enzyme $\cdot O_2 \cdot o$-diphenol complex, and $E \cdot H_2O \cdot B$ is a ternary enzyme $\cdot H_2O \cdot o$-benzoquinone complex.

methods give the same results (Table 9.9) when true initial velocities are determined (< 5% conversion of substrate to product).

Results of experiments showing the effects of chlorogenic acid and O_2 concentrations on the initial velocities are shown in Figs. 9.6 and 9.7. The *y*-intercepts on the slopes can be replotted to determine the true $K_m(O_2)$, $K_m(DP)$, and V_{max} ($= k_{cat}[E]_0$) values (Whitaker 1994).

MECHANISM OF ACTION

Role of Copper

Polyphenol oxidases are copper-containing proteins. Evidence is quite strong that the active site of *N. crassa* polyphenol oxidase contains two copper atoms, each of which is liganded to three histidine residues (Fig. 9.9; Lerch 1983; Huber et al. 1985). Figure 9.4 shows that eight other polyphenol oxidases have at least five conserved histidine residues in similar positions to that of *N. crassa* polyphenol oxidase. Therefore, any mechanism of action for polyphenol oxidases must include the role of the coppers.

Proposed overall mechanism. The proposed mechanism of action for *N. crassa* polyphenol oxidase (Lerch 1983; Solomon et al. 1992) appears to fit the data for most polyphenol oxidases (Fig. 9.10). The proposed mechanisms of hydroxylation (Eq. 1) and dehydrogenation (Eq. 2) reactions are presented separately, but linked by a common *oxy*polyphenol oxidase intermediate. Proposed intermediates in the *o*-diphenol oxidation pathway are shown in Fig. 9.10A. O_2

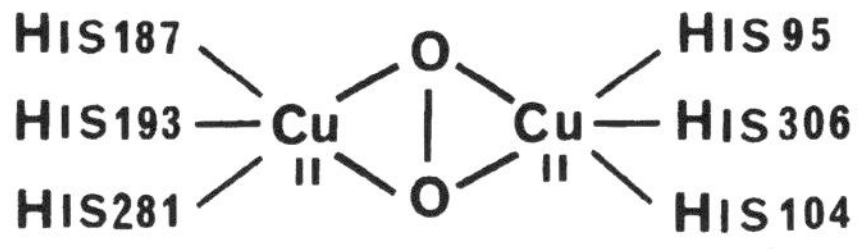

$$
\begin{array}{c}
\text{HIS}187 \diagdown \qquad \text{O} \qquad \diagup \text{HIS}95 \\
\text{HIS}193 - \text{Cu} \quad | \quad \text{Cu} - \text{HIS}306 \\
\text{HIS}281 \diagup \quad {}^{\text{II}} \quad \text{O} \quad {}^{\text{II}} \diagdown \text{HIS}104
\end{array}
$$

Figure 9.9. Coordination of copper to six histidine residues in the active site of *Neurospora crassa* polyphenol oxidase (Lerch 1983).

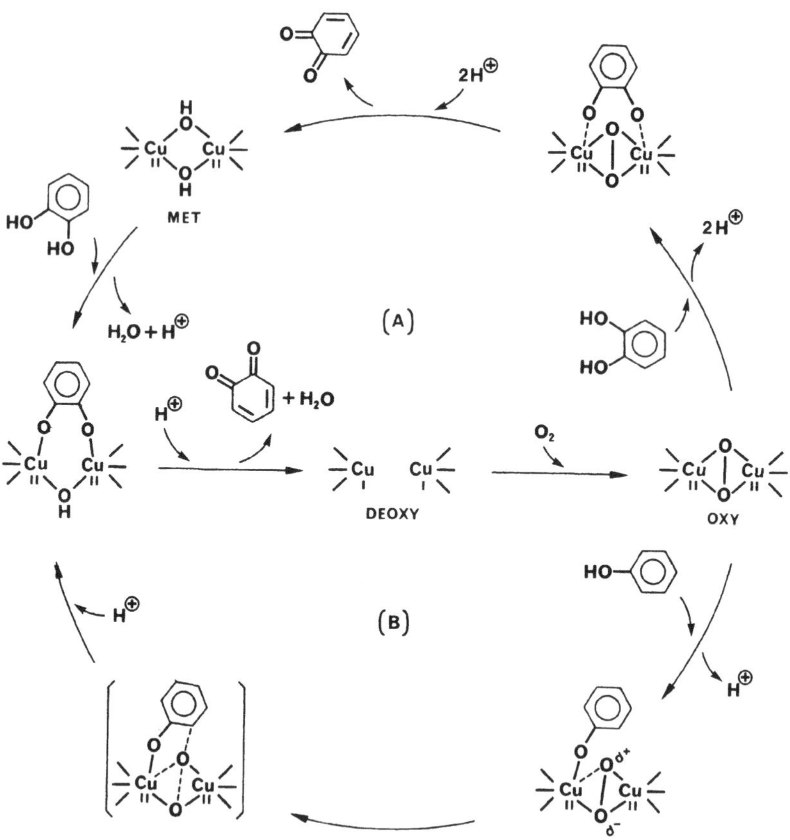

Figure 9.10. Proposed kinetic scheme depicting the mechanisms of oxidation of *o*-diphenol (catechol; top (A) and monophenol (phenol; bottom (B) for *Neurospora crassa* polyphenol oxidase. (Adapted from Lerch 1983 and Solomon et al. 1992.)

is bound first to the two Cu(I) groups of *deoxy*polyphenol oxidase to give *oxy*polyphenol oxidase in which the O_2 has the characteristics of a peroxide (Solomon et al. 1992). The two Cu(II)s groups of *oxy*polyphenol oxidase then bind the two hydroxyl groups of catechol, replacing the two water molecules or OH groups, to form the $O_2 \cdot$ catechol $\cdot$ enzyme complex. The catechol is oxidized to benzoquinone, leaving the enzyme as *met*polyphenol oxidase. Another molecule of catechol binds to *met*polyphenol oxidase, and is oxidized to benzoquinone, in the process reducing the enzyme to *deoxy*polyphenol oxidase, thereby completing the cycle. Two catechol molecules are oxidized in a complete cycle, consuming two atoms of O_2. However, the mechanisms of the two reactions appear to be different, with O_2 accounting for the oxidation of catechol in step 1 and the two Cu(II)s providing the oxidation of catechol in step 2. Further details and evidence for the proposed reaction are presented below.

The mechanism of monophenol hydroxylation by polyphenol oxidase is shown in Fig. 9.10B. In vitro, the reaction begins with *met*polyphenol oxidase (top of scheme in Fig. 9.10A). Although Wilcox et al. (1985) indicated that the "resting" form of *N. crassa* polyphenol oxidase is a mixture of >85% *met* and <15% *oxy*, it is the general experience that resting polyphenol oxidase has no activity on monophenols initially unless BH_2 is added (Eq. 1; see previous section in this chapter, Reactions Catalyzed). BH_2, which has a unique binding site (catechol with K_d of ~$1 \times 10^{-7} M$), converts the *met*polyphenol oxidase to the *deoxy* form (with oxidation of BH_2), which can then bind O_2 to give *oxy*polyphenol oxidase. The *oxy* form then binds the monophenol via one of the Cu(II)s, displacing the water molecule. In the subsequent step, the steric orientation of the bound monophenol and O_2 are altered to place the *o*-position of the monophenol adjacent to the second Cu(II) in *oxy*polyphenol oxidase, so that the *o*-position is hydroxylated by the bound -O-O- moiety. The initial product of the monophenol is catechol bound to both Cu(II)s. While still bound it is oxidized to *o*-benzoquinone, and the enzyme is reduced to the *deoxy* form and can again bind O_2 without cycling through the *met* form. It is interesting, and surprising, that the step(s) involving oxidation of the *o*-diphenol resembles neither of the two steps in oxidation of *o*-diphenols in Fig. 9.10A. Therefore, Fig. 9.10 provides three separate mechanisms of oxidation of *o*-diphenols by polyphenol oxidases!

Evidence for mechanism. Evidence for the mechanisms of oxidation of monophenols and *o*-diphenols by *N. crassa* and *Agaricus bisporus* polyphenol oxidases are based on spectral changes (Mason 1965; Lerch 1983), the effects of several substrate analogs (Wilcox *et al.* 1985), on the use of $^{18}O_2$ (Mason et al. 1955), essentiality of copper for activity of the enzyme (Kubowitz 1938; Keilin and Mann 1938), on raman spectroscopy (Eickman et al. 1978), and resemblance to hemocyanins (Himmelwright et al. 1980; Solomon et al. 1992; Solomon and Lowery 1993).

Evidence for the role of copper in catalysis of oxidation of monophenols and *o*-diphenols came early on (Kubowitz 1938; Keilin and Mann 1938). It was shown first that the addition of $CuSO_4$ to plant extracts containing polyphenol oxidase increased the observed activity. Later, it was shown that the copper can be removed in the presence of reducing compounds, including ascorbate, leading to complete inactivation of the enzyme, and that the activity can be restored by incubation of the inactive polyphenol oxidase with $CuSO_4$ (Dawson and Mallette 1945). Initially, researchers reported that *A. bisporus* and *N. crassa* polyphenol oxidases contained a mononuclear copper active site (Fling et al. 1963; Gutteridge and Robb 1975). However, based on determination of the primary structure of *N.*

crassa polyphenol oxidase, and resemblance of a sequence of amino acids to an amino acid sequence of the O_2-binding segment of hemocyanin, and from EPR studies, evidence accumulated that the active site must contain a binuclear copper complex. Subsequent studies have shown that the binuclear copper complex exists in three forms, the *met-*, the *oxy-* and the *deoxy-* (Eq. 3).

$$ \tag{3} $$

Evidence for *met*polyphenol oxidase came from lack of an EPR signal, on changes during reaction with H_2O_2, and from absorbance and circular dichroism properties attributable to the Cu(II) *d-d* transitions (Himmelwright et al. 1980) and from specific reaction of the *met* form with certain inhibitors, such as azide and mimosine (Wilcox et al. 1985), and the ability to form a mixed-valence oxidation state [Cu(I)-Cu(II)] (Wilcox et al. 1985).

Evidence for d*eoxy*polyphenol oxidase was obtained from lack of an EPR signal, by analogy to *deoxy*hemocyanin where the copper was shown to be Cu(I) (Brown et al. 1980), and by treatment of *met*polyphenol oxidase with sodium nitrite and ascorbic acid to give a mixed-valence oxidation state [Cu(I)-Cu(II)] that could then be substituted with different anions (Himmelwright et al. 1980).

*Oxy*polyphenol oxidase can be produced from *met*polyphenol oxidase in the presence of reducing compounds (ascorbic acid, hydroxylamine, dithionite, *o*-diphenols) and hydrogen peroxide in the presence of O_2 (Gutteridge and Robb 1975). The circular dichroism and absorbance spectra are very similar to those of *oxy*hemocyanin (Lerch 1983). Subsequent raman spectroscopic studies support evidence for the *oxy* form and its structure (Eickman et al. 1978). EXAFS (extended X-ray absorption fine structure) study of *N. crassa oxy*polyphenol oxidase showed that the Cu(II)-Cu(II) distances are 3.6 A, somewhat larger than for the Cu(II)-Cu(II) distance in the *met* form (Brown et al. 1980). Other evidence indicates that the oxygen-oxygen bond in the *oxy* form is longer than in O_2, and resembles the bond in hydrogen peroxide (Solomon et al. 1992). Investigations of inorganic and bioinorganic copper-dioxygen complexes have also added to structural knowledge of the active site of polyphenol oxidases (Kitajima and Moro-oka 1994).

Another intermediate reaction occurs during polyphenol oxidase-catalyzed oxidation of *o*-diphenols, and perhaps monophenols. In the oxidation of catechol to *o*-benzoquinone, there is a relatively slow irreversible inactivation of polyphenol oxidase (Nelson and Dawson 1944; Ingraham 1959; Brooks and Dawson 1966). It was initially concluded that inactivation was due to reaction of the *o*-benzoquinone with an essential ε-amino group of a lysyl residue of the enzyme (Wood and Ingraham 1965). Later, it was shown that inactivation was due to a free-radical-catalyzed fragmentation of one or more of the active site histidine residues, which explains the loss of histidine and release of copper (Nelson and Dawson 1944; Dietler and Lerch 1982; Golan-Goldhirsh and Whitaker 1984). Inactivation of *N. crassa* polyphenol oxidase is due to loss of His306 in the active site (Fig. 9.9; Dietler and Lerch 1982). Golan-Goldhirsh et al. (1992) subsequently showed that ascorbic acid and copper mimic this reaction not only with polyphenol oxidase, but also with ovalbumin, Kunitz trypsin inhibitor, and bovine serum albumin. Golan-Goldhirsh and Whitaker (1984) calculated that inactivation of mushroom polyphenol oxidase occurs at the rate of approximately one in 5,000 turnovers of the enzyme. Therefore, it appears that there is a free-radical intermediate produced in the enzyme-catalyzed reaction with *o*-diphenols (Eq. 4) and that the free radical inactivates the enzyme (reaction inactivation mechanism). It has been impossible until recently to detect the semiquinone radical directly, possibly because it is converted rapidly to the *o*-quinone (rate-determining step).

Spin resonance stabilization NMR techniques have shown that semiquinones are formed during the polyphenol oxidase-catalyzed reaction (Peter et al. 1985; Korytowski et al. 1987). Sugumaran et al. (cited in Sugumaran 1988) have observed free radicals in insect cuticle formation.

Products formed. The mechanisms for monophenol and *o*-diphenol oxidation by polyphenol oxidases indicate that *o*-benzoquinone is the final product of the enzyme (Fig. 9.10). In vivo and in vitro, further reactions occur because of high reactivity of the *o*-benzoquinone, because of further involvement of polyphenol oxidase, or because of involvement of other enzymes.

The biosynthetic pathways of pheomelanin and eumelanin synthesis from tyrosine in humans are shown in Fig. 9.11. Polyphenol oxidase acts at three points in the reaction: (1) the hydroxylation of tyrosine to dihydroxyphenylalanine (DOPA); (2) the oxidation of DOPA to DOPA quinone; and (3) the oxidation of

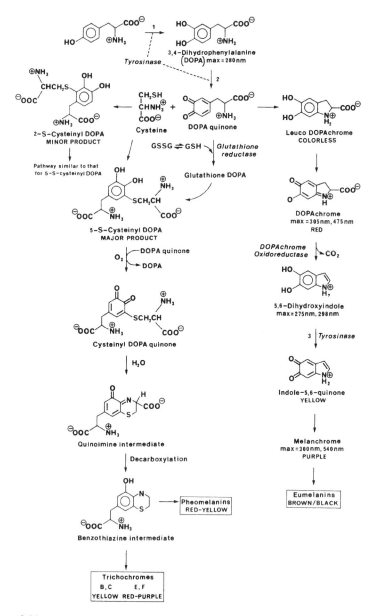

Figure 9.11. Proposed biochemical pathway of eumelanin and pheomelanin biosynthesis in humans from the substrate tyrosine by polyphenol oxidase. Polyphenol oxidase catalyzes the two reactions labelled 1 and the third reaction labelled 3. See text for further details. From Witkop. 1985. Clinics in Dermatology, vol. 2, p. 72. With permission. Copyright 1985 J. B. Lippincott Company.

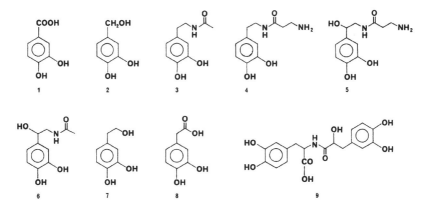

Figure 9.12. Structures of some catecholic sclerotizing precursors in insects (beetles with exoskeleton). The compounds are: (1), 3,4-dihydroxybenzoic acid; (2), 3,4-dihydroxybenzyl alcohol; (3), N-acetyldopamine; (4), N-β-alanyldopamine; (5), N-β-alanylnorepinephrine; (6), N-acetylnorepinephrine; (7), 3,4-diphenethyl alcohol; (8), 3,4-dihydroxyphenylacetic acid; and (9), N-(3,4-dihydroxyphenyllactyl)-dihydroxyphenylalanine. From Sugumaran. 1988. Adv. Insect Physiol. 21, p. 185. With permission. Copyright 1988 Academic Press Limited.

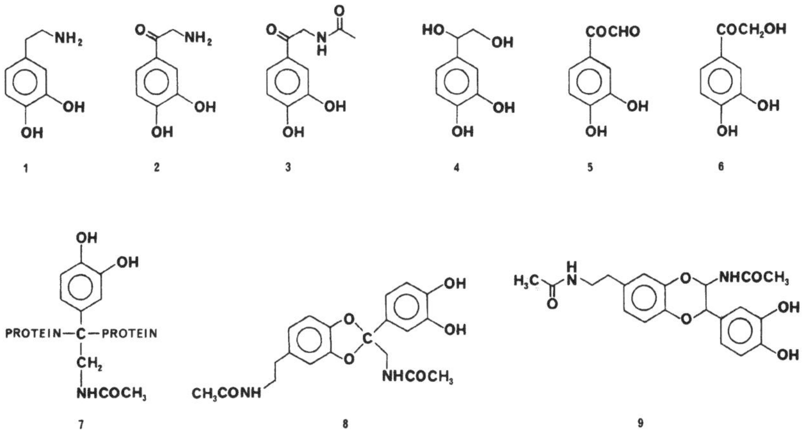

Figure 9.13. Structures of some catecholic products associated with β-sclerotization in insects (beetles with exoskeletons). They are: (1), dopamine; (2), 2-amino-3′,4′-dihydroxyacetophenone; (3), N-acetylarterenone; (4), 3,4-dihydroxyphenylglycol; (5), 3,4-dihydroxyphenylglyoxal; (6), 2-hydroxy-3′,4′-dihydroxyacetophenone; (7), crosslinked structure of N-acetyldihydroxyphenylalanine with proteins (via tyrosine OH?); (8), proposed structure of acid (1 M HCl) hydrolysis dimeric product from sclerotized cuticular tissue; and (9), established structure of acid hydrolysis dimeric product from sclerotized cuticular tissue. From Sugumaran. 1988. Adv. Insect. Physiol. 21, p. 206. With permission. Copyright 1988 Academic Press Limited.

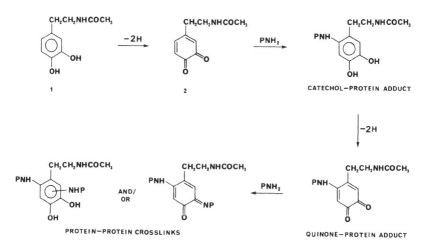

Figure 9.14. Quinone tanning hypothesis for schlerotization of insect exoskeleton. N-Acetyldopamine (1) (or other sclerotizing catechols, see Fig. 9.12) secreted by the epidermal cells is oxidized to the corresponding quinone (2) by the cuticular polyphenol oxidase. Compound 2 undergoes nonenzymatic Michael-1,4-addition reaction with cuticular proteins (P) forming catechol-protein adducts. These adducts, following further oxidation and coupling to another protein, yield protein-protein crosslinks. From Sugumaran. 1988. Adv. Insect Physiol. 21, p. 191. With permission. Copyright 1988 Academic Press Limited.

5,6-dihydroxyindole to indole-5,6-quinone, provided DOPAchrome oxidoreductase is present to convert DOPAchrome to 5,6-dihydroxyindole. In this pathway, the final product is eumelanin. If DOPAchrome oxidoreductase is absent, pheomelanin is formed via a separate pathway, in which either cysteine or reduced glutathione is required.

Polyphenol oxidase is essential for the tanning (sclerotization) of insect cuticles (Sugumaran 1988). Much research has been done on the mechanism of sclerotization in insects, expecially by Sugumaran and his associates. The structures of some of the catecholic sclerotizing precursors and catecholic products are shown in Figs. 9.12 and 9.13, respectively, and a quinone tanning hypothesis is proposed in Fig. 9.14. Reaction of the intermediate quinone methide with cuticular components (including side chains of proteins) is shown in Fig. 9.15. A remarkably large number of reactions occur and polyphenol oxidase appears to be involved in more than just the initial oxidation of the *o*-diphenol to quinone methide or quinone (Sugumaran 1988).

In only a few cases has the chemistry of the reactions catalyzed by polyphenol oxidase of higher plants been elucidated beyond the *o*-benzoquinone step, primarily because many of the products are insoluble. In some cases, subsequent

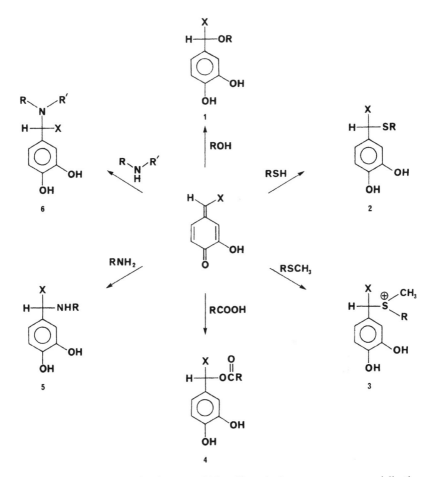

Figure 9.15. Reaction of quinone methide with cuticular components, especially the protein side chains of serine and threonine (1), cysteine (2), methionine (3), carboxyl groups (4), amino groups (5), and histidine and N-terminal proline (6). From Sugumaran. 1988. Adv. Insect Physiol. 21, p. 216. With permission. Copyright 1988 Academic Press Limited.

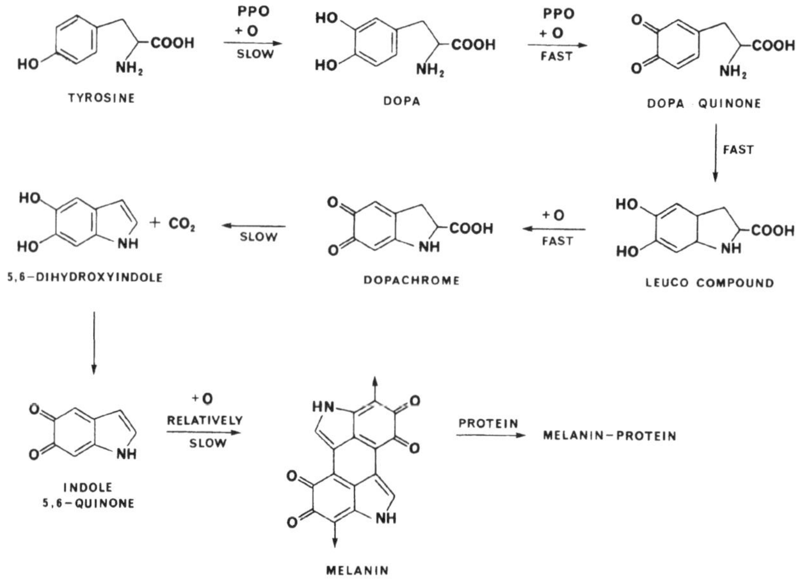

Figure 9.16. Oxidation of tyrosine by plant polyphenol oxidase. The polyphenol oxidase catalyzes the oxidation of tyrosine to DOPA and to DOPA quinone. The remainder of the reactions leading to melanin formation and its reaction with proteins are nonenzymatic. From Lerner. 1953. Adv. Enzymol. 14, p. 76. With permission. Copyright 1953 John Wiley and Sons, Inc.

reactions, thought to be nonenzymatic, lead to further active products. The polyphenol oxidase-catalyzed oxidation of tyrosine (Fig. 9.16; Lerner 1953) appears to follow the same pathway as shown for eumelanin formation in humans (Fig. 9.11). The products through indole-5,6-quinone (Fig. 9.16) are soluble in in vitro systems and have been studied.

o-Benzoquinones react with other phenolic compounds by Michael addition to the 4-position of the o-benzoquinones, often to give intensely colored products that range from yellow, red, blue, green, and black (Wong et al. 1971B). These types of products are very important in the discoloration of fruits and vegetables and lead to modification of the ε-amino group of lysyl residues of proteins (Matheis and Whitaker 1984). The proposed reactions (Matheis and Whitaker 1984) are very similar to those found in the sclerotization of insect cuticles (Fig. 9.15). Those interested in the chemistry of polyphenol oxidase-catalyzed browning in fruits and vegetables must extend our knowledge of the reactions to the level achieved for human and insect polyphenol oxidase-catalyzed reactions.

REFERENCES

BERNAN, V.; FILPULA, D.; HERBER, W.; BIBB, M.; and KATZ, E. 1985.The nucleotide sequence of the tyrosinase gene of *Streptomyces antibioticus* and characterization of the gene product. *Gene 37,* 101–110.

BOUCHILLOUX, S.; MCMAHILL, P.; and MASON, H. S. 1963. The multiple forms of mushroom tyrosinase. Purification and molecular properties of the enzymes. *J. Biol. Chem. 238,* 1699–1707.

BROOKS, D. W., and DAWSON, C. R. 1966. Aspects of tyrosinase chemistry. In: *The Biochemistry of Copper,* J. Peisach, P. Aisen, and W. E. Blumberg, eds., Academic Press, New York, pp. 343–357.

BROWN, J. M.; POWERS, L.; KINCAID, B.; LARRABEE, J. A.; and SPIRO, T. G. 1980. Structural studies of the hemocyanin active site. 1. Extended X-ray absorption fine structure (EXAFS) analyses. *J. Am. Chem. Soc. 102,* 4210–4216.

CARY, J. W.; LAX, A. R.; and FLURKEY, W. H. 1992. Cloning and characterization of cDNAs coding for *Vicia faba* polyphenol oxidase. *Plant Mol. Biol. 20,* 245–253.

CLELAND, W. W. 1963A. The kinetics of enzyme-catalyzed reactions with two or more substrates or products. I. Nomenclature and rate equations. *Biochim. Biophys. Acta 67,* 104–137.

———. 1963B. The kinetics of enzyme-catalyzed reactions with two or more substrates or products. II. Inhibition: Nomenclature and theory. *Biochim. Biophys. Acta 67,* 173–187.

———. 1963C. The kinetics of enzyme-catalyzed reactions with two or more substrates or products. III. Prediction of initial velocity and inhibition patterns by inspection. *Biochim. Biophys. Acta 67,* 188–196.

DAWSON, C. R., and MALLETTE, M. F. 1945. Copper proteins. *Adv. Prot. Chem. 2,* 179–248.

DIETLER, C., and LERCH, K. 1982. Reaction inactivation of tyrosinase. In: *Oxidases and Related Redox Systems,* T. E. King, H. S. Mason, and M. Morrison, eds., Pergamon Press, New York, pp. 305–317.

EICKMAN, N. C.; SOLOMON, E. I.; LARRABEE, J. A.; SPIRO, T. G.; and LERCH, K. 1978. Ultraviolet resonance Raman study of oxytyrosinase. Comparison with oxy-hemocyanins. *J. Am. Chem. Soc. 100,* 6529–6531.

FLING, M.; HOROWITZ, N. H.; and HEINEMANN, S. F. 1963. The isolation and properties of crystalline tyrosinase from *Neurospora. J. Biol. Chem. 238,* 2045–2053.

FLURKEY, W. H. 1994. Isolation and properties of broad bean polyphenol oxidase. Abstract 91 (AGFD), 208th Am. Chem. Soc. National Meeting, Washington, D.C., August 21–25.

GAYKEMA, W. P. J.; HOL, W. G. J.; VEREIJKEN, J. M.; SOETER, N. M.; BAK, H. J.; and BEINTEMA, J. J. 1984. 3.2 A Structure of the copper-containing, oxygen-carrying protein *Panulirus interruptus* haemocyanin. *Nature 309,* 23–29.

GOLAN-GOLDHIRSH, A.; OSUGA, D. T.; CHEN, A. O.; and WHITAKER, J. R. 1992. Effect of ascorbic acid and copper on proteins and other polymers. In: *The Bioorganic*

Chemistry of Enzymatic Catalysis: An Homage to Myron L. Bender, V. T. D'Souza and J. Feder, eds., CRC Press, Boca Raton, pp. 61–76.

GOLAN-GOLDHIRSH, A., and WHITAKER, J. R. 1984. k_{cat} inactivation of mushroom polyphenol oxidase. *J. Mol. Catal. 32,* 141–147.

GUTTERIDGE, S., and ROBB, D. 1975. Catecholase activity of *Neurospora* tyrosinase. *Eur. J. Biochem. 54,* 107–116.

HIMMELWRIGHT, R. S.; EICKMAN, N. C.; LUBIEN, C. D.; LERCH, K.; and SOLOMON, E. I. 1980. Chemical and spectral studies of the binuclear copper active site of *Neurospora* tyrosinase: Comparison to hemocyanin. *J. Am. Chem. Soc. 102,* 7339–7344.

HUBER, M.; HINTERMANN, G.; and LERCH, K. 1985. Primary structure of tyrosinase from *Streptomyces glaucescens. Biochemistry 24,* 6038–6044.

HUNT, M. D.; EANNETTA, N. T.; YU, H.; NEWMAN, S. M.; and STEFFENS, J. C. 1993. cDNA cloning and expression of potato polyphenol oxidase. *Plant Mol. Biol. 21,* 59–68.

INGRAHAM, L. L. 1959. Polyphenol oxidase at low pH values. In: *Pigment Cell Biology,* M. Gordon, ed., Academic Press, New York, pp. 609–617.

JOLLEY, R. L., JR.; ROBB, D. A.; and MASON, H. S. 1969A. The multiple forms of mushroom tyrosinase. Association-dissociation phenomena. *J. Biol. Chem. 244,* 1593–1599.

JOLLEY, R. L., JR.; NELSON, R. M.; and ROBB, D. A. 1969B. The multiple forms of mushroom tyrosinase. *J. Biol. Chem. 244,* 3251–3257.

KEILIN, D., and MANN, T. 1938. Polyphenol oxidase: Purification, nature and properties. *Proc. Royal Soc. Ser B 125,* 187–204.

KITAJIMA, N., and MORO-OKA, Y. 1994. Copper-dioxygen complexes. Inorganic and bioinorganic perspectives. *Chem. Rev. 94,* 737–757.

KORYTOWSKI, W.; SARNA, T.; KALYANARAMAN, B.; and SEALEY, R. C. 1987. Tyrosinase-catalyzed oxidation of dopa and related catechol (amines): A kinetic electron spin resonance investigation using spin-stabilization and spin label oximetry. *Biochim. Biophys. Acta 924,* 383–392.

KUBOWITZ, F. 1938. Cleavage and resynthesis of polyphenoloxidase and of hemocyanin. *Biochem. Z. 299,* 32–57.

KWON, B. S.; HAQ, A. K.; POMERANTZ, S. H.; and HALABAN, R. 1987. Isolation and sequence of a cDNA clone for human tyrosinase that maps at the mouse *c*-albino locus. *Proc. Natl. Acad. Sci. USA 84,* 7473–7477.

LERCH, K. 1978. Amino acid sequence of tyrosinase from *Neurospora crassa. Proc. Natl. Acad. Sci. USA 75,* 3635–3639.

———. 1983. *Neurospora* tyrosinase: Structural, spectroscopic and catalytic activity. *Mol. Cell. Biochem. 52,* 125–138.

LERCH, K., and ETTLINGER, L. 1972. Purification and characterization of a tyrosinase from *Streptomyces glaucescens. Eur. J. Biochem. 31,* 427–437.

LERNER, A. B. 1953. Metabolism of phenylalanine and tyrosine. *Adv. Enzymol. 14,* 73–128.

MACRAE, A. R., and DUGGLEBY, R. G. 1968. Substrates and inhibitors of potato tuber phenolase. *Phytochemistry 7,* 855–861.

MASON, H. S. 1965. Oxidases. *Ann. Rev. Biochem. 34,* 595–634.

MASON, H. S.; FOWLKS, W. B.; and PETERSON, E. W. 1955. Oxygen transfer and electron transport by the phenolase complex. *J. Am. Chem. Soc. 77,* 2914–2915.

MATHEIS, G., and WHITAKER, J. R. 1984. Modification of proteins by polyphenol oxidase and peroxidase and their products. *J. Food Biochem. 8,* 137–162.

MAYER, A. M.; HAREL, E.; and BEN-SHAUL, R. 1966. Assay of cathechol oxidase. A critical comparison of methods. *Phytochemistry 5,* 783–789.

NELSON, J. M., and DAWSON, C. R. 1944. Tyrosinase. *Adv. Enzymol. 4,* 99–152.

OLAH, A. F., and MUELLER, W. C. 1981. Ultrastructural localization of oxidative and peroxidative activities in a carrot suspension cell culture. *Protoplasmia 106,* 231–248.

OSUGA, D; VAN DER SCHAAF, A.; and WHITAKER, J. R. 1994. Control of polyphenol oxidase activity using a catalytic mechanism. In: *Protein Structure-Function Relationships in Foods,* R. Y. Yada, R. L. Jackman, and J. L. Smith, eds., Blackie Academic & Professional, Glasgow, Scotland, pp. 62–88.

PETER, M. G.; STEGMANN, H. B.; DAO-BA, H.; and SCHEFFLER, K. 1985. Detection of semiquinone radicals of *N*-acyldopamines in aqueous solution. *Z. Naturforsch. 40c,* 535–538.

QUEVEDO, W. C., JR. 1971. Genetic regulation of pigmentation in mammals. In: *Bio logy of Normal and Abnormal Melanocytes,* T. Kawamura and T. B. Fitzpatrick, eds., University Park Press, Baltimore, MD, pp. 99–115.

RIVAS, N. J., and WHITAKER, J. R. 1973. Purification and some properties of two polyphenol oxidases from Bartlett pears. *Plant Physiol. 52,* 501–507.

ROBB, D. A.; SWAIN, T.; and MAPSON, L. W. 1966. Substrates and inhibitors of the activated tyrosinase of broad bean (*Vicia faba L.*). *Photochemistry 5,* 665–675.

ROBINSON, S. P., and DRY, I. B. 1992. Broad bean leaf polyphenol oxidase is a 60-kilodalton protein susceptible to proteolytic cleavage. *Plant Physiol. 99,* 317–323.

RODRIQUEZ, M. O., and FLURKEY, W. H. 1992. A biochemistry project to study mushroom tyrosinase: Enzyme localization, isoenzymes, and detergent activation. *J. Chem. Educ. 69,* 767–769.

SAUL, S. J., and SUGUMARAN, M. 1987. Protease mediated prophenoloxidase activation in the hemolymph of the tobacco hornworm, *Manduca sexta. Arch. Insect Biochem. Physiol. 5,* 1–11.

———. 1988. Prophenoloxidase activation in the hemolymph of *Sarcophaga bullata* larvae. *Arch. Insect Biochem. Physiol. 7,* 91–103.

SCHARTAU, W.; EYERLE, F.; REISINGER, P.; GEISERT, H.; STORZ, H.; and LINZEN, B. 1983. Hemocyanin in spiders. XIX. Complete amino acid sequence of subunit *d* from *Eurypelma californicum* hemocyanin, and comparison to chain *e. Hoppe-Seyler's Z. Physiol. Chem. 364,* 1383–1409.

SCHNEIDER, H. J.; DREXEL, R.; FELDMAIER, G.; and LINZEN, B. 1983. Hemocyanins in spiders. XVIII. Complete amino-acid sequence of subunit *e* from *Eurypelma californicum* hemocyanin. *Hoppe-Seyler's Z. Physiol. Chem. 364,* 1357–1381.

SCHOENBEIN, C. F. 1856. On ozone and oronic actions in mushrooms. *Phil. Mag. 11,* 137–141.

SHAHAR, T.; HENNIG, N.; GUTFINGER, T.; HAREVEN, D.; and LIFSCHITZ, E. 1992. The tomato 66.3-kD polyphenoloxidase gene: Molecular identification and developmental expression. *The Plant Cell 4,* 135–147.

SHERMAN, T. O.; VAUGHN, K. C.; and DUKE, S. O. 1991. A limited survey of the phylogenetic distribution of polyphenol oxidase. *Phytochemistry 30,* 2499–2506.

SHIBAHARA, A.; TOMITA, Y.; SAKAKURA, T.; NAGER, C.; CHAUDHURI, B.; and Müller, R. 1986. Cloning and expression of cDNA encoding mouse tyrosinase. *Nucl. Acids Res. 14,* 2413–2427.

SILVERS, W. K. 1979. *The Coat Colors of Mice: A Model for Mammalian Gene Action and Interaction.* Springer-Verlag, New York.

SOLOMON, E. I.; BALDWIN, M. J.; and LOWREY, M. D. 1992. Electronic structures of active sites in copper proteins: Contributions to reactivity. *Chem. Rev. 92,* 521–542.

SOLOMON, E. I., and LOWERY, M. D. 1993. Electronic structure contributions to function in bioinorganic chemistry. *Science 259,* 1575–1581.

SUGUMARAN, M. 1988. Molecular mechanisms for cuticular sclerotization. *Adv. Insect Physiol. 21(9),* 179–231.

———. 1990. Prophenoloxidase activation and insect immunity. In: *Defense Molecules,* J. J. Marchalonis and E. L. Reinisch, eds., Wiley-Liss, New York, pp. 47–62.

———. 1994. Regulation of phenoloxidase activity in insects. Abstract 93 (AGFD), 208th Am. Chem. Soc. Nat. Meeting, Washington, D.C., August 21–25.

SUGUMARAN, M.; HENNIGAN, B.; and O'BRIEN, J. 1987. Tyrosinase catalyzed protein polymerization as an in vitro model for quinone tanning of insect cuticle. *Arch. Insect Biochem. Physiol. 6,* 9–25.

SZENT-GYÖRGYI, A., and VIETORISZ, K. 1931. Function and significance of polyphenol oxidase from potatoes. *Biochem. Z. 233,* 236–239.

VAUGHN, K. C. 1987. Polyphenol oxidase. In: *Handbook of Plant Cytochemistry,* vol. 1, K. C.Vaughn, ed., CRC Press, Boca Raton, pp. 159–162.

VAUGHN, K. C., and DUKE, S. O. 1981. Tissue localization of polyphenol oxidase in sorghum. *Protoplasmia 108,* 319–327.

WHITAKER, J. R. 1994. *Principles of Enzymology for the Food Sciences,* Marcel Dekker, New York, pp. 184–192.

WILCOX, D. E.; PORRAS, A. G.; HWANG, Y. T.; LERCH, K.; WINKLER, M. E.; and SOLOMON, E. I. 1985. Substrate analogue binding to the coupled binuclear copper active site in tyrosinase. *J. Am. Chem. Soc. 107,* 4015–4027.

WITKOP, C. J., JR. 1979. Depigmentations of the general and oral tissues and their genetic foundations. *Ala. J. Med. Sci. 16,* 331–343.

———. 1985. Inherited disorders of pigmentation. In: *Genodermatoses: Clinics in Dermatology,* vol. 2, R. M. Goodman, ed., J. M. Lippincott, Philadelphia, pp. 70–134.

WONG, T. C.; LUH, B. S.; and WHITAKER, J. R. 1971A. Isolation and characterization of polyphenol oxidases of clingstone peach. *Plant Physiol. 48,* 19–23.

———. 1971B. Effect of phloroglucinol and resorcinol on the clingstone peach polyphenol oxidase-catalyzed oxidation of 4-methyl catechol. *Plant Physiol. 48,* 24–30.

WOOD, B. J. B., and INGRAHAM, L. L. 1965. Labelled tyrosinase from labelled substrate. *Nature 205,* 291–292.

Chapter 10

Glucose Oxidase

Glucose oxidase (β-D-glucose: oxygen 1-oxidoreductase, EC 1.1.3.4) catalyzes the oxidation of β-D-glucose to δ-D-glucono-1,5-lactone coupled with the reduction of O_2 to H_2O_2 (Fig. 10.1). The enzyme is found in a number of fungal sources, including *Aspergillus oryzae*, *Penicillium notatum*, *Penicillium glaucum*, *Phanerochaete chrysosporium*, and *Talaromyces flavins*, but most studies have been done on the enzyme purified from *Aspergillus niger* and *Penicillium amagasakiense*.

GENERAL CHARACTERISTICS

The enzymes from different fungi share certain similarities in that all are dimers composed of two identical subunits, have 2 moles of firmly bound FAD per mole of protein, and contain 11–20% carbohydrates. Glucose oxidase from *Aspergillus niger* is a dimer of MW of 160 kD with a pH optimum of 5.5 and a pI of 4.2. Molecular weights of 150 kD (Pazur and Kleppe 1964; Nakamura and Fujiki 1968), and 186 kD (Swoboda and Massey 1965) have also been reported. The *Penicillium* enzyme shares many similarities with only minor variations (Nakamura and Fujiki 1968). The enzyme cross-reacts with antiserum against the *Aspergillus* enzyme (Hayashi and Nakamura 1976). The activation energy is nearly the same for the two enzymes (24 kcal/mol), although ΔH and ΔS are different (ΔH = 85 and 119 kcal/mol, and ΔS = 173 and 282 kcal/mol for the *Aspergillus* and *Penicillium* enzyme, respectively).

Each enzyme subunit contains one disulfide bond and one free sulfhydryl. The two subunits are noncovalently associated; only the dimer form is active (Tsuge et al. 1975; Jones et al. 1982; Ye and Combes 1989). ^{31}P-NMR shows a covalent phosphorus residue located near the surface of the enzyme molecule. The disubstituted phosphorus forms a phosphodiester linkage between two amino acid residues similar to disulfide links stabilizing proteins (James et al. 1981).

FLAVIN COFACTOR

The FAD cofactor in glucose oxidase has its pyrophosphates buried in the protein, and is tightly bound to the enzyme with a dissociation constant $K_d > 10^{-10}$ (James et al. 1981). Similar to many flavoproteins, the protein-bound flavin interacts with sulfite (SO_3^-/HSO_3^-) to form a flavin-sulfite adduct (Swoboda and Massey 1966) that is catalytically inactive. The sulfite binds at the N5 position of the flavin molecule in a reversible equilibrium with a $K_d = 2.3 \times 10^{-4}$ M (Muller and Massey 1969). The recombinant enzyme secreted by a transformed *Saccharomyces cerevisiae* gives a K_d value of $1.1 \times 10^{-4} M$ (Frederick et al. 1990), slightly lower than that of the wild type glucose oxidase. In the oxidized enzyme, the flavin isoalloxazine ring is not planar, with hydrogen bonds at positions N3-H, N5-H, and C2-O (Sanner et al. 1991). In the reduced state, the N1 is anionic, possibly stabilized by a positively charged amino acid side chain (Fig. 10.2). Strong hydrogen bonding occurs at positions N3-H, C2-O, and C4-O.

CARBOHYDRATE MOIETY

The total carbohydrate content of glucose oxidase has been reported in a range of 11–20% (w/w) (Swoboda and Massey 1965; Nakamura and Fujiki 1968; Tsuge et al. 1975; Hayashi and Nakamura 1976) with mannose as the most abundant (~80%) component (Hayashi and Nakamura 1981). The carbohydrate moiety

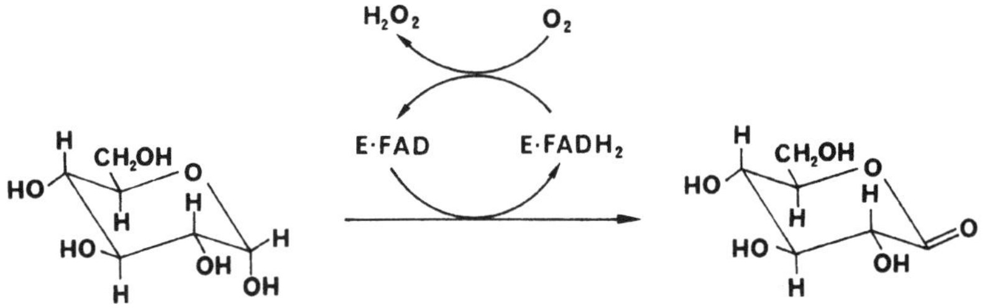

Figure 10.1. Conversion of β-D-glucose to β-δ-gluconolactone by glucose oxidase.

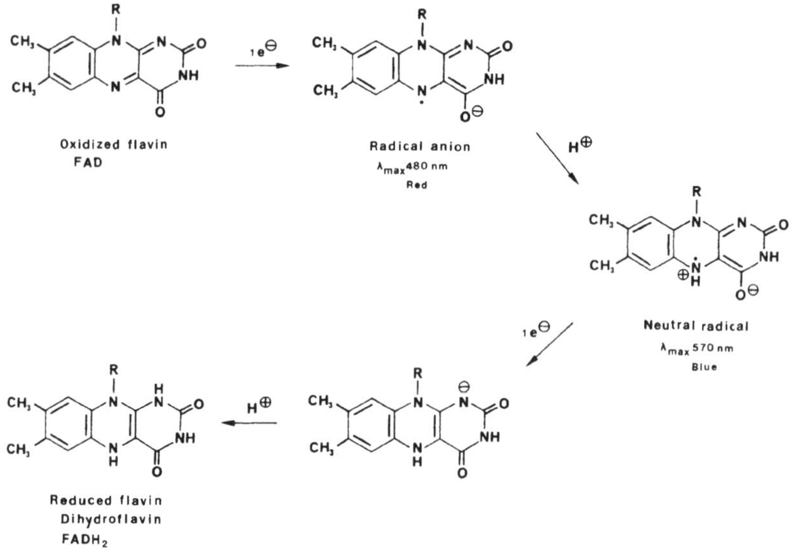

Figure 10.2.   The oxidized and the reduced forms of flavin.

contributes no effect to the thermal stability of the protein. Periodate-treated enzyme with 60% of its original carbohydrate deleted shows thermal inactivation parameters similar to those of the native enzyme (Nakamura et al. 1976). In the presence of denaturing agents, however, the treated enzyme shows a decrease in thermal stability. Similar studies using endo-β-*N*-acetylglucosaminidase to cleave specifically the N-linked carbohydrate chains on the *Aspergillus niger* enzyme also confirm that the effect of deglycosylation (70% deleted) is minimum on catalytic activity, and thermal and pH stabilities (Takegawa et al. 1989). Because both O- and N-glycosylation exist in the molecule, the possible effect of threonyl and seryl carbohydrate chains also needs to be considered. Glucose oxidase from *Aspergillus niger* deglycosylated by 95% with endoglucosidase H and α-mannosidase shows no significant changes in the circular dichroism spectra, and other physical properties such as thermal, stability, pH, and temperature optima (Kalisz et al. 1991). Studies on unglycosylated enzyme produced by *Penicillium amagasakiense* in the presence of tunicamycin (an antibiotic that inhibits N-glycosylation) also indicate that the carbohydrate moiety plays no significant role in the enzyme function (Kim and Schmid 1991). The primary structure of N-linked sugar chains consists of: ± Manα1->3Manα1->3(Manα1->6)Manα1->6(± Manα1->3Manα1->3)Manβ1->4GlcNAcβ1->4GlcNAc (Takegawa et al. 1991). O-Glycosylation consists of mainly single mannose residues.

AMINO ACID SEQUENCE

The amino acid sequence of the *Aspergillus niger* enzyme is known as deduced from the encoding cDNA gene and the genomic gene (Fig. 10.3) (Kriechbaum et al. 1989; Frederick et al. 1990). The enzyme subunit consists of 583 amino acids, with three Cys residues and eight potential N-linked glycosylation sites with a typical -Asn-X-Thr/Ser- sequence. The amino acid sequence is 26% identical with that of alcohol oxidase with four regions of high homology. The N-terminal segment (residues 20–50) is also homologous with a number of other flavoenzymes—glutathione reductase, *p*-hydroxybenzoate hydroxylase, and lipoamide dehydrogenase. A short region (residues 101–126) of the enzyme are found to be similar with sequences in α-amylase, glucoamylase, and glucose dehydrogenase (Sierks et al. 1992). Another segment at the C-terminal end exhibits sequence similarity with the active site regions of alcohol oxidases and three of the disulfide oxidoreductases—glutathione reductase, lipoamide dehydrogenase, and mercuric reductase (Frederick et al. 1990). The secondary structure of the N-terminal homologous segment is indicative of a β-α-β motif that may be involved in flavin binding. Crystallization and preliminary x-ray diffraction studies of deglycosylated glucose oxidase from *Aspergillus niger* (Kalisz et al. 1990) and *Penicillium amagasakiense* (Hendle et al. 1992) have been reported. The three-dimensional structure is known (Hecht et al. 1993).

The enzyme monomer can be divided into two domains between which the flavin site is located inside a deep pocket (Fig. 10.4). The domain that binds FAD consists of a 5-stranded β sheet (A1–A5) layered between a 3-stranded β sheet (B1–B3) and 3 α helices (H1, H6, H13). Strands A3, A2, and helix H1 form the βαβ motif for FAD binding. The second domains consists of a large 6-stranded antiparallel β sheet (C1–C6) covered by three helices (H8, H9, H11) on the outer surface, and another three helices (H5, H10, H12) on both sides. The β sheet (C1–C6) forms one side of the active site pocket containing the flavin. Two β-turn-β motifs (residues 75–98, 432–455) located at one end of the pocket provide most of the contacts of the dimer interface. In the dimer the FAD binding channel is blocked by the contact segment. Conformational change of this "lid" is required before FAD binding. Extensive hydrogen bonding exists in the flavin-enzyme interaction, involving primarily the ribose and pyrophosphate groups. At least five of the potential N-glycosylation sites are known to be Asn89, Asn161, Asn168, Asn355, and Asn388.

ENZYME SPECIFICITY

In addition to β-D-glucose, glucose oxidase catalyzes the irreversible oxidation of a number of aldoses, including mannose, xylose, and galactose. Compared with the oxidation of β-D-glucose, the reaction rates for these sugars are ex-

```
        MQTLLVSSLVVSLAAALPHYIRSNGIEASLLTDPKDVSGR
 19     TVDYIIAGGGLTGLTTAARLTENPNISVLVIESGSYESDR
 59     GPIIEDLNAYGDIFGSSVDHAYETVELATNNQTALIRSGN
 99     GLGGSTLVNGGTWTRPHKAQVDSWETVFGNEGWNWDNVAA
139     YSLQAERARAPNAKQIAAGHYFNASCHGVNGTVHAGPRDT
179     GDDYSPIVKALMSAVEDRGVPTKKDFGCGDPHGVSMFPNT
219     LHEDQVRSDAAREWLLPNYQRPNLQVLTGQYVGKVLLSQN
259     GTTPRAVGVEFGTHKGNTHNVYAKHEVLLAAGSAVSPTIL
299     EYSGIGMKSILEPLGIDTVVDLPVGLNLQDQTTATVRSRI
339     TSAGAGQGQAAWFATFNETFGDYSEKAHELLNTKLEQWAE
379     EAVARGGFHNTTALLIQYENYRDWIVNHNVAYSELFLDTA
419     GVASFDVWDLLPFTRGYVHILDKDPYLHHFAYDPQYFLNE
459     LDLLGQAAATQLARNISNSGAMQTYFAGETIPGDNLAYDA
499     DLSAWTEYIPYHFRPNYHGVGTCSMMPKEMGGVVDNAARV
539     YGVQGLRVIDGSIPPTQMSSHVMTVFYAMALKISDAILED
579     YASMQ    583
```

Figure 10.3. Amino acid sequence of *Aspergillus niger* glucose oxidase (605 amino acids r deduced from cDNA sequencing, 583 residues for the mature protein, signal peptide = 22 amino acids, Frederick et al.1990). The N-terminal amino acid is indicated by double lines.

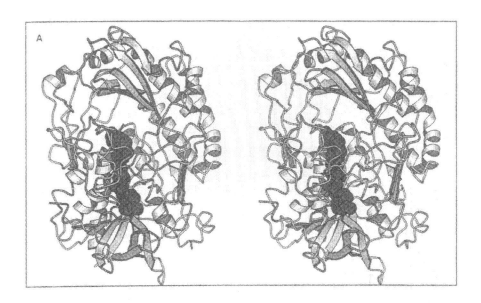

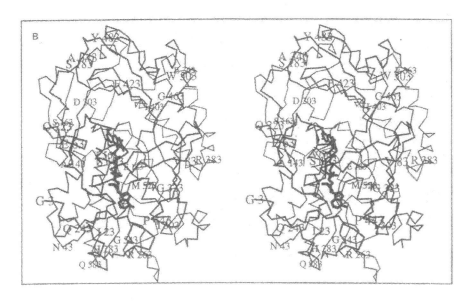

Figure 10.4. (A) A stereo drawing of the glucose oxidase monomer from *Aspergillus niger*. (B) C_α backbone of the enzyme. The orientation of the molecule is identical in both (A) and (B). (From Hecht et al. 1993. *J. Mol. Biol. 229,* 159, with permission. Copyright 1993 Academic Press, Ltd. Figure courtesy of Dr. H. J. Hecht, Gesellschaft für Biotechnologische Forschung mbH, Germany.)

tremely low. The enzyme shows high anomeric specificity in that the activity is ~160 times higher for the β-D-glucose as compared with the corresponding α-anomer, as confirmed by NMR studies (Feather 1970). The enzyme reaction also shows a significantly higher rate with D (C1) conformer (with the OH groups and the CH_2OH group in equatorial position). Furthermore, the stereochemistry is conserved in that the product is also a D conformer (Feather 1970). Replacing C1-OH with H abolishes the enzyme activity entirely. Substitution at C2 through C6 in the substrate reduces the catalytic efficiency of the enzyme. The values of k_{cat}/K_m = 8.58, 3.57, 1.62, 0.17, and 0.19 mM^{-1} s^{-1} for glucose, 2-deoxy, 3-deoxy, 4-deoxy, and 6-deoxy-D-glucose, respectively (Sierks et al. 1992). Changes in activation energy due to substitution of individual OH groups may suggest that these OH groups are important for hydrogen-binding interactions with amino acid side chains.

REACTION MECHANISMS

The glucose oxidase-catalyzed reaction consists of two half-reactions:

(1) Reductive half-reaction in which E-FAD (oxidized form of the enzyme) is reduced to E-FADH$_2$ (reduced form of the enzyme) and the substrate S$_1$ (β-D-glucopyranose used in Fig. 10.1) is oxidized to the product P$_1$ (δ-D-gluconolactone).

(2) Oxidative half-reaction in which E-FADH$_2$ is oxidized to the reduced enzyme (E-FAD) and the electron acceptor O$_2$ is reduced to H$_2$O$_2$.

Spectrophotometric studies of pH-dependent kinetics and isotope labelling experiments (Gibson et al. 1964; Bright and Gibson 1967) provide detailed kinetic parameters of glucose oxidation. The reaction catalyzed by glucose oxidase follows a ping-pong mechanism. For the overall reaction, V_{max} = 1150 M s^{-1}, K_m (glucose) = 0.11 M, K_m (O$_2$) = 0.48 M, k_{cat} = 1,200 s^{-1} (glucose, 27°C, *Aspergillus niger* enzyme) (Gibson et al. 1964; Bright and Appleby 1969). Enzyme-product complexes exist before dissociation, and the dissociation step (k_5) is rate limiting (for substrate = β-D-glucose).

$$E\text{-}FAD + S \underset{k_{-1}}{\overset{k_1}{\rightleftharpoons}} E\text{-}FAD \cdot S \overset{k_2}{\rightarrow} E\text{-}FADH_2 \cdot P \overset{k_3}{\rightarrow} E\text{-}FADH_2 + P$$

$$E\text{-}FADH_2 + O_2 \overset{k_{ox}}{\rightarrow} E\text{-}FAD \cdot H_2O_2 \overset{k_5}{\rightarrow} E\text{-}FAD + H_2O_2$$

Some studies argue against the formation of an enzyme (oxidized form)-H$_2$O$_2$ compound (E-FAD · H$_2$O$_2$). In this scheme, the conversion of enzyme-substrate to the reduced enzyme and product is the first-order rate-determining step in the overall reaction (Nakamura and Ogura 1962; 1968). For the *Aspergillus niger* enzyme, $k_2 = 320$ s^{-1}, and $k_{ox} = 9.4 \times 10^{-5}$ M^{-1} s^{-1} (glucose, 25°C). For the *Penicillium amagasakiense* enzyme, $k_2 = 550$ s^{-1}, and $k_{ox} = 1.4 \times 10^{-6}$ M^{-1} s^{-1} (Nakamura and Ogura 1962). The conversion of enzyme-substrate to the reduced enzyme and product is the first-order rate-determining step in the overall reaction (Nakamura and Ogura 1968).

$$\text{E-FAD} + \text{S} \underset{k_{-1}}{\overset{k_1}{\rightleftharpoons}} \text{E-FAD} \cdot \text{S} \overset{k_2}{\rightarrow} \text{E-FADH}_2 + \text{P}$$

$$\text{E-FADH}_2 + \text{O}_2 \overset{k_{ox}}{\rightarrow} \text{E-FAD} + \text{H}_2\text{O}_2$$

The two half-reactions are pH dependent. The pH profiles for k_1 and k_{ox} are sigmoidal with $pK_1 = 3.0$ and $pK_2 = 6.9$, respectively (Bright and Appleby 1969). The substrate binds only with a basic form of the enzyme, which requires the acidic group ($pK_1 = 3.0$) unprotonated. For the reaction of O_2 with the reduced enzyme, the process is acid catalyzed in that the enzyme must have a group with $pK_2 = 6.9$ protonated. The ionization groups are suggested to be carboxyl (pK_1) and histidine imidazole (pK_2). The dissociation constant K_2 is found to be sensitive to ionic strength. An increase in the ionic strength from 0.025 to 0.225 causes a sharp decrease of 0.9 pH unit in pK_2, and a 3-fold increase of k_{ox} (Voet et al. 1981). The intrinsic pK of the acidic group is 6.7, consistent with the group being a histidine residue.

The reduction-reoxidation of the flavin cofactor of glucose oxidase can be described by a single-step 2e$^-$ (hydride) transfer, or a two-step 1e$^-$ transfer mechanism.

Hydride Transfer

In this mechanism, a two-electron reduction of flavin involves a hydride transfer generated from a C1-H bond breaking in the substrate (glucose in Fig. 10.5). The process can be visualized as promoted by general base catalysis, with interaction between the glucose C1-OH to a side chain group of the enzyme (Bright and Gibson 1967; Weibel and Bright 1971; Massey and Hemmerich 1980). The general base, originally assigned to a carboxylate group, has now been identified as His559 (Hecht et al. 1993). The characteristic pH dependence of this reaction step is linked to Glu412, which is hydrogen bonded to His559. Protonation of Glu412 disrupts the hydrogen bond, resulting in shifting of the side chain to a position partially blocking the active site. The oxidative half-reaction involves the formation of a flavin-OOH adduct at the C10 position of the flavin (Massey and Hemmerich 1980). The acidic residue required for this reaction is His516 (Hecht et al. 1993) which, in its protonated form, stabilizes the negative charge localized at the N1 position of the flavin (Sanner et al. 1991).

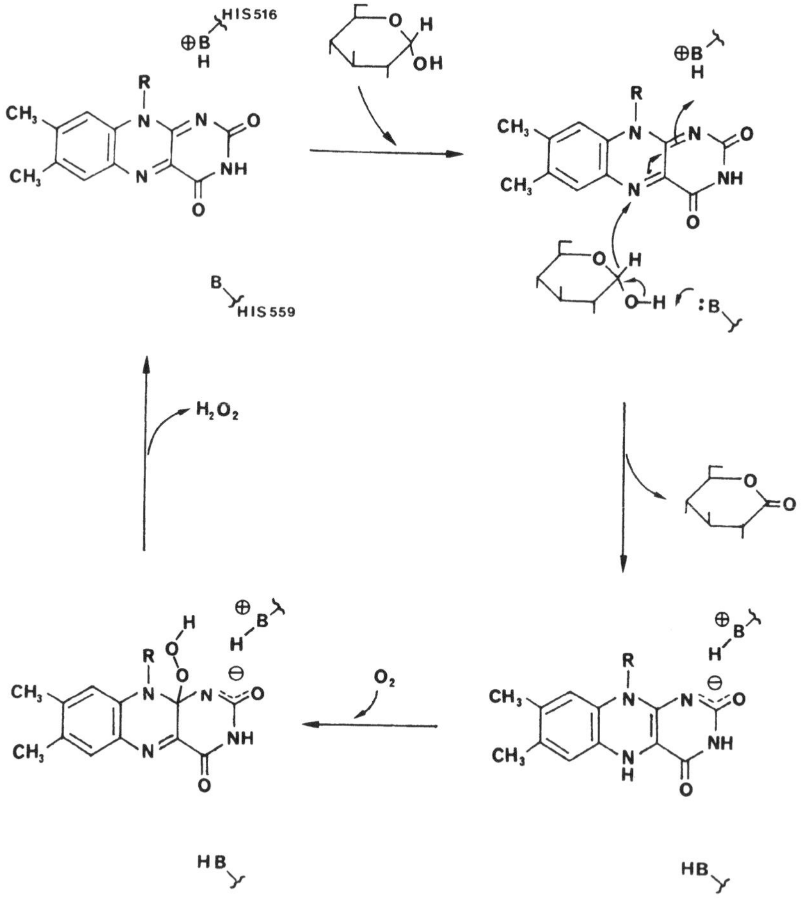

Figure 10.5. Reaction mechanism of glucose oxidase involving hydride transfer (Bright and Gibson 1967; Weibel and Bright 1971; Massey and Hemmerich 1980).

A hydride transfer mechanism is supported by the following findings. (1) Electron spin resonance experiments show no evidence of a radical intermediate in the glucose oxidase system (Nakamura and Ogura 1962; Gibson et al. 1964). (2) The presence of superoxide radical (O_2^-) cannot be detected by electron paramagnetic resonance experiments (Massey et al. 1969; Weibel and Bright 1971). (3) The reduction of E-FAD is accompanied by a substantial substrate deuterium isotope effect. No primary isotope effect is observed for both the reduction and oxidation reactions (Bright and Gibson 1967).

One-Electron Transfer

An alternative to the hydride transfer involves a two-step one-electron transfer giving a radical pair intermediate that is then transformed to the products (Chan and Bruice 1977). Reduction of glucose oxidase (E-FAD) using a glucose analog, furoin (which is an α-hydroxy-carbonyl compound), yields a semidione radical (furil). The formation of a semiquinone intermediate in this model is in direct contrast with the absence of detectable radicals in experiments using natural substrates. This difference is attributed to the fact that the $\Delta G°$ for the formation of radical intermediates for many substrates are close to the ΔG for the conversion of E-FAD· to E-FADH$_2$. The low free energy barrier for the second electron transfer means that the radical is less likely to be detected and the initial one-electron transfer becomes rate determining. In the case of using furoin as a substrate, the formation of the semidione intermediate has a smaller $\Delta G°$, creating a higher free energy barrier for the second electron transfer, and rendering the radical stable enough to be detected.

The reaction of oxygen with E-FADH$_2$ also proceeds with radical formation, based on the study using nitroxide as a model of $^3O_2$. This process of oxygen being a two-step one-electron acceptor generates superoxide anion, (O_2^-) and flavin semiquinone (E-FADH·).

A one-electron transfer mechanism can be summarized by Fig. 10.6. The

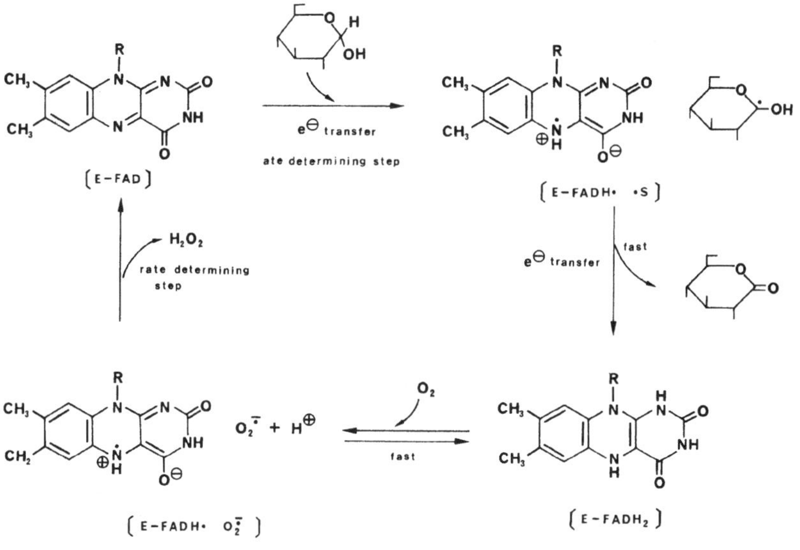

Figure 10.6. Reaction mechanism of glucose oxidase involving one-electron transfer, (Chan and Bruice 1977).

oxidized glucose oxidase (E-FAD) is reduced to the semiquinone by the substrate, forming a radical pair intermediate (E-FADH· ·S). The second electron transfer yields the reduced enzyme (E-FADH$_2$) and the product (δ-D-gluconolactone). Likewise, the oxidative half-reaction proceeds via a radical pair between E-FADH· and O$_2^-$. Radical pair intermediates do not give EPR signals, in agreement with experimental results where no free radicals were detected in glucose oxidase systems using natural substrates.

The argument against a one-electron transfer process is based on the study indicating that it is kinetically unfavorable to transfer a second electron in the reduction reaction in which one-electron reductants are used (Stankovich et al. 1978). The difference in the redox potential between the two one-electron transfer steps is small; $\Delta E_m = 2$ mV and 40 mV at pH 5.3 and pH 9.3, respectively. The presence of a kinetic barrier against the transfer of the second electron is further indicated by the fact that, in spite of the small potential difference of the two separate electron transfer steps, the rate of the second electron transfer is ~10 times slower than the first electron transfer when one-electron reductants are used. In addition, the small potential difference makes it thermodynamically unfavorable to the stabilization of a glucose oxidase radical, and favors a hydride transfer. Finally, the one-electron mechanism does not provide adequate consideration of the role of the active site residues as depicted by the pH dependence of the reaction.

REFERENCES

BENTLEY, R. 1963. Glucose oxidase. *The Enzymes*, 2d ed., 7, 575–586.

BRIGHT, H. J., and APPLEBY, M. 1969. The pH dependence of the individual steps in the glucose oxidase reaction. *J. Biol. Chem. 244*, 3625–3634.

BRIGHT, H. J., and GIBSON, Q. H. 1967. The oxidation of 1-deuterated glucose by glucose oxidase. *J. Biol. Chem. 242*, 994–1003.

CHAN, T. W., and BRUICE, T. C. 1977. One and two electron transfer reactions of glucose oxidase. *J. Am. Chem. Soc. 99*, 2387–2389.

FEATHER, M. S. 1970. A nuclear magnetic resonance study of the glucose oxidase reaction. *Biochim. Biophys. Acta 220*, 127–128.

FREDERICK, K. R.; TUNG, J.; EMERICK, R. S.; MASIARZ, F. R.; CHAMBERLAIN, S. H.; VASAVADA, A.; ROSENBERG, S.; CHAKRABORTY, S.; SCHOPTER, L. M.; and MASSEY, V. 1990. Glucose oxidase from *Aspergillus niger*. Cloning, gene sequence, secretion from *Saccharomyces cerevisiae* and kinetic analysis of a yeast-derived enzyme. *J. Biol. Chem. 265*, 3793–3802.

GIBSON, Q. H.; SWOBODA, B. E. P.; and MASSEY, V. 1964. Kinetics and mechanism of action of glucose oxidase. *J. Biol. Chem. 239*, 3927–3934.

HAYASHI, S., and NAKAMURA, S. 1976. Comparison of fungal glucose oxidase. *Biochim. Biophys. Acta 438*, 37–48.

————. 1981. Multiple forms of glucose oxidase with different carbohydrate compositions. *Biochim. Biophys. Acta 657*, 40–51.

HECHT, H. J.; KALISZ, H. M.; HENDLE, J.; SCHMID, R. D.; and SCHOMBURG, D. 1993. Crystal structure of glucose oxidase from *Aspergillus niger* refined at 2.3 A resolution. *J. Mol. Biol. 229*, 153–172.

HENDLE, J.; HECHT, J.-J.; KALISZ, H. M.; SCHMID, R. D.; and SCHOMBURG, D. 1992. Crystallization and preliminary x-ray diffraction studies of a deglycosylated glucose oxidase from *Penicillium amagasakiense. J. Mol. Biol. 223*, 1167–1169.

JAMES, T. L.; EDMONDSON, D. E.; and HUSAIN, M. 1981. Glucose oxidase contains a disubstituted phosphorus residue. Phosphorus-31 nuclear magnetic resonance studies of the flavin and nonflavin phosphate residues. *Biochemistry 20*, 617–621.

JONES, M. N.; MANLEY, P.; and WILKINSON, A. 1982. The dissociation of glucose oxidase by sodium *n*-dodecyl sulphate. *Biochem. J. 203*, 285–291.

KALISZ, H. M.; HECHT, H.-J.; SCHOMBURG, D.; and SCHMID, R. D. 1990. Crystallization and preliminary x-ray diffraction studies of a deglycosylated glucose oxidase from *Aspergillus niger. J. Mol. Biol. 213*, 207–209.

————. 1991. Effects of carbohydrate depletion on the structure, stability and activity of glucose oxidase from *Aspergillus niger. Biochim. Biophys. Acta 1080*, 138–142.

KIM, J. M., and SCHMID, R. D. 1991. Comparison of *Penicillium amagasakiense* glucose oxidase purified as glyco- and aglyuco-proteins. *FEBS Microbiol. Lett. 78*, 221–226.

KRIECHBAUM, M.; HEILMANN, H. J.; WIENTJES, F. J.; HAHN, M.; JANY, K.-D.; GASSEN, H. G.; SHARIF, F.; and ALAEDDINOGLU, G. 1989. Cloning and DNA sequence analysis of the glucose oxidase gene from *Aspergillus niger* NRRL-3. *FEBS Lett. 255*, 63–66.

MASSEY, V., and HEMMERICH, P. 1980. Active-site probes of flavoproteins. *Biochem. Soc. Trans. 8*, 246–257.

MASSEY, V.; MULLER, F.; FELDBERG, R.; SCHUMAN, M.; SULLIVAN, P. A.; HOWELL, L. G.; MAYHEW, S. G.; MATTHEWS, R. G.; and FOUST, G. P. 1969. The reactivity of flavoproteins with sulfite. Possible relevance to the problem of oxygen reactivity. *J. Biol. Chem. 244*, 3999–4006.

MASSEY, V.; STRICKLAND, S.; MAYHEW, S. G.; HOWELL, L. G.; ENGEL, P. C.; MATTHEWS, R. G.; SHUMAN, M.; and SULLIVAN, P. A. 1969. *Biochem. Biophys. Res. Comm. 36*, 891.

MULLER, F., and MASSEY, V. 1969. Flavin-sulfite complexes and their structures. *J. Biol. Chem. 244*, 4007–4016.

NAKAMURA, S., and FUJIKI, S. 1968. Comparative studies on the glucose oxidases of *Aspergillus niger* and *Penicillium amagasakiense. J. Biochem. 63*, 51–58.

NAKAMURA, S.; HAYASHI, S.; and KOGA, K. 1976. Effect of periodate oxidation on the structure and properties of glucose oxidase. *Biochim. Biophys. Acta 445*, 294–308.

NAKAMURA, S., and OGURA, Y. 1962. Kinetic studies on the action of glucose oxidase. *J. Biochem. 52*, 214–220.

————. 1968. Action mechanism of glucose oxidase of *Aspergillus niger. J. Biochem. 63*, 308–316.

PAZUR, J. H., and KLEPPE, K. 1964. The oxidation of glucose and related compounds by glucose oxidase from *Aspergillus niger. Biochemistry 3*, 578–583.

SANNER, C.; MACHEROUX, P.; RUTERJANS, H.; MULLER, F.; and BACHER, A. 1991. ^{15}N- and ^{13}C-NMR investigations of glucose oxidase from *Aspergillus niger. Eur. J. Biochem. 196*, 663–672.

SIERKS, M. R.; BOCK, K.; REFN, S.; and SVENSSON, B. 1992. Active site similarities of glucose dehydrogenase, glucose oxidase, and glucoamylase probed by deoxygenated substrates. *Biochemistry 31*, 8972–8977.

STANKOVICH, M. T.; SCHOPFER, L. M.; and MASSEY, V. 1978. Determination of glucose oxidase oxidation-reduction potentials and the oxygen reactivity of fully reduced and semiquinoid forms. *J. Biol. Chem. 253*, 4971–4979.

SWOBODA, B. E. P., and MASSEY, V. 1965. Purification and properties of the glucose oxidase from *Aspergillus niger. J. Biol. Chem. 240*, 2209 2215.

————. 1966. On the reaction of the glucose oxidase from *Aspergillus niger* with bisulfite. *J. Biol. Chem. 241*, 3409–3416.

TAKEGAWA, K.; FUJIWARA, K.; IWAHARA, S.; YAMAMOTO, K.; and TOCHIKURA, T. 1989. Effect of deglycosylation of N-linked sugar chains on glucose oxidase from *Aspergillus niger. Biochem. Cell. Biol. 67*, 460–464.

TAKEGAWA, K.; KONDO, A.; IWAMOTO, H.; FUJIWARA, K.; HOSOKAWA, Y.; KATO, I.; HIROMI, K.; and IWAHARA, S. 1991. Novel oligomannose-type sugar chains derived from glucose oxidase of *Aspergillus niger. Biochemistry International 25*, 181–190.

TSUGE, H.; NATSUAKI, O.; and OHASHI, K. 1975. Purification, properties, and molecular features of glucose oxidase from *Aspergillus niger. J. Biochem. 78*, 835–843.

VOET, J. G.; COE, J.; EPSTEIN, J.; MATOSSIAN, V.; and SHIPLEY, T. 1981. Electrostatic control of enzyme reactions: Effect of ionic strength on the pK_a of an essential acidic group on glucose oxidase. *Biochemistry 20*, 7182–7185.

WEIBEL, M. K., and BRIGHT, H. J. 1971. The glucose oxidase mechanism. Interpretation of the pH dependence. *J. Biol. Chem. 246*, 2734–2744.

YE, W.-N., and COMBES, D. 1989. The relationship between the glucose oxidase subunit structure and its thermostability. *Biochim. Biophys. Acta 999*, 86–93.

Chapter 11

Horseradish Peroxidase

Horseradish peroxidase (donor:hydrogen-peroxide oxidoreductase, EC 1.11.1.7) (HRP) belongs to a family of proteins with ferriprotoporphyrin IX as a prosthetic group. These heme enzymes function to either activate dioxygen for incorporation into the substrate (oxygenase activity) or use peroxides for oxidation of the substrate (peroxidase activity). Peroxidase is found widely distributed in higher plants (horseradish, turnip, fig sap, tobacco, potato) and microorganisms (yeast cytochrome c). The present discussion focuses on the well-studied horseradish peroxidase.

GENERAL CHARACTERISTICS

Horseradish peroxidase consists of many isozymes; a total of 42 has been reported (Hoyle 1977). The major isozymes, A, B, and C have pI of 6.1, 6.9, and 8.9, respectively. The complete primary structure of isozyme C has been described (Welinder 1976, 1979). The enzyme consists of a prosthetic group (ferriprotoporphyrin IX), 2 calcium atoms, and 308 amino acid residues with 4 disulfide bonds (Cys11–91, Cys44–49, Cys97–30, Cys177–209) and a calculated MW of 33,890 based on amino acid composition. The carbohydrate moieties are comprised of N-acetylglucosamine, mannose, fucose, and xylose, attached to an Asn-X-Ser/Thr sequence mostly in the C-terminal half of the molecule. The molecular weight of HRP-C, counting the carbohydrate composition, is close to 44 kD. Six isozymes, E1 to E6, with extremely high pI values (10.6 for E1 and E2, and $\geq$ 12 for E3–E6), have also been isolated (Aibara et al. 1981). The molecular weights of these basic isozymes are slightly lower than the neutral isozymes, because of the variation in the contents of carbohydrate moiety. The HRP-E5 whose crystal structure has been recently analyzed, has a MW of 36 kD, consisting of 306 amino acid residues, 2 glucosamines, and 8 other sugars.

The native enzyme (HRP-C) has the ferric iron coordinated to the four nitrogens of the pyrrole ring of the protoporphyrin IX (Fig. 11.1). The fifth coordination is occupied by an imidazole ligand, the proximal His170. The sixth coordination is vacant (Lanir and Schejter 1975; Spiro et al. 1979). The heme pocket is more buried and less accessible to solvent than that found in myoglobin (La Mar and de Ropp 1979). The regions of the proximal His170 and the distal His42 residues are highly conserved not only in plants, but also in microbial peroxidases.

HRP-C contains 2 moles of calcium per mole of enzyme with one of the calcium atoms being essential for correct folding conformation (Haschke and Friedhoff 1978; Ogawa et al. 1979). This calcium at the high-affinity binding site is responsible for maintaining the protein structure around the heme group (Morishima

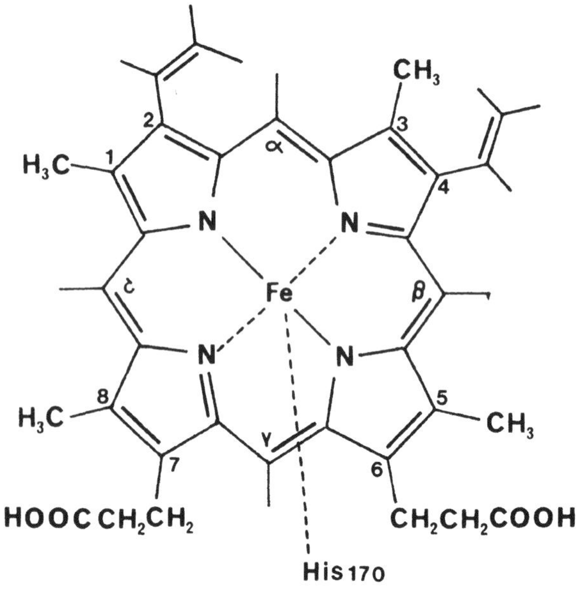

Figure 11.1.   Horseradish peroxidase has the ferric iron coordinated to the four nitrogens of the pyrrole ring of the protoporphyrin IX.

et al. 1986). Studies on the recombinant enzyme also suggest that Ca^{++} is critical for correct folding and activity (Smith et al. 1990). Removal of Ca^{++} causes changes in the heme distal and proximal structures. These conformational changes affect the rate constant of the reduction process (HRP-II + $AH_2 \rightarrow$ HRP + AH·), resulting in the decrease of enzyme activity (Table 11.1) (Shiro et al. 1986).

Table 11.1. Rate Constants of HRP in the Enzymatic Cycle[a]

Enzymes	Rate Constant $(M^{-1} s^{-1})$			Specific Activities (%)	
	$k_1 \times 10^{-7}$	$k_7 \times 10^{-4}$	$k_4 \times 10^{-2}$	Calculation	Experiment[b]
Native HRP	1.6 ± 0.1	1.2 ± 0.3	8.1 ± 1.6	100	100
Ca^{++}-free HRP	1.4	1.5 ± 0.2	3.6 ± 0.6	47	40
Reconstituted HRP	1.6 ± 0.1	1.3 ± 0.2	7.6 ± 2.1	94	100

From Shiro et al. 1986 *J. Biol. Chem.* **261**, 9383, with permission. Copyright 1986 American Society for Biochemistry and Molecular Biology.

[a]Conditions: pH 7, 22°C, 1 mM PIPES.NaOH, [*p*-aminobenzoic acid] = 0.07–1 mM.

[b]At steady state by using *o*-aminophenol as a substrate.

AMINO ACID SEQUENCES

The presence of different genes for the major HRP isozymes has been documented. The HRP-C1 sequence deduced from cDNA nucleotides is identical to the primary structure determined by amino acid sequencing (Fig. 11.2) (Fujiyama et al. 1988). Two other cDNA clones (HRP-C2, HRP-C3) also code for 308 amino

```
HRP C1           ELTPTFYDNSCPNVSNIV
HRP E5           E-RPD--SRT--S-FN-I
TUP              Q-TTN--STS---LLST-
TOB    MSFLRFVGAILFLVAIFGASNAQ-SAT--DTT----TS--

HRP C1   19   RDTIVNELRSDP RIAASILRLHFHDCFVNGCDASILLDN
HRP E5   19   KNVIVDELQTD- -IA----R---H----R---A-----T
TUP      19   KSGVKSAVSSQ- -MG----R-F-H----N---G-----D
TOB      19   R GVMDQRQRTDA-AG-K-IR---H----N---G-----
```

Figure 11.2. Comparison of amino acid sequences of plant peroxidases: HRP C1 (horseradish peroxidase C1, Welinder 1979), HRP E5 (horseradish peroxidase E5, Morita et al. 1991), TUP (turnipperoxidase 7, Mazza and Welinder 1980), TOB (tobaccoanionic peroxidase isozyme pI 4.5–6.5, 324 amino acids deduced from cDNA sequencing, the mature protein with 302 residues, signal peptide = 22 amino acids, Lagrimini et al. 1987).

```
HRP C1    58    TTSFRTEKDAFG NANSARGFPVIDRMKAAVESACP RTV
HRP E5    58    SK--R-----AP -V------N---RM-T-L-R--- RT-
TUP       58    -S--TG-QN- GP-R------T--NDI-S-V-K--- GV-
TOB       57    -DGTQ----- PANVG-G--DIV-DI-T-L-NV-- GV-

HRP C1    96    SCADLLTIAAQQSVTLAGGPSWRVPLGRRDSLQAFLDLAN
HRP E5    96    ----I-T--SQI--L-S-----A-PL-----VE-FFDL--
TUP       96    ----I-A--ARD--VQL---N-N-KV----AKT-SQAA--
TOB       94    ----I-AL-SEIG-V-AK----Q-LF--K--LT-NRSG--

HRP C1   136    ANLPAPFFTLPQLKDSFRNVGLNRSSDLVALSGG**H**TFGKN
HRP E5   136    TAL-S--F--A--KKA-AD--LNRPS-------G**H**---RA
TUP      136    SNI-A-SMSIS--ISS-SA--LS TR-M-----A**H**-I-QS
TOB      134    SDI-S--E--AVMIPQ-TNK-MDLT -------A**H**---RA

HRP C1   176    QCRFIMD RLYNFSNTGLPDPTLNTTYLQTLRGLCP LNG
HRP E5   176    R-LFVTA -----NGTNR--P-LNPSY-AD--RL-- RNG
TUP      175    R-VNFRA -V--ETNINAAFA        TLRQ-SCPRAAGS
TOB      173    R-GTFEQ --F--NGSGN--L-VDATF-QT-QGI--QGGN

HRP C1   214    NLSALVDFDLRTPTIFDNKYYVNLEEQKGLIQSDQELFSS
HRP E5   214    -GTV-VNF-VM--NT---QFYT--RNGK--I--------T
TUP      208    GDAN-APL-INSATS---S-FK--MAQR--LH---V-- N
TOB      212    -GNTFTNL-IS--ND---D-FT--QSNQ--L-T------T

HRP C1   254    PNATDTIPLVRSFANSTQTFFNAFVEAMDRMGNITPLTGT
HRP E5   254    PGA -T-PL-NLYSSNTLS--GA-AD--IR---LR-----
TUP      247    GGST-S- -RGYSN-PSS-NSD-AA--IK--D-S----S

TOB      252    SGSATIAI -NRYAG-QTQ--DD-VSS-IKL---S-----
HRP C1   294    QGQIRLNCRVVNSNS    308
HRP E5   293    Q-E--QN-RV--SR    306
TUP      285    S-E--KV-GKT-    296
TOB      291    N-Q--TD-KR--    302
```

Figure 11.2   (Continued).

acid residues with 91–94% homology among the isozymes. Comparison of HRP-C with other plant peroxidases shows the following sequence homologies: turnip HRP isozyme 49%, tobacco HRP anionic enzyme 50%, and potato HRP anionic enzyme 36% (Fujiyama et al. 1990). HRP-E5 contains 306 amino acids with a pyrrolidonecarboxylic acid at the N-terminus. The sequence of HRP-E5 shows a 69.6% and 47.0% homology with HRP-C and turnip HRP-7, respectively (Morita et al. 1991). The catalytic residues of HRP C1—Arg38 and His42—are conserved in all the peroxidases (Fig. 11.3).

```
                              Distal Histidine                      Proximal Histidine

HRP C1               31  RIAASILRLHFHDCFVNGCDASILLD       155  VGLNRSSDLVALSGGHTFGKNQCR
HRP C2               31  RIA-SILR-H-H----N---A-I---       155  V--DRPS-L-A---GH---KNQCR
HRP C3               31  RIA-SLLR-H-H----R---A-I---       155  V--NRPS-L-A---GH---RAQCQ
HRP E5               31  RIA-SILR-H-H----R---A-I---       155  V--NRPS-L-A---GH---RARCL
TURNIP P1                IG-SLIR-H-H----N---G-L---             V--N TT-V-V---AH---R
TURNIP P2                IG-SLIR-H-H----K---G-L---             V--N TT-V-V---AH---R
TURNIP P3                IG-SLIR-H-H----N---A-I---             V--N TN-L-A---AH---R
TURNIP P7            31  RMG-SILR-F-H----N---G-I---       155  V--S TR-M-A---AH-I-QSRCV
TOBACCO              31  RAG-KIIR-H-H----N---G-I---       153  K-MD LT-L-A---AH---RARCQ
POTATO               86  RMG-SLIR-H-H----D---GGI---       211  DKNFTLREM-A-A-AH-V-FARCS
FUNGUS Lig1          36  AEAHESIR-V-H-SIAISPAMEAQGK       161  A-EFDELEL-WMLSAHSVAAVNDV
FUNGUS Lig2          36  AEAHEALRMV-H-SIAISPKLQSQGK       162  A-GFDEIET-W-LSAHSIAAANDV
FUNGUS Lig3          36  AEAHESIR-V-H-AIAISPAMEPQAS       160  A-EFDELEL-WMLSAHSVAAANDI
YEAST Cyt C          41  GYGPVLVR-AWHTSGTWDKHDNTGGS       160  QR-NMDREV-A-M-AHAL-KTHLK
E. coli              96  SYAGLFIRAMWHGAGTYRSIDGRGGA       255  M-MNDEETVALIA-GH-L-KTHGA
B.stearothermophilus 90  HYGPLFIRMAWHSAGTYRIGDGRGGA       247  M-MNDEETVALIA-GH---KAHRG
```

Figure 11.3. Conserved regions of amino acid sequences of peroxidases showing the distal and proximal His residues (From Fujiyama et al. 1990. *Gene 89,* 167, with permission. Copyright 1990 Elsevier Science Publishers, B.V.)

THREE-DIMENSIONAL STRUCTURE

The crystal structure of HRP-C predicted from a comparison with cytochrome c peroxidase suggests that the enzyme consists of mainly 10 α helices (Fig. 11.4) (Welinder 1985). The enzyme is folded into two domains: domain I formed by helices A, B, C, and D (residues 1–144), and domain II of helices F, G, H, I, and J (residues 158–289). Two antiparallel β pairs are found in domain II. The folding pattern of HRP-C is almost identical to that of the yeast enzyme, although the latter is not a glycoprotein, and does not bind to Ca^{++}. All the eight sites of glycosylation in HRP-C are located in turns near the surface of the molecule.

The structure of the basic isozyme, HRP-E5, analyzed at 3.1-A resolution, is presented in Fig. 11.5 (Morita et al. 1991, 1993). The enzyme contains the corresponding helices A–J, very similar to that of cytochrome c peroxidase. However, the shorter N-terminus in HRP-E5 is not extended to the far side of domain I. One of the disulfides bridges helices F and G, holding the proximal His170 in stable ligation to the heme iron. Both the N- and the C-terminal segments are also held in position by disulfide bonds. Two glycosylation sites have the sequence -Asn-Gly-Thr-, similar to sites 4 and 6 in HRP-C. In both the predicted structure of HRP-C and the crystal structure of HRP-E5, the heme is located at the interface between domains I and II.

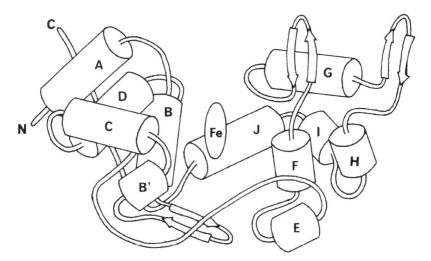

Figure 11.4. Structural model of horseradish peroxidase C predicted from a comparison with cytochrome c peroxidase. The two domains of the peroxidase enclosing the central hemin prosthetic group are pulled apart for clarity and simplicity. (From Welinder 1985. *Eur. J. Biochem. 151,* 500, with permission. Copyright 1985 European Journal of Biochemistry.)

Figure 11.5. Schematic model of the three-dimensional structure of horseradishperoxidase isozyme E5 at 3.1-A resolution analysis. (From Morita et al. 1993. In: *Plant Peroxidases: Biochemistry and Physiology,* p. 3, with permission. Copyright 1993 University of Geneva.)

GENERAL REACTION MECHANISM

The general mechanism of HRP involves two reaction steps (Fig. 11.6).

(1) A two-electron oxidation of the native ferric enzyme to compound I intermediate (HRP-I), with the peroxide substrate cleaved at the O-O bond.

(2) A two one-electron reduction of HRP-I by electron donor substrates to the native enzyme via compound II intermediate (HRP-II).

The common substrates for steps 1 and 2 are H_2O_2 and aromatic compounds, respectively. The rate constant for the formation of HRP-I using H_2O_2 as the substrate is $1.6 \pm 0.1 \times 10^7 \, M^{-1} \, s^{-1}$. Using *p*-nitroperbenzoic acid, the apparent second-order rate constant $= 3.7 \pm 0.2 \times 10^7 \, M^{-1} \, s^{-1}$ (Adediran and Dunford 1983). The reduction process occurs at a low rate constant: $1.2 \pm 0.3 \times 10^4 \, M^{-1} \, s^{-1}$ for HRP-I to HRP-II, and $8.1 \pm 1.6 \times 10^2 \, M^{-1} \, s^{-1}$ for HRP-II to native HRP, using *p*-aminobenzoic acid as the electron donor (Shiro et al. 1986). At

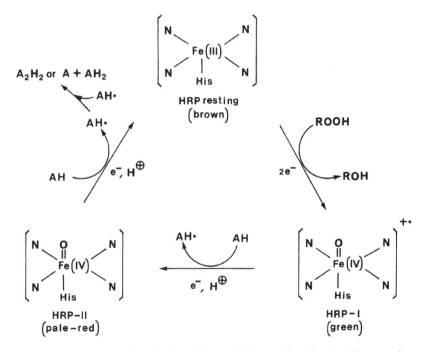

Figure 11.6. The general mechanism of horseradish peroxidase showing the conversion of the native enzyme to HRP-I and HRP-II. Details described in text.

high concentrations of most donors, the reduction of HRP-II to HRP is the rate-determining step.

THE IONIZATION GROUPS

Kinetic studies of HRP-C indicate that these reaction steps are pH dependent, with pK values of 4.2 and 8.6 for the formation of HRP-I and the reduction of HRP-II, respectively (Cotton and Dunford 1973; Dunford et al. 1978; Kato et al. 1984; Dunford and Adeniran 1986). The ionization with pK = 4.2 is found in three isozymes, A2 (acidic), C1 (neutral), and E5 (basic), although the effect on the rate constant for the formation of HRP-I differs. For the second ionization of pK = 8.6, the isozymes A2 and E5 exhibit pK values of 7.3 and 6.6, respectively (Kato et al. 1984). The ionization group responsible for these two steps is attributed to the distal His42 (Kato et al. 1984; Sitter et al. 1985; Penner-Hahn et al. 1986; Bhattachyaryya et al. 1992). The change in the pK value of the distal His may be attributed to the interactive ionization of the neighboring charge groups and the heme iron oxidation state (Sachs et al. 1971; Tanokura et al. 1976; Valentine et al. 1979). This is exemplified in the case of papain in which the essential His159 has its pK shifted from 8.5 to 4.0 depending on the ionization of the other catalytic residue Cys25 (Lewis et al. 1981). The importance of a carboxylate group (Asp34) has been initially assigned to the first step with pK = 4.2. However, comparison of the structural model of HRP heme environment with that of cytochrome c peroxidase suggests that the catalytically important residues His42 and Arg38 are conserved (Fig. 11.2) (Sakurada et al. 1986; Thanabal et al. 1987).

OXIDATION OF HRP TO HRP-I AND REDUCTIVE CLEAVAGE OF PEROXIDE

In the two-electron oxidation of the native ferric enzyme to HRP-I the peroxide substrate is reductively cleaved at the oxygen bond with the distal His42 acting as a general acid-base catalyst (Fig. 11.7).

Reaction Mechanism

Binding of the peroxide (H_2O_2 used in Fig. 11.7) to the ferric iron is accompanied by coordination between the peroxide α oxygen and the heme Fe(III). Hydrogen bonding of the α proton to His42 further increases the acidity of the proton. This is followed by the migration of the negative charge from the O_α to O_β, as assisted by stabilization via the positively charged guanidium group of Arg38. Finally, deprotonation occurs with the α proton transferred to the β oxygen via the imidazole, resulting in the heterolytic cleavage of the O_α–O_β bond (Schon-

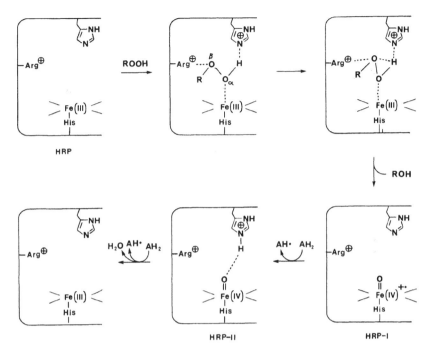

Figure 11.7. Reaction mechanism of horseradish peroxidase showing the pathways involving the two-electron oxidation of the native enzyme to HRP-I and the two-step one-electron reduction of HRP-I to HRP-II. Details described in text.

baum and Lo 1972; Dunford et al. 1978; Dunford and Araiso 1979; Poulos and Kraut 1980; Chance et al. 1984; Finzel et al. 1984; Behere et al. 1985). The departure of H_2O leaves a transient $Fe(III)$-O^+ structure, which acquires a double bond oxo-iron species formed by an electron transfer from the $Fe(III)$ together with a second electron from the rearrangement in the porphyrin π electron system. The incorporation of one oxygen from the oxidizing peroxide to HRP has been confirmed by isotope labelling (Hager et al. 1972) and kinetic studies (Adediran and Dunford 1983).

The presence of the positively charged Arg38 and the proximal His plays a dominant role in defining the catalytic mechanism of HRP (Finzel et al. 1984; Chance et al. 1986A). The heme-containing protein myoglobin, which lacks a corresponding charge group on the distal side of the heme, fails to polarize the O-O bond although it binds H_2O_2 to form myoglobin peroxide. Substitution of Arg38 for Lys causes a 1,000-fold decrease in the rate of formation of compound I (Smith et al. 1992).

The proximal His170 in the native enzyme is hydrogen bonded to an adjacent carboxylate side chain in the protein (La Mar and de Ropp 1982; La Mar et al. 1982; Teraoka et al. 1983). In HRP-I and HRP-II, the His N_δ-H is deprotonated with the proton transferred to the carboxylate group (de Ropp et al. 1985). The imidazolate thus formed strengthens the iron-imidazole bond, and serves to "push" electrons towards the iron and heme π electron system. In the oxidation of HRP to HRP-I, this facilitates the electron transfer from the porphyrin to the oxygen and the formation of Fe(IV). In both HRP-I and HRP-II, such interaction enhances stabilization of the high iron oxidation state.

The Nature of HRP-I

The structural aspect of HRP-I has been extensively studied by Mössbauer, EPR (Moss et al. 1969; Schulz et al. 1979), NMR (Dolphin et al. 1971; Groves et al. 1981), raman (Oertling and Babcock 1985; Salehi et al. 1986; Palaniappan and Terner 1989), and x-ray absorption (Penner-Hahn et al. 1983). HRP-I is consistent with a heme structure of O=Fe(IV)-porphyrin π-cation radical with S = 3/2. The HRP-I is therefore two oxidation equivalents above the native ferric enzyme—one oxidation equivalent located in the metal as [Fe(IV)=O] structure (low spin, S = 1), and the other in the porphyrin ring as a π cation (S 1/2). The Fe-O bond length has been confirmed to be 1.64 A by x-ray absorption studies, characteristic of a double bond (Chance et al. 1984; Penner-Hahn 1986). The π cation radical has also been detected by ESR (electron spin resonance) investigation in enzymatic one-electron oxidation of metal-free porphyrin (Morehouse et al. 1989). The green color of HRP-I comes from the porphyrin radical. Abstraction of an electron from the porphyrin occurs in the nearly degenerate highest occupied a_{2u} π orbital (Fujita et al. 1983; Rutter et al. 1983). Some studies have suggested that the cation radical is not entirely localized on the porphyrin, but follows delocalization to the proximal His, providing stabilization to the HRP-I structure (Fujita et al. 1983; Thanabal et al. 1988). However, this suggestion is anomalous with other results supporting a porphyrin π cation radical (Ogura and Kitagawa 1987; Paeng and Kincaid 1988; Palaniappan and Terner 1989).

REDUCTION OF HRP-I BY ELECTRON DONOR SUBSTRATES

The 2-electron oxidation of HRP-I is reversible via formation of the compound II intermediate (HRP-II) in a two-step one-electron reduction process. The porphyrin π cation radical first accepts an electron from a donor substrate, because the ferryl iron is relatively more stable. This is then followed by a second electron transfer from another donor to the Fe(IV) to yield the native enzyme.

HRP-II is one oxidation equivalent above the native ferric enzyme, with the porphyrin filled by donor substrate (AH_2). The heme structure has been described as an oxo-iron complex, Fe(IV)=O, like HRP-I (La Mar et al. 1983; Terner et al. 1985; Penner-Hahn et al. 1983, 1986), and also as a single bonded species, Fe(IV)–OH (Chance et al. 1984, 1986A, 1986B; Chang et al. 1993). The discrepancy lies in the assignment of the bond length of Fe-O in HRP-II. A bond length of 1.64 ± 0.03 A is consistent with a double bond character, whereas a distance of 1.93 ± 0.02 A is more indicative of a single bonded model.

The reactivity of HRP-II is linked to an ionization of pK 8.5. This is attributed to hydrogen bonding between the oxyferryl group and the distal His42 (Sitter et al. 1985; Evangelista-Kirkup et al. 1986). Deprotonation of the His at the basic side of the pK disrupts the hydrogen bonding, resulting in decrease of the reaction rate. Ionization of the distal His causes a conformational change that results in a distortion of the proximal His170 (Teraoka and Kitagawa 1980, 1981; Chang et al. 1993). The significance of this structural implication of the heme-link ionization of HRP-II and its precise role in the catalytic reaction need further investigation.

Reduction of HRP-I and HRP-II by AH_2 can be summarized as follows: (1) An electron is transferred from the substrate to the porphyrin cation radical of HRP-I, and a proton to His42. (2) The HRP-II thus formed has the oxoferryl hydrogen bonded to the protonated His42, which can then also pick up an electron from a second AH_2 molecule.

REACTIONS OF THE SUBSTRATES IN THE REDUCTION STEPS OF HRP-I AND HRP-II

Binding Site

Binding of an aromatic donor involves a hydrophobic interaction with Tyr185 and hydrogen bonding with Arg183, all in the vicinity of the porphyrin 8-CH_3. The aromatic ring of a number of substrates including indolepropionic acid, *p*-cresol, resorcinol, 2-methoxy-4-methyl phenol, and benzhydroxamic acid are located ~8.5–12.0 A perpendicular to the heme iron, and the latter three ~4 from the 8-CH_3 protons of the porphyrin (Sakurada et al. 1986; Thanabal et al. 1987; Fidy et al. 1989; Saxena et al. 1990). Therefore, HRP differs from the monooxygenases in that substrates react only with the heme edge and are not accessible for interaction with the ferryl oxygen. The Tyr185 is located in a flexible region of the HRP, which may allow the accommodation of various substrates, accounting for the low specificity of the enzyme.

Interaction of substrate with the heme porphyrin edge, around the δ-meso-carbon and 8-CH_3 group, is further confirmed by the inhibitory effect of alkylation on enzyme activity by alkyl- and phenylhydrazines (Ator and Montellano 1987; Ator et al. 1987).

Reaction Chemistry of Substrates

Alkylation by hydrazine compounds involves the addition of diazene, the oxidized product of hydrazine, to the δ-meso-carbon (Fig. 11.8). In the first step, transfer of an electron from the diazene to HRP-I produces the diazenyl radical and results in the conversion of HRP-I to HRP-II. Elimination of N_2 from the diazenyl radical yields the alkyl (or phenyl) radical, which then adds to the δ-meso-carbon of HRP-II. In the case of phenylhydrazine, addition also occurs at the 8-CH$_3$ of HRP-II, because of the higher reactivity of the phenyl radical. The alkyl (or phenyl) radical formed can couple to form termination products.

The mechanism of inhibition of HRP by cyclopropanone hydrate also suggests meso-alkylation, although this is preceded by H-abstraction at the ferryl oxygen and the C1-C2 ring opening of the substrate to yield a free alkyl radical of propionic acid, which is then added to the heme edge (Wiseman et al. 1982). It has also been argued that the cyclopropanone hydrate is oxidized by electron transfer to the heme edge directly without going through the interaction with the iron oxygen (Ator and Montellano 1987).

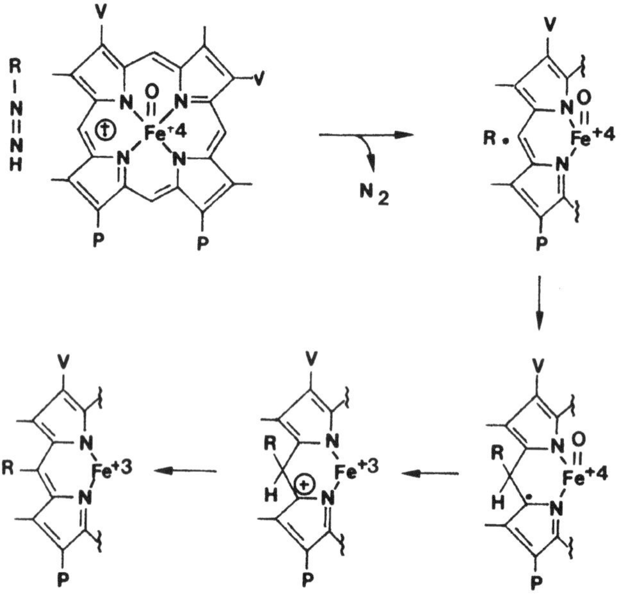

Figure 11.8. The proposed 4 mechanism for the inactivation of horseradish peroxidase by alkylhydrazines. R = phenylethyl, ethyl, or methyl. (From Ator et al. 1987. *J. Biol. Chem. 262,* 14959, with permission. Copyright 1987 American Society for Biochemistry andMolecular Biology.)

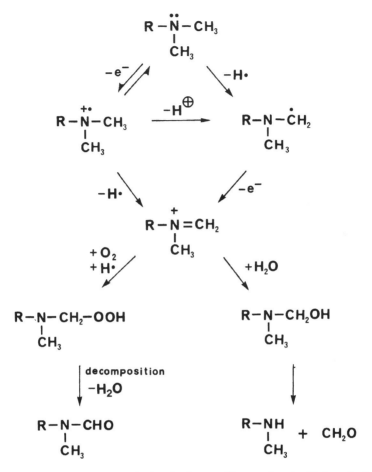

Figure 11.9. The proposed mechanism for the peroxidase-catalyzed oxidation of arylamines. (From Miwa et al. 1983. *J. Biol. Chem. 258,* 14446, with permission. Copyright 1993 American Society for Biochemistry and Molecular Biology.)

HRP-catalyzed oxidation of thiocyanate also proceeds via one electron transfer with the binding of thiocyanate to HRP near 1- and 8-CH_3 groups. The SCN · radical formed can also combine to form oxidation products $(SCN)_2$ (Modi et al. 1989, 1990, 1991).

In the peroxidase-catalyzed oxidation of secondary and tertiary arylamines, N-dealkylation occurs (Fig. 11.9) (Miwa et al. 1983). Reaction of arylamine with HRP-I involves hydrogen atom abstraction (H ·) from the methyl group, forming an N-alkyl-centered radical. Further oxidation of this radical by HRP-II yields an iminium cation, which reacts with H_2O to form a carbinolamine. Alternatively, the N-alkyl-centered radical reacts with O_2 to form a hydroperoxide, which, upon decomposition, forms the carbinolamine (Griffin and Ting 1978; Kedderis et al. 1986). It has been suggested that the demethylation reaction may actually involve formation of a nitrogen-centered cation radical by electron transfer, followed by hydrogen atom abstraction to form the iminium cation (van der Zee et al. 1989). It is obvious from the the demethylation process and the previously described reactions that the initial formation of the substrate radicals is enzyme catalyzed involving the reduction of HRP-I and/or HRP-II, whereas the rest of the pathway consists of nonenzymatic radical reactions.

The oxidation of indole-3-acetic acid by HRP also includes a radical intermediate with O_2 to form hydroperoxide and its decomposition to aldehyde (Fig. 11.10). In this case, the indole-3-methyl radical is formed by a one-electron abstraction in the decarboxylation step (Kobayashi et al. 1984).

The oxidation of substituted phenols and aromatic amines by HRP-I and HRP-II has been extensively investigated. The effects of substituent in aromatic molecules on the reaction rate follow the Hammett equation. For *m*- and *p*-monosubstituted phenols reacting with HRP-I and HRP-II, the *r* values are -6.9 and -4.6, respectively (Dunford and Adeniran 1986). Therefore, electron-donating groups enhance the reaction rate, and HRP-II is relatively less sensitive to the electron-donating substituent. HRP-I, being a porphyrin π-cation radical, is more electron deficient and is readily reduced by the substrate. Furthermore, the mechanism of oxidation for phenols (or anilines) by HRP-I involves a simple

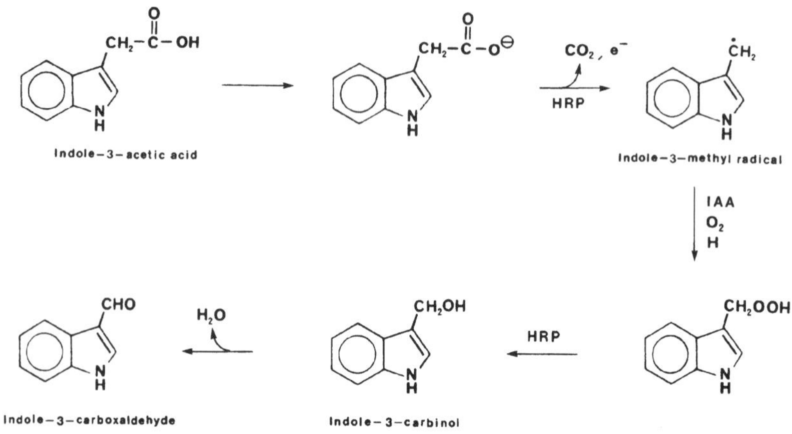

Figure 11.10. Oxidation of indole-3-acetic acid by horseradish peroxidase. (From Kobayashi et al. 1984. *Biochemistry 23*, 4596, with permission. Copyright 1984 American Chemical Society.)

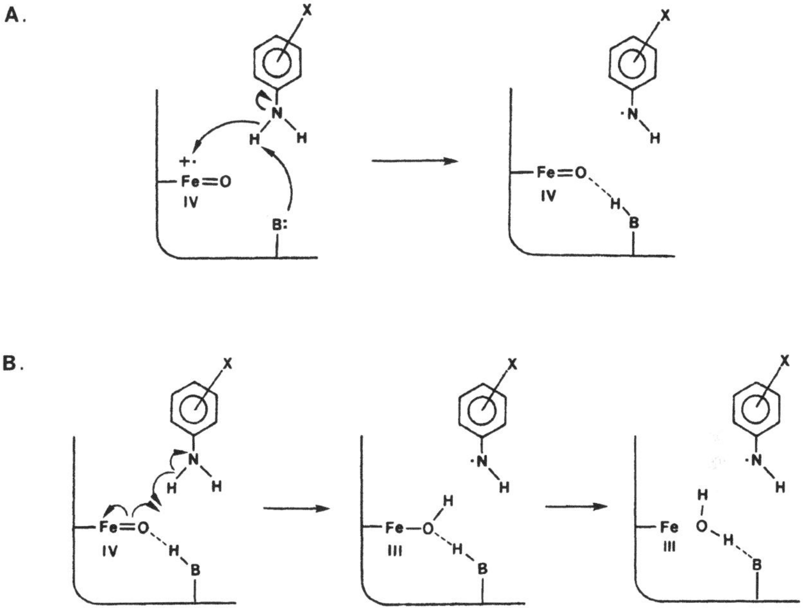

Figure 11.11. Oxidation of phenols by (A) HRP-I via a one-electron transfer, (B) HRP-II via electron and proton transfers. (From Huang and Dunford 1990. *Can. J. Chem. 68,* 2162, with permission. Copyright 1990 National Research Council of Canada.)

one-electron transfer from the amines to the π-cation radical porphyrin (Fig. 11.11A). In the case of HRP-II, electron transfer is to the Fe(IV), and in addition, two protons are transferred to combine with the oxo oxygen to form a H_2O molecule (Fig. 11.11B). Thus the latter reaction involves more bond breaking and forming, a more complex reaction that does not depend solely on the electron-donating substituent (Dunford and Adeniran 1986; Bohne et al. 1987; Huang and Dunford 1990; Sakurada et al. 1990). The substrate radical formed from the substrate may be stabilized by the π-conjugation involving a cation radical (Huang and Dunford 1991).

CATALATIC ACTIVITY

In the absence of the usual electron donors, the HRP-H_2O_2 system behaves as a catalase, converting H_2O_2 to H_2O + O_2 (Adediran and Lambeir 1989; Nakajima and Yamazaki 1987). H_2O_2 reacts with HRP-I to form HRP-I $\cdot$ H_2O_2 complex and by a two-electron transfer, yields HRP and O_2 with a second-order rate constant of $5 \times 10^2 \, M^{-1} \, s^{-1}$ (Fig. 11.12). Alternatively, the HRP-I·H$_2$O$_2$ complex formation may lead to the enzyme inactive form, with inactivation constant $3.92 \pm 0.06 \times 10^{-3} \, s^{-1}$. That H$_2O_2$ behaves as a suicide substrate suggests in the peroxidase pathway that a competition between the reductant (donor) substrate and H$_2$O$_2$ for HRP-I may protect the enzyme from inactivation (Arnao et al. 1990A, B).

A minor pathway involves a one-electron transfer to generate HRP-II and a superoxide anion (O$_2^-$). Reaction between HRP-II with H$_2$O$_2$ forms HRP-III Fe(II)O$_2$ (also known as oxyperoxidase), with an apparent second-order rate constant of 25 $M^{-1} \, s^{-1}$. HRP-III decomposes to the native enzyme and O$_2^-$ with a first-order reaction rate of $2.2 \times 10^{-3} \, s^{-1}$ (Nakajima and Yamazaki 1987). Some studies suggests the involvement of addition of O$_2^-$ to HRP-II in HRP-III formation (Fe(III)O$_2^-$ <-> Fe(II)O$_2$). A recent electron spin resonance study also favors direct addition of H$_2$O$_2$ to HRP-II (Moore et al. 1992), although the exact mechanism remains unclear.

In the normal reaction cycle, a high ratio of [donor substrate]/[H$_2$O$_2$] favors the HRP-I and HRP-II cycle, because of the higher reaction rate of reduction of

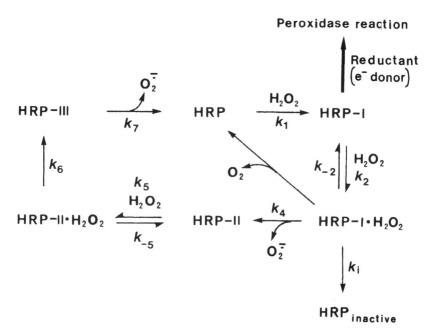

Figure 11.12. The catalatic pathway involving the formation of HRP-III in the absence of reductant substrates. (From Arnao et al. 1990B. *Biochim. Biophys. Acta 1041*, 45, with permission. Copyright 1990 Elsevier Science Publisher B.V.)

HRP-I by the donor substrate. Therefore, the peroxidase reaction constitutes the dominant pathway.

HYDROXYLATION

HRP, in the presence of dihydroxyfumaric acid (DHFA) and O_2, catalyzes the hydroxylation of a number of aromatic compounds. H_2O_2 or other peroxide substrates are not required for this reaction. It is now apparent that HRP-III is also involved.

The initial step is nonenzymatic and involves the reaction of DHFA with O_2 in the formation of DHFA radical and O_2^- (Fig. 11.13). The reaction of O_2^- with HRP yields HRP-III (Fe(III)O_2^-). Oxidation of another molecule of DHFA by HRP-III results in the formation of hydroxyl radical ($\cdot$OH), which is responsible for hydroxylation (Dordick et al. 1986).

However, $\cdot$OH can also be generated nonenzymatically via a dismutation

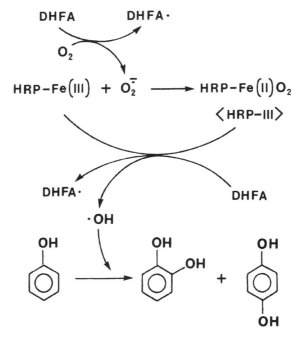

Figure 11.13. Horseradish peroxidase-catalyzed hydroxylation of aromatic compounds involving HRP-III. (From Dordick et al. 1986. *Biochemistry 25,* 2950, with permission. Copyright 1986 American Chemical Society.)

of O_2^- ($2O_2^-$ + $2H^+$ → H_2O_2 + O_2) to form H_2O_2, which then interacts to form the hydroxyl radical (O_2^- + H_2O_2 → HO· + OH^- + O_2) according to the Haber-Weiss reaction. The H_2O_2 generated in this process can be used by HRP in the oxidation of DHPA to DHPA radical, which can then react with O_2 to yield more O_2^-. In this case, the concentration of the enzyme is the determining factor and depends on the rate of breakdown of HRP-III (Tamura and Yamazaki 1972; Halliwell 1977).

For the hydroxylation of 2,4,6-trimethylphenol to 2,6-dimethyl-4-hydroxy-methylphenol, the OH group introduced at the 4-CH_3 carbon is accomplished by the addition of H_2O to the quinone methide or via the addition of O_2 to the substrate radical (Fig. 11.14) (de Montellano et al. 1987). But this reaction requires H_2O_2 and therefore must involve HRP-I and HRP-II which differs from the HRP-DHFA system.

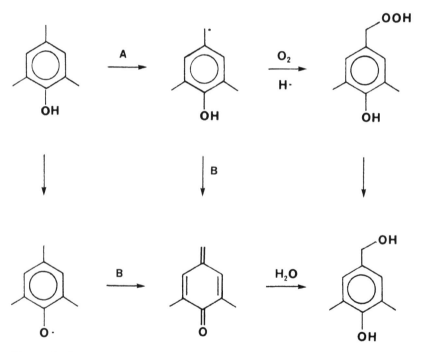

Figure 11.14. Hydroxylation L of 2,4,6-trimethylphenol by peroxidase-catalyzed oxygeninsertion reactions. (From de Montellano et al. 1987. *J. Biol. Chem. 262*, 11645, with permission. Copyright 1987 American Society for Biochemistry and Molecular Biology.)

CROSSLINKING

The formation of crosslinking products in proteins treated with peroxidase-H_2O_2 has been shown to occur via the formation of dityrosines with biphenyl C-C bonds. This type of crosslinking is also found in vivo in a number of structural proteins: elastin, collagen, silk, keratin, etc. (Aeschbach et al. 1976; Amado et al. 1984; Matheis and Whitaker 1984). Isodityrosine has been detected in the hydroxyproline-rich protein, extensin, found in plant cell walls. The crosslink in this case is a dimer of tyrosine linked via a diphenyl ether bond (Fry 1984). The biological function of this post-translational modification is to crosslink the glycoprotein into a network together with other cell wall components.

REFERENCES

Adediran, S. A., and Dunford, H. B. 1983. Structure of horseradish peroxidase compound I. Kinetic evidence for the incorporation of one oxygen atom from the oxidizing substrate into the enzyme. *Eur. J. Biochem. 132*, 147–150.

ADEDIRAN, S. A., and LAMBEIR, A.-M. 1989. Kinetics of the reaction of compound II of horseradish peroxidase with hydrogen peroxide to form compound III. *Eur. J. Biochem. 186*, 571–576.

AESCHBACH, R.; Amado, R.; and Neukom, H. 1976. Formation of dityrosine crosslinks in proteins by oxidation of tyrosine residues. *Biochim. Biophys. Acta. 439*, 292–301.

AIBARA, S.; KOBARYASHI, T.; and MORITA, Y. 1981. Isolation and properties of basic isoenzymes of horseradish peroxidase. *J. Biochem. 90*, 489–496.

AMADO, R.; AESCHBACH, R.; and NEUKOM, H. 1984. Dityrosine: *In vitro* production and characterization. *Methods in Enzymology 107*, 377–388.

ARNAO, M. B.; ACOSTA, M.; DEL RIO, J. A.; and GARCIA-CANOVAS, F. 1990A. Inactivation of peroxidase by hydrogen peroxide and its protection by a reductant agent. *Biochim. Biophys. Acta 1038*, 85–89.

ARNAO, M. B.; ACOSTA, M.; DEL RIO, J. A.; VARON, R.; and GARCIA-CANOVAS, F. 1990B. A kinetic study on the suicide inactivation of peroxidase by hydrogen peroxide. *Biochim. Biophys. Acta. 1041*, 43–47.

ATOR, M. A., and DE MONTELLANO, P. R. O. 1987. Protein control of prosthetic heme activity. Reaction of substrates with the heme edge of horseradish peroxidase. *J. Biol. Chem. 262*, 1542–1551.

ATOR, M. A.; DAVID, S. K.; and DE MONTELLANO, P. R. O. 1987. Structure and catalytic mechanism of horseradish peroxidase. *J. Biol. Chem. 262*, 14954–14960.

BEHERE, D. V.; GONZALEZ-VERGARA, E.; and GOFF, H. M. 1985. Unique cyanide nitrogen-15 nuclear magnetic resonance chemical shift values for cyano-peroxidase complexes. Relevance to the heme active-site structure and mechanism of peroxide activation. *Biochim. Biophys. Acta 832*, 319–325.

BHATTACHYARYYA, D. K.; BANDYOPADHYAY, U.; and BANERJEE, R. K. 1992. Chemical and kinetic evidence for an essential histidine in horseradish peroxidase for iodide oxidation. *J. Biol. Chem. 267,* 9800–9804.

BOHNE, C.; MACDONALD, D. I.; and DUNFORD, H. B. 1987. Transient state kinetics of the reactions of isobutyraldehyde with compounds I and II of horseradish peroxidase. *J. Biol. Chem. 262,* 3572–3578.

CHANCE, B.; POWERS, L.; CHING, Y.; POULOS, T.; SCHONBAUM, G. R.; YAMAZAKI, I.; and PAUL, K. G. 1984. X-Ray absorption studies of intermediates in peroxidase activity. *Arch. Biochem. Biophys. 235,* 596–611.

CHANCE, M.; POWERS, L.; KUMAR, C.; and CHANCE, B. 1986A. X-Ray absorption studies of myoglobin peroxide reveal functional differences between globins and heme enzymes. *Biochemistry 25,* 1259–1265.

CHANCE, M.; POWERS, L.; POULOS, T.; and CHANCE, B. 1986B. Cytochrome c peroxidase compound ES is identical with horseradish peroxidase compound I in iron-ligand distances. *Biochemistry 25,* 1266–1270.

CHANG, C. S.; YAMAZAKI, I.; SINCLAIR, R.; KHALID, S.; and POWERS, L. 1993. pH dependence of the active site of horseradish peroxidase compound II. *Biochemistry 32,* 923–928.

COTTON, M. L., and DUNFORD, H. B. 1973. Studies on horseradish peroxidase. XI. On the nature of compounds I and II as determined from the kinetics of the oxidation of ferrocyanide. *Can. J. Chem. 51,* 582–587.

DE MONTELLANO, P. R. O.; CHOE, Y. S.; DEPILLIS, G.; and CATALANO, C. E. 1987. Structure-mechanism relationships in hemoproteins. *J. Biol. Chem. 262,* 11641–11646.

DE ROPP, J. S.; THANABAL, V.; LA MAR, G. N. 1985. NMR evidence for a horseradish peroxidase state with a deprotonated proximal histidine. *J. Am. Chem. Soc. 107,* 8268–8270.

DOLPHIN, D.; FORMAN, A.; BORG, F. D. C.; FAJER, B. J.; and FELTON, R. H. 1971. Compounds I of catalase and horse radish peroxidase: π-cation radicals. *Proc. Natl. Acad. Sci. USA 68,* 614–618.

DORDICK, J.; KLIBANOV, A. M.; and MARLETTA, M. A. 1986. Horseradish peroxidase catalyzed hydroxylations: Mechanistic studies. *Biochemistry 25,* 2946–2951.

DUNFORD, H. B., and ADENIRAN, A. J. 1986. Hammett $\rho\sigma$ correction for reactions of horseradish peroxidase compound II with phenols. *Arch. Biochem. Biophys. 251,* 536–542.

DUNFORD, H. B., and ARAISO, T. 1979. Horseradish peroxide. XXXVI. On the difference between peroxidase and metmyoglobin. *Biochem. Biophys. Res. Comm. 89,* 764–768.

DUNFORD, H. B.; HEWSON, W. D.; and STEINER, H. 1978. Horseradish peroxidase. XXIX. Reactions in water and deuterium oxide: Cyanide binding, compound I formation, and reactions of compounds I and II with ferrocyanide. *Can. J. Chem. 56,* 2844–2852.

EVANGELISTA-KIRKUP, R.; SMULEVICH, G.; and SPIRO, T. G. 1986. Alternative carbon monoxide binding modes for horseradish peroxidase studied by resonance raman spectroscopy. *Biochemistry 25,* 4420–4425.

FIDY, J.; PAUL, K.-G.; and VANDERKOOI, J. M. 1989. Differences in the binding of aromatic substrates to horseradish peroxidase revealed by fluorescence line narrowing. *Biochemistry 28*, 7531–7541.

FINZEL, B. C.; POULOS, T. L.; and KRAUT, J. 1984. Crystal structure of yeast cytochrome c peroxidase refined at 1.7-A resolution. *J. Biol. Chem. 259*, 13027–13036.

FRY, S. C. 1984. Isodityrosine, a diphenyl ether cross-link in plant cell wall glycoprotein: Identification, assay, and chemical synthesis. *Methods in Enzymology 107*, 388–397.

FUJITA, I.; HANSON, L. K.; WALKAR, F. A.; and FAJER, J. 1983. Models for compounds I of peroxidases: Axial ligand effects. *J. Am. Chem. Soc. 105*, 3296–3300.

FUJIYAMA, K.; TAKEMURA, H.; SHIBAYAMA, S.; KOBAYASHI, K.; CHOI, J.-K.; SHINMYO, A.; TAKANO, M.; YAMADA, Y.; and OKADA, H. 1988. Structure of the horseradish peroxidase isozyme C genes. *Eur. J. Biochem. 173*, 681–687.

FUJIYAMA, K.; TAKEMURA, H.; SHINMYO, A.; OKADA, H.; and TAKANO, M. 1990. Genomic DNA structure of two new horseradish-peroxidase-encoding genes. *Gene 89*, 163–169.

GRIFFIN, B. W., and TING, P. L. 1978. Mechanism of N-demethylation of aminopyrine by hydrogen peroxide catalyzed by horseradish peroxidase, metmyoglobin, and protohemin. *Biochemistry 17*, 2206–2211.

GROVES, J. T.; HAUSHALTER, R. C.; NAKAMURA, M.; NEMO, T. E.; and EVANS, B. J. 1981. High-valent iron-porphyrin complexes related to peroxidase and cytochrome P-450. *J. Am. Chem. Soc. 103*, 2884–2886.

HALLIWELL, B. 1977. Generation of hydrogen peroxide, superoxide and hydroxyl radicals during the oxidation of dihydroxyfumaric acid by peroxidase. *Biochem. J. 163*, 441–448.

HASCHKE, R. H., and FRIEDHOFF, J. M. 1978. Calcium-related properties of horseradish peroxidase. *Biochem. Biophys. Res. Comm. 80*, 1039–1042.

HASHIMOTO, S.; TATSUNO, Y.; and KITAGAWA, T. 1986. Resonance raman evidence for oxygen exchange between the Fe^{IV}=O heme and bulk water during enzymic catalysis of horseradish peroxidase and its relation with the heme-linked ionization. *Proc. Natl. Acad. Sci. USA 83*, 2417–2421.

HOYLE, M. C. 1977. High resolution of peroxidase-indoleacetic acid oxidase isoenzymes from horseradish by isoelectric focusing. *Plant Physiol. 60*, 787–793.

HUANG, J., and DUNFORD, H. B. 1990. Oxidation of substituted anilines by horseradish peroxidase compound II. *Can. J. Chem. 68*, 2159–2163.

———. 1991. One-electron oxidative activation of 2-aminofluorene by horseradish peroxidase compounds I and II: Special and kinetic studies. *Arch. Biochem. Biophys. 287*, 257–262.

KAPUT, J.; GOLTZ, S.; and BLOBEL, G. 1982. Nucleotide sequence of the yeast nuclear gene for cytochrome c peroxidase precursor. *J. Biol. Chem. 257*, 15054–15058.

KATO, M.; AIBARA, S.; MORITA, Y.; NAKATANI, H.; and HIROMI, K. 1984. Comparative studies on kinetic behavior of horseradish peroxidase isoenzymes. *J. Biochem. 95*, 861–870.

KEDDERIS, G. L.; RICKERT, D. E.; PANDEY, R. N.; and HOLLENBERG, P. F. 1986. [18]O studies of the peroxidase-catalyzed oxidation of N-methylcarbazole. *J. Biol. Chem. 261*, 15910–15914.

LAGRIMINI, L. M.; BURKHART, W.; MOYER, M.; and ROTHSTEIN, S. 1987. Molecular cloning of complementary DNA encoding the lignin-forming peroxidase from tobacco: Molecular analysis and tissue-specific expression. *Proc. Natl. Acad. Sci. USA 84*, 7542–7546.

KOBAYASHI, S.; SUGIOKA, K.; NAKANO, H.; NAKANO, M.; and TERO-KUBOTA, S. 1984. Analysis of the stable end products and intermediates of oxidative decarboxylation of indole-3-acetic acid by horseradish peroxidase. *Biochemistry 23*, 4589–4597.

LA MAR, G. N., and DE ROPP, J. S. 1979. Assignment of exchangeable proximal histidine resonances in high-spin ferric hemoproteins: Substrate binding in horseradish peroxidase. *Biochem. Biophys. Res. Comm. 90*, 36–41.

———. 1982. Proton NMR characterization of the state of protonation of the axial imidazole in reduced horseradish peroxidase. *J. Am. Chem. Soc. 104*, 5203–5206.

LA MAR, G. N.; DE ROPP, J. S.; CHACKO, V. P.; SATTERLEE, J. D.; and ERMAN, J. E. 1982. Axial histidyl imidazole non-exchangeable proton resonances as indicators of imidazole hydrogen bonding in ferric cyanide complexes of heme peroxidases. *Biochim. Biophys. Acta. 708*, 317–325.

LA MAR, G. N.; DE ROPP, J. S.; LATOS-GRAZYNSKI, L.; BALCH, A. L.; JOHNSON, R. B.; SMITH, K. M.; PARISH, D. W.; and CHENG, R. 1983. Proton NMR characterization of the ferryl group in model heme complexes and hemeproteins: Evidence for the $Fe^{IV}=O$ group in ferryl myoglobin and compound II of horseradish peroxidase. *J. Am. Chem. Soc. 105*, 782–787.

LANIR, A., and SCHEJTER, A. 1975. Nuclear magnetic resonance evidence for the absence of iron coordinated water in horseradish peroxidase. *Biochem. Biophys. Res. Comm. 62*, 199–203.

LEWIS, S. D.; JOHNSON, F. A.; and SHAFER, J. A. 1981. Effect of cysteine-25 on the ionization of histine-159 in papain as determined by proton nuclear magnetic resonance spectroscopy. Evidence for a His-159-Cys-25 ion pair and its possible role in catalysis. *Biochemistry 20*, 48–51.

MATHEIS, G., and WHITAKER, J. R. 1984. Peroxidase-catalyzed cross linking of proteins. *J. Protein Chem. 3*, 35–48.

MAZZA, G., and WELINDER, K. G. 1980. Covalent structure of turnip peroxidase. *Eur. J. Biochem. 108*, 481–489.

METODIEWA, D., and DUNFORD, H. B. 1989. The reactions of horseradish peroxidase, lactoperoxidase, and myeloperoxidase with enzymatically generated superoxide. *Arch. Biochem. Biophys. 272*, 245–253.

MIWA, G. T.; WALSH, J. S.; KEDDERIS, G. L.; and HOLLENBERG, P. F. 1983. The use of intramolecular isotope effects to distinguish between deprotonation and hydrogen atom abstraction mechanisms in cytochrome P-450- and peroxidase-catalyzed N-demethylation reactions. *J. Biol. Chem. 258*, 14445–14449.

MODI, S.; BEHERE, D. V.; and MITRA, S. 1991. Horseradish peroxidase catalyzed oxidation of thiocyanate by hydrogen peroxide: comparison with lactoperoxidase-catalyzed oxidation and role of distal histidine. *Biochim. Biophys. Acta 1080*, 45–50.

————. 1989. Interaction of thiocyanate with horseradish peroxidase. *J. Biol. Chem.* *264*, 19677–19684.

MODI, S.; SAXENA, A. K.; BEHERE, D. V.; and MITRA, S. 1990. Binding of thiocyanate and cyanide to manganese (III)-reconstituted horseradish peroxidase: A ^{15}N nuclear magnetic resonance study. *Biochim. Biophys. Acta 1038*, 164–171.

MOORE, K. L.; MORONNE, M. M.; and MEHLHORN, R. J. 1992. Electron spin resonance study of peroxidase activity and kinetics. *Arch. Biochem. Biophys. 299*, 47–59.

MOREHOUSE, K. M.; SIPE, H. J.; and MASON, R. P. 1989. The one-electron oxidation of porphyrins to porphyrin pi-cation radicals by peroxidases: An electron spin resonance investigation. *Arch. Biochem. Biophys. 273*, 158–164.

MORISHIMA, I.; KURONO, M.; and SHIRO, Y. 1986. Presence of endogenous calcium ion in horseradish peroxidase. *J. Biol. Chem. 261*, 9391–9399.

MORITA, Y.; FUNATSU, J.; and MIKAMI, B. 1993. X-Ray crystallographic analysis of horseradish peroxidase E5. In: *Plant Peroxidases: Biochemistry and Physiology*, K. G. Welinder, S. K. Rasmussen, C. Penel, H. Greppin, eds., University of Geneva, 1993, pp. 1–4.

MORITA, Y.; MIKAMI, B.; YAMASHITA, H.; LEE, J. Y.; AIBARA, S.; SATO, M.; KATSUBE, Y.; and TANAKA, N. 1991. Primary and crystal structures of horseradish peroxidase isozyme E5. In: *Biochemical, Molecular and Physiological Aspects of Plant Peroxidases*, J. Lobarzewski, H. Greppin, C. Penel, and Th. Gasper, eds., University of Geneva, 1991, pp. 81–88.

MOSS, T. H.; EHRENBERG, A.; and BEARDEN, A. J. 1969. Mössbauer spectroscopic evidence for the electronic configuration of iron in horseradish peroxidase and its peroxidase derivatives. *Biochemistry 8*, 4160–4162.

NAKAJIMA, R., and YAMAZAKI, I. 1987. The mechanism of oxyperoxidase formation from ferryl peroxidase and hydrogen peroxide. *J. Biol. Chem. 262*, 2576–2581.

OGURA, T., and KITAGAWA, T. 1987. Device for simultaneous measurements of transient raman and absorption spectra of enzymic reactions: application to compound I of horseradish peroxidase. *J. Am. Chem. Soc. 109*, 2177–2179.

OERTLING, W. A., and BABCOCK, G. T. 1985. Resonance raman scattering from horseradish peroxidase compound I. *J. Am. Chem. Soc. 107*, 6406–6407.

————. 1988. Time-resolved and static resonance raman spectroscopy of horseradish peroxidase intermediates. *Biochemistry 27*, 3331–3338.

OGAWA, S.; SHIRO, Y.; and MORISHIMA, I. 1979. Calcium binding by horseradish peroxidase C and the heme environmental structure. *Biochem. Biophys. Res. Comm. 90*, 674–678.

PAENG, K.-J., and KINCAID, J. R. 1988. The resonance raman spectrum of horseradish peroxidase compound I. *J. Am. Chem. Soc. 110*, 7913–7915.

PALANIAPPAN, V., and TERNER, J. 1989. Resonance raman spectroscopy of horseradish peroxidase derivatives and intermediates with excitation in the near ultraviolet. *J. Biol. Chem. 264*, 16046–16053.

PENNER-HAHN, J. E.; EBLE, K. S.; MCMURRY, T. J.; RENNER, M.; BALCH, A. L.; GROVES, J. T.; DAWSON, J. H.; and HODGSON, K. O. 1986. Structural characterization of horseradish peroxidase using EXAFS spectroscopy. Evidence for $Fe=O$ ligation in compounds I and II. *J. Am. Chem. Soc. 108*, 7819–7825.

PENNER-HAHN, J. E.; McMURRY, T. J.; RENNER, M.; LATOS-GRAZYNSKY, L.; EBLE, K. S.; DAVIS, I. M.; BALCH, A. L.; GROVES, J. T.; DAWSON, J. H.; and HODGSON, K. O. 1983. X-Ray absorption spectroscopic studies of high valent iron porphyrins. *J. Biol. Chem. 258*, 12761–12764.

POULOS, T. L., and KRAUT, J. 1980. The stereochemistry of peroxidase catalysis. *J. Biol. Chem. 255*, 8199–8205.

RUTTER, R.; VALENTINE, M.; HENDRICH, M.; HAGER, L.; and DEBRUNNER, P. 1983. Chemical nature of the porphyrin π cation radical in horseradish peroxidase compound I. *Biochemistry 22*, 4769–4774.

SACHS, D. H.; SCHECHTER, A. N.; and COHEN, J. S. 1971. Nuclear magnetic resonance titration curves of histidine ring protons. *J. Biol. Chem. 246*, 6576–6580.

SAKURADA, J.; SEKIGUCHI, R.; SATO, K.; and HOSOYA, T. 1990. Kinetic and molecular orbital studies on the rate of oxidation of monosubstituted phenols and anilines by horseradish peroxidase compound II. *Biochemistry 29*, 4093–4098.

SAKURADA, J.; TAKAHASHI, S.; and HOSOYA, T. 1986. Nuclear magnetic resonance studies on the spatial relationship of aromatic donor molecules to the heme iron of horseradish peroxidase. *J. Biol. Chem. 261*, 9657–9662.

SALEHI, A.; OERTLING, W. A.; BABCOCK, G. T.; and CHANG, C. K. 1986. One-electron oxidation of the porphyrin ring of cobaltous octaethylporphyrin ($Co^{II}OEP$). Absorption and resonance raman spectral characteristics of the $Co^{II}OEP^+\cdot ClO_4$-π-cation radical. *J. Am. Chem. Soc. 108*, 5630–5631.

SAXENA, A.; MODI, S.; BEHERE, D. V.; and MITRA, S. 1990. Interaction of aromatic donor molecules with manganese (III) reconstituted horseradish peroxidase: Proton nuclear magnetic resonance and optical difference spectroscopic studies. *Biochim. Biophys. Acta 1041*, 83–93.

SCHONBAUM, G. R., and CHANCE, B. 1976. Catalase. *The Enzymes 13*, 363–408.

SCHONBAUM, G. R., and LO, S. 1972. Interaction of peroxidases with aromatic peracids and alkyl peroxides. *J. Biol. Chem. 247*, 3353–3360.

SCHULZ, C. E.; DEVANEY, P. W.; WINKLER, H.; DEBRUNNER, P. G.; DOAN, N.; CHIANG, R.; RUTTER, R.; and HAGER, L. P. 1979. Horseradish peroxidase compound I: Evidence for spin coupling between the heme iron and a "free" radical. *FEBS Lett. 103*, 102–105.

SHIRO, Y.; KURONO, M.; and MORISHIMA, I. 1986. Presence of endogenous calcium ion and its functional and structural regulation in horseradish peroxidase. *J. Biol. Chem. 261*, 9382–9390.

SITTER, A. J.; RECZEK, C. M.; and TERNER, J. 1985. Heme-linked ionization of horseradish peroxidase compound II monitored by the resonance raman $Fe(IV)\!\!=\!\!O$ stretching vibration. *J. Biol. Chem. 260*, 7515–7522.

SMITH, A. T.; SANDERS, S. A.; GRESCHIK, H.; THORNELEY, R. N. F.; BURKE, J. F.; and BRAY, R. C. 1992. Probing the mechanism of horseradish peroxidase by site-directed mutagenesis. *Biochem. Soc. Trans. 20*, 340–345.

SMITH, A. T.; SANTAMA, N.; DACEY, S.; EDWARDS, M.; BRAY, R. C.; THORNELEY, R. N. F.; and BURKE, J. F. 1990. Expression of synthetic gene for horseradish peroxidase C in *Escherichia coli* and folding and activation of the recombinant enzyme with Ca^{++} and heme. *J. Biol. Chem. 265*, 13335–13343.

SPIRO, T. G.; STRONG, J. D.; and STEIN, P. 1979. Porphyrin core expansion and doming in heme proteins. New evidence from resonance raman spectra of six-coordinate high-spin iron (III) heme. *J. Am. Chem. Soc. 101,* 2648–2655.

TAMURA, M., and YAMAZAKI, I. 1972. Reactions of the oxyform of horseradish peroxidase. *J. Biochem. 71,* 311–319.

TANOKURA, M.; TASUMI, M.; and MIYAZAWA, T. 1976. ^{1}H Nuclear magnetic resonance studies of histidine-containing di- and tripeptides. Estimation of the effects of charged groups on the pK_a value of the imidizole ring. *Biopolymers 15,* 393–401.

TERAOKA, J.; JOB, D.; MORITA, Y.; and KITAGAWA, T. 1983. Resonance raman study of plant tissue peroxidase. Common characteristics in iron coordination environments. *Biochim. Biophys. Acta 747,* 10–15.

TERAOKA, J., and KITAGAWA, T. 1980. Resonance raman study of the heme-linked ionization in reduced horseradish peroxidase. *Biochem. Biophys. Res. Comm. 93,* 694–700.

———. 1981. Structural implication of the heme-linked ionization of horseradish peroxidase probed by the Fe-histidine stretching raman line. *J. Biol. Chem. 256,* 3969–3977.

TERNER, J.; SITTER, A. J.; and RECZEK, C. M. 1985. Resonance raman spectroscopic characterizations of horseradish peroxidase. Observation of the $Fe^{IV}=O$ stretching vibration of compound II. *Biochim. Biophys. Acta 828,* 73–80.

THANABAL, V.; DE ROPP, J. S.; and LA MAR, G. N. 1987. Identification of the catalytically important amino acid residue resonances in ferric low-spin horseradish peroxidase with nuclear overhauser effect measurements. *J. Am. Chem. Soc. 109,* 7516–7525.

THANABAL, V.; LA MAR, G. N.; and DE ROPP, J. S. 1988. Nuclear overhauser effect study of the heme crevice in the resting state and compound I of horseradish peroxidase: Evidence for cation radical delocalization to the proximal histidine. *Biochemistry 27,* 5400–5407.

VALENTINE, J. S.; SHERIDAN, R. P.; ALLEN, L. C.; and KAHN, P. C. 1979. Coupling between oxidation state and hydrogen bond conformation in heme proteins. *Proc. Natl. Acad. Sci. USA 76,* 1009–1013.

VAN DER ZEE, J.; DULING, D. R.; MASON, R. P.; and ELING, T. E. 1989. The oxidation of N-substituted aromatic amines by horseradish peroxidase. *J. Biol. Chem. 264,* 19828.

WELINDER, K. G. 1976. Covalent structure of the glycoprotein horseradish peroxidase (EC1.11.1.7). *FEBS Lett. 72,* 19–23.

———. 1979. Amino acid sequence studies of horseradish peroxidase. Amino and carboxyl termini, cyanogen bromide and tryptic fragments, the complete sequence, and some structural characteristics of horseradish peroxidase C. *Eur. J. Biochem. 96,* 483–502.

———. 1985. Plant peroxidases. Their primary, secondary and tertiary structures, and relation to cytochrome c peroxidase. *Eur. J. Biochem. 151,* 497–504.

WISEMAN, J. S.; NICHOLS, J. S.; and KOLPAK, M. X. 1982. Mechanism of inhibition of horseradish peroxidase by cyclopropanone hydrate. *J. Biol. Chem. 257,* 6328–6332.

Chapter 12

Catalase

Catalase (hydrogen peroxide:hydrogen-peroxide oxidoreductase, (EC 1.11.1.6) occurs ubiquitously in aerobic organisms. The enzyme functions to protect cells from the toxic effects of hydrogen peroxide by catalyzing predominantly the reaction $[2H_2O_2 \rightarrow 2H_2O + O_2]$. The enzyme exists as a tetramer; each subunit consisting of a single polypeptide associated with a prosthetic group of high-spin ferric protoporphyrin IX. Catalase is relatively stable between pH 3 and 10. Denaturation by dissociation occurs at extreme pHs in the presence of detergents.

AMINO ACID SEQUENCES

The primary structure of the bovine liver catalase (BLC) subunit consists of 506 amino acid residues (Fig. 12.1). The four Cys residues in each subunit do not form disulfides in the tetrameric molecule (Schroeder et al. 1982). Association of the subunits involves mainly hydrophobic interactions. In addition to BLC, amino acid sequences of the following catalases have been deduced from their corresponding cDNA nucleotides: human (Bell et al. 1986; Quan et al. 1986), rat (Furuta et al. 1986), yeast (Hartig and Ruis 1986), sweet potato (Sakajo et al. 1987), corn (Bethards et al. 1987), *Escherichia coli* (Triggs-Raine et al. 1988), cotton seed (Ni et al. 1990), *Drosophila* (Orr et al. 1990), mouse (Shaffer et al. 1990), tomato (Inamine and Baker 1990), and pea (Isin and Allen 1991).

```
BLcat          ADNRDPASDQMKHWKEQRAAQKPDVLTTGGGNPVGDKLN
Mcat           MS-S--------KQ------S-RP------G---I-----
HMcat          MA-S--------QH------A-KA------A---V-----


BLcat    40    SLTVGPRGPLLVQDVVFTDEMAHFDRERIPERVVHAKGAG
Mcat     40    IM-A-S--------------------------H-----
HMcat    40    VI-V-P--------------------------H-----


BLcat    80    AFGYFEVTHDITRYSKAKVFEHIGKRTPIAVRFSTVAGES
Mcat     80    ------------R-----------R--------------
HMcat    80    ------------K-----------K--------------


BLcat   120    GSADTVRDPRGFAVKFYTEDGNWDLVGNNTPIFFIRDALL
Mcat    120    -------------------------------------AI-
HMcat   120    -------------------------------------PI-


BLcat   160    FPSFIHSQKRNPQTHLKDPDMVWDFWSLRPESLHQVSFLF
Mcat    160    -------------------------------------T---
HMcat   160    -------------------------------------S---


BLcat   200    SDRGIPDGHRHMDGYGSHTFKLVNADGEAVYCKFHYKTDQ
Mcat    200    ------------N------------D--------------
HMcat   200    ------------N------------N--------------


BLcat   240    GIKNLSVEDAARLAHEDPDYGLRDLFNAIATGNYPSYTLY
Mcat    240    -----P-GE-G--AQ------L--------N-N---W-F-
HMcat   240    -----S-ED-A--SQ------I--------T-K---W-F-


BLcat   280    IQVMTFSEAEIFPFNPFDLTKVWPHGDYPLIPVGKLVLNR
Mcat    280    ------KE--T-------------K-------------K
HMcat   280    ------NQ--T-------------K-------------R


BLcat   320    NPVNYFAEVEQLAFDPSNMPPGIEPSPDKMLQGRLFAYPD
Mcat    320    -----------M-------------P-----------Y--
HMcat   320    -----------I-------------A-----------Y--


BLcat   360    THRHRLGPNYLQIPVNCPYRARVANYQRDGPMCMMDNQGG
Mcat    360    -----------Q---------------------H-----
HMcat   360    -----------H---------------------Q-----
```

Figure 12.1. Comparison of amino acid sequences of catalases: BLcat (bovine liver catalase subunit, Schroeder et al. 1982), Mcat (mouse liver, sequence deduced from cDNA sequencing, Shaffer et al. 1990), HMcat (human kidney, 527 amino acids deduced from cDNA sequencing, 526 residues for the mature protein, Bell et al. 1986).

```
BLcat   400  APNYYPNSFSAPEHQPSALEHRTHFSGDVQRFNSANDDNV
Mcat    400  ---------S---Q-R-----SVQCAVD-K---S--E---
HMcat   400  ---------G---Q-P-----SIQYSGE-R---T--D---

BLcat   440  TQVRTFYLKVLNEEQRKRLCENIAGHLKDAQLFIQKKAVK
Mcat    440  ----T--TK-----E----------------L--------
HMcat   440  ----A--VN-----Q----------------I--------

BLcat   480  NFSDVHPEYGSRIQALLDKYNEEKPKN   506
Mcat    480  --TD--PD--AR---------A-----AIHTYTQAGSHMA
HMcat   480  --TE--PD--SH---------A---------FV-S---L-

BLcat
Mcat    520  AKGKANL   526
HMcat   520  -RE----   526
```

Figure 12.1   (Continued).

THREE-DIMENSIONAL STRUCTURES

Each bovine liver catalase subunit is composed of four domains (Fig. 12.2) (Murthy et al. 1981). The first domain comprises the N-terminal of 75 residues that forms an arm (with two helices, α1 and α2) interacting with the neighboring subunits. The second domain, which consists of residues 76–320, forms a β barrel with two topologically similar, 4-stranded antiparallel sheets connected by three helices (α3, α4, and α5). Residues from this region constitute the distal side and the hydrophobic cavity of the heme. The third domain (residues 321–436) forms the outer layer of the subunit with the proximal ligand Tyr357 located in the α9 helix. The fourth domain is the C-terminal portion of a four-helical region (α10–13) located mostly on the external part of the molecule. Each subunit also contains one molecule of NADPH that is tightly bound near the C-terminal domain, and strongly protected from oxidation. The NADPH molecule acquires a conformation of a right-handed helix. The function of this cofactor has not been clearly established (Kirkman and Gaetani 1984; Fita and Rossman 1985; Fita et al. 1986). It has been proposed that the catalase-bound NADPH prevents the enzyme from forming the active form of compound II via a 2e transfer to a compound II precursor (Hillar and Nicholls 1992).

The BLC tetrameric enzyme acquires a dumbell shape, with a dimension of 105 × 15 A. The heme is buried at a distance of 20 A from the center of the protein. The heme is accessible to substrates through a hydrophobic channel formed between the β barrels and the C-terminal segments. Hydrophobic residues include Val115, Ala116, Pro128, Phe152, Phe153, Phe163, Ile164, and Leu198. The heme ferric iron is coordinated to the four nitrogens of the pyrrole ring of

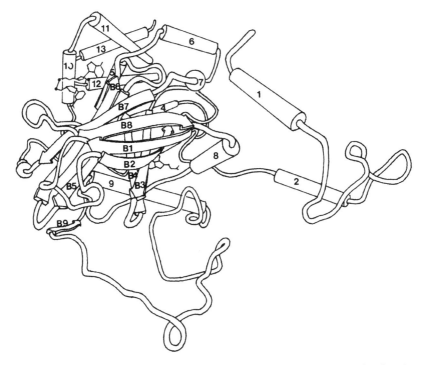

Figure 12.2. Diagrammatic view of one subunit of bovine liver catalase showing the bound heme and NADP groups. (From Melik-Adamyan et al. 1986. *J. Mol. Biol. 188,* 65, with permission. Copyright 1986 Academic Press, Ltd.)

protoprophyrin IX, with the fifth ligand occupied by Tyr357. Located on the distal side are His74 and Asn147, which are catalytically important residues.

Although its complete amino acid sequence has not yet been determined, fungal catalase from *Penicillium vitale* (PVC) has been crystallized and its structure refined by computer simulation (Vainshtein et al. 1986). PVC subunits contain structural domains very similar to those of BLC. Approximately 68% and 91% of the residues in PVC and BLC, shows structural homology (Melik-Adamyan et al. 1986). PVC contains additional residues in the C-terminal region with an α/β structure (a five-parallel stranded β sheet surrounded by four helices) (Fig. 12.3). This domain has a topology very similar to that of flavodoxin. The function of this flavodoxin-like domain is not known, as it does not bind FMN.

The catalase enzyme from the bacterium *Micrococcus lysodeikiticus,* has many similarities to BLC (Murshudov et al. 1992). It is a tetramer with a dimension of 66 × 90 × 93 A. Each subunit consists of at least three domains. The

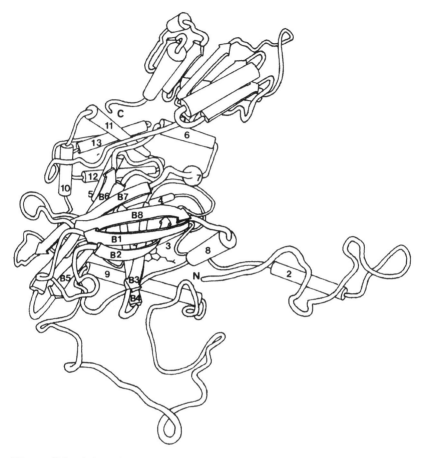

Figure 12.3. Schematic representation of *Penicillium vitale* catalase showing the L bound heme. (From Melik-Adamyan et al. 1986. *J. Mol. Biol. 188,* 65, with permission. Copyright 1986 Academic Press, Ltd.)

N-terminal helix (residues 3–58) forms an arm interacting with the neighboring subunit. The heme-containing domain (residues 59–348) contains a β barrel of eight antiparellel β strands, connected by eight α helices typical of an α/β topology. The third small domain contains four helices (residues 422–497) connecting to the α/β domain by a large loop that lies on the surface of the enzyme molecule. The binding of NADPH by MLC remains to be established.

E. coli contains two catalases: HP-I (hydroperoxidase I) is a tetramer with both catalatic and peroxidatic activities, and HP-II has an unusual hexameric structure and contains a chlorin (dihydroporphyrin) as the prosthetic group (Loewen and Switala 1986; Mulvey et al. 1988; Chiu et al. 1989). Catalases with a manganese active center have been found in bacteria, including *Thermus thermophilus*, *Lactobacillus plantarum*, and *Thermoleophilum album* (Khangulov et al. 1990; Waldo et al. 1991).

REACTION MECHANISM

The present discussion will focus primarily on the extensively studied BLC. The overall reaction involves the enzyme reacting with two moles of H_2O_2 to give water and oxygen as the final products. The chemistry of the reaction mechanism implies that H_2O_2 is involved in both oxidizing and reducing activities, with a two-electron transfer in each of the processes. This is shown in the following equations.

(1) Oxidation of native ferric enzyme to catalase compound I, with the substrate H_2O_2 reduced to H_2O.

$$En\text{-}Fe(III) + H_2O_2 \rightarrow En\text{-}Fe(IV)\text{=}O^{+\cdot} + H_2O$$

(2) Reduction of compound I back to the native ferric enzyme, with a second molecule of H_2O_2 oxidized to O_2.

$$En\text{-}Fe(IV)\text{=}O^{+\cdot} + H_2O_2 \rightarrow En\text{-}Fe(III) + O_2$$

Formation of Compound I and Reduction of H_2O_2

The first reaction is a two-electron oxidation of the ferric enzyme to compound I which contains a π-cation radical structure of $[O\text{=}Fe(IV)\text{-}porphyrin]^{+\cdot}$ similar to that of HRP-I. The rate constant for the formation of compound I with H_2O_2 as substrate = $6 \times 10^6\ M^{-1}\ s^{-1}$. Other peroxides, such as alkyl or acyl peroxides, can also serve as substrates, although with much lower rate constants (Table 12.1). The products in this case are the corresponding alcohols [En-Fe(III) + R-OOH $\rightarrow$ En-Fe(IV)=O$^{+\cdot}$ + ROH]. The low reaction rates exhibited by these other substrates may be explained by the fact that these molecules have a comparatively restricted access to interaction with the heme.

Binding of peroxide begins with displacement of the water molecules in the heme cavity (Fig. 12.4). The substrate is then positioned for the following interactions: (1) hydrogen bonding of the α proton with $N_{\varepsilon2}$ of His74, which forms a charge relay system with Ser113 and the propionic carboxylate chain on pyrrole II, (2) interaction between the α oxygen and the heme iron, (3) interaction between the β oxygen and the amide side chain of Asn147. These interactions cause an increase in the acidity of the α proton, polarization of the O_α-O_β bond, and con-

Table 12.1.   Rate Constants of Catalase in the Enzymatic Cycle

Substrate	Rate Constants ($M^{-1} s^{-1}$)	
	k_1 Formation of compound I	k_4' Reduction of compound I
H_2O_2/H_2O_2	6×10^6	3.5×10^7
CH_3OOH/CH_3OH	8.5×10^6	830
CH_3CH_2OOH/CH_3ch_2OH	2×10^4	1020
$CH_3(CH_2)_2OOH/CH_3(CH_2)_2OH$	5.5×10^3	6.5
$HOCH_2OOH/HOCH_2OH$	3×10^4	1200

From Schonbaum and Chance (1976).

sequently transfer of the α proton to the β oxygen. Protonation facilitates the interaction between O_α and the iron reinforced by electron transfer from the heme Fe(III) and the porphyrin π electron system. The development of a positively charged heme lowers the basicity of the imidazole of His74 (which lies parallel to the heme) and, in turn, assists the transfer of the α proton to the β oxygen.

The reaction mechanism has many similarities to that of HRP with a few differences. (1) Both catalase and HRP contain a distal His (74 and 42, respec-

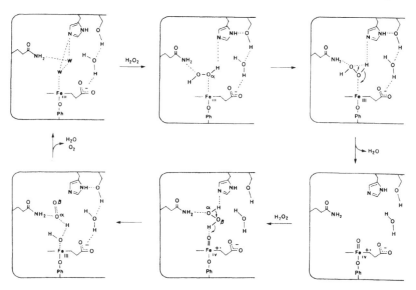

Figure 12.4.   Reaction mechanism of catalase showing the formation and subsequent reduction of compound I. (From Fita and Rossman 1985).

tively) with the imidazole $N_{\epsilon 2}$ close to the heme iron and the $N_{\delta 1}$ hydrogen bonded to another residue (Ser113 in catalase). The His74 in catalase, however, is positioned parallel to the heme approximately 3.5 A above the pyrrole III ring. (2) The imidazole group of the distal His in both enzymes acts in the transfer of the α proton from O_α to O_β of the peroxide substrate, H_2O_2. (3) In both cases, polarization of the O_α-O_β bond is enhanced by interaction with side chain groups. In catalase, Asn147 replaces Arg48, which is highly conserved in HRP and other peroxidases. (4) Compound I of both enzymes contains a π-cation radical of the heme prosthetic group, with the heme iron having an oxo-iron [Fe(IV)=O] structure (Dolphin et al. 1971). (5) Both enzymes have the proximal residue in compound I deprotonated. In catalase, the heme iron coordinates with the phenolate oxygen of Tyr357, which forms two hydrogen bonds with Arg353. In HRP, the His170-$N_{\delta 1}$ is deprotonated with the proton transferred to an adjacent carboxylate side chain.

Reduction of Compound I

Catalase compound I reacts with a second molecule of H_2O_2 to complete the catalitic cycle, giving back the native enzyme with generation of O_2 and H_2O.

Because compound I has a positively charged heme, the imidazole of His74, which lies parallel above the heme, becomes less nucleophilic and less likely to deprotonate the α proton as occurred in the first reaction. Reduction of compound I therefore occurs with the abstraction of the β hydrogen (most likely as hydride) by the heme ferryl oxygen (Dounce 1983). This hydride transfer allows redistribution of electrons in the heme iron and the porphyrin π system, and development of a double bond character at O_α-O_β. Because compound I acquires an Fe(III) state, the nucleophilicity of the imidazole of His74 increases, and a weakening of the O_α-H bond occurs. Transfer of the α proton thus occurs via the imidazole of His74, from the α oxygen to the heme ferryl oxygen.

Catalase Compound I

Although compound I of catalase consists of a π-cation radical heme structure similar to that of HRP-I, the two reaction mechanisms show a marked difference. HRP-I undergoes a two-step one-electron reduction via the formation of compound II. As described earlier, HRP-I reacts with electron donors at the heme edge. Apparently, in catalase, the phenoxylate oxygen of the proximal Tyr provides more electron density to the Fe heme system than the His-$N_{\epsilon 2}$ in HRP. This allows electrons to add to both the iron and the heme instead of the preferential addition occurred in the HRP enzyme (Chance 1984). Furthermore, H_2O_2 is a very poor substrate for HRP-I and HRP-II (k_{cat} 4.0 s^{-1} and 9 $\times$ 10^4 s^{-1} for HRP and catalase, respectively, at pH 7, 20°C, 10^{-2} M H_2O_2). In contrast, catalase compound I undergoes a single two-electron reduction to yield the native enzyme and in the process, the substrate H_2O_2 is oxidized to O_2 and H_2O with $k_4' = 3.5$

$\times$ 10^7 M^{-1} s^{-1}. H_2O_2 is directed to interact with the ferryl oxygen of compound I, instead of the heme edge (Chuang et al. 1989). Unlike HRP-I, catalase is inactivated by reaction with phenylhydrazine due to the formation of iron-phenyl complex (Ortiz de Montellano and Kerr 1983).

Catalase Compound II

Catalase compound I can be converted to compound II via a one-electron reduction by certain substrates, such as ferrocyanide, ascorbic acid, and phenols. Compared to HRP-II, catalase compound II is a poor oxidant and regarded as an inactive form (Nicholls and Schonbaum 1963). There is marked structural similarity between catalase compound II and HRP-II. The heme structure of catalase compound II is described as Fe(IV)=O, with the oxygen hydrogen bonded to the distal His74 (corresponding to His42 in HRP-II), although the fifth ligand is a Tyr phenoate ion instead of a His imidazole. Catalase compound II also has an Fe=O frequency identical to that of HRP-II (Chuang et al. 1989). The difference in the reactivities of compound II in the two enzymes may be related to the way the protein moiety controls the accessibility and orientation of the substrates.

PEROXIDATIC ACTIVITY

Catalase compound I also reacts with other electron donor substrates such as alkyl or acyl alcohols, but with significantly lower rate constants (Table 12.1). In these reactions, instead of H_2O acting as the hydrogen donor, the alcohols are oxidized to the corresponding aldehydes.

$$En\text{-}Fe(III) + H_2O_2 \rightarrow En\text{-}Fe(IV)=O^{+\cdot} + H_2O$$
$$En\text{-}Fe(IV)=O^{+\cdot} + RCH_2OH \rightarrow En\text{-}Fe(III) + RCHO + H_2O$$

The overall reaction [$H_2O_2 + RCH_2OH \rightarrow RCHO + 2H_2O$] exhibits a peroxidase-catalyzed process in that the electrons for the reduction of H_2O_2 are furnished by a second substrate, in this case, an alcohol (although here, a two-electron step reduction of compound I is involved). Here, the first step involves the formation of compound I with the reduction of H_2O_2 to H_2O as discussed in the catalatic mechanism. The difference is seen in the second step.

In the case of alcohols as the hydrogen donor, the rate constant k_4' for the reduction of compound I is $\sim 1 \times 10^3$ $M^{-1}s^{-1}$ for methanol and ethanol, significantly lower than that of H_2O_2 (Table 12.1). For the oxidation of H_2O_2 in the catalatic cycle, the two hydrogen and two oxygen atoms are equivalent (H:O:O:H). In the oxidation of alkyl or acyl alcohols, the orientation of the substrate molecule becomes more restricted and consequently, the reaction is stereospecific with the pro-R hydrogen selectively abstracted (Corrall et al. 1974) similar to that of alcohol dehydrogenase.

It is noted that in the absence of hydrogen donors, compound I also reacts with organic hydroperoxides to yield compound II and a radical that catalyzes destruction of the enzyme (Jones et al. 1968; Pichorner et al. 1993).

$$E\text{-}Fe(III) + ROOH \rightarrow E\text{-}Fe(IV)\text{=}O^{+\cdot} + ROH$$
$$E\text{-}Fe(IV)\text{=}O^{+\cdot} + ROOH \rightarrow Fe(IV)\text{=}O + ROO\cdot + H^+$$

REFERENCES

BELL, G. I.; NAJARIAN, R. C.; MULLENBACK, G. T.; and HALLEWELL, R. A. 1986. cDNA sequence coding for human kidney catalase. *Nucl. Acids Res. 14*, 5561–5562.

BETHARDS, L. A.; SKADSEN, R. W.; and SCANDALIOS, J. G. 1987. Isolation and characterization of a cDNA clone for the *Cat2* gene in maize and its homology with other catalases. *Proc. Natl. Acad. Sci. USA 84*, 6830–6834.

CHANCE, B.; POWERS, L.; CHING, Y.; POULOS, T.; SCHONBAUM, G. R.; YAMAZAKI, I.; and PAUL, K. G. 1984. X-ray absorption studies of intermediates in peroxidase activity. *Arch. Biochem. Biophys. 235*, 596–611.

CHIU, J. T.; LOEWEN, P. C.; SWITALA, J.; GENNIS, R. B.; and TIMKOVICH, R. 1989. Proposed structure for the prosthetic group of the catalase HPII from *Escherichia coli. J. Am. Chem. Soc. 111*, 7046–7050.

Cuhuang, W.-J.; HELDT, J.; and VAN WART, H. E. 1989. Resonance raman spectra of bovine liver catalase compound II. *J. Biol. Chem. 264*, 14209–14215.

CORRALL, R. J. M.; RODMAN, H. M.; MARGOLIS, J.; and LANDAU, B. R. 1974. Stereospecificity of the oxidation of ethanol by catalase. *J. Biol. Chem. 249*, 3181–3182.

DOLPHIN, D.; FORMAN, A.; BORG, D. C.; FAJER, J.; and FELTON, R. H. 1971. Compounds I of catalase and horseradish peroxidase: π-cation radicals. *Proc. Natl. Acad. Sci. USA 68*, 614–618.

DOUNCE, A. L. 1983. A proposed mechanism for the catalase action of catalase. *J. Theor. Biol. 105*, 553–567.

FITA, I., and ROSSMANN, M. G. 1985. The NADPH binding site on beef liver catalase. *Proc. Natl. Acad. Sci. USA 82*, 1604–1608.

FITA, I.; SILVA, A. M.; MURTHY, M. R. N.; and ROSSMANN, M. G. 1986. The refined structure of beef liver catalase at 2.5 A resolution. *Acta Cryst. B42*, 497–515.

FURATA, S.; HAYASHI, H.; HIJIKATA, M.; MIYAZAWA, S.; OSUMI, T.; and HASHIMOTO, T. 1986. Complete nucleotide sequence of cDNA and deduced amino acid sequence of rat liver catalase. *Proc. Natl. Acad. Sci. USA 83*, 313–317.

HARTIG, A., and RUIS, H. 1986. Nucleotide sequence of the *Saccharomyces cerevisiae* CTTa gene and deduced amino-acid sequence of yeast catalase T. *Eur. J. Biochem. 160*, 487–490.

HILLAR, A., and NICHOLLS, P. 1992. A mechanism for NADPH inhibition of catalase compound II formation. *FEBS Lett. 314*, 179–182.

INAMINE, G. S., and BAKER, J. E. 1990. The catalase genes of tomato: Molecular analysis and sequence. *J. Cell Biochem. 14E*, 349.

ISIN, S. H., and ALLEN, R. D. 1991. Isolation and characterization of a pea catalase cDNA. *Plant Mol. Biology 17*, 1263–1265.

JONES, P.; SUGGETT, A.; and PAIN, R. H. 1968. Sub-unit structure of catalase compound II. *Nature 217*, 1050.

JOUVE, H.-M.; GOUET, P.; BOUDJADA, N.; BUISSON, G.; KAHN, R.; and DUEE, R. 1991. Crystallization and crystal packing of *Proteus mirabilis* PR catalase. *J. Mol. Biol. 221*, 1075–1077.

KHANGULOV, S. V.; BARYNIN, V. V.; VOEVODSKAYA, N. V.; and GREBENKO, A. I. 1990. ESR spectroscopy of the binuclear cluster of manganese ions in the active center of Mn-catalase from *Thermus thermophilus*. *Biochim. Biophys. Acta 1020*, 305–310.

KIRKMAN, H. N., and GAETANI, G. F. 1984. Catalase: A tetrameric enzyme with four tightly bound molecules of NADPH. *Proc. Natl. Acad. Sci. USA 81*, 4343–4347.

LOEWEN, P. C., and SWITALA, J. 1986. Purification and characterization of catalase HPII from *Escherichia coli* K12. *Biochem. Cell Biol. 64*, 638–648.

MELIK-ADAMYAN, W. R.; BARYNIN, V. V.; VAGIN, A. A.; BORISOV, V. V.; VAINSHTEIN, B. K.; FITA, I.; MURTHY, R. N.; and ROSSMANN, M. G. 1986. Comparison of beef liver and *Penicillium vitale* catalases. *J. Mol. Biol. 188*, 63–72.

MULVEY, M. R.; SORBY, P. A.; TRIGGS-RAINE, B. L.; and LOEWEN, P. C. 1988. Cloning and physical characterization of *katE* and *katF* required for catalase HPII expression in *Escherichia coli*. *Gene 73*, 337–345.

MURSHUDOV, G. N.; MELIK-ADAMYAN, W. R.; GREBENKO, A. I.; BARYNIN, V. V.; VAGIN, A. A.; VAINSHTEIN, B. K.; DAUTER, Z.; and WILSON, K. S. 1992. Three-dimensional structure of catalase from *Micrococcus lysodeikticus* at 1.5 Å resolution. *FEBS Lett. 312*, 127–131.

MURTHY, M. R. N.; REID, T. J. III; SICIGNANO, A.; TANAKA, N.; and ROSSMANN, M. G. 1981. Structure of beef liver catalase. *J. Mol. Biol. 152*, 465–499.

NI, W.; TURLEY, R. B.; and TRELEASE, R. N. 1990. Characterization of a cDNA encoding cottonseed catalase. *Biochim. Biophys. Acta 1049*, 219–222.

NICHOLLS, P., and SCHONBAUM, G. R. 1963. Catalases. *The Enzymes 8*, 147–225.

ORITZ DE MONTELLANO, P. R., and KERR, D. E. 1983. Inactivation of catalase by phenylhydrazine. *J. Biol. Chem. 258*, 10558–10563.

ORR, E. C.; BEWLEY, G. C.; and ORR, W. C. 1990. cDNA and deduced amino acid sequence of *Drosophila* catalase. *Nucl. Acids Res. 18*, 3663.

PICHORNER, H.; JESSNER, G.; and EBERMANN, R. 1993. tBOOH acts as a suicide substrate for catalase. *Arch. Biochem. Biophys. 300*, 258–264.

QUAN, F.; KORNELUK, R. G.; TROPAK, M. B.; and GRAVEL, R. A. 1986. Isolation and characterization of the human catalase gene. *Nucl. Acid Res. 14*, 5321–5335.

REID, T. J. III; MURPHY, M. R. N.; SICIGNANO, A.; TANAKA, N.; MUSICK, W. D. L.; and ROSSMANN, M. G. 1981. Structure and heme environment of beef liver catalase at 2.5 Å resolution. *Proc. Natl. Acad. Sci. USA 78*, 4767–4771.

SAKAJO, S.; NAKAMURA, K.; and ASAHI, T. 1987. Molecular cloning and nucleotide sequence of full-length cDNA for sweet-potato catalase mRNA. *Eur. J. Biochem. 165*, 437–442.

SCHONBAUM, G. R., and CHANCE, B. 1976. Catalase. *The Enzymes 13*, 363–408.

SCHROEDER, W. A.; SHELTON, J. R.; SHELTON, J. B.; ROBERSON, B.; APELL, G.; FANG, R. S.; and VENTURA, J. B. 1982. The complete amino acid sequence of bovine liver catalase and the partial sequence of bovine erythrocyte catalase. *Arch. Biochem. Biophys. 214*, 397–421.

SHAFFER, J. B.; PRESTON, K. E.; and SHEPARD, B. A. 1990. Nucleotide and deduced amino acid sequences of mouse catalase: Molecular analysis of a low activity mutant. *Nucl. Acids Res. 18*, 4941.

SHARMA, K. D.; ANDERSON, L. A.; LOEHR, T. M.; TERNER, J.; and GOFF, H. M. 1989. Comparative spectral analysis of mammalian, fungal, and bacterial catalases. *J. Biol. Chem. 264*, 12772–12779.

TORMO, J.; FITA, I.; SWITALA, J.; and LOEWEN, P. C. 1990. Crystallization and preliminary x-ray diffraction analysis of catalase HPII from *Escherichia coli. J. Mol. Biol. 213*, 219–220.

TRIGGS-RAINE, B. L.; DOBLE, B. W.; MULVEY, M. R.; SORBY, P. A.; and LOEWEN, P. C. 1988. Nucleotide sequence of *katG*, encoding catalase HP I of *Escherichia coli. J. Bacteriol. 170*, 4415–4419.

VANISHTEIN, B. K.; MELIK-ADAMYAN, W. R.; BARYNIN, V. V.; VAGIN, A. A.; GREHENKO, A. I.; BORISOV, V. V.; BARTELS, K. S.; FITA, I.; and ROSSMANN, M. G. 1986. Three-dimensional structure of catalase from *Penicillium vitale* at 2.0 A resolution. *J. Mol. Biol. 188*, 49–61.

WALDO, G. S.; FRONKO, R. M.; and PENNER-HAHN, J. E. 1991. Interaction and reactivation of manganese catalase: Oxidation-state assignments using x-ray absorption spectroscopy. *Biochemistry 30*, 10486–10490.

Chapter 13

Xylose Isomerase

Xylose isomerase (D-xylose ketol-isomerase, EC 5.3.1.5), also known as glucose isomerase, catalyzes the reversible isomerization of D-xylose and D-glucose to their respective ketoses, D-xylulose and D-fructose, in the presence of divalent metal ions (Fig. 13.1). The enzyme is essentially a "xylose" isomerase; isomerization of glucose is not the primary function in physiological systems.

Various microorganisms grow on hemicellulose, which comprises 40% of the plant biomasss. In bacterial systems the xylose from degradation of hemicellulose is converted to xylulose by xylose isomerase and phosphorylated to xylulose-5-phosphate, which is then channelled to the pentose phosphate pathway. Most yeasts, however, require two enzymes in a two-step process to convert xylose to xylulose—xylose reductase using NADPH for reducing xylose to xylitol, and NAD$^+$-linked D-xylitol dehydrogenase for the oxidation of xylitol to xylulose. There is considerable interest in introducing a bacterial xylose isomerase gene into yeast cells to achieve direct conversion of biomass (Alexander 1986; Chan et al. 1989). An alternative approach involves cloning the genes for the ethanol pathway into E. coli to achieve a complete conversion of xylose to ethanol in a single stage process (Beall et al. 1991).

The most important utilization of xylose isomerase is recognized in the corn sweetener industry. For the industrial production of high-fructose corn syrup, the enzyme is immobilized in various ways, including whole cell entrapment, covalent crosslinking, and adsorption on cellular materials. The enzymes used in these

α–D–XYLOPYRANOSE   α–D–XYLULOFURANOSE

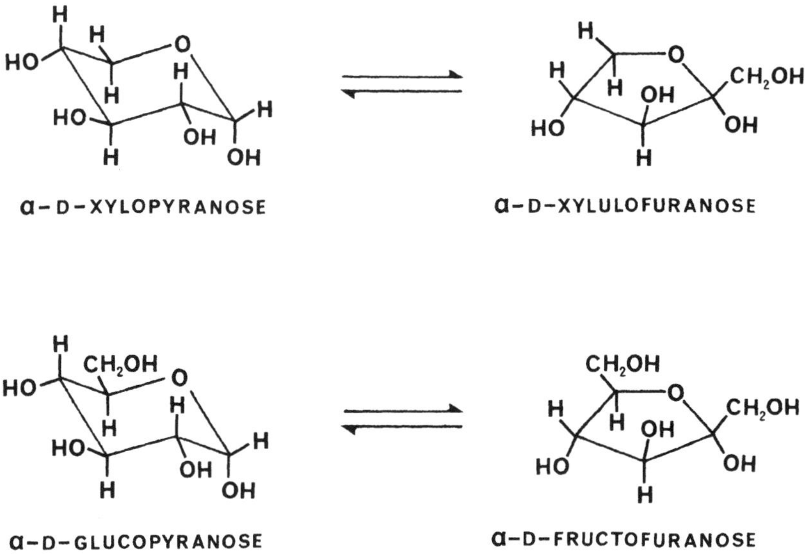

α–D–GLUCOPYRANOSE   α–D–FRUCTOFURANOSE

Figure 13.1. Reversible isomerization of D-xylose and D-glucose to their respective ketoses.

processes are usually products of microorganisms obtained from sophisticated screening procedures to optimize thermal stability and operation conditions (Antrim and Auterinen 1986). The properties of some of these commercially available immobilized enzyme systems are presented in Table 13.1.

GENERAL CHARACTERISTICS

The enzyme is a homomeric tetramer with varying molecular weight depending on the microbial source: 180 kD for *Streptomyces griseofuscus* (Kasumi et al. 1981), 165 kD for *Streptomyces albus* (Hogue-Angeletti 1975), 120 kD for *Streptomyces olivochromogenes*, 130 kD for *Bacillus stearothermophilus* (Suekane et al. 1978), 183 kD for *Lactobacillus xylosus* (Yamanaka and Takahara 1977), 185 kD for *Arthrobacter* strain N.R.R.L. B3728 (Smith et al. 1991), 196 kD for *Thermus aquaticus* HB8 (Lehmacher and Bisswanger 1990A), and 200 kD for *Clostridium thermohydrosulfuricum* (Dekker 1991A) and for *Thermus thermophilus* (Dekker et al. 1991B). The *E. coli* isomerase is a dimer of subunits with a MW of 44 kD (Tucker et al. 1988).

Table 13.1. Properties of Commercially Available Immobilized Glucose Isomerase

	Maxazyme GI-Immob. Gist Brocades	Sweetzyme Q NOVO Industri	Optisweet 22 Miles Kali-Chemie	Taka-Sweet[a] Miles Lab, Inc.
Trade name & company				
Enzyme source	*Actinoplanes missouriensis*	*Bacillus coagulans*	*Streptomyces rubiginosus*	*Flavobacterium arborescens*
Immobilization method	Whole cells entrapped in gelatin, crosslinked with glutaraldehyde	Whole cells crosslinked with glutaraldedehyde	Purified enzyme covalently bound on silica	Bound on nonviable materials
Product form	Wet spherical particles	Granulated dried cylindrical particles	Suspension of spherical particles	Granular
Reactant (reported)	D-xylose, D-ribose, D-allose, D-galactose, D-glucose	D-xylose, D-ribose, D-glucose	D-glucose	D-glucose
T (maximum activity)	90°C	83°C	—	—
T (recommended)	60°C	60°C	60°C	60°C
pH range	6–9	5–9	—	6.5–8.0
pH (recommended)	7.5	8.2	7.5	7.5
Cation requirement	$3 \times 10^{-3}\ M$ Mg^{++}	$4 \times 10^{-4}\ M$ Mg^{++}	$5 \times 10^{-3}\ M$ Mg^{++}	Mg^{++}
Inhibitors	Ag$^+$, Ni^{++}, Hg^{++}, Fe^{++}, Zn^{++}, Ca^{++}, Al^{+++}	Cu^{++}, Ni^{++}, Ga^{++}, Zn^{++}, Ba^{++}, Hg^{++}, As^{++}	Ca^{++}	Ca^{++}
Half-lifetime (at recommended conditions)	1,300 hr	1,050 hr	1,200 hr	—
Productivity at half lifetime (kg dry product/kg dry enzyme preparation)	2,430	1,430	13,000	—
Molecular weight	80,000	100,000	160,000	—
Enzyme content of dry enzyme preparation	6–7%	~12%	7–8%	—
Activity (mol·gm^{-1}·hr^{-1})	0.044	0.116	0.617	0.009–0.015
Apparent turnover numbers (s^{-1})	50	30	360	—

From Makkee et al. 1985. *Starch/Stärke* 37, 234 with permission. Copyright 1985 VCH Verlagsgesellschaft mbH.
[a]Anon 1983.

Thermal and pH Stability

Xylose isomerase is relatively heat stable. The temperature optimum is 80°C for *Streptomyces olivochromogenes* and *Bacillus stearothermophilus* (Suekane et al. 1978), and *Streptomyces* sp. (Takasaki et al. 1969). The *Bacillus* enzyme retains its activity after incubation for 30 min at 75°C, whereas the *Streptomyces* enzyme loses its activity above 55°C. The *Lactobacillus* enzyme is stable at 60°C for 30 min (Yamanaka and Takahara 1977). The pH optimum for xylose isomerases is generally in the slightly alkaline range (pH 8.0–8.5 for *Streptomyces*, pH 7.5 for *Lactobacillus*, 7.5–8.0 for *Bacillus* enzyme). In general, the enzyme is stable over a wide pH range of 4.0–11.0.

Divalent Metal Cations

The activity of xylose isomerase is dependent on divalent cations. The *Streptomyces* enzyme requires Mg^{++} or Co^{++} for activity. Other metal ions (Fe^{++}, Mn^{++}, Ba^{++}, Ni^{++}, Ca^{++}, Zn^{++}, and Cu^{++}) have relatively little effect (Takasaki et al. 1969). For the *Streptomyces violaceoruber* enzyme, the highest activity is obtained with Mg^{++}, to a lesser degree with Co^{++}, Mn^{++} and Fe^{++} have slight effect, and Ni^{++}, Ca^{++}, Zn^{++}, Cu^{++}, and Hg^{++} show no activation (Callens et al. 1986). The *Arthrobacter* (strain N.R.R.L. B3728) enzyme requires Mg^{++}, Co^{++}, or Mn^{++} for activity, with Mg^{++} being the most efficient (Smith et al. 1991). The effect of metal (Mg^{++}, Co^{++}, Mn^{++}) activation on the kinetic parameters for the interconversion of xylose-xylulose and glucose-fructose for five xylose isomerases (*Streptomyces violaceoruber, Streptomyces* sp., *Lactobacillus xylosus, Lactobacillus brevis, and Bacillus coagulans*) are presented in Table 13.2. For the *Streptomyces* enzymes, the highest catalytic efficiency (k_{cat}/K_m) occurs with activation by Mg^{++}. For the other three enzymes studied, Mn^{++} shows higher activation than Mg^{++} or Co^{++} for the pentoses, whereas Co^{++} is preferred for the isomerization of hexoses (van Bastelaere et al. 1991).

Metal ion plays a major role in stabilizing the molecular structure of xylose isomerases. The Co^{++}-enzyme of *Streptomyces griseofuscus* shows increasing stability toward heat and acid inactivation, and denaturants, such as GuHCl and EDTA-urea (Kasumi et al. 1982A,B), but Mg^{++} tends to enhance the denaturing effect. The stability of *Arthrobacter* isomerase to heat and denaturant is enhanced by the binding of Mg^{++} or Co^{++}. The half-life of inactivation at 80°C, pH 7 is 3.5 min, and 5.5 hr and >18 hr for the apo-, Mg^{++}- and Co^{++}-enzymes, respectively (Rangarajan et al. 1992).

Substrate Specificity

Although the enzyme isomerizes D-glucose and in some cases, D-ribose, other than D-xylose, the K_m is lowest and V_{max} is highest when xylose is the

Table 13.2. Kinetic Parameters for D-Xylose/D-Xylulose, and D-Glucose/D-Fructose

Substrate	k_{cat} (s^{-1})			K_m (mM)			k_{cat}/K_m (M^{-1} s^{-1})		
	Mg^{++}	Co^{++}	Mn^{++}	Mg^{++}	Co^{++}	Mn^{++}	Mg^{++}	Co^{++}	Mn^{++}
D-Xylose									
S. violaceoruber	8.2	3.62	2.39	3.2(1.2)	3.3(1.2)	9.1(3.4)	2563(6833)	1097(3017)	263(703)
Streptomyces sp.	8.4	3.91	3.07	3.1(1.1)	3.2(1.2)	7.3(2.7)	2710(7636)	1222(3258)	421(1137)
L. xylosus	2.19	2.65	7.0	3.5(1.3)	3.7(1.4)	5.1(1.9)	626(1685)	716(1893)	1373(3684)
L. brevis	0.273	0.73	7.2	0.555(0.205)	0.70(0.26)	2.2(0.81)	492(1332)	1043(2808)	3273(8888)
B. coagulans	1.29	1.25	7.0	0.903(0.334)	0.38(0.14)	1.71(0.63)	1429(3862)	3289(8929)	4094(11111)
D-Xylulose									
S. violaceoruber	5.3	8.5	0.68	0.35(0.061)	1.2(0.21)	0.59(0.10)	15143(86885)	7083(40476)	1150(6800)
Streptomyces sp.	7.8	5.6	1.3	0.37(0.065)	0.9(0.16)	0.43(0.075)	21081(120000)	6222(35000)	3023(17333)
L. xylosus	1.03	0.99	7.2	0.22(0.039)	0.20(0.035)	0.62(0.11)	4682(26410)	4950(28286)	11613(65455)
L. brevis	0.39	0.19	9.6	0.11(0.019)	0.024(0.0042)	0.69(0.12)	3545(20526)	7917(45238)	13913(80000)
B. coagulans	1.4	0.80	11	0.14(0.025)	0.033(0.0058)	0.40(0.070)	10000(56000)	26667(137931)	27500(157143)
D-Glucose									
S. violaceoruber	2.35	1.43	0.300	166(64)	339(130)	1039(400)	14.2(37)	4.2(11.0)	0.3(0.75)
Streptomyces sp.	2.57	1.63	0.56	145(56)	374(144)	950(365)	17.7(46)	4.4(11.3)	0.59(1.53)
L. xylosus	0.420	2.01	—	493(190)	574(221)	>1000(>385)	0.85(2.2)	3.5(9.1)	—
L. brevis	0.240	1.22	1.10	373(143)	412(158)	656(252)	0.64(1.68)	3.0(7.7)	1.68(4.4)
B. coagulans	0.462	2.14	0.243	145(56)	189(73)	168(65)	3.2(8.3)	11.3(29)	1.45(3.7)
D-Fructose									
S. violaceoruber	1.96	4.02	0.278	175(15.9)	900(82)	880(80)	11.3(124)	4.5(49)	0.32(3.5)
Streptomyces sp.	2.69	4.20	0.428	130(11.8)	1070(97)	752(68)	21(228)	3.9(43)	0.57(6.3)
L. xylosus	0.073	0.406	—	98(8.9)	135(12.3)	>1000(>90)	0.74(8.2)	3.0(33)	—
L. brevis	0.253	0.55	0.65	397(36)	232(21.1)	424(39)	0.64(7.0)	2.4(26)	1.53(16.7)
B. coagulans	0.266	0.70	0.135	100(9.1)	67(6.1)	99(9.0)	2.7(29)	10.4(115)	1.36(15.0)

From van Bastelaere et al. 1991. *Biochem. J.* 278, 289, with permission. Copyright 1991 The Biochemical Society and Portland Press.

Average standard error: k_{cat} = 2% (D-xylose), 9% (D-xylulose), 2% (D-glucose and D-fructose); K_m = 5% (D-xylose), 12% (D-xylulose), 5% (D-glucose and D-fructose). The values in parentheses are for the reactive form. The anomeric composition: (1) D-xylose at 31°C = 36.5% α-D-xylopyranose (reactive form), 63% β-D-xylopyranose, <1% D-xylofuranose (α + β; (2) D-xylulose at 35°C = 17.5% α-D-xylulose (reactive form), 62.3% β-D-xylulose; (3) D-glucose at 31°C = 38% α-D-glucopyranose (reactive form), 62% β-D-glucopyranose; D-fructose at 36°C = 9% α-D-fructofuranose (reactive form), 31% β-D-fructofuranose, <1% α-D-fructopyranose, 57% β-D-fructopyranose.

substrate. For the *Bacillus coagulans* enzyme, $K_m = 1.1 \times 10^{-3}, 9 \times 10^{-2}$, and $7.7 \times 10^{-2} M$, and $V_{max} = 1.1, 0.52$, and 0.25 mg/min/mg at 40°C, pH 7.0, for xylose, glucose, and ribose, respectively (Danno 1970). Similar results are observed for the *Streptomyces olivochromogenes* enzyme in that $K_m = 0.25$ and $0.033 M$, and $V_{max} = 5.33$ and 21.84 µmol/min/mg for glucose and xylose (at 60°C, pH 7.5, 20 mM Mg^{++}, 3 mM Co^{++}) (Suekane et al. 1978). For the reverse reaction with fructose and xylulose, the K_m values are 0.20 and $0.018 M$, respectively. Kinetic studies of xylose isomerases from five different sources (*Streptomyces*, *Lactobacillus*, and three *Bacillus* sp.) provide further support that pentoses are better substrates than the hexoses with significantly lower K_m (2 to 3 orders of difference) and slightly higher k_{cat}. The ketoses show lower K_m values compared with the corresponding aldoses (Table 13.2) (van Bastelaere et al. 1991). The observed differences in the catalytic efficiencies, k_{cat}/K_m, are largely the result of the differences in the Michaelis constants for the reaction.

Inhibitors

The sugar alcohols of the corresponding substrates—D-xylitol, D-sorbitol, and D-mannitol—are competitive inhibitors, with $K_i = 2.5 \times 10^{-3}, 2.9 \times 10^{-2}$, and $7 \times 10^{-2} M$, respectively, when D-glucose, D-xylose, or D-ribose is used as a substrate (*Bacillus coagulans* enzyme at 40°C, pH 7.0) (Danno 1970). For *Lactobacillus xylosus*, the enzyme is competitively inhibited by xylitol with a K_i of $7 \times 10^{-3} M$, whereas arabitol and ribitol exhibit no inhibition (Yamanaka and Takahara 1977). For the *Arthrobacter* enzyme, xylitol and sorbitol show inhibition with $K_i = 3 \times 10^{-4}$ and $6.5 \times 10^{-3} M$, respectively. In some cases, the inhibitions are not entirely competitive, because there is a decrease in k_{cat}. It has been postulated that the additional OH group in the sugar alcohols imposes steric effects on the catalytic activity of adjacent subunits of the enzyme molecule (Smith et al. 1991). The binding affinity of these polyols is usually higher than that of their respective substrates. In the inhibition study of the xylose isomerase from *Streptomyces violaceoruber*, the dissociation constant K_d for xylitol and sorbitol on enzyme-Mg^{++} complexes is 4.6×10^{-4} and $4.5 \times 10^{-3} M$, respectively, in comparison with the substrates D-xylose ($K_m = 2.8 \times 10^{-3} M$) and D-glucose ($K_m = 1.5 \times 10^{-1} M$) (Callens et al. 1986).

AMINO ACID SEQUENCES

The cDNA sequences are known for xylose isomerases from *Bacillus subtilis* (Wilhelm and Hollenberg 1985), *Escherichia coli* (Schellenberg et al. 1984; Lawlis et al. 1984), *Ampullariella* sp. strain 3876 (Saari et al. 1987), *Streptomyces violaceoniger* (Drocourt et al. 1988), *Actinoplanes missouriensis* (Amore and Hollenberg 1989), *Clostridium thermosulfurogenes* (Lee et al. 1990), *Clostridium thermohydrosulfuricum* (Dekker et al. 1991A), *Arthrobacter* strain N.R.R.L. B3728 (Loviny-Anderton et al. 1991), *Thermus thermophilus* (Dekker et al. 1991B), and *Lactobacillus brevis* (Bor et al. 1992). The *Streptomyces, Ampullariella*, and *Arthrobacter* sequences show a high sequence similarity of ~65%. *Bacillis* and *E. coli* share ~50% similarity. The *Clostridium* enzymes share a higher homlogy with *Bacillus* and *E. coli* (50–70%) than with the former (22–24%) (Lee et al. 1990). Xylose isomerases show several highly conserved regions suggesting their possible roles in the catalytic process and/or in metal binding (Fig. 13.2).

```
SV  XI
Art XI
Amp XI
TT  XI
CT  XI                       MEYFKNVPQIKYEGPKSNNPYSFKFYNPEEVI
BS  XI                       MAQSHSSSVNYFGSVNKVVFEGKASTNPLAFKYYNPQEVI
EC  XI                       MQAYFDQLDRVRYEGSKSSNPLAFRHYNPDELV


SV  XI          MSFQPTPEDKFTFGLW  TVGWQGRDPFGDAT   RPALDP
Art XI          MSVQPTPADH-TFGL-  -VGWT-A-P--V--   -KNLDP
Amp XI          MSLQATPDDK-SFGL-  -VGWQAR-A--D--   --VLDP
TT  XI          MYEPKPEHR-TFGL-   TVGNV-R-P--D-V   -ERLDP
CT  XI    33    DGKTMEEHLR-SIAY-H-FTAD-T-Q--K--MQ--WNHYT
BS  XI    41    GGKTMKEHLR-SIAY-H-FTAD-T-V--A--MQ--WDHYK
EC  XI    34    LGKRMEEHLR-AACY-H-FCWN-A-M--VGAFN--WQQPG


SV  XI    36               VETVQRLAELGAYGVTFHDDDLIPFGS
Art XI    36               V-AVHK-AELG-YGIT-H-N-LI-FDA
Amp XI    36               I-AVHK-AEIG-YGVT-H-D-LV-F-A
TT  XI    21               VYVVHK-AELG-YGVNLH-E-LI-RGT
CT  XI    73    DPMDIAKARVEAAF-FFDK-   N-PYFC-H-R-IA-E-D
BS  XI    81    G MDLARARVEAAF-MFEK-   D-PFFA-H-R-IA-E-S
EC  XI    74    EALALAKRKADVAF-FFHK-   HVPFYC-H-V-VS-E-A
```

Figure 13.2. Comparison of amino acid sequences of xylose isomerases: SV (*Streptomyces violaceoniger* (Drocourt et al. 1988), Art (*Arthrobacter* strain B3728 (Loviny-Anderton et al. 1991), Amp (*Ampullariella* strain 3876, Saari et al. 1987), TT (*Thermus thermophilus*, Dekker et al. 1991B), CT (*Clostridium thermohydrosulfuricum* (Dekker et al. 1991A), BS (*Bacillus subtilis* (Wilhelm and Hollenberg 1985), *Escherichia coli* (Schellenberg et al. 1984). All the amino acid sequences are deduced from cDNA sequencing. The N-terminal amino acids are indicated by double lines.

```
SV  XI   63   SDTER   ESHIKRFRQALDATGMTVPMATTNLFTHPVF
Art XI   63   TEA-R   EKILGDFNQA-KDT-LK-PMV-T---SH-V-
Amp XI   63   DAATR   DGIVAGFSKA-DET-LI-PMV-T----H-V-
TT  XI   48   PPQ-R   DQIVRRFKKA-DET-LK-PMV-A---SD-A-
CT  XI  110   TLR-TNKNLDTIVAMIKDY-KTSKTK-LWG-A---SN-R-
BS  XI  117   TLK-TNQNLDIIVGMIKDYMRDSNVKLLWN-A-M--N-R-
EC  XI  111   SLK-YINNFAQMVDVLAGKQEES-VKLLWG-A-C--N-RY


SV  XI   99   KDGGFTANDRDVRRYALRKTIRNIDLAAELGAKTYVAWGG
Art XI   99   KD-GF-SN-RSIRRF-LAKVLHNIDLAA-M--ETF-M---
Amp XI   99   KD-GF-SN-RS-RR--IRKVLRQMDLGA----KTL-L---
TT  XI   84   KD-AF-SP-PW-RA--LRKSLETMDLGA----EIY-V-P-
CT  XI  150   VH-AS-SCNAD-FA-SAAQVKKALEITK---GENY-F---
BS  XI  157   VH-AA-SCNAD-FA--AAQVKKGLETAK----ENY-F---
EC  XI  151   GA-AA-NP-PE-FSW-ATQVVTAMEATHK--GENY-L---

SV  XI  139   REGAESGGAKDVRDALDRMKEAFDLLGEYVTAQGYDLRFA

Art XI  139   ---S-YDGSK-LAAA--RMREGVDTAAG-IKDK--NLRIA
Amp XI  139   ---A-YDSAK-VGAA--RYREALNLLAQ-SEDQ--GLP-A
TT  XI  124   ---A-VEATGKARKVW-WVREALNFMAA-AEDQ--GYR-A
CT  XI  190   ---Y-TLLNT-MEFE--NFARFLHMAVD-AKEI-FEGQ-L
BS  XI  197   ---Y-TLLNT-LKFE--NLARFMHMAVD-AKEIE-TGQ-L
EC  XI  191   ---Y-TLLNT-LRQEREQLGRFMQMVVEHKHKI-FQGTLL

SV  XI  179   IEPKPNEPRGDILLPTVGHALAFIERLERPELYGVNPEVG
Art XI  179   L----N--RGDIFLPTVGHGL--IEQLEHGDIVGL-P-TG
Amp XI  179   -----N--RGDILLPTAGHAI--VQELERPELFGI-P-TG
TT  XI  164   L----N--RGDIYFATVGSML--IHTLDRPERFGL-P-FA
CT  XI  230   -----K--TKHQYDFDVANVL--LRKYDLDKYFKV-I-AN
BS  XI  237   -----K--TTHQYDTDAATTI--LKQYGLDNHFKL-L-AN
EC  XI  231   -----Q--TKHQYDYDAATVYG-LKQFGLEKEIKL-I-AN

SV  XI  219   HEQMAGLNFPHGIAQALWAGKLFHIDLNGQSGIKYDQDLR
Art XI  219   -EQM--LN-T-GI-Q-LWAEK-FH--L-GQRGIKY-QD-V
Amp XI  219   -EQMSNLN-TQGI-Q-LWHKK-FH--L-GQHGPKF-QD-V
TT  XI  204   -ETM-GLN-VHAV-Q-LDA-K-FH--L-DQRMSRF-QD-R
CT  XI  270   -ATL-FHD-Q-ELRY-RIN-V-GS--A-TG      -ML-G
BS  XI  277   -ATL--HT-E-ELRM-RVH-L-GSV-A-QG      HPL-G
EC  XI  271   -ATL--HS-H-EI-T-IAL-LFGSV-A-RG      -AQ-G
```

Figure 13.2 (Continued).

```
SV  XI   259   FGAGDLRAAFWLVDLLESA      GYE GPRHFDFKPPRT
Art XI   259   FGHGDLTSAFFTVD-LENGFPNGGPKYT-PRH--Y-PS-T
Amp XI   259   FGHGDLLNAFSLVD-LENGPDGGPAYD -PRH--Y-PS-T
TT  XI   244   FGSENLKAAFFLVD--E      SSGYQ -PRH--AHAL-T
CT  XI   305   WDTDQFPTDIRMTT-AMYEVIKMGGFDK-GLN--A-VR-A
BS  XI   312   WDTDEFPTDLYSTT-AMYEILQNGGLGS-GLN--A-VR-S
EC  XI   306   WDTDQFPNSVEENA-VMYEILKAGGFTT-GLN--A-VR-Q


SV  XI   293   E DFDGVWASAEGCMRNYLILKERAAAFRADPEVQEALRA
Art XI   299   D GYDGVWDSAKAN-SMYLLLKER-LAFRA-PECQEAMKT
Amp XI   298   E DFDGVWISAKDNIRMYLLLKER-KAFRA-PEVQAALAE
TT  XI   278   E DEEGVWAFARGC-RTYLILKER-EAFRE-PEVKELLAA
CT  XI   345   SFEPEDLFLGHIAG-DAFAKGFKV-YKLVK-RVFDKFIE
BS  XI   352   SFEPDDLVYAHIAG-DAFARGLKV-HKLIE-RVFEDVIQ
E.C.     346   STDKYDLFYGHIGA-DTMALALKI-ARMIE-GELDKRIA


SV  XI   332   ARLDQLAQPT   AADGLEALLADRTAFEDFDVEAA AARA
Art XI   338   SGVFELGETTLNAGESAADLMNDSASFAGFDAEAA AERN
Amp XI   337   SKVDELRTPTLNPGETYADLLADRSAFEDYDADAVGAKG
TT  XI   317   YYQEDPAALA   LLGPYSREKAEALKRAELPLEA  KRRR
CT  XI   384      ERYASYKDGIGADIVSGKADFRSLEKYALERSQI VN
BS  XI   391      HRYRSFTEGIGLEITEGRANFHTLEQYALNNKTIK N
EC  XI   385   QRYSGWNSELGQQILKGQMSLADLAKYAQEHHLSPVH


SV  XI   369   AWPFERLDQLAMDHLLGARG 388
Art XI   377   FA FIR-NQLAIEHL-GSR 394
Amp XI   376    YGFVK-NQLAIDHL-GAR 393
TT  XI   353   GYALER-DQLAVEYL-GVRG 372
CT  XI   420   KSGRQE-LESILNQY-FAE 438
BS  XI   427   ESGRQERLKPILNQ 440
EC  XI   422   QSGRQEQLENLVNHY-FDK 440
```

Figure 13.2   (Continued).

THREE-DIMENSIONAL STRUCTURES

The tertiary structures of xylose isomerase have been investigated for *Streptomyces olivochromogenes* (Farber et al. 1988, 1989; Lavie et al. 1994), *Streptomyces violaceoniger* (Glasfeld et al. 1988), *Streptomyces rubiginosus* (Carrell et al. 1984, 1989; Whitlow et al. 1991), *Arthrobacter* B3728 (Akins et al. 1986; Henrick et al. 1989), *Actinoplanes missouriensis* (Jenkins et al. 1992; Lambeir et al. 1992; van Tilbeurgh et al. 1992), and *Clostridium thermosulfurogenes* (Lee et al. 1990). Each subunit of the enzyme consists of two structural domains. Domain I contains eight strands of parallel β pleated sheet connected by segments of α helices antiparallel to the strands forming an $(\alpha/\beta)_8$ barrel of ~320 amino acids. Domain II is the C-terminal arm consisting of five helices linked by a random coil, forming a large loop (~65 amino acids) away from the domain. Domain I of one subunit interacts with domain II of another subunit to form a dimer (Fig. 13.3). The interactions between two subunits are concentrated at the interface

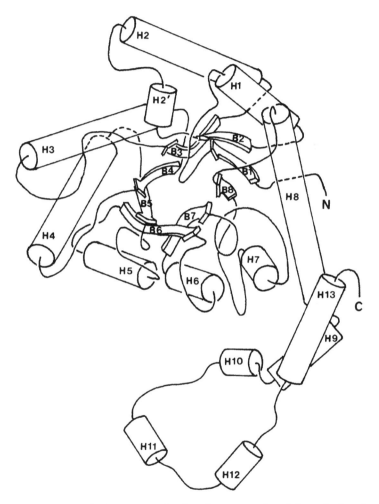

Figure 13.3. The subunit structure of xylose isomerase, viewed down the barrel axis from the carboxyl end of the sheet. (From Henrick et al. 1989. *J. Mol. Biol. 208,* 138, with permission. Copyright 1989 Academic Press, Ltd.)

between helices 4, 5, 6 of the $(\alpha/\beta)_8$ barrel, which consists mostly of highly hydrophobic amino acid residues (Farber et al. 1989). The tetramer consists of two dimers aligned in a 2-fold symmetry, better described as a dimer of dimers: dimer A-B associated with dimer A*-B*. The dimer-dimer interaction is relatively weaker than the association of monomer in the formation of dimer. Dissociation of tetramer yields three possible dimers, A-A*, A-B, and A-B*. Only the latter is fully active because it preserves the interface residues of the active site (Rangarajan et al. 1992).

The active site is located close to the C-terminal end of the β strands, and lies within the loops connecting these β strands. The site is buried in a deep pocket (amphipathic in nature) in the interior, close to the 2-fold axis of the tetramer molecule. A cluster of aromatic side chains (Trp15, Phe25, Phe93, and Trp136 in the *Arthrobacter* enzyme) interact with the carbon backbone of the sugar substrate. A number of acidic and polar residues form an array of ligands with the metal cations and a network of hydrogen bonds with the hydroxyl groups of the substrate.

THE NATURE OF METAL CATIONS

Two Metal Binding Sites

All xylose isomerases contain two metal binding sites, with varying affinities. Spectroscopic studies on the *Streptomyces rubiginosus* enzyme reveal two metal-binding sites per monomer—one with high affinity for Mg^{++} but weak binding to Co^{++}, and the other binding in reverse order. Both sites need to be occupied by Co^{++} for full activation. The Co^{++} affinity site has an octahedral symmetry, while the other site is penta-coordinated. The former contains nitrogen ligands, in addition to carboxylate oxygens (Sudfeldt et al. 1990).

Similar high- and low-affinity sites for different metals have been confirmed by kinetic studies of xylose isomerase from *Thermus aquaticus* HB8 (Lehmacher and Bisswanger 1990B). One site has high affinity for Mg^{++}, Co^{++}, and Mn^{++}, whereas a low-affinity site binds these metals weakly. In a study on the *Bacillus coagulans* enzyme, two sites are identified, showing preferential affinity for different metals—one site with high affinity for Mn^{++}, and the other site for Co^{++} (Marg and Clark 1990). Enzyme activity toward xylose is high when Mn^{++} is coordinated to the first site. For the substrate fructose, both sites need to be occupied by Co^{++}.

The *Streptomyces violaceoruber* enzyme contains two metal binding sites per monomer. Site 1 binds Co^{++} with a binding constant K_b of $> 3.3 \times 10^6$ M^{-1}, and site 2 with a value of $4 \times 10^4 M^{-1}$ (Callens et al. 1988). Similar results have also been obtained with Mg^{++}. The site 1 metal enhances the structural stability of the enzyme, and the site 2 metal is needed for activity. The formation of the active ternary complex involves the binding of cations to both sites. Sites 1 and 2 bind Mg^{++} and Co^{++} with different affinities. Site 1 has high affinity for Mg^{++}, but low affinity for Co^{++}; a reverse order of binding affinity is evidenced for site 2.

A study on the binding characteristics of Mn^{++}, Co^{++}, and Mg^{++} with xylose isomerases from five microbial sources (*Streptomyces violaceoruber, Streptomyces* sp., *Lactobacillus xylosus, Lactobacillus brevis,* and *Bacillus coagulus*) confirms the general requirements of 2 moles of metal per mole of monomer for catalytic function. The enzymes are active only when the two sites are both metal bound (van Bastelaere et al. 1992). The dissociation constants of the three metal ions for the two metal binding sites of these enzymes indicate metal binding in the order of $Mn^{++} > Co^{++} > Mg^{++}$ (Table 13.3). The K_d values are identical for Mn^{++} binding to both sites in all five enzymes. For the binding of Co^{++}, xylose isomerases from the *Streptomyces* have identical K_d valves for both sites. In contrast, the other three enzymes have K_d values for site 2 greater than site 1 by an order of magnitude. All five enzymes show high and low affinity for Mg^{++} with K_d differences between site 1 and site 2 of 1 to 2 orders of magnitude.

In most literature, the metal-binding site that primarily contributes to structural stability of the enzyme is designated as site 1. The metal site that is mainly involved in the catalytic process is designated as site 2. In the crystal structure, metal site 2 is the one closest to C1 and C2 of the substrate as illustrated in Fig. 13.7B.

Metal Coordination

The two metal ions assume different geometry in ligand coordination. In the *Streptomyces rubiginosus* enzyme, the site 1 metal has an octahedral coordination with the side chain oxygen of Glu181, Glu217, Asp245, Asp287, and C3-O and C5-O of the xylose substrate intermediate which exists in an open-chain form. The site 2 metal ligands with Glu217, Asp255, Asp257, His220, and a H_2O molecule. The following substrate bindings also occur: C4-O to Asp287, C5-O to H_2O bound to metal site 2, C1-O and C2-O to Thr90. The metal ions stabilize the substrate in its open-chain form, and in general, position the active site residues in the correct orientation (Carrell et al. 1989). In another study on the same enzyme, the xylose substrate (in its extended form) C2-O and C4-O are coordinated directly to the site 1 metal, and C1-O is liganded to the site 2 metal. The other OH groups are hydrogen bonded with neighboring amino acid side chains. The geometry of site 1 remains the same after substrate binding, but site 2 assumes hepta-coordination (Whitlow et al. 1991). In the *Streptomyces olivochromogenes* enzyme, the metal ion is bound to Glu180, Asp286, and His219 (Farber et al. 1989).

The changes of coordination of metal sites upon substrate binding are also observed in the *Actinoplanes missouriensis* enzyme. Site 1 changes from tetrahedral to octahedral geometry; whereas site 2 remains octahedral in all cases

Table 13.3. Dissociation Constants for Mn^{++}, Co^{++}, and Mg^{++} for the Two Metal Binding Sites in Xylose Isomerase from Several Microbial Sources

D-Xylose isomerase	Preincubation conditions	Mn^{++}		Co^{++}		Mg^{++}	
		K_{d1}	K_{d2}	K_{d1}	K_{d2}	K_{d1}	K_{d2}
Streptomyces sp.	35°C, pH 8.0	4×10^{-8}	8×10^{-8}	4×10^{-7}	4×10^{-7}	$\leq 10^{-6}$	6.2×10^{-5}
S. violaceoruber	35°C, pH 8.0	8×10^{-8}	8×10^{-8}	6×10^{-7}	6×10^{-7}	$\leq 10^{-6}$	7.4×10^{-5}
L. brevis	0°C, ph 7.5	6×10^{-8}	8×10^{-8}	6×10^{-8}	4×10^{-7}	Very low activity	
L. xylosus	35°C, pH 7.5	2×10^{-8}	6×10^{-8}	4×10^{-8}	6×10^{-7}	$\leq 10^{-6}$	8×10^{-5}
B. coagulans	35°C, pH 7.0	2×10^{-8}	2×10^{-7}	4×10^{-8}	4×10^{-7}	$\leq 10^{-6}$	6×10^{-5}

From Van Bastelaere et al. 1992. *Biochem. J. 286*, 734, with permission. Copyright 1992 The Biochemical Society and Portland Press.

K_{d1} = dissociation constant for the high-affinity (structural) metal site; K_{d2} = dissociation constant for the low-affinity (catalytic) metal site.

although its position shifts depending on the metal and the substrate (Jenkins et al. 1992).

For the *Arthrobacter* enzyme, the site 1 metal is coordinated with Glu180, Glu216, Asp244, Asp292, and C2-O and C4-O of the substrate in its extended form. The metal at site 2 is octahedrally bound to Glu216, Asp254, Asp256, His219, and an H_2O molecule (Collyer et al. 1990). The substrate C1-O and C5-O are linked to Lys182 and His53, respectively. In addition, there is an extensive network of hydrogen bonds via H_2O molecules interacting with a number of conserved amino acids (Henrick et al. 1989).

THE ACTIVE SITE HISTIDINE

The involvement of a single-active site His residue has been suggested by kinetic studies of ethyloxycarbonylation of His residues (Vangrysperre et al. 1988, 1989A). It has also been demonstrated that the enzyme-substrate complex is resistant to inactivation by His modification, whereas the Co^{++} or Mn^{++} bound enzyme does not have the same effect (Gaikwad et al. 1988). Chemical modification of several bacterial enzymes with Woodward's reagent suggests that a carboxylate residue may also contribute to the catalytic activity of xylose isomerase (Vangrysperre et al. 1989B). Peptide mapping reveals that the His and carboxylate groups are in close proximity (Vangrysperre et al. 1990). The result is consistent with the crystallographic structure of the *Streptomyces rubiginosus* enzyme showing the reactive His54 hydrogen bonded to Asp57 (Carrell et al. 1989).

Discrepancies exist regarding the precise functional role of the essential His in the catalytic mechanism. In the *Streptomyces rubiginosus* enzyme, the essential His54 has been assumed to be located adjacent to the C1-C2 bond of the open-chain substrate, acting as a general base that assists in proton abstraction (based on *cis*-enediol intermediate mechanism). However, kinetic studies of the *Streptomyces rubiginosus* enzyme with the His54 modifed chemically implicate that the His is not involved in the isomerization step (Vangrysperre et al. 1990). A later study of the same microbial enzyme structure suggests that the His54 acts in transferring the C1-OH proton to the ring oxygen in the ring-opening process (Whitlow et al. 1991). For the *Arthrobacter* enzyme, the active site His53 is believed to be positioned near the C4-C5 bond, and not above the C1-C2 bond (Collyer and Blow 1990). The C1-C2 bond in this case is oriented towards the metal sites and a number of hydrophobic residues (Trp136, Phe93, and Phe25) in the immediate environment. In this case, His53 is positioned to act as a catalytic base for the ring opening of the substrate. Studies by protein engineering suggest, however, that the His does not act as a catalytic base for the ring-opening step, but rather functions to stabilize the open-chain substrate by hydrogen bonding with the C5-OH (Lee et al. 1990) or to direct the anomeric specificity of the enzyme (Lambeir et al. 1992). (See section later in this chapter, The Ring-Opening Step).

STEREOCHEMISTRY OF THE REACTION

Xylose isomerase is specific only for the α-D-pyranose anomer of the substrates xylose and glucose and the α-D-furanose form of fructose (Fig. 13.1). The interconversion is also stereospecific, where a substrate of α-D-glucopyranose (with C2-*H labelled) gives (R)-(1*H)-α-D-fructofuranose as the product (Makkee et al. 1984). Similarly, in the reverse reaction, only the C1-pro-*R* hydrogen atom of α-D-fructofuranose is transferred to the C2 position to form the (S)-(2H)-α-D-glucopyranose. Steric restrictions in the active site of the enzyme prevent a β-pyranose or β-furanose from productive binding (according to the hydride shift mechanism described below) (Collyer et al. 1992).

MECHANISMS OF ISOMERIZATION

Two reaction mechanisms have been postulated to account for the reaction pathway of isomerization: (1) proton transfer involving a *cis*-enediol intermediate and (2) hydride shift mediated by metal ion. The general chemistry aspects of these two types of reactions are presented in Fig. 13.4.

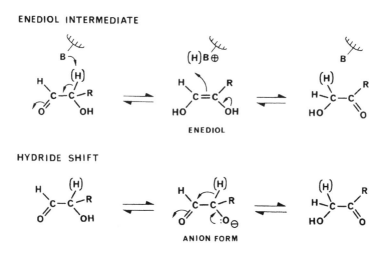

ENEDIOL INTERMEDIATE

HYDRIDE SHIFT

Figure 13.4. Mechanism for the isomerization of an aldose to a ketose via an enediol intermediate or a hydride shift. (From Collyer and Blow 1990. *Proc. Natl. Acad. Sci. USA* *87*, 1363, with permission. Copyright 1990 The National Academy of Sciences.)

Cis-enediol Intermediate

In the mechanism initially proposed by Schray and Rose (1971) for isomerases, aldose-ketose interconversion proceeds via a cis-enediol intermediate (Fig. 13.5). In the forward reaction, abstraction of the proton from C2 and its transfer to C1 is base-catalyzed and occurs from the same face of the cis-enediol. The abstraction of a proton from C2 is assisted by a H_2O molecule or an electrophilic residue (e.g. lysine ε-NH_3^+) acting as an acid catalyst in polarizing the carbonyl at C1. This molecule also serves as a proton donor (to C1-O) or acceptor (from C2-OH) in the enolization step, which results in ring opening between C1 and the ring oxygen. The cis-arrangement of the enediol intermediate allows suprafacial transfer of the proton at C2-C1, requiring almost no conformational changes of the substrate. Mechanistically, the assumption of the *cis*-eclipsed conformation requires the substrate to be in a skew boat configuration.

In the model proposed by Makkee et al. (1984), the metal ion coordinates the C1-OH and C2-OH of the substrate, which exists in a pseudo-cyclic cis-enediol (Fig. 13.6). In this case, it is the metal that assists H^+ abstraction from C1 or C2. For ring closure, the C5-O becomes oriented for attack of the carbonyl C1 from the opposite direction of the protonation step (at C2). Consequently, the reactive substrates are both in the α-D-form (i.e., α-D-glycopyranose <-> α-D-fructofuranose).

Crystal analysis of the *Streptomyces rubiginosus* enzyme reveals close interactions between C1 of the substrate and the essential His54, suggesting His54 as the base catalyst. Polarization of the carbonyl group in the substrate is mediated by a H_2O molecule (held in place by Thr90 and Thr91) linking the C1-OH and C2-OH oxygens (Carrell et al. 1989).

The mechanism involving a 1,2-enediol intermediate has been well established for other aldose-ketose isomerases (Straus et al. 1985; Alagona et al. 1986). However, there are features that distinguish xylose isomerase from the other enzymes. Xylose isomerase requires divalent metal ions for activity and stability, and has a higher K_m for substrate and approximately 1/1,000 the k_{cat} of triosephosphate isomerase. The ^3H exchange ratio for D-glyceraldehyde 3-phosphate to dihydroxyacetone is 50 compared to 10^{-4} for the xylose-xylulose isomerization reaction. The high isotope exchange shown in triosephosphate isomerase is consistent with a mechanism involving proton transfer and enolization; however, the striking absence of exchange between the transfer and solvent hydrogen in xylose isomerase-catalyzed reaction is hard to explain if the same mechanism is applied. Harris and Feather (1975) suggest that the abstracted proton remains associated in partial bonding with the base and the developing π-electron cloud between C1-C2. Isotope exchange will not be observed if a rapid proton transfer occurs in an active site that is shielded from solvent access. To investigate the possible exis-

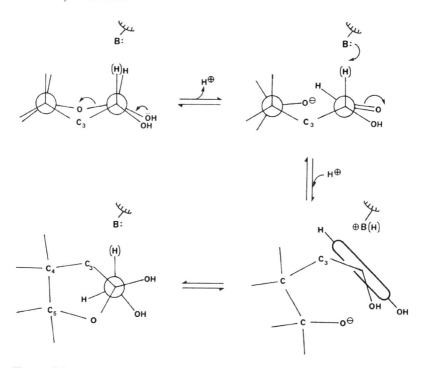

Figure 13.5. Mechanism for aldose-ketose isomerization by a cis-enediol intermediate. (From Schray and Rose 1971. *Biochemistry 10*, 1061, with permission. Copyright 1971 American Chemical Society.)

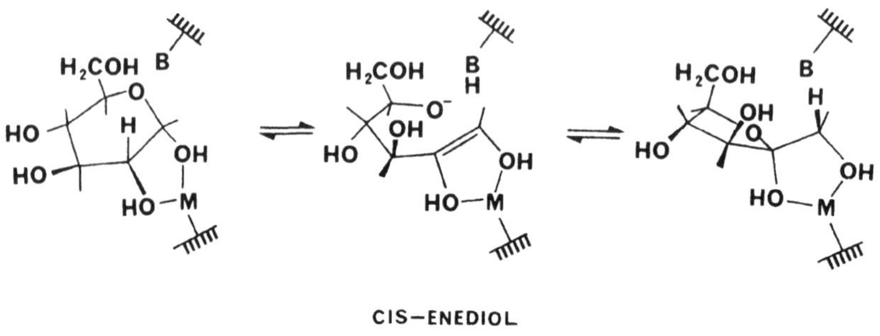

CIS—ENEDIOL

Figure 13.6. Mechanism of aldose-ketose isomerization via metal-mediated proton transfer. (From Makkee et al. 1984. *J. Royal Netherlands Chem. Soc. 103*, 363, with permission. Copyright 1984 Elsevier Science Publishers B.V.)

tence of such an effect, Allen et al. (1994A) have conducted isotope-exchange experiments at elevated temperatures that should affect positively the rate of solvent exchange if it occurs. The results do not support the presence of a shielding effect.

1,2-Hydride Shift

An alternative mechanism involving hydride transfer has received increased attention. A hydride shift mechanism would lead to no isotope exchange at the C1- or C2-hydrogen (Topper 1957; Harris and Feather 1975, Farber et al. 1989; Allen et al. 1994A,B). The primary features of this mechanism are: (1) The bound cyclic substrate undergoes ring opening to form an extended form. (2) Isomerization is achieved by a hydride shift mediated by a metal ion (rate-limiting step). (3) An anion intermediate is involved.

The proposal based on the crystallographic studies of *Streptomyces olivochromogenes* enzyme (Farber et al. 1989) suggest that the ring-opening step involves His219 as the catalytic base. Hydride transfer proceeds via a cyclic transition state involving coordination of the metal ion with the carbonyl groups. In a later study by the same group, ring opening was demonstrated by measuring the release of H_2S using 1-thioglucose as the substrate (Allen et al. 1994A)

A more detailed mechanism based on the *Arthrobacter* enzyme has been described by Collyer et al. (1990) and Whitlow et al. (1991). The initial step of ring opening of the substrate is catalyzed by the His53-Asp56 relay system, in which the C1-OH is transferred for protonation of the ring oxygen (Fig. 13.7A). The initial binding of the pyranose substrate (in its cyclic form) to the metal ion at site 1 via C3-OH and C4-OH is replaced by the binding of C2-OH and C4-

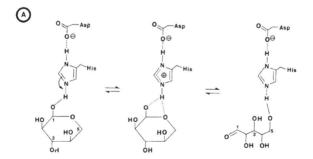

Figure 13.7. The proposed hydride shift mechanism of xylose isomerase. Schematic diagram illustrating (A) His-catalyzed ring opening of the aldose. (B) The metal-catalyzed stereospecific hydride transfer. (From Collyer et al. 1990. *J. Mol. Biol. 212,* 230, with permission. Copyright 1990 Academic Press, Ltd.)

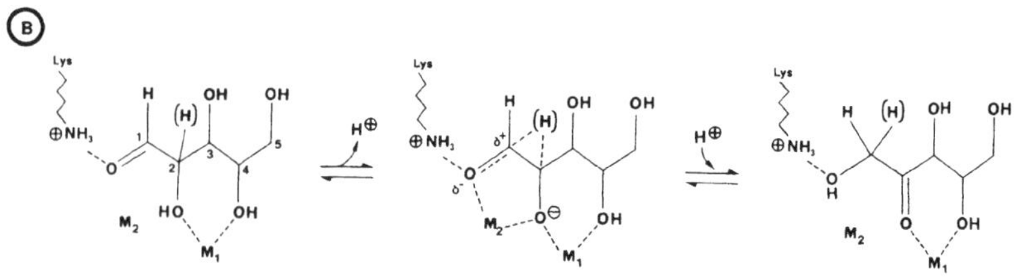

Figure 13.7   (Continued).

OH in the extended form of the substrate. Chemical modification studies on the hydroxyl group at various positions suggest that the C4-OH is important for enzyme action (Bock et al. 1983).

The next step involves the abstraction of the proton from C2-OH, in which the H_2O molecule bound to the site 2 metal serves as the proton acceptor. The resulting anionic intermediate is stabilized by the metal cations at sites 1 and 2 by coordination via C1-O and C2-O, and also by hydrogen bonding with Lys183 via the C1 carbonyl oxygen (Fig. 13.7B). The role of Lys183 in the isomerization step is supported by site-directed mutation studies (Lambeir et al. 1992)

Consequently, the C-O bonds at C1 and C2 are polarized, and in the transition state, hydride ion originating from C2 is partially associated with C1 and C2. In the enzyme, the hydride ion is located in a strongly hydrophobic environment inaccessible to solvent. A transfer of the hydride ion to C1 and disengagement of the metal sites from the coordination complex converts the intermediate to a stable ketone. The final step includes ring closure to yield the ketose reaction product.

THE RING-OPENING STEP

The role of the essential His as a base catalyst in ring opening has not been firmly established. In a site-directed mutagenesis study of the *Clostridium thermosulfurogenes* enzyme, substitution of the essential His101 by Glu, Gln, Asp, Asn retains 10–16% of the wild type activity, indicating that the essential His may not act as a simple base in the catalysis (Lee et al. 1990). Furthermore, the specific activity of the wild type enzyme decreases below pH 6.5 where the His is positively charged, suggesting that His101 is not involved in electrostatic bonding in the ring-opening step, but rather is hydrogen bonded with C5-OH to stabilize the substrate. The mechanism proposed by Lee et al. (1990) differs largely in the role of the essential His, which, in this case, forms a hydrogen bond with the C5-OH in stabilizing the open-chain form of the xylose substrate. The rest of the reaction steps, following the general scheme, involve a metal-ion induced C2-C1 hydride shift in an aldose-ketose isomerization. Bond rotation at C3-C4, followed by hemiketal formation yields the ketose product. Results from site-directed muta-genesis studies of *Actinoplanes missouriensis* also indicate that the essential His54 does not act in ring opening, but interacts with the C1-OH and C5-OH of the cyclic and open-chain substrate, respectively (Lambeir et al. 1992).

THE CATALYTIC BASE IN ISOMERIZATION

A major catalytic step subsequent to ring opening involves the transfer of a proton from C2-OH to C1-O with a concomitant hydride shift from C2 to C1. Crystal-lographic study of the *Streptomyces rubiginosus* enzyme under high resolution (Whitlow et al. 1991) suggests that the transfer of C1-OH proton to C2-O is mediated by a H_2O molecule acting as a proton donor. In this step, a H_2O molecule coordinated to the site 1 metal cation hydrogen-bonded Asp257, serves as the proton acceptor of the C2-OH proton by transferring its own proton to Asp257. The resulting anion intermediate is stabilized by the metal ion. A hydride shift follows that transfers the hydrogen from C2 to C1 (Fig. 13.8). The role of an Asp side chain in assisting the C1-OH to C2-O proton transfer is also identified in mutation studies of the *Actinoplanes missouriensis* enzyme (van Tilbeurgh et al. 1992). Another study suggests that a site 2 metal-bound hydroxide ion acts as a base in the abstraction of the C2-OH proton. The metal-bound water thus formed can protonate C1-O accompanied by the hydride shift (Allen et al. 1994B, Lavie et al. 1994).

THE ACTION OF METAL CATIONS

All the xylose isomerases with known crystal structures show a shift of coordi-nation in the binding of cyclic or extended substrates. The site 1 metal coordinates

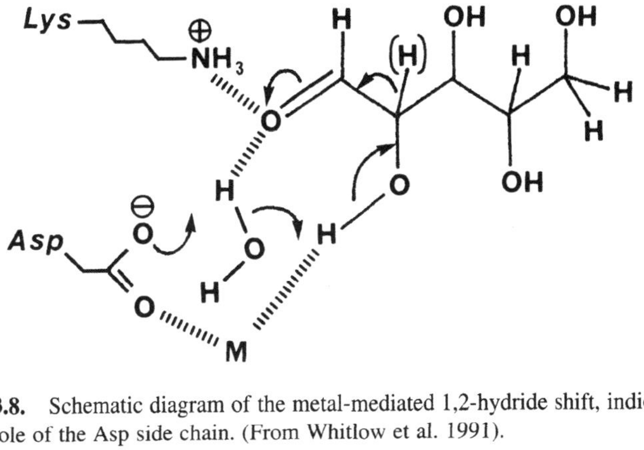

Figure 13.8. Schematic diagram of the metal-mediated 1,2-hydride shift, indicating the possible role of the Asp side chain. (From Whitlow et al. 1991).

with C3-O and C4-O in cyclic form, and C2-O and C4-O in the extended form (Collyer et al. 1990). The site 2 metal shifts in position (a distance of 1.8 A apart) upon substrate binding to ligand with C1-O and C2-O (Jenkins et al. 1992; Lavie et al. 1994). The metal ions at both sites are required for full enzyme activity. Mutation of active site residues that possibly causes perturbation of metal binding has been shown to decrease k_{cat} significantly (Lambeir et al. 1992), and also the thermal stability of the enzyme (Cha et al. 1994). The site 2 metal cation plays a major role in assisting proton transfer from C1-O to C2 carbonyl oxygen that involves either a general base catalysis (Whitlow et al. 1991; van Tilbeurgh et al. 1992) or direct participation of a metal-bound hydroxide ion (Allen et al. 1994B). (See previous section on the catalytic Base.) The anion at C2 is stabilized by both sites 1 and 2 metal cations. Site 2 metal ion coordination to C1-O together with charge interaction of a lysyl ε-amino group polarizes the carbonyl bond at C1. The development of a partial positive charge at the carbonyl carbon facilitates the hydride ion transfer (Collyer et al. 1990; Lavie et al. 1994).

REFERENCES

AKINS, J.; BRICK, P.; JONES, H. B.; HIRAYAMA, N.; SHAW, P.-C.; and BLOW, D. M. 1986. The crystallization of glucose isomerase from *Arthrobacter* B3728. *Biochim. Biophys. Acta 874*, 375–377.

ALAGONA, G.; GHIO, C.; and KOLLMAN, P. A. 1986. Simple model for the effect of Glu165 → Asp165 mutation on the rate of catalysis in triose phosphate isomerase. *J. Mol. Biol. 191*, 23–27.

ALEXANDER, N. J. 1986. Genetic manipulation of yeasts for ethanol production from xylose. *Food Technol. 40*, 99–103.

ALLEN, K. N.; LAVIE, A.; FARBER, G. K.; GLASFELD, A.; PETSKO, G. A.; and RINGE, D. 1994A. Isotope exchange plus substrate and inhibition kinetics of D-xylose isomerase do not support a proton-transfer mechanism. *Biochemistry 33*, 1481–1487.

ALLEN, K. N.; LAVIE, A.; GLASFELD, A.; TANADA, T. N.; GERRITY, D. P.; CARLSON, S. C.; FARBER, G. K.; PETSKO, G. A.; and RINGE, D. 1994B. Role of the divalent metal ion in sugar binding, ring opening, and isomerization by D-xylose isomerase: Replacement of a catalytic metal by an amino acid. *Biochemistry 33*, 1488–1494.

AMORE, R., and HOLLENBERG, C. P. 1989. Xylose isomerase from *Actinoplanes missouriensis*: Primary structure of the gene and the protein. *Nucl. Acids Res. 17*, 7515.

ANON. 1983. Taka-Sweet, immobilized glucose isomerase for high fructose syrup production. Miles Laboratories, Inc., Biotech Products Division, Elkhart, IN 46516.

ANTRIM, R. L., and AUTERINEN, A.-L. 1986. A new regenerable immobilized glucose isomerase. *Starch/Starke 38*, 137–142.

BATT, C. A.; JAMIESON, A. C.; and VANDEYAR, M. A. 1990. Identification of essential histidine residues in the active site of *Escherichia coli* xylose (glucose) isomerase. *Proc. Natl. Acad. Sci. USA 87*, 618–622.

BEALL, D. S.; OHTA, K.; and INGRAM, L. O. 1991. Parametric studies of ethanol production from xylose and other sugars by recombinant *Escherichia coli*. *Biotechnol. Bioengineer. 38*, 296–303.

BLUNDELL, T. L.; JENKINS, J. A.; SEWELL, B. T.; PEARL, L. H.; COOPER, J. B.; TICKLE, I. J.; VEERAPANDIAN, B.; and WOOD, S. P. 1990. X-ray analysis of aspartic proteinases. The three-dimensional structure at 2.1 Å resolution of endothiapepsin. *J. Mol. Biol. 211*, 919–941.

BOCK, K.; MELDAL, M.; MEYER, B.; and WIEBE, L. 1983. Isomerization of D-glucose with glucose-isomerase. A mechanistic study. *Acta Chem. Scand. B37*, 101–108.

BOR, Y.-C.; MORAES, C.; LEE, S.-P.; CROSBY, W. L.; SINSKEY, A. J.; and BATT, C. A. 1992. Cloning and sequencing the *Lactobacillus brevis* gene encoding xylose isomerase. *Gene 114*, 127–131.

CALLENS, M.; KERSTERS-HILDERSON, H.; van OSPTAL, O.; and DE BRUYNE, C. K. 1986. Catalytic properties of D-xylose isomerase from *Streptomyces violaceoruber*. *Enzyme Microbiol. Technol. 8*, 696–700.

CALLENS, M.; TOMME, P.; KERSTERS-HILDERSON, H.; CORNELIS, R.; VANGRYSPERRE, W.; and DeBRUYNE, K. 1988. Metal ion binding to D-xylose isomerase from *Streptomyces vilaceoruber*. *Biochem. J. 250*, 285–290.

CARRELL, H. L.; GLUSKER, J. P.; BURGER, V.; MANFRE, F.; TRITSCH, D.; and BIELLMANN, J.-F. 1989. X-ray analysis of D-xylose isomerase at 1.9 Å: Native enzyme in complex with substrate and with a mechanism-designed inactivator. *Proc. Natl. Acad. Sci. USA 86*, 4440–4444.

CARRELL, H. L.; RUBIN, B. H.; HURLEY, T. J.; and GLUSKER, J. P. 1984. X-ray crystal structure of D-xylose isomerase at 4-Å resolution. *J. Biol. Chem. 259*, 3230–3236 (1984).

CHA, J.; CHO, Y.; WHITAKER, R. D.; CARRELL, H. L.; GLUSKER, J. P.; KARPLUS, P. A.; and BATT, C. A. 1994. Perturbing the metal site in D-xylose isomerase. *J. Biol. Chem. 269*, 2687–2694.

CHAN, E.-C.; UENG, P. P.; and CHEN, L. F. 1989. Metabolism of D-xylose in *Schizosaccharomyces pombe* cloned with a xylose isomerase gene. *Appl. Microbiol. Biotechnol. 31*, 524–528.

COLLYER, C. A., and BLOW, D. M. 1990. Observations of reaction intermediates and the mechanism of aldose-ketose interconversion by D-xylose isomerase. *Proc. Natl. Acad. Sci. USA 87*, 1362–1366.

COLLYER, C. A.; GOLDBERG, J. D.; VIEHMANN, H.; BLOW, D. M.; RAMSDEN, N. G.; FLEET, G. W. J.; MONTGOMERY, F. J.; and GRICE, P. 1992. Anomeric specificity of D-xylose isomerase. *Biochemistry 31*, 12211–12218.

COLLYER, C. A.; HENRICK, K.; and BLOW, D. M. 1990. Mechanism for aldose-ketose interconversion by D-xylose isomerase involving ring opening followed by a 1,2-hydride shift. *J. Mol. Biol. 212*, 211–235.

Danno, G.-I. 1970. Studies on D-glucose-isomerizing enzyme from *Bacillus coagulans*, strain HN-68. *Agric. Biol. Chem. 34*, 1805–1814.

DEKKER, K.; YAMAGATA, H.; SAKAGUCHI, K.; and UDAKA, S. 1991A. Xylose (Glucose) isomerase gene from the thermophile *Clostridium thermohydrosulfuricum*; Cloning, sequencing, and expression in *Escherichia coli. Agric. Biol. Chem. 55*, 221–227.

―――. 1991B. Xylose (Glucose) isomerase gene from the thermophile *Thermus thermophilus*: Cloning, sequencing and comparison with other thermostable xylose isomerase. *J. Bacteriol. 173*, 3078–3083.

DROCOURT, D.; BEJAR, S.; CALMELS, T.; REYNES, J. P.; and TIRABY, G. 1988. Nucleotide sequence of the xylose isomerase gene from *Streptomyces violaceoniger. Nucl. Acids Res. 16*, 9337.

FARBER, G. K.; GLASFELD, A.; TIRABY, G.; RINGE, D.; and PETSKO, G. A. 1989. Crystallographic studies of the mechanism of xylose isomerase. *Biochemistry 28*, 7289–7297.

FARBER, G. K.; MACHIN, P.; ALMO, S. C.; PETSKO, G. A.; and HAJDU, J. 1988. X-ray Laue diffraction from crystals of xylose isomerase. *Proc. Natl. Acad. Sci. 85*, 112–115.

FEATHER, M. S.; DESHPANDE, V.; and LYBYER, M. J. 1970. Anomeric specificity during some isomerase reactions. *Biochem. Biophys. Res. Comm. 38*, 859–863.

GAIKWARD, S. M.; PAWAR, H. S.; VARTAK, H. G.; and DESHPANDE, V. V. 1989. Streptomyces glucose/xylose isomerase has a single active site for glucose and xylose. *Biochem. Biophys. Res. Comm. 159*, 457–463.

GLASFELD, A.; FARBER, G. K.; RINGE, D.; MARCEL, T.; DROCOURT, D.; TIRABY, G.; and PETSKO, G. A. 1988. Characterization of crystals of xylose isomerase from *Streptomyces violaceoniger. J. Biol. Chem. 263*, 14612–14613.

HANSON, K. R., and ROSE, I. A. 1975. Interpretations of enzyme reaction stereospecificity. *Acc. Chem. Res. 8*, 1–10.

HARRIS, D. W., and FEATHER, M. S. 1975. Studies on the mechanism of the interconversion of D-glucose, D-mannose, and D-fructose in acid solution. *J. Am. Chem. Soc. 97*, 178–181.

HENRICK, K.; COLLYER, C. A.; and BLOW, D. M. 1989. Structure of D-xylose isomerase from *Arthrobacter* strain B3728 containing the inhibitors xylitol and D-sorbitol at 2.5A and 2.3A resolution, respectively. *J. Mol. Biol. 208*, 129–157.

HOGUE-ANGELETTI, R. A. 1975. Subunit structure and amino acid composition of xylose isomerase from *Streptomyces albus. J. Biol. Chem. 250*, 7814–7818.

JENKINS, J.; JANIN, J.; REY, F.; CHIADMI, M.; VAN TILBEURGH, H.; LASTERS, I.; DE MAEYER, M.; VAN BELLE, D.; STANSSENS, P.; MRABET, N. T.; SNAUWAERT, J.; MATTHYSSENS, G.; and LAMBEIR, A.-M. 1992. Protein engineering of xylose (glucose) isomerase from *Actinoplanes missouriensis*. 1. Crystallography and site-directed mutagenesis of metal binding sites. *Biochemistry 31*, 5449–5458.

KASUMI, T.; HAYASHI, K.; TSUMURA, N.; and TAKAGI, T. 1981. Physiochemical characterization of glucose isomerase from *Streptomyces griseofuscus* S-41. *Agric. Biol. Chem. 45*, 1087–1095.

KASUMI, T.; HAYASHI, K.; and TSUMURA, N. 1982A. Roles of magnesium and cobalt in the reaction of glucose isomerase from *Streptomyces griseofuscus* S-41. *Agric. Biol. Chem. 46*, 21–30.

————. 1982B. Role of cobalt in stabilizing the molecular structure of glucose isomerase from *Streptomyces griseofuscus* S-41. *Agric. Biol. Chem. 46*, 31–39.

LAMBEIR, A.-M.; LAUWEREYS, M.; STANSSENS, P.; MRABET, N. T.; SNAUWAERT, J.; VAN TILBEURGH, H.; MATTHYSSENS, G.; LASTERS, I.; DE MAEYER, M.; WODAK, S. J.; JENKINS, J.; CHIADMI, M.; and JANIN, J. 1992. Protein engineering of xylose (glucose) isomerase from *Actinoplanes missouriensis*. 2. Site-directed mutagenesis of the xylose binding site. *Biochemistry 31*, 5459–5466.

LAVIE, A.; ALLEN, K. N.; PETSKO, G. A.; and RINGE, D. 1994. X-Ray crystallographic structures of D-xylose isomerase-substrate complexes position the substrate and provide evidence for metal movement during catalysis. *Biochemistry 33*, 5469–5480.

LAWLIS, V. B.; DENNIS, M. S.; CHEN, E. Y.; SMITH, D. H.; and HENNER, D. J. 1984. Cloning and sequencing of the xylose isomerase and xylulose kinase gene of *Escherichia coli. Appl. Environ. Microbiol. 47*, 15–21.

LEE, C.; BAGDASARIAN, M.; MENG, M.; and ZEIKUS, J. G. 1990. Catalytic mechanism of xylose (glucose) isomerase from *Clostridium thermosulfurogenes. J. Biol. Chem. 265*, 19082–19090.

LEHMACHER, A., and BISSWANGER, H. 1990A. Isolation and characterization of an extremely thermostable D-xylose isomerase from *Thermus aquaticus* HB8. *J. Gen. Microbiol. 136*, 679–686.

————. 1990B. Comparative kinetics of D-xylose and D-glucose isomerase activities of the D-xylose isomerase from *Thermus aquaticus* HB8. *Biol. Chem. Hoppe-Seyler 371*, 527–536.

LOVINY-ANDERTON, T.; SHAW, P.-C.; SHIN, M.-K.; and HARTLEY, B. S. 1991. D-Xylose (D-glucose) isomerase from *Arthrobacter* strain N.R.R.L. B3728. Gene cloning, sequence and expression. *Biochem. J. 277*, 263–271.

MAKKEE, M.; KIEBOOM, P. G.; and VAN BEKKUM, H. 1984. Glucose isomerase-catalyzed D-glucose-D-fructose interconversion: Mechanism and reaction species. *Recl. Trav. Chim. Pays-Bas 103*, 361–364.

————. 1985. Glucose isomerase and its behavior under hydrogenation conditions. *Starch/Starke 37*, 232–241.

MARG, G. A., and CLARK, D. S. 1990. Activation of glucose isomerase by divalent cations: Evidence for two distinct metal-binding sites. *Enzyme Microb. Technol. 12*, 367–373.

MENG, M.; LEE, C.; BAGDASARIAN, M.; and ZEIKUS, G. 1991. Switching substrate preference of thermophilic xylose isomerase from D-xylose to D-glucose by redesigning the substrate binding pocket. *Proc. Natl. Acad. Sci. USA 88*, 4015–4019.

MRABET, N. T.; VAN DEN BROECK, A.; VAN DE BRANDE, I.; STANSSENS, P.; LAROCHE, Y.; LAMBEIR, A.-M.; MATTHIJSSENS, G.; JENKINS, J.; CHIADMI, M.; VAN TILBEURGH, H.; REY, F.; JANIN, J.; QUAX, W. J.; LASTERS, I.; DE MAEYER, M.; and WODAK, S. J. 1992. Arginine residues as stabilizing elements in proteins. *Biochemistry 31*, 2239–2253.

RANGARAJAN, M.; ASBOTH, B.; and HARTLEY, B. S. 1992. Stability of *Arthrobacter* X-xylose isomerase to denaturants and heat. *Biochem. J. 285*, 889–898.

SAARI, G. C.; KUMAR, A. A.; KAWASAKI, G. H.; INSLEY, M. Y.; and O'HARA, P. J. 1987. Sequence of the *Ampullariella* sp. strain 3876 gene coding for xylose isomerase. *J. Bacteriol. 169*, 612–618.

SCHELLENBERG, G. D.; SARTHY, A.; LARSON, A. E.; BACKER, M. P.; CRABB, J. W.; LIDSTROM, M.; HALL, B. D.; and FURLONG, C. E. 1984. Xylose isomerase from *Escherichia coli* characterization of the protein and the structural gene. *J. Biol. Chem. 259*, 6826–6832.

SCHRAY, K. J., and ROSE, I. A. 1971. Anomeric specificity and mechanism of two pentose isomerase. *Biochemistry 10*, 1058–1062.

SMITH, C. A.; RANGARAJAN, M.; and HARTLEY, B. S. 1991. D-xylose (D-glucose) isomerase from *Arthrobacter* strain N.R.R.L. B3728. Purification and properties. *Biochem. J. 277*, 255–261.

STRAUS, D.; RAINES, R.; KAWASHIMA, E.; KNOWLES, J. R.; and GILBERT, W. 1985. Active site of triosephosphate isomerase: In vitro mutagenesis and characterization of an altered enzyme. *Proc. Natl. Acad. Sci. USA. 82*, 2272–2276.

SUDFELDT, C.; SCHAFFER, A.; KAGI, J. H. R.; BOGUMIL, R.; SCHULZ, H.-P.; WULFF, S.; and WITZEL, H. 1990. Spectroscopic studies on the metal-ion-binding sites of Co^{++}-substituted D-xylose isomerase from *Streptomyces rubiginosus. Eur. J. Biochem. 193*, 863–871.

SUEKANE, M.; TAMURA, M.; and TOMIMURA, C. 1978. Physico-chemical and enzymatic properties of purified glucose isomerases from *Streptomyces olivochromogenes* and *Bacilus stearothermophilus. Agric. Biol. Chem. 42*, 909–917.

TAKASAKI, Y.; KOSUGI, Y.; and KANBAYASHI, A. 1969. Studies on sugar-isomerizing enzyme. Purification, crystallization and some properties of glucose isomerase from *Streptomyces* sp. *Agric. Biol. Chem. 33*, 1527–1534.

TOPPER, Y. J. 1957. On the mechanism of action of phosphoglucose isomerase and phosphomannose isomerase. *J. Biol. Chem. 225*, 419–425.

TUCKER, M. Y.; TUCKER, M. P.; HIMMEL, M. E.; GROHMANN, K.; and LASTICK, S. M. 1988. Properties of genetically overproduced *E. coli* xylose isomerase. *Biotechnol. Lett. 10*, 79–84.

VAN BASTELAERE, P.; VANGRYSPERRE, W.; and KERSTERS-HILDERSON, H. 1991. Kinetic studies of Mg^{2+}-, Co^{2+}- and Mn^{2+}- activated D-xylose isomerases. *Biochem. J. 278*, 285–292.

VAN BASTELAERE, P. B. M.; CALLENS, M.; VANGRYSPERRE, W. A. E.; and KERSTERS-HILDERSON, H. L. M. 1992. Binding characteristics of Mn^{2+}, Co^{2+}, and Mg^{2+} ions with several D-xylose isomerases. *Biochem. J. 286*, 729–735.

VANGRYSPERRE, W.; CALLENS, M.; KERSTERS-HILDERSON, H.; and DEBRUYNE, K. 1988. Evidence for an essential histidine residue in D-xylose isomerases. *Biochem. J. 250*, 153–160.

VANGRYSPERRE, W.; AMPE, C.; KERSTERS-HILDERSON, H.; and TEMPST, P. 1989A. Single active-site histidine in D-xylose isomerase from *Streptomyces violaceoruber. Biochem. J. 263*, 195–199.

VANGRYSPERRE, W.; KERSTERS-HILDERSON, H.; CALLENS, M.; and DEBRUYNE, C. K. 1989B. Reaction of Woodward's reagent K with D-xylose isomerase. *Biochem. J. 260*, 163–169.

VANGRYSPERRE, W.; VAN DAMME, J.; VANDEKERCKHOVE, J.; DE BRUYNE, C. K.; CORNELIS, R.; and KERSTERS-HILDERSON, H. 1990. Localization of the essential histidine and carboxylate group in D-xylose isomerase. *Biochem. J. 265*, 699–705.

VAN TILBEURGH, H.; JENKINS, J.; CHIADMI, M.; JANIN, J.; WODAK, S. J.; MRABET, N. T.; and LAMBEIR, A.-M. 1992. Protein engineering of xylose (glucose) isomerase from *Actinoplanes missouriensis*. 3. Changing metal specificity and the pH profile by site-directed mutagenesis. *Biochemistry 31*, 5467–5471.

WHITLOW, M.; HOWARD, A. J.; FINZEL, B. C.; POULOS, T. L.; WINBORNE, E.; and GILLILAND, G. L. 1991. A metal-mediated hydride shift mechanism for xylose isomerase based on the 1.6 A *Streptomyces rubiginosus* structures with xylitol and D-xylose. *Proteins: Structure, Function, and Genetics 9*, 153–173.

WILHELM, M., and HOLLENBERG, C. P. 1985. Nucleotide sequence of the *Bacillus subtilis* xylose isomerase gene: extensive homology between the *Bacillus* and *Escherichia coli* enzyme. *Nucl. Acids. Res. 13*, 5717–5721.

YAMANAKA, K., and TAKAHARA, N. 1977. Purification and properties of D-xylose isomerase from *Lactobacillus xylosus*. *Agric. Biol. Chem. 41*, 1909–1915.

Index

Portland Community College

Made in the USA
Lexington, KY
03 February 2011